ACS SYMPOSIUM SERIES **629**

Chemical Applications of Density-Functional Theory

Brian B. Laird, EDITOR
University of Kansas

Richard B. Ross, EDITOR
PPG Industries

Tom Ziegler, EDITOR
University of Calgary

Developed from a symposium sponsored
by the Division of Physical Chemistry and the
Division of Computers in Chemistry
at the 209th National Meeting
of the American Chemical Society,
Anaheim, California
April 2–6, 1995

American Chemical Society, Washington, DC 1996

Library of Congress Cataloging-in-Publication Data

Chemical applications of density-functional theory / Brian B. Laird, editor, Richard B. Ross, editor, Tom Ziegler, editor.

   p.   cm.—(ACS symposium series, ISSN 0097–6156; 629)

"Developed from a symposium sponsored by the Division of Physical Chemistry and the Division of Computers in Chemistry at the 209th National Meeting of the American Chemical Society, Anaheim, California, April 2–6, 1995."

Includes bibliographical references and indexes.

ISBN 0–8412–3403–5

1. Density functionals—Congresses.

I. Laird, Brian B., 1960–   . II. Ross, Richard B., 1958–   . III. Ziegler, Tom, 1945–   . IV. American Chemical Society. Division of Physical Chemistry. V. American Chemical Society. Division of Computers in Chemistry. VI. American Chemical Society. Meeting (209th: 1995: Anaheim, Calif.) VII. Series.

QD462.6.D46C48   1996
541.2′8—dc20   96–14753
                                                          CIP

This book is printed on acid-free, recycled paper.

Foreword

THE ACS SYMPOSIUM SERIES was first published in 1974 to provide a mechanism for publishing symposia quickly in book form. The purpose of this series is to publish comprehensive books developed from symposia, which are usually "snapshots in time" of the current research being done on a topic, plus some review material on the topic. For this reason, it is necessary that the papers be published as quickly as possible.

Before a symposium-based book is put under contract, the proposed table of contents is reviewed for appropriateness to the topic and for comprehensiveness of the collection. Some papers are excluded at this point, and others are added to round out the scope of the volume. In addition, a draft of each paper is peer-reviewed prior to final acceptance or rejection. This anonymous review process is supervised by the organizer(s) of the symposium, who become the editor(s) of the book. The authors then revise their papers according to the recommendations of both the reviewers and the editors, prepare camera-ready copy, and submit the final papers to the editors, who check that all necessary revisions have been made.

As a rule, only original research papers and original review papers are included in the volumes. Verbatim reproductions of previously published papers are not accepted.

ACS BOOKS DEPARTMENT

Contents

FURTHER APPLICATIONS AND THEORETICAL DEVELOPMENT
IN ELECTRONIC STRUCTURE DENSITY-FUNCTIONAL THEORY

INDEXES

Preface

THE USE OF DENSITY-FUNCTIONAL THEORY (DFT) to approach problems of a chemical nature has increased rapidly in recent years. Traditionally, research exploiting this powerful theoretical tool has been divided into two main areas: (1) the calculation of electronic structures by quantum chemists and solid-state physicists, and (2) the study of phase transitions and inhomogeneous fluids by statistical mechanicians. Although these two approaches share a common origin and fundamental justification in the theorems of Peter Hohenberg, Walter Kohn, and David Mermin, they have evolved more or less independently.

In the calculation of electronic structure, DFT methods are increasingly being used by researchers as a standard tool and alternative to traditional ab initio schemes in which the wave function is expanded as a sum of Slater determinants. The methodology is now employed by many groups, including those historically active in DFT and those previously using traditional ab initio methods. DFT methodology is also being used as a modeling tool in several industrial companies as well as in academic research. Furthermore, much effort has gone into development of the methodology itself.

The use of DFT methods to predict the structure and thermodynamics of inhomogeneous fluids and solids near the melting point is not as well developed an area as the electronic structure calculations, but it is becoming a standard research technique. The method has been successful in the study of solid–liquid phase transitions, solid–liquid interfaces, elastic constants of crystals near the melting point, and fluids confined in pores or capillaries. Although primarily developed for inhomogeneous fluids, the method has been used to generate new ways of studying bulk fluids themselves. With important exceptions, the systems that are studied are approached classically, in contrast to the inherently quantum electronic structure calculations.

To our knowledge, the symposium upon which this book is based was the first large-scale common forum for the two areas. It involved scientists from around the world who are carrying out state-of-the-art research in areas of application as well as methodology development. Both industry and academia were well represented.

Through its dual emphasis on basic methodology and applications, it is hoped that this book will become a valuable reference for experts in the field as well as a useful introduction to the current state of DFT research for students and researchers new to the field.

Acknowledgments

We gratefully acknowledge the Petroleum Research Fund, PPG Industries, and Biosym, Inc., for their support in the form of travel grants for selected participants, as well as the American Chemical Society Divisions of Physical Chemistry and Computers for their financial support of the symposium.

BRIAN B. LAIRD
Department of Chemistry
University of Kansas
Lawrence, KS 66045

RICHARD B. ROSS
PPG Industries
P.O. Box 9
Allison Park, PA 15101

TOM ZIEGLER
Department of Chemistry
University of Calgary
2500 University Drive Northwest
Calgary
Alberta T2N 1N4
Canada

November 16, 1995

Chapter 1

Density-Functional Methods in Chemistry: An Overview

Brian B. Laird[1], Richard B. Ross[2], and Tom Ziegler[3]

[1]Department of Chemistry, University of Kansas, Lawrence, KS 66045
[2]PPG Industries, P.O. Box 9, Allison Park, PA 15101
[3]Department of Chemistry, University of Calgary,
2500 University Drive Northwest, Calgary, Alberta T2N 1N4, Canada

Density-functional theory (DFT), in its various forms, has become an important research tool for chemists, physicists and materials scientists. Its development in recent years has proceeded along two (largely independent) paths. The first, and unarguably the most well-trodden, concerns the use of DFT in systems of many electrons. Such methods have undergone tremendous growth over the last decade and are successfully challenging traditional wavefunction-based methods as the technique of choice in large-scale quantum chemistry calculations. In a second direction, DFT has found much success in constructing theories of inhomogeneous fluids and phase transitions, both important sub-fields of statistical mechanics. In contrast to the problem of electronic structure, which is inherently quantum mechanical in nature, applications in statistical mechanics have been, with a few exceptions, classical.

Although density-functional methods for many-electron systems and their classical counterparts used in statistical mechanical studies have evolved more or less independently over the years, they, in fact, have very similar structure and origins. Modern density-functional theory, for both quantum and classical systems, has its fundamental roots in the theorems of Hohenberg and Kohn [1]. For an N-particle system interacting with a given interparticle interaction, the Hamiltonian and thus the ground-state wavefunction and energy are completely determined by specification of the external field $\phi(\mathbf{r})$. In other words, the ground state energy is a functional of $\phi(\mathbf{r})$. In an elegant and simple proof, Hohenberg and Kohn, showed in 1964 that there is a one-to-one correspondence between external field $\phi(\mathbf{r})$ and the single-particle density $\rho(\mathbf{r})$ and that as consequence it is possible to write the total ground-state energy as a functional of $\rho(\mathbf{r})$,

$$E[\rho] = E_0[\rho] + \int d\mathbf{r}\phi(\mathbf{r})\rho(\mathbf{r}); .\qquad(1)$$

Here $E_0[\rho]$ is a functional that is independent of the external potential, $\phi(\mathbf{r})$, i.e., it is a universal functional for a given interparticle interaction. Hohenberg

0097–6156/96/0629–0001$15.00/0

and Kohn also proved a second theorem which provides an energy variational principle. They showed that for any trial density $\bar{\rho}(\mathbf{r})$ that satisfies $\int \rho(\mathbf{r})\, d\mathbf{r} = N$,

$$E[\bar{\rho}] \geq E_g \,, \tag{2}$$

where E_g is the true ground-state energy. The equality in equation 2 holds only when $\bar{\rho}(\mathbf{r})$ is the true ground-state single-particle density.

If the functional $E_0[\rho]$ were known for a system of interacting electrons, then equations 1 and 2 would allow the calculation of the ground-state energy and electron density for any multielectron system in an arbitrary external field. The HK theorem establishes that such a functional exists, but gives no prescription for its determination; therefore, the goal of researchers in this field is the development of accurate approximate functionals.

The Hohenberg-Kohn theorems apply specifically to the ground state energy and therefore apply strictly only at zero absolute temperature. Statistical mechanics applications require the extension of these theorems to non-zero temperature. This was accomplished by Mermin [2], who showed that for a system at fixed temperature, T, chemical potential, μ, and external single-particle potential, $v(\mathbf{r})$, there exists a functional $\mathcal{F}[\rho(\mathbf{r})]$, independent of $v(\mathbf{r})$ and μ, such that the functional

$$\Omega[\rho(\mathbf{r})] = \mathcal{F}[\rho(\mathbf{r})] + \int d\mathbf{r}[v(\mathbf{r}) - \mu] \tag{3}$$

is a minimum for the the correct equilibrium density $\rho(\mathbf{r})$ subject to the external potential. The value of Ω at this minimum is the grand potential. (For systems restricted to constant density, i.e. canonical ensemble, one minimizes functional $\mathcal{F}$ itself to give the equilibrium Helmholtz potential.)

The rest of this overview is organized as follows: In the next two sections we will discuss the various approximations and applications of DFT as applied to electronic structure and classical statistical mechanics, respectively. In the last section, the current state-of-the-art of DFT applications to industrial research are reviewed.

Density-Functional Theory for Many Electron Systems

The basic notion in quantum mechanical density-functional theory of many electron systems is that the energy of an electronic system can be expressed in terms of its density. This notion is almost as old as quantum mechanics and dates back to the early work by Thomas [3], Fermi [6], Dirac [4], and Wigner [5]. The theory by Thomas and Fermi is a true density-functional theory since all parts of the energy, kinetic as well as electrostatic, are expressed in terms of the electron density. The Thomas-Fermi method, although highly approximate, has been applied widely in atomic physics as a conceptually useful and computational expedient model.

The Hartree-Fock-Slater or $X\alpha$ method was one of the first DFT-based schemes to be used in studies on electronic systems with more than one atom. The $X\alpha$ theory has its origin in solid-state physics. The method emerged from the work of J.C. Slater [7] who in 1951 proposed to represent the exchange-correlation potential by a function which is proportional to the 1/3 power of the electron density. This approximation evolved out of the need to develop techniques that were able to handle solids within a reasonable time frame. DFT based methods have been predominant in solid-state physics since the pioneering work by Slater [7] and Gáspár [8]. Slater [7] has given a vivid account of how the $X\alpha$ method evolved during the 1950's and 1960's, with reference to numerous applications up to 1974.

The Thomas-Fermi method as well as the $X\alpha$ scheme were at the time of their inceptions considered as useful models based on the notion that the energy of an electronic system can be expressed in terms of its density. As mentioned in the introduction, a formal proof of this notion came in 1964 when it was shown by Hohenberg and Kohn [1] that the ground-stateenergy of an electronic system is uniquely defined by its density, although the exact functional dependence of the energy on density remains unknown. This important theorem has later been extended by Levy [9]. Of further importance was the derivation by Kohn and Sham [10] of a set of one-electron equations from which one in principle could obtain the exact electron density and thus the total energy. The work of Hohenberg, Kohn and Sham has rekindled much interest in methods where the energy is expressed in terms of the density. In particular the equations by Kohn and Sham have served as a starting point for new approximate DF methods. These schemes can now be considered as approximations to a rigorous theory rather than just models.

The Kohn-Sham Equation. The total energy of an n-electron system can be written [10] without approximations as

$$E_{el} = -\frac{1}{2}\sum_i \int \phi_i(\mathbf{r_1})\nabla^2\phi_i(\mathbf{r_1})\,d\mathbf{r_1} + \sum_A \frac{Z_A}{|\mathbf{R}_A - \mathbf{r_1}|}\rho(\mathbf{r_1})\,d\mathbf{r_1}$$
$$+\frac{1}{2}\int \frac{\rho(\mathbf{r_1})\rho(\mathbf{r_2})}{|\mathbf{r_1} - \mathbf{r_1}|}\,d\mathbf{r_1}\,d\mathbf{r_2} + E_{XC}\,. \tag{4}$$

The first term in equation 4 represents the kinetic energy of n non- interacting electrons with the same density $\rho(\mathbf{r_1}) = \sum_i^{occ}\phi_i(\mathbf{r_1})\phi_i(\mathbf{r_1})$ as the actual system of interacting electrons. The second term accounts for the electron- nucleus attraction and the third term for the Coulomb interaction between the two charge distributions $\rho(\mathbf{r_1})$ and $\rho(\mathbf{r_2})$. The last term contains the exchange-correlation energy, E_{XC}. The exchange-correlation energy can be expressed in terms of the spherically averaged exchange-correlation hole functions [11, 12] $\rho_{XC}^{\gamma\tau}(\mathbf{r}_1, s)$ as

$$E_{XC} = -\frac{4\pi}{2} \sum_{\gamma} \sum_{\tau} \rho_1^{\gamma}(\mathbf{r}_1) \rho_{XC}^{\gamma\tau}(\mathbf{r}_1, s) \, d\mathbf{r}_1 \, s^2 \, ds \,, \tag{5}$$

where the spin indices γ and τ both run over α-spin as well as β-spin and $s = |\mathbf{r}_1 - \mathbf{r}_2|$.

The one-electron orbitals, of equation 4 are solutions to the set of one-electron Kohn-Sham equations [10]

$$[-\frac{1}{2}\nabla_1^2 + \frac{Z_A}{|\mathbf{R}_A - \mathbf{r}_1|} + \int \frac{\rho(\mathbf{r}_2) \, d\mathbf{r}_2}{|\mathbf{r}_1 - \mathbf{r}_2|} + V_{XC}]\phi_i(\mathbf{r}_1) = \hat{h}_{KS}\phi_i(\mathbf{r}_1) = \epsilon_i\phi_i(\mathbf{r}_1) \,, \tag{6}$$

where the exchange-correlation potential V_{XC} is given as the functional derivative of E_{XC} with respect to the density [10]:

$$V_{XC} = \frac{\delta E_{XC}[\rho]}{\delta\rho} \,. \tag{7}$$

The hole function $\rho_{XC}^{\gamma\tau}(\mathbf{r}_1, s)$ contains all information about exchange and correlation between the interacting electrons as well as the influence[10] of correlation on the kinetic energy. The interpretation of $\rho_{XC}^{\gamma\tau}(\mathbf{r}_1, s)$ is that an electron at $\mathbf{r}_1$ to a larger or smaller extent will exclude other electrons from approaching within a distances s. The extent of exclusion or screening increases with the magnitude of $\rho_{XC}^{\gamma\tau}(\mathbf{r}_1, s)$.

The first generation of DFT based methods took their hole from the homogeneous electron gas. The Hartree-Fock-Slater method (HFS) [13] treats only exchange whereas the local density approximation (LDA) [14] takes into account both exchange and correlation. The HFS and LDA schemes are often referred to as the local methods. The second generation of DFT based theories acknowledge the fact that the density in molecules is far from homogeneous by introducing correction terms based on the electron density gradients. These theories (LDA/NL) due to Langreth and Mehl [15], Becke [16] and Perdew [17] are referred to as nonlocal. Nonlocal corrections are essential for a quantitative estimate of bond energies [18] as well as metal-ligand bond distances [18]. They are also of importance for other properties[18]. Both LDA and LDA/NL suffer from the fact that they allow an electron to interact with itself to some degree. A number of authors have recently suggested ways in which the "self-interaction" error can be eliminated in a computationally efficient way [19, 20]. These new methods form the vanguard for the third generation of DFT based schemes. An account of the formal developments in DFT since 1964 and its applications to chemistry has been reviewed [18, 21]. Newer developments in the theory are discussed in this volume by Baerends, et al. (Chapter 2) and Gross, et al. (Chapter 3).

Practical Implementations.   A practical DFT based calculation is in many ways similar to a traditional HF treatment in that the final outcome is a set of molecular orbitals. These functions are referred to as Kohn-Sham (KS) orbitals [10] and are often expanded in terms of a basis set as in the traditional Linear Combination of Atomic Orbitals (LCAO) approach.

In the earliest implementation of DFT applied to molecular problems, Johnson [22] used scattered-plane waves as a basis and the exchange-correlation energy was that of the LDA method, see Table 1. This SW-$X\alpha$ method assumes in addition that the Coulomb potential is spherical around each atom; i.e., the muffin-tin approximation. This approximation is well suited for solids, for which the SW-$X\alpha$ method originally was developed. However, it is less appropriate in molecules where the potential around each atom might be far from spherical. The SW-$X\alpha$ method is computationally expedient compared to standard *ab initio* techniques and has been used with considerable success to elucidate the electronic structure in complexes and clusters of transition metals. However, the use of the muffin-tin approximation precludes accurate calculations of total energies. A more recent method in which scattered plane waves are used without the muffin-tin approximation is presented by Springborg, *et al.* in Chapter 8.

The first implementations of self-consistent DFT, without recourse to the muffin-tin approximation, are due to Ellis and Painter [23] (DVM), Baerends, *et al.* [24], (ADF), Sambe, and Felton [25], Dunlap, *et al.* [26], (LCGT-LDA) as well as Gunnarson, *et al.* [27] (LCMTO). A further step forward was taken when Ziegler [28], Dunlap [26] and Jones [27] introduced methods that allowed total energies to be calculated accurately within the DFT framework. Other implementations [29-34] and refinements have also appeared more recently. The available DFT program packages, ADF [24], DeMon [33], DMOL [32], DGAUSS [31], GAUSSIAN/DFT [35] and CADPAC [36] are now nearly as user-friendly as their *ab initio* counterparts, although much work remains to be done. A unique approach has lately been taken by Becke [37] in which the KS-orbitals are found directly without basis sets. This approach, which seems promising, was first applied to diatomic molecules and more recently to polyatomics.

The use of plane waves as basis functions has received renewed interest in connection with the recent pioneering work by Carr and Parrinello (CP) [38] . These authors have been able to study molecular dynamics from first principles by evaluating the required forces "on the go" from DFT calculations. The method is best suited for plane wave basis sets. Blöchl [39] has eliminated some of the drawbacks of plane waves by introducing projector augmented plane waves (CP-PAW). The CP-PAW method has great potential in studies of organometallic kinetics. Blöchel provides an account of the CP-PAW method in Chapter 4.

The various SCF-schemes based on DFT are attractive alternatives to conventional *ab initio* methods in studies on large size molecules since the computational effort increases as n^3 with the number of electrons, n, as opposed to between n^4 and n^7 for post- HF methods of similar accuracy. The scope of density-functional based methods has further been enhanced to include pseudo-potentials [18], relativistic-effects [18], as well as energy gradients of use in geometry op-

Table 1: **The development of DFT as a practical tool in chemical research**

Year	Comments	Reference
1966	First chemical application of DFT using the muffin-tin approximation - UV-spectra and photoelectron spectroscopy	[22]
1973-1976	First implementations of DFT without use of the muffin-tin approximation - UV-spectra and photoelectron spectroscopy	[23]-[25]
1977	The implementation of accurate algorithms for total energy calculations - geometry, bond energies, vibrational spectra	[26]-[28]
1983-1988	The introduction of non-local density functionals - accurate bond energies and geometries	[15]-[17]
1985	The development of first-principles molecular-dynamics methods	[38, 39]
1988	Implementation of analytical energy gradients - automated geometry optimization	[18]
1989	Force-field calculations based on analytical energy gradients - accurate vibrational spectra for large molecules	[18]
1990	The determination of transition-state structures	[18]
1990	The Divide and Conquer method of Yang	[46]
1993	Accurate calculations of NMR and ESR parameters	[43]-[45]
1994	Analytical second derivatives	[35]-[42]

timization [18] and analytical second derivatives [39-42] with respect to nuclear displacements. One of the implementations of analytical second derivatives are described by Jacobsen, *et al.* in Chapter 11.

Future outlook for Electronic Structure DFT. DFT is now a well established method after a bumpy and exhilarating path through the 1970's and 1980's. Many of its previous opponents have become lukewarm practitioners or even staunch supporters, and it has been implemented into a number of major program packages such as ADF, DeMon, DMOL, DGAUSS, GAUSSIAN/DFT and CADPAC.

Much still has to be done in terms of improving the existing functionals. Applications to excited states have to be explored further [18], and new applications to properties such as NMR shifts [44] and ESR [45] hyperfine tensor are emerging. Other new developments are the first principles molecular dynamics method by Carr and Parrinello [38] which has the potential to introduce temperature into the description of molecular behavior; the divide-and-conquer scheme by Yang [46] which allows one to describe large systems; as well as several recent inclusions of solvation effects [47] into the DFT formalism. Solvation is discussed here by Truong, *et al.*, (Chapter 6), whereas St-Amant deals with the divide-and conquer scheme in Chapter 5. Combining DFT with molecular mechanics is the subject of Chapter 10 by Merz, *et al.* All of the new methods are bound to benefit chemistry to the same degree as the development of DFT over the past decade, and it is likely that DFT will become as indispensable a research tool in chemistry as any of the major spectroscopic techniques.

Stephens, *et al.*, deals with an application of DFT to vibrational spectroscopy in Chapter 7. Sosa discuss DFT calculations on the hydrogen bond in Chapter 9.

Density-Functional Methods in Statistical Mechanics

At nonzero temperature, density-functional theory is a procedure for the determination of the free energy associated with a given spatially dependent single-particle density, $\rho(\mathbf{r})$. That is, the free energy is determined as a functional of $\rho(\mathbf{r})$. The equilibrium free energy and microscopic density can then be found by minimizing this functional for a given external field $\phi(\mathbf{r})$ over the space of single-particle densities consistent with the ensemble under study. For readers seeking more information than is contained in this brief overview, a detailed description of basic classical DFT and its mathematical justifications is presented by Evans [49].

Fundamental Theory. The application of density-functional theory to classical statistical mechanics was pioneered by Stillinger and Buff [50] and Liebowitz and Percus [51] and developed into its present form by Saam and Ebner [52]. The universal functional $\mathcal{F}[\rho]$ from Equation 3 can be written as the sum of an ideal part, $\mathcal{F}_{id}[\rho]$, and an excess part, $\mathcal{F}_{ex}[\rho]$, due to the interparticle interactions:

$$\mathcal{F}[\rho] = \mathcal{F}_{id}[\rho] + \mathcal{F}_{ex}[\rho] . \tag{8}$$

In contrast to quantum density functionals, the the exact ideal part of the functional for monatomic systems is known explicitly, and is given by

$$\beta\mathcal{F}_{id}[\rho] = \int d\mathbf{r}\rho(\mathbf{r})\{\ln[\Lambda^3\rho(\mathbf{r})] - 1\} , \tag{9}$$

where Λ is the thermal wavelength and $\beta = (kT)^{-1}$. (Note that, because of bonding constraints, the ideal part of the functional for polyatomic molecules is not as straightforward and must be either approximated by a theoretical model or calculated via computer simulation - see the chapters in this volume by McCoy, Yethiraj or Keirlik, *et al.*, for more information.) The excess part is, in general, unknown; therefore, the central task of classical DF theory is to provide a suitable approximation for this quantity.

To this end, classical density-functional theories usually begin by defining the n-body direct correlation functions, $c^{(n)}(\mathbf{r}_1, ..., \mathbf{r}_n; [\rho])$, in terms of the functional derivatives of $\mathcal{F}_{ex}[\rho]$:

$$c^{(n)}(\mathbf{r}_1, ..., \mathbf{r}_n; [\rho]) = -\frac{\delta^{(n)} \beta \mathcal{F}_{ex}[\rho]}{\delta\rho(\mathbf{r}_1)...\rho(\mathbf{r}_n)} . \qquad (10)$$

For arbitrary $\rho(\mathbf{r})$, these correlation functions, like $\mathcal{F}[\rho]$ itself, are unknown, except in the homogeneous density (liquid) limit, where, due to the advances in liquid-state theory over the past three decades, they can be determined for $n \leq 2$. This known information about the homogeneous ($\rho(\mathbf{r}) = \rho_0$) state is then used to construct the functional $\mathcal{F}_{ex}[\rho]$.

In general, most such approximations fall into two classes: a) methods based on functional Taylor-series expansions about some reference state - usually the homogeneous phase, and (b) non-perturbative methods based on the concept of weighted density functionals.

The Taylor series approach was outlined by Saam and Ebner [52]. In this approximation, the free energy of the inhomogeneous phase (here, the crystal) is expanded in a functional Taylor expansion about a reference liquid density $\rho_0(\mathbf{r})$. This expansion is subsequently truncated at second order to yield

$$\begin{aligned} \beta \mathcal{F}_{ex}[\rho] &= \beta \mathcal{F}_{ex}(\rho_0(\mathbf{r})) - \int d\mathbf{r}_1 c_l^{(1)}(\rho_0(\mathbf{r}_1)[\rho(\mathbf{r}_1) - \rho_0(\mathbf{r}_1)] \qquad (11) \\ &\quad - \int \int d\mathbf{r}_1 d\mathbf{r}_2\, c_l^{(2)}(\mid \mathbf{r}_1 - \mathbf{r}_2 \mid; \rho_0(\mathbf{r})) \\ &\quad \times [\rho(\mathbf{r}_1) - \rho_0(\mathbf{r}_1)][\rho(\mathbf{r}_2) - \rho_0(\mathbf{r}_2)] + ... \end{aligned}$$

Generally, information about $c^{(2)}$ is only available for homogeneous liquids, so the reference density is usually taken to be constant.

Initiated by Tarazona [53], weighted-density-functional (WDF) methods are modifications of the usual Local Density Approximation for inhomogeneous systems. In the LDA, the free energy density at a point $\mathbf{r}$ in a system with inhomogeneous single-particle density $\rho(\mathbf{r})$ is given by the free energy per particle of a *homogeneous* system, evaluated at the value of the single-particle density at point $\mathbf{r}$. However, for very strongly inhomogeneous systems such as a crystal, the LDA breaks down. To remedy this, the local density averaged over a small region using a weighting function $w(\mid \mathbf{r}_1 - \mathbf{r}_2 \mid; \hat{\rho})$ to create a coarse-grained or "weighted" density $\hat{\rho}(\mathbf{r})$:

$$\hat{\rho}(\mathbf{r}_1) = \int d\mathbf{r}_2 \rho(\mathbf{r}_2)\, w(\mid \mathbf{r}_1 - \mathbf{r}_2 \mid; \hat{\rho}(\mathbf{r}_1)) . \qquad (12)$$

The functional is then assumed to have the same form as the LDA, but with a homogeneous free energy evaluated at the weighted density instead of the local density. The free-energy functional is then given by

$$\beta \mathcal{F}_{ex}[\rho] = \int d\mathbf{r}\, \beta f_0(\hat{\rho}(\mathbf{r}))\rho(\mathbf{r}) , \qquad (13)$$

where $f_0(\rho)$ is the excess Helmholtz free energy of the homogeneous phase. The functional is then completely specified by the choice of the weighting function w. The task of a successful weighted density-functional theory is to choose a weighting function that leads to a good description of the structure and thermodynamics of the inhomogeneous phase. For example, in the weighted density approximation (WDA) of Curtin and Ashcroft [54], the weighting function $w(|\mathbf{r}_1 - \mathbf{r}_2|; \hat{\rho})$ is chosen such that both the free energy and the liquid pair correlation function are exactly reproduced in the limit of a homogeneous density. In this sense, the WDF theories are similar to the Taylor series methods in that they both require only the two-particle direct-correlation function of the liquid phase as input.

Applications of Classical DFT Perhaps the most common use for classical DFT is in the determination of solid-liquid coexistence; i.e. freezing [55-57]. The idea of using data from the liquid phase to determine crystal free energies has its origin in the work of Kirkwood and Monroe [58], but modern DFT approaches are due to Ramakrishnan and Yousoff [59], whose theory was later reformulated into the language of classical density functionals by Haymet and Oxtoby [60]. After a functional has been chosen, an equilibrium freezing DFT calculation proceeds as follows. First, the periodic single-particle density of the crystal is parametrized so that the minimization of the free-energy functional can be performed. The most general parametrization for a given lattice type is the Fourier series

$$\rho(\mathbf{r}) = \rho_L[1 + \eta + \sum_{\{\mathbf{k}\}} \mu(\mathbf{k})e^{i\mathbf{k}\cdot\mathbf{r}}], \tag{14}$$

where ρ_L is the bulk liquid density, η is the fractional density change on freezing, $\mathbf{k}$ represents the set of reciprocal lattice vectors (RLV's) corresponding to the particular lattice type under study, and $\rho_L\mu(\mathbf{k})$ is the Fourier component of the density corresponding to the wavevector $\mathbf{k}$. A simpler but less general parametrization that is commonly used expresses the density as a sum of Gaussian peaks centered at the lattice sites

$$\rho(\mathbf{r}) = \left(\frac{\alpha}{\pi}\right)^{3/2} \sum_i \exp\left(-\alpha \mid \mathbf{r} - \mathbf{R}_i \mid^2\right), \tag{15}$$

where the $\mathbf{R}_i$ are the real space lattice vectors, and α measures the width of the Gaussian peaks. For fcc systems, the Gaussian parametrization has been found to give almost identical freezing results as the more complicated, but more general Fourier expansion [61]. After parametrization of the crystal density, the free energy (grand or Helmholtz) is minimized in such a way as to ensure the thermodynamic conditions of phase coexistence are satisfied, that is, the pressure, temperature, and chemical potential of the crystal phase equals that of the liquid phase. The $\rho(r)$ at the minimum is the equilibrium crystal density.

Because of the important role that packing considerations play in the solid-liquid phase transition and because accurate data is available in analytic form

[62, 63] for the liquid structure and thermodynamics, the hard-sphere system has been the primary benchmark for the evaluation of new DFT approaches to freezing (For a complete list of references, see Reference [57]). Although differing in quantitative accuracy, the results from the various methods are qualitatively quite similar in that they all predict coexistence densities that are roughly in agreement with the simulation results.

For the above reason, the use of the hard-sphere system as the standard to test newly developed theories was, in hindsight, unfortunate as it is only when one considers systems with long-range potentials and non-close-packed crystal structures that the principal flaws in the density functionals emerge. For example, the WDA of Curtin and Ashcroft [54] gives very good results for hard-sphere systems and systems with purely repulsive interactions that freeze into an fcc crystal, but the full theory fails when attractive forces are introduced, as in the Lennard-Jones system [64] The most accurate theory for hard-sphere freezing, the Generalized Effective Liquid Approximation (GELA) of Lutsko and Baus [65] is most extreme in that it gives results for hard spheres that are nearly identical with simulations, but completely fails for any other system [66, 64]. This suggests that the current density-functional theories are doing a relatively good job at describing the entropy of a system (hard-sphere freezing is purely entropic), but are lacking in their description of the energy. Also, neither the second-order functionals nor the standard weighted-density methods have been successful at describing the liquid-bcc transition in systems such as the repulsive inverse-sixth (or fourth) power potential [66, 67].

For systems such as LJ fluids [68] and C_{60} [69] which have attractive interactions, there has been some success in in combining DFT methods with perturbation theory. In these problems, the attractive part of the potential is included as a perturbation on a hard-sphere reference state. (For a similar approach to the phase diagram of colloidal systems see chapter 21 in the present volume by Tejero). It is also possible to construct similar perturbative approaches to systems with long-ranged repulsions [70, 71].

Another important area of classical DFT research is in constructing theories for inhomogeneous fluids, i.e., calculating the response of a fluid to an external field or confining geometry. It is often difficult to apply lessons from the above freezing theories to such systems since it is sometimes the case that methods that completely fail for freezing give very good results for inhomogeneous fluids [72]. Some of the major areas of research include fluids near a wall, fluids in pores, liquid-vapor interfaces and wetting phenomena. These subjects have a long and very interesting history and it would be impossible to do them justice in this short overview - for an excellent review of this aspect of DFT see Reference [73]. In Chapter 12, Evans uses similar ideas to explore the relationship between the decay of correlations in bulk fluids and in fluids near a liquid-vapor interface.

In a related application, it is also possible to use DFT in construction of theories to describe the structure and thermodynamics of solid-liquid interfaces.

This application, which can be regarded as a hybrid between DFT freezing methods and those for inhomogeneous fluids, is very important due to the paramount role played by the interface in near-equilibrium crystal growth processes [74]. There has been much recent success for systems with short-range interactions (see Reference [74] and Chapter 15 in this volume by Marr and Gast), but because of the intimate relationship between theories of interfacial structure and the ability of a density-functional theory to reproduce the coexistence properties of a system, theoretical study on crystal-liquid interfaces is contingent upon development of a more generally applicable density-functional approximation. Density-functional theory can also be used in a similar way to develop theories of nucleation by providing a way to calculate the free energy of a nucleating droplet [75].

Future Outlook for DFT in Statistical Mechanics.   Perhaps, the single biggest challenge today in classical density-functional theory is development of more generally applicable functional methods, especially ones that work well for systems with attractive or long-range interactions. Neither the second-order functionals nor the standard weighted-density methods have been successful at describing the liquid-bcc transition in systems such as the repulsive inverse-sixth (or fourth) power potential [66, 67]. Perturbation theories have potential to this end, but more fundamental approaches are probably needed. For an introduction into two approaches of functional development that differ from the standard methods presented in this overview see the chapters 13 and 14 in this volume by Percus and Rosenfeld, respectively.

It would seem that, given the problems encountered in the application of classical DFT techniques to simple systems, their application to the complex systems of interest to industry such as polymers would futile. On the contrary, the special nature of macromolecular systems makes them ideal candidates. For this reason, such applications will be a growth area in classical DFT technology, especially considering the enormous technological importance of such materials. For further insight into this important topic, please see the following chapters in this volume: Chapter 16 by Keirlik, Phan and Rosinberg, Chapter 17 by McCoy and Nath, Chapter 18 by McMullen, and Chapter 19 by Yethiraj.

Although most applications of density-functional theory to statistical mechanics are classical, this is not a necessary restriction. Extension of current techniques to quantum systems at finite temperature will be an important area of future research. It is here that cross-fertilization from electronic structure DFT will have the greatest impact. Chapter 20 by Rick, McCoy and Haymet explores such issues in the context of developing a freezing theory for helium.

The use of density-functional theory in statistical mechanical applications is not as well-developed and mature an area as electronic structure DFT. Consequently, there is much room for improvement, but the successes so far indicate that the method has much promise as a general method in statistical mechanics.

Industrial Applications of Density-Functional Theory

Applications of density-functional molecular-orbital theory in an environment such as in industrial research laboratories have been increasing. Applications of theoretical and computational chemistry in industry have been established in pharmaceutical research and more recently in diversified industrial areas such as automotive, chemicals, coatings, glass, materials, petroleum, and polymers.

The theoretical methods commonly used fall into two general categories: classical and quantum mechanical. Classical methods, as implemented in molecular mechanics and dynamics programs, have been used widely for example in the pharmaceutical industry in the study of macromolecules [76]. More recently, classical tools have been applied to materials-oriented areas such as polymers [77], catalysts [78, 79] and zeolites [78, 80].

While useful for many applications, classical methods do not directly account for electronic effects which is required for many additional applied studies. To account directly for electronic effects, a molecular-orbital (MO) theory-based method (such as density-functional theory) is required.

There are many examples of applications for MO theory in an applied environment. For example, the calculation of bond dissociation energies can aid in understanding degradation of polymers and other materials. The calculation of proton and electron affinities can aid in understanding relative acidities of industrially important compounds.

Additional examples of applications include the characterization of the kinetics and thermodynamics of chemical reactions. Resulting quantities such as activation energies and heats of reaction can potentially be used to design more optimal products with better properties or enhanced processing characteristics. For example in a thermodynamically controlled endothermic reaction, heats of reaction for a series compounds can be compared to determine which compounds will likely result in reactions with the lowest energy requirements.

Molecular-orbital theory studies can also be employed to predict spectra such as for nuclear magnetic resonance and infrared spectroscopy. Prediction of spectra can aid in the interpretation of experimental spectra, confirming reaction products, and in identifying unknown compounds.

In addition, MO theory can be used to characterize electron distributions in compounds. The electron distributions can then be analyzed to identify quantities such as relative areas of positive and negative charge. The resulting information can be interpreted, for example, to aid in the prediction of which regions of molecules are more or less prone to nucleophillic or electrophillic attack.

MO theory methods can also be used to optimize and aid in determining structures for industrially important organometallic compounds in cases for which classical force field parameters are not well developed or unavailable. The resulting structural information can be used to aid in understanding the mechanisms of interaction of organometallic compounds in chemical reactions or in more complex composite or mixed systems.

In addition to single property applications, a combination of MO theory tools can be used to characterize important industrial processes. For example, electron charge distributions, molecular orbital plots, relevant bond dissociation energies and/or proton or electron affinities and other applicable quantities can be calculated for a series of compounds used in the same industrial process. The calculated results can be then compared to each other as well as to experiment to potentially increase understanding of which quantities are most important for good experimental performance.

These examples are small segment of the wide array of applied research studies which can be approached with MO theory. For additional discussion of quantities potentially useful for applied research see for example the monograph *Experiments in Computational Organic Chemistry* by Hehre, Burke, Shusterman, and Pietro [86] as well as chapters that have appeared in Volumes 1-6 of the series Reviews in Computational Chemistry edited by Lipkowitz and Boyd [82].

MO theory methods used commonly in an applied environment include: semiempirical methods [83-85]; Hartree-Fock-based *ab initio* methods [86]; and the subject of this proceedings volume, DFT *ab initio* methods.

Semiempirical methods are useful for example for predicting structures of organic molecules, heats of reaction for classes of compounds, and trends in UV-visible absorption spectra produced by substituent changes. These methods are the fastest of the three methods typically scaling on the order of n^2 with the number of electrons, n. However, parameterization of semiempirical methodologies can lead to difficulties if parameters are not available for an element of interest. A second difficulty can be the comparison of quantities such as heats of reaction if the values differ by small amounts that are on the same order of magnitude as errors in the methodology. In addition, while semiempirical methods are applicable for many studies, some applications require accuracy which is higher than that possible from the methodologies.

Two first-principles alternatives to the semiempirical methodologies are the Hartree-Fock (HF)-based and density-functional theory methods. In Hartree-Fock-based methodology an unparameterized self-consistent field (SCF) calculation is carried out. If electron correlation effects are important, the SCF study is followed by systematic post Hartree-Fock treatments [87]. CPU time required for HF-based methodology, including post electron correlation treatment, increases at between n^4 and n^7 with the number of electrons n as discussed earlier. The extensive CPU requirements for the methodology renders it most useful for studies on small molecules and clusters.

As discussed earlier, density-functional methods have been established for many applications, can be of similar accuracy to post-HF treatments, and scale as n^3. In addition, since electron correlation is included in part in the formalism, practical application of the methodology is a single-step calculation rather than the two or more steps required in a post HF-based treatment. In addition, commercial packages have been developed to study periodic systems [88] which complement those developed and discussed earlier to study non-periodic molecular or cluster systems. The comparable accuracy, decreased computational requirements, reduced wall clock and researcher interaction time as well as the development of commercial periodic packages make DFT methodology an attractive ab initio alternative for the study of large systems as well as for the faster study of smaller systems. Largely for these reasons, DFT methods are growing in use in applied environments.

The use of density-functional methods in an applied environment has been illustrated by several presentations at this symposium. Presentations were made by industrial theoreticians from a diverse array of industries including automotive, chemicals, coatings, computers, software, petroleum, and polymers. Application areas discussed included: copper-substituted zeolite catalysts; iron porphyrn catalysts; amide hydrogen bonding and hydrolysis in the context of commercial industrial polymers; characterization of polysulfide additives; and characterization of metal fragments. The last two topics are discussed in detail in chapters in this book by Anne Chaka, John Harris and Xiao-Ping Li and by Rick Ross, Bill Kern, Shaoping Tang, and Art Freeman.

In addition to the presentations by industrial scientists, scientists from academia, governmental, and scientific software research laboratories discussed the use of DFT methods for many additional applications. The topics of discussion included: redox potentials; iron-sulfur clusters in the context of proteins; reaction potential energy surfaces and transition states; structures and vibrational frequencies; surface applications and reactions; enzyme reaction mechanisms; chemical reactions on the $Si(100)2x1$ surface; the calculation of NMR spectroscopic parameters; prediction of diradical singlet/triplet gaps; comparison of methods to predict heats of reaction, geometries, and barrier heights; fast density-functional methods; and molecular and materials design. Chapters are included in this book on many of these topics including the last five which are discussed in contributions by George Schreckenbach, Ross Dickson, Yosadara Ruiz-Morales and Tom Ziegler; Myong Lim, Sharon Worthington, Frederic Dulles, and Chris Cramer; Jon Baker, Max Muir, Jan Andzelm, and Andrew Scheiner; John Harris, Xiao-Ping Li, and Jan Andzelm; and by Erich Wimmer, respectively. The prediction of these properties have potential applications in industrial as well as governmental academic and software development laboratories where applied research is carried out.

The wide array of applications discussed in this symposium illustrates the motivation for the increasing use of DFT methods in an applied research environment. As the applicability of DFT methods to predict additional properties and as further theoretical improvements occur, the use of the methodology in an applied environment will increase at an even faster rate.

Literature Cited

[1] Hohenberg, P., Kohn, W. *Phys.Rev.* **1964**, *136*, A864.
[2] Mermin, N. D. *Phys. Rev.* **1965**, *137*, A1441.
[3] Thomas, L.H. *Proc.Camb.Phil.Soc.* **1927**, *23*, 542.
[4] Fermi, E. *Z. Phys.* **1928**, *48*, 73.
[5] Dirac, P.A.M. *Cambridge Phil.Soc.* **1930**, *26*, 376.
[6] Wigner, E.P. *Phys.Rev.* **1934**, *46*, 1002.
[7] Slater, J.C. *The Self-Consistent Field for Molecules and Solids*: Quantum Theory of Molecules and Solids, Vol. 4. New York, 1974, McGraw Hill.
[8] Gaspar, R. *Acta Phys. Acad.Sci.Hung.* **1954**, *3*, 263.
[9] Levy, M. *Proc.Natl. Acad.Sci. USA* **1979**, *76*, 6062.
[10] Kohn, W.; Sham, L.J. *Phys.Rev.* **1965**, *140*, A1133.
[11] McWeeney, R.; Sutcliffe, B.T. *Methods of Molecular Quantum Mechanics*; Academic Press: New York, 1969.
[12] Luken,W.L.; Beratan, D.N., *Theoret. Chim.Acta.* **1982**, *61*, 265.
[13] Gunnarsson, O; Lundquist, L. *Phys.Rev.* B **1974**, *10*, 1319.
[14] Slater, J.C. *Adv. Quantum Chem.* **1972**, *6*, 1.
[15] Langreth, D.C.; Mehl, M.J. *Phys.Rev.* B **1983**, *28*, 1809.
[16] Becke, A.D. *Phys.Rev.* A **1988**, *33*, 2786.
[17] Perdew, J.P. *Phys. Rev.* B **1986**, *33*, 8822. Also see the erratum, Phys.Rev. B **1986**, *34*, 7046.
[18] Ziegler, T. *Chem Rev.* **1991**, *91*, 651.
[19] Becke, A.D. *J.Chem.Phys.* **1993**, *98*, 1372.
[20] van Leeuwen, R.; Baerends, E.J. *Phys.Rev.* A **1994**, *49*, 2421.
[21] Parr R.G.; Yang W., *Density-Functional Theory of Atoms and Molecules*; Oxford University Press, New York, 1989.
[22] Johnson, K.H. *J.Chem.Phys.* **1966**, *45*, 3085.
[23] Ellis D.E.; Painter, G.S. *Phys.Rev.* B **1970**, *2*, 2887.
[24] Baerends, E.J.; Ellis, D.E.; Ros, P. *Chem.Phys.* **1973**, *2*, 41.
[25] Sambe H.; Felton R.H. *J.Chem.Phys.* **1975**, *62*, 1122.
[26] Dunlap, B.I.; Connolly, J.W.D.; Sabin J.F., *J.Chem.Phys.* **1977**, *71*, 3396.
[27] Gunnarson, O.; Harris, J.; Jones, R.O *Phys. Rev.* A **1977**, *15*, 3027.
[28] Ziegler, T.; Rauk, A. *Theor.Chim.Acta.* **1977**, *46*, 1.
[29] Wimmer, E; Freeman, A.; Fu, C.L; Cao, S.H.; Delley, B. In *Supercomputer Research in Chemistry and Chemical Engineering*; Jensen, K.F.; Truhlar,D.G., Eds.; ACS Symposium Series No. 353; American Chemical Society: Washington, D.C., 1987; p 49.
[30] Andzelm, J.; Wimmer, E.; Salahub, D.R. In *Spin Density Functional Approach to the Chemistry of Transition Metal Clusters*, D.R. Salahub; M.C.Zerner, Eds. ACS Symposium Series No. 394; American Chemical Society: Washington, D.C., 1989; p.229.
[31] DGauss available from Cray Research, Inc., Eagan, MN.
[32] DMol available from Biosym/Molecular Simularions, Inc., San Diego, CA.
[33] DeMon is developed by Salahub, *et al.* at University of Montreal, Canada.
[34] NUMOL is developed by Becke at Queens University, Kingston, Canada.

[35] Johnson, B.G.; Gonzales, C.A.; Gill, P.M.W.; Pople, J.A. *Chem.Phys.Lett.* **1994**, *221*, 100.
[36] Handy, N.C.; Tozer, D.J.; Muirray, C.W.; Laming; G.J.; Amos, R.D.; *Isr.J.Chem.* **1993**, *33*, 331.
[37] Becke, A.D.; Dickson, R.M. *J.Chem.Phys.* **1990**, *92*, 3610.
[38] Carr, R.; Parrinello M. *Phys.Rev.Lett.* **1985**, *55*, 2471.
[39] Blöchl, P.E. *Phys.Rev. B* **1994**, *50*, 17953.
[40] Komornicki, A.; Fitzgerald, G. *J.Chem.Phys.* **1993**, *98*, 1398.
[41] Johnson, B.G.; Fisch, M.J. *J.Chem.Phys.* **1994**, *100*, 7429.
[42] Berces, A.; Dickson, R.M.; Fan, L.; Jacobsen, H.; Swerhone, D.; Ziegler, T. *Computer Physics Communications*, submitted.
[43] Malkin, V.G.; Malkina, O.L.; Salahub, D.R. *Chem.Phys,Lett.* **1993**, *204*, 80.
[44] Schreckenbach, G.; Ziegler, T. *J.Phys.Chem,* **1995**, *99*, 606.
[45] Eriksson, L.A.; Wang, J.; Boyd, *Chem.Phys.Lett.* **1993**, *211*, 88.
[46] Yang, W. *Phys.Rev. A* 1991, 44,7823.
[47] Chen, J.L.; Noodleman, L.; Bashford, D. *J.Phys.Chem.* **1994**, *98*, 11059.
[48] Ruiz-Lopez, M.F.; Bohr, F.; Martins-Costa, M.T.C. *Chem.Phys.Lett.* **1994**, *221*, 109.
[49] Evans, R. *Adv. Phys.* **1979**, *28*, 143.
[50] Stillinger, F.H.; Buff, F.P. *J. Chem. Phys.* **1962** *37*, 1.
[51] Liebowitz, J.L; Percus, J.K. *J. Math. Phys.* **1963** *4*, 116.
[52] Saam, W.F.; Ebner, C. *Phys. Rev. A* **1977** *15*, 2566.
[53] Tarazona, P. *Mol. Phys.* **1984**, *52*, 81.
[54] Curtin, W.A.; Ashcroft, N.W. *Phys. Rev. A* **1985**, *32*, 2909.
[55] Haymet, A. D. J. *Annu. Rev. Phys. Chem.* **1987**, *38*, 89.
[56] Baus, M. *J. Phys. Condens. Matter* **1987**, *2*, 2111.
[57] Singh, Y. *Phys. Rep.* **1991**, *207*, 351.
[58] Kirkwood, J.G.; Monroe, E. *J. Chem. Phys.* **1951** *9*, 514.
[59] Ramkrishnan, T. V.; Yussouff, R. *Phys. Rev. B* **1979**, *19*, 2775.
[60] Haymet, A. D. J.; Oxtoby, D. W. *J. Chem. Phys.* **1981**, *74*, 2559.
[61] Laird, B. B.; McCoy, J. D.; Haymet, A. D. J. *J. Chem. Phys.* **1987**, *87*, 5449.
[62] Wertheim, M.S. *Phys. Rev. Lett* **1963** *10*, 321.
[63] Thiele, E. *J. Chem. Phys.* **1963** *39*, 474.
[64] de Kuiper, W.L.; Vos, W.L.; Barrat, J.L.; Hansen, J.-P.; Shoulen, J.A. *J. Chem. Phys.* **1990** *93*, 5187.
[65] Lutsko, J.F.; Baus, M. *Phys. Rev. Lett.* **1990** *64*, 761.
[66] Laird, B. B.; Kroll, D. M. *Phys. Rev. A* **1990**, *42*, 4810.
[67] Pastore, G; Barrat, J.L.; Hansen, J.P., Waisman E.M. *J. Chem. Phys.* **1987** *86*, 6360.
[68] Curtin, W.A.; Ashcroft, N.W. *Phys. Rev. Lett.*, **1986** *56* 2775.
[69] Mederos, L.; Navascués, G. *Phys. Rev. E* **1994** *50*, 1301.
[70] Salgi, P.; Rajagopalan, R. *Langmuir* **1991** *7*, 1383.
[71] Ludsko, J.F.; Baus, M. *J. Phys. Conds. Matter* **1991** *3*, 6547.
[72] Laird, B. B.; Kroll, D. M. *Phys. Rev. A* **1990**, *42*, 4806.
[73] Evans, R. In *Fundamentals of Inhomogeneous Fluids*; Henderson, D., Ed.; Marcel Dekker: New York, NY, 1992, pp. 85-176.

[74] Laird, B.B.; Haymet, A.D.J. *Chemical Reviews* **1992** *92*, 1819.

[75] Oxtoby, D.W. In *Fundamentals of Inhomogeneous Fluids*; Henderson, D., Ed.; Marcel Dekker: New York, NY, 1992, pp. 407-442.

[76] Balbes, L.M.; Mascarella, S.W.; Boyd, D.B. In *A Perspective of Modern Methods in Computer-Aided Drug Design*; Lipkowitz, K.B.; Boyd, D.B., Eds.; Reviews in Computational Chemistry; VCH Publishers: New York, NY, 1994, Vol. 5; pp 337-380.

[77] Galiatsatos, V. In *Computational Methods for Modeling Polymers: An Introduction*; Lipkowitz, K.B.; Boyd, D.B., Eds.; Reviews in Computational Chemistry; VCH Publishers: New York, NY, 1995, Vol. 6; pp 149-208.

[78] Landis, C.R.; Root, D.M.; Cleveland, T. In *Molecular Mechanics Force Fields for Modeling Inorganic and Organometallic Compounds*; Lipkowitz, K.B.; Boyd, D.B., Eds.; Reviews in Computational Chemistry; VCH Publishers: New York, NY, 1995, Vol. 6; pp 73-136.

[79] Vercauteren, D.P.; Leherte, L.; Vanderveken, D.J.; Horsley, J.A.; Freeman, C.M.; Derouane, E.G. In *Molecular Modeling and Molecular Graphics of Sorbates in Molecular Sieves*; Joyner, R.W.; van Santen, R.A. Eds. ; Elementary Reaction Steps in Heterogeneous Catalysis; Kluwer Academic Publishers: Amsterdam, 1993; pp. 389-401.

[80] Newsome, J.M. in *Zeolite Structural Problems from a Computational Perspective*; von Ballmoos, R.; Higgins, J.B.; Treacy, M.M.J. Eds.; Proceedings from the Ninth International Zeolite Conference, Butterworth-Heinemann: Stoneham, MA, 1993, Vol. 1; pp 127-141.

[81] Hehre, W.J.; Burke, L.D.; Shusterman, A.J.; Pietro, W.J. *Experiments in Computational Organic Chemistry*; Wavefunction: Irvine, CA, 1993.

[82] Lipkowitz, K.B.; Boyd, D.B., Eds.; Reviews in Computational Chemistry; VCH Publishers: New York, NY, 1990-1995, Vol. 1-6.

[83] Zerner, M.C. In *Semiempirical Molecular Orbital Methods*; Lipkowitz, K.B.; Boyd, D.B., Eds.; Reviews in Computational Chemistry; VCH Publishers, Inc.: New York, NY, 1991, Vol. 2; pp 313-366.

[84] Stewart, J.J.P. In *Semiempirical Molecular Orbital Methods*; Lipkowitz, K.B.; Boyd, D.B., Eds.; Reviews in Computational Chemistry; VCH Publishers: New York, NY, 1989, Vol. 1; pp 45-82.

[85] Stewart, J.J.P. *J. Comp. Aided Mol. Design*, **1990**, 4, 1.

[86] Hehre, W.J.; Radom, L.; Schleyer, P.v.R.; Pople, J.A. *Ab Initio Molecular Orbital Theory*; Wiley: New York, NY, 1986.

[87] Bartlet, R.J.; Stanton, J.F. In *Applications of Post-Hartree-Fock Methods: A Tutorial*; Lipkowitz, K.B.; Boyd, D.B., Eds.; Reviews in Computational Chemistry; VCH Publishers: New York, NY, 1995, Vol. 5; pp. 65-170.

[88] For example the programs DSOLID, ESOCS, and PLANEWAVE from Biosym Technologies, Inc./Molecular Simulations, Inc., San Diego, CA.

Density-Functional Methods in Electronic Structure Calculation

Chapter 2

Effective One-Electron Potential in the Kohn–Sham Molecular Orbital Theory

Evert Jan Baerends, Oleg V. Gritsenko, and Robert van Leeuwen

Afdeling Theoretische Chemie, Vrije Universiteit, De Boelelaan 1083, 1081 HV, Amsterdam, Netherlands

Density functional theory has received great interest mostly because of the accurate bonding energies and related properties (geometries, force constants) it provides. However, the Kohn-Sham molecular orbital method, that is almost exclusively used, is more than a convenient tool to generate the required electron density. The effective one-electron potential in the Kohn-Sham equations is intimately related to the physics of electron correlation. We demonstrate that it is useful to break down the exchange-correlation part of the potential into a part that is directly related to the total energy (the hole potential or screening potential) and a socalled response part that is related to "response" of the exchange-correlation hole to density change. The latter part is poorly represented by the generalized gradient approximation, explaining why this approximation yields accurate total energies but fails for simple orbital related quantities such as the HOMO orbital energy. A simple modelling of the response potential is proposed.
We stress the usefulness of Kohn-Sham orbitals in chemistry, both quantitatively (by providing in principle the exact electron density, to be used in density functionals for the energy) and qualitatively (for use in qualitative MO theory).

Although density-functional methods have been around for some time, such as the Thomas-Fermi method and the $X\alpha$ method, a rigorous foundation has only been given with the formulation of the Hohenberg-Kohn theorems [1]. In the second place, the Kohn-Sham one-electron model [2] has endowed density functional theory with a very expedient and at the same time very successful method for practical applications. The use of both the old and the new density functional methods has been stimulated by their providing relatively high accuracy for relatively low cost. In particular the so-called generalized gradient approximations (GGA) [3, 4, 5] are clearly a major step forward (much more so in fact than adding electron gas correlation effects in the local-density approximation (LDA) to the exchange-only LDA or Xα method). In chemistry, applications almost invariably use the Kohn-Sham molecular orbital model. The exchange-correlation functional $E_{XC}[\rho]$, which is defined in the context of the Kohn-Sham model, has been the primary focus of theoretical work directed towards the immediate goal of obtaining highly accurate energies from Kohn-Sham calculations. In this line of research, the Kohn-Sham one-electron model has been viewed primarily as a convenient method to generate accurate electron densities, which can be used in some approximate $E_{XC}[\rho]$ to obtain the total energy. The theoretical status of the Kohn-Sham model has received comparatively little attention, as may be evident from the frequently voiced opinion that the Kohn-Sham orbitals are just a means to generate electron densities but do not have any physical meaning themselves. However, this is a far too restricted view on the Kohn-Sham model. As we will demonstrate, the effective potential of the Kohn-Sham model has an intimate connection with the physics of the exchange and correlation effects in atoms and molecules. The Kohn-Sham orbitals thus represent electrons that move in a potential that is certainly as realistic as the Hartree-Fock "potential" and indeed has some advantages. There is no reason to believe that the Kohn-Sham orbitals are any less "physical" or useful than the Hartree-Fock orbitals and they may and have indeed been used quite succesfully in qualitative MO explanations that are so typical for present-day chemistry.

The physics embodied in the effective one-electron potential of the Kohn-Sham model leads to certain requirements that have to be fulfilled by these potentials. Well-known ones are the $-1/r$ behaviour for $r \to \infty$ and the finiteness at the nucleus. Other properties, such as certain invariance properties [6], special behaviour at atomic shell boundaries [7, 8, 9] and at the bond midpoint [10, 11] have also been identified. It is a rather stringent test for approximations to the exchange-correlation energy $E_{XC}[\rho]$, that the exchange-correlation potential v_{XC} that may be derived from it by functional differentiation, obeys these requirements. However, in order to obtain a complete assessment of the quality of trial $E_{XC}[\rho]$ and their functional derivatives, it would be necessary to obtain the exact v_{XC} at all points in space. Several procedures have been published [12, 6, 13] to generate the Kohn-Sham potential that belongs to a given density, in the sense that the occupied eigenfunctions of that potential produce the given density. Application of the Hohenberg-Kohn theorem to the Kohn-Sham system of non-interacting electrons proves that the Kohn-Sham potential so obtained is unique. A detailed study of the potentials derived from the current GGA's for $E_{XC}[\rho]$ shows that, even if these GGA's are quite succesful for the energy, their potentials exhibit important deficiencies. This knowledge may be helpful when devising improved functionals for the energy.

One may also wonder if it is possible and useful to model Kohn-Sham potentials as density functionals directly. Since Kohn-Sham potentials can now be obtained for a variety of systems (the first applications to molecules are just appearing [14]), there will be more complete data to judge proposals for model potentials than there is for functionals for the energy. The latter are usually judged only by their performance for the energy, which is an integral over all space of the energy density, in which local deficiencies may have cancelled. Locally different energy densities may lead to (almost) identical energies, and given their non-uniqueness it is hard to judge different proposals for energy densities. The Kohn-Sham potential on the other hand is a unique, local function of $\vec{r}$.

These considerations naturally lead to the question: is it possible to determine the energy of a system, given its Kohn-Sham potential? This is indeed the case, when one is prepared to perform a line-integral in the space of densities on which $E_{XC}[\rho]$ and $v_{XC}[\rho]$ are defined [15]. This, however, requires knowledge of the KS potential for all ρ's along the line. It is maybe more practical to study and model $E_{XC}[\rho]$ and $v_{XC}[\rho]$ simultaneously, since it is possible to identify a part of the Kohn-Sham potential that can serve as energy density and thus leads directly to the energy, and a part that serves to add features to the potential that are required to generate the exact density, but that do not enter the energy calculation.

It is the purpose of this contribution to summarize our present understanding of the Kohn-Sham potentials. The paper is structured as follows. In section 2 we introduce a method to calculate exact KS potentials from exact (in practice, highly accurate) electron densities. A comparison is made between such accurate potentials and the GGA potentials, highlighting important deficiencies of the GGA potentials. In section 3 the KS potential is analyzed and in particular its relationship established with traditional concepts in the theory of electron correlation such as density matrices and Fermi and Coulomb correlation holes [16]. In the last section we specialize to the exchange-only case and discuss the breaking up of the KS potential into a screening part, directly related to the potential of the exchange hole charge density, and a response part, related to the "response" of the hole to density change. Characteristic features of these components of the potential are used to model them accurately.

2 Exact and GGA potentials in atoms

We will use the procedure outlined in ref. [6] to generate Kohn-Sham potentials from a given density. The given density is the best available density, usually from an extensive configuration interaction calculation.

If we multiply the Kohn-Sham equations

$$(-\frac{1}{2}\nabla^2 + v_s(\vec{r}))\phi_i(\vec{r}) = \epsilon_i\phi_i(\vec{r}) \tag{1}$$

from which the density is obtained as:

$$\sum_i^N |\phi_i(\vec{r})|^2 = \rho(\vec{r}) \tag{2}$$

(N is the number of electrons in the system), by ϕ_i^* and sum over i, we obtain after dividing by ρ:

$$v_s(\vec{r}) = \frac{1}{\rho(\vec{r})} \sum_i^N \frac{1}{2}\phi_i^*(\vec{r})\nabla^2\phi_i(\vec{r}) + \epsilon_i|\phi_i(\vec{r})|^2 \tag{3}$$

We now define an iterative scheme using this equation. We want to calculate the potential corresponding to the density ρ. Suppose that at some stage in the iteration we have calculated orbitals ϕ_i^o with eigenvalues ϵ_i^o and density ρ^o and potential v^o. In the next step we define the new potential:

$$v^n(\vec{r}) = \frac{1}{\rho(\vec{r})} \sum_i^N \frac{1}{2} \phi_i^{o*}(\vec{r}) \nabla^2 \phi_i^o(\vec{r}) + \epsilon_i^o |\phi_i^o(\vec{r})|^2$$

$$= \frac{\rho^o(\vec{r})}{\rho(\vec{r})} v^o(\vec{r}) \tag{4}$$

Using this potential we calculate new orbitals and a new density and define in the same way a new potential. This procedure is continued until the density calculated from the orbitals is the same as the given density. In practice until:

$$\max_{\vec{r}} |1 - \frac{\rho^o(\vec{r})}{\rho(\vec{r})}| < \epsilon \tag{5}$$

with ϵ a given threshold. In practice, several precautions have to be taken to ensure proper convergence. For instance, one should take care to keep the prefactor in the last term of equation 4 in each iteration within an acceptable range:

$$1 - \delta < \frac{\rho^o(\vec{r})}{\rho(\vec{r})} < 1 + \delta \tag{6}$$

for example with $\delta = 0.05$. It may also be expedient to split off the external potential $v(\vec{r})$ (in our case always the nuclear attraction) and apply the "update procedure" only to the electronic part of the potential, i.e. the Hartree potential and the exchange-correlation potential:

$$v_s = v(\vec{r}) + v_{el}(\vec{r}) \tag{7}$$

where

$$v_{el}(\vec{r}) = \int \frac{\rho(\vec{r}_1)}{|\vec{r} - \vec{r}_1|} d\vec{r}_1 + v_{XC}(\vec{r}) \tag{8}$$

and v is the external potential, in our case the nuclear potential. The scheme is not guaranteed to converge as there are densities which are not non-interacting v-representable. However if the density is non-interacting v-representable and if the procedure converges then its limit is unique as guaranteed by the Hohenberg-Kohn theorem applied to a non-interacting electron system [1].

We show in fig. 1a the exchange-correlation potential obtained in this way from the accurate density for Ne obtained by Bunge and Esquivel [17] using configuration interaction calculations in a large Slater type basis set. Comparison is made to a GGA potential, obtained by functional differentiation of an exchange-correlation energy with Becke's [5] gradient corrections to the electron gas exchange energy density and Perdew's [3] gradient correction to the correlation energy density. The electron gas (LDA) energy densities have been used in the Vosko-Wilk-Nusair parametrization [18]. The picture demonstrates that the GGA potential is deficient both asymptotically (it decays more rapidly for $r \to \infty$ than the required Coulmbic long range $-1/r$) and close to the nucleus, where it exhibits an erroneous though weak Coulombic ($-k/r$, $k \approx 0.02$) singularity [6]. One may observe that the LDA potential does not differ much from the GGA potential for medium and large r (it also is not correct asymptotically), and at the nucleus the LDA potential is actually in better agreement with the exact v_{xc}, having similar slope and being only slightly too little attractive.

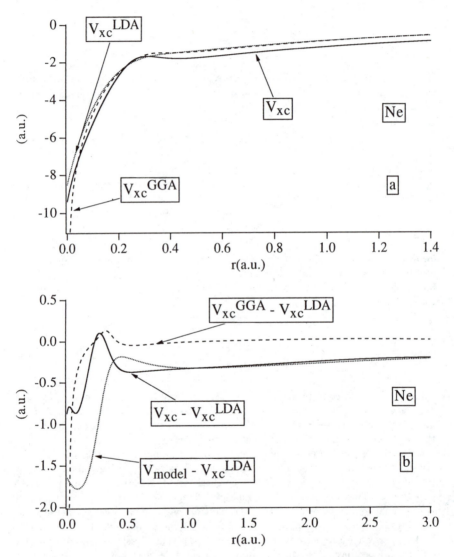

Figure 1. Exchange-correlation potentials for the neon atom:
a) total potentials and b) non-local components of the
potentials.

The fact that the Becke-Perdew gradient corrections, which provide quite important corrections to the energy density, fail to correct the LDA potential is very clearly visible in fig. 1b, where the difference $v_{XC} - v_{LDA}$ is compared to the sum of Becke and Perdew correction potentials. Asymptotically there is virtually no correction and in the inner region the GGA potential deviates strongly from $v_{XC} - v_{LDA}$, although it does make an attempt to model the peak in v_{XC} at the boundary between 1s and 2s shells. The model potential introduced in ref. [6] provides good asymptotic behaviour but is clearly deficient in the inner region and also in the border region between 1s and 2s.

The too rapid decay of the LDA and GGA potentials in the asymptotic region has the important consequence that the one-electron energy of the highest occupied orbital is not equal to the exact ionisation energy, as it should be [19], but is far too high. The error is in fact quite large, in the order of .25 hartree (5 - 6 eV), as may be seen in table I. It is also clear from this table that the nonlocal or gradient corrections do not provide any improvement. We may therefore expect the electron density to be too diffuse, which may have consequences for properties that depend on the outer regions such as dipole moments and (hyper)polarizabilities and in particular properties related their derivatives such as infrared and Raman intensities.

It is clear from these results that there is no reason to expect that the GGA approximation to the KS potential will generate a better density than the LDA. The general practice, particularly in solid state calculations, to use the LDA potential during the SCF and to apply the GGA only after convergence, in order to calculate the GGA functional for the energy with the LDA density, seems to be justified. The deficiency of the GGA potential has been also reported for a model of two interacting electrons in an external harmonic potential [20], which can be solved analytically for certain parameter values.

We will in the next section break down the KS potential in physically meaningful parts, which will allow us to see the relationship with the physics of electron correlation and to identify the causes for the discrepancy between the exact and approximate (both LDA and GGA) potentials.

3 Kohn-Sham potentials and electron correlation

The exact total energy of a system may be written as

$$E = \langle T \rangle + \langle V \rangle + \frac{1}{2} \int \frac{\Gamma(1,2)}{r_{12}} d1 d2 \tag{9}$$

Here T is the kinetic energy, V is the potential energy in the external field and Γ is the (diagonal part of the) two-electron density matrix. $\Gamma(1,2)$ is the probability to find an electron with coordinates 1 (at position $\vec{r}_1$ with spin s_1) and simultaneously another electron with coordinates 2 (at position $\vec{r}_2$ with spin s_2). It embodies the physics of the correlation phenomenon, which is exhibited clearly by splitting off the uncorrelated (independent particle) probability, which is just the product of the one-

Table   I.Ionisation energies and electron affinities (ionisation energies of the negative ions) from the highest occupied Kohn-Sham orbital energies: HF - the Hartree-Fock approxima- tion, LDA - the local density approximation, BP - a combination of Becke's [5] approximation for exchange and Perdew's [3] approximation for correlation, MODEL - the model potential of ref. [6], EXPT - the experimental data

ATOM	HF	LDA	BP	MODEL	EXPT
H	0.500	0.234	0.280	0.440	0.500
He	0.918	0.571	0.585	0.851	0.903
Be	0.309	0.206	0.209	0.321	0.343
Ne	0.850	0.490	0.496	0.788	0.792
Ar	0.591	0.381	0.380	0.577	0.579
Kr	0.524	0.346	0.344	0.529	0.517
Xe	0.457	0.310	0.308	0.474	0.446
ION	HF	LDA	BP	MODEL	EXPT
F^-		-0.097	-0.099	0.128	0.125
Cl^-		-0.022	-0.023	0.140	0.133
Br^-		-0.008	-0.009	0.140	0.124
I^-		+0.005	+0.004	0.139	0.112
MOLEC	HF	LDA	BP	MODEL	EXPT
N_2	0.622	0.328	0.322	0.557	0.573
F_2	-	0.339	0.334	0.607	0.582
CO	0.551	0.334	0.336	0.529	0.515

electron probability densities $\rho(1)$ and $\rho(2)$

$$\Gamma(1,2) = \rho(1)\rho(2) + \Gamma_{XC}(1,2) \tag{10}$$

The exchange-correlation part of Γ describes how the probability to find a second electron is modified by the presence of the first one. This may be cast in the language of an exchange-correlation hole surrounding the electron at 1 by using the definition of the conditional probability $\Gamma(1,2)/\rho(1)$ to find an electron at 2 when one is known to be at 1:

$$\rho^{cond}(2|1) = \rho(2) + \frac{\Gamma_{XC}(1,2)}{\rho(1)} \tag{11}$$

The probability of the other electrons to be at 2 is the unconditional probability $\rho(2)$ plus the exchange-correlation hole $\Gamma_{XC}(1,2)/\rho(1)$ that the electron at 1 digs around itself in the unconditional density $\rho(2)$. The electron-electron part of the total energy may accordingly be written as

$$
\begin{aligned}
\langle W \rangle &= \frac{1}{2}\int \frac{\rho(1)\rho(2)}{r_{12}}d1d2 + \frac{1}{2}\int \frac{\Gamma_{XC}(1,2)}{r_{12}}d1d2 \\
&= \frac{1}{2}\int \rho(1)\int \frac{\rho(2)}{r_{12}}d2d1 + \frac{1}{2}\int \rho(1)\int \frac{\Gamma_{XC}(1,2)}{\rho(1)}\frac{1}{r_{12}}d2d1 \\
&= \frac{1}{2}\int \rho(1)V_{Hartree}(1)d1 + \frac{1}{2}\int \rho(1)v^{hole}(1)d1 \tag{12} \\
&= E_{Hartree} + W_{XC} \tag{13}
\end{aligned}
$$

It is customary to write the conditional electron density in terms of the pair-correlation factor $g(1,2)$

$$\rho^{cond}(2|1) = g(1,2)\rho(2) = \rho(2) + \frac{\Gamma_{XC}(1,2)}{\rho(1)} \tag{14}$$

so that the hole density may be written as

$$\rho^{hole}(2|1) = \frac{\Gamma_{XC}(1,2)}{\rho(1)} = (g(1,2) - 1)\rho(2) \tag{15}$$

and

$$
\begin{aligned}
W_{XC} &= \frac{1}{2}\int \rho(1)\int \frac{\rho^{hole}(2|1)}{r_{12}}d2d1 \\
&= \frac{1}{2}\int \frac{\rho(1)(g(1,2)-1)\rho(2)}{r_{12}}d1d2 \tag{16}
\end{aligned}
$$

The definition of the exchange-correlation part W_{XC} of the electron-electron interaction energy is convenient for the definition of the exchange-correlation energy E_{XC} that features in Kohn-Sham theory. When a Kohn-Sham calculation has been performed, an alternative way to write the total energy would be

$$E = T_s[\rho] + \int \rho(1)v(1)d1 + \frac{1}{2}\int \frac{\rho(1)\rho(2)}{r_{12}}d1d2 + E_{XC} \tag{17}$$

Since the KS potential is uniquely determined by the density, so are the solutions of the one-electron Kohn-Sham equations. Therefore, the kinetic energy of the electrons described by Kohn-Sham orbitals is a functional of the density:

$$T_s[\rho] = \sum_i^N \int \phi_i^*[\rho](1)(-\frac{1}{2}\nabla^2)\phi_i[\rho](1)d1 \qquad (18)$$

This follows of course also immediately from the Hohenberg-Kohn theorem, since this theorem implies that the wavefunction of a nondegenerate system is a functional of the density, and therefore every expectation value is, including that of the kinetic energy. The theorem holds for systems with arbitrary electron interaction, therefore also for the non-interacting Kohn-Sham system. So the first three terms in the r.h.s. of eq.(17), as well as the l.h.s. are defined and this equation therefore defines the socalled exchange-correlation energy E_{XC}. We emphasize the well-known fact that E_{XC} is different from the traditional quantum chemical definition of the exchange-correlation energy as the sum of the Hartree-Fock exchange energy plus the correlation energy, the latter being traditionally defined as the difference between the exact and Hartree-Fock energies. If we compare the equations (17) and (9) for the exact total energy, it is clear that

$$\begin{aligned} E_{XC}[\rho] &= T[\rho] - T_s[\rho] + W_{XC} \\ &= T_{XC} + W_{XC} \end{aligned} \qquad (19)$$

where we have used the fact that also the exact kinetic energy $\langle T \rangle$ is a functional of ρ, and have written the difference between the exact kinetic energy and the KS kinetic energy as the exchange-correlation contribution to T (also often simply referred to as the correlation part of the kinetic energy, T_c). It is to be noted that the Kohn-Sham exchange-correlation energy consists of a kinetic part and a pure exchange-correlation part of the electron-electron interaction energy. In contrast, the traditional definition $E_X^{HF} + E_{corr}$ contains - sometimes sizable - corrections to the electron-nuclear and Hartree energies due to the difference $\Delta\rho$ between exact and Hartree-Fock densities:

$$\begin{aligned} \Delta\rho(1) &= \rho(1) - \rho^{HF}(1) \\ E_X^{HF} + E_{corr} &= E_X^{HF} + E - E^{HF} \\ &= T[\rho] - T^{HF} \\ &\quad + \int \Delta\rho(1)v(1)d1 \\ &\quad + \int \frac{\Delta\rho(1)\rho(2)}{r_{12}}d1d2 + \frac{1}{2}\int \frac{\Delta\rho(1)\Delta\rho(2)}{r_{12}}d1d2 \\ &\quad + W_{XC} \end{aligned} \qquad (20)$$

The Kohn-Sham definition has the advantage that it only consists of the exchange-correlation corrections to the kinetic energy $(T[\rho]-T_s)$ and electron-electron interaction energy (W_{XC}) and is not "cluttered" by other terms. These other terms such as the correlation correction to the electron-nuclear energy and the corrections to the Hartree energy are often quite large, see table 5.1 in ref. [16]. For instance, for the N_2 molecule, the correlation correction $\int \Delta\rho v d\vec{r}$ to the electron-nuclear energy is -13.8 eV, to be compared to the total correlation energy of -11.1 eV and -11.0 eV for the correlation correction to the electron-electron interaction energy. The traditional definition has of course the operational advantage that the reference Hartree-Fock energy does not contain unknowns and can be calculated to virtually arbitrary accuracy. This evidently is not the case for the Kohn-Sham system, E_{XC} and its functional derivative v_{xc} being only known approximately. Obtaining them exactly is equivalent to a full solution of the many-electron problem.

We are now in a position to consider the physical interpretation of the Kohn-Sham potential $v_{xc}(\vec{r}) = \delta E_{XC}/\delta\rho(\vec{r})$. It is possible to take the above equation(19) as a starting point, but it is also possible to incorporate the kinetic energy part in an expression that formally is similar to eq.(16), but in which the pair-correlation factor has been redefined by the so-called coupling-constant integration [21, 22, 23, 24]. The coupling constant integrated hole is described by the "average" pair correlation factor $\bar{g}(1,2)$, in terms of which E_{XC} may be written as

$$E_{XC} = \frac{1}{2}\int \frac{\rho(1)(\bar{g}(1,2)-1)\rho(2)}{r_{12}}d2d1$$
$$= \frac{1}{2}\int \rho(1)\bar{v}_{scr}(1)d1 \qquad (21)$$

The screening (or the hole) potential $\bar{v}_{scr}$ is now due to an average exchange-correlation hole. It is referred to as the screening potential [7], since the exchange-correlation effects embodied in $\bar{g}$ may be considered as screening effects on the full electron-electron interaction $1/r_{12}$. The considerations concerning exchange-correlation holes, on which present day approximations for E_{XC} are based, use almost always the coupling constant integrated form, either implicitly (e.g. by referring to electron gas calculations of exchange-correlation that employ $\bar{g}$) or explicitly (cf. the explicit introduction of the $\lambda = 0$ limit ("exact exchange") by Becke [26]). We will use both expressions.

The Kohn-Sham potential

$$v_s(\vec{r}) = v(\vec{r}) + V_{Hartree}(\vec{r}) + v_{xc} \qquad (22)$$

is related to E_{XC} since

$$v_{xc} = \frac{\delta E_{XC}}{\delta\rho(\vec{r})} \qquad (23)$$

Using eq.(21) we may write v_{xc} as

$$v_{xc}(3) = \frac{\delta E_{XC}}{\delta\rho(3)} = \bar{v}_{scr}(3) + \bar{v}_{scr}^{response}(3) \qquad (24)$$

where

$$\bar{v}_{scr}^{response}(3) = \frac{1}{2}\int \rho(1)\frac{\delta\bar{g}(1,2)}{\delta\rho(3)}\rho(2)d1d2 \qquad (25)$$

Eq. 24 demonstrates the physical nature of the components of v_{xc} and therefore of the Kohn-Sham potential: the most important part of v_{xc} is just the potential due to the averaged exchange-correlation hole. It is interesting to note that this part of v_{xc}

is directly related to the exchange-correlation energy according to equation (21). In other words, the XC energy density is just onehalf times the screening potential:

$$E_{XC} = \int \rho(\vec{r})\epsilon_{xc}(\vec{r})d\vec{r} = \frac{1}{2}\int \rho(\vec{r})\bar{v}_{scr}(\vec{r})d\vec{r} \qquad (26)$$

This relation between the exchange-correlation energy and the screening potential part of v_{xc} may be exploited when modelling exchange-correlation functionals for the energy and potential simultaneously. Of course v_{xc} also contains an additional term, which we have called the response part [7]. It is a measure of the sensitivity of the pair-correlation factor to density variations. These density variations may be understood in the following way. If the density changes to $\rho + \delta\rho$, also assumed v-representable, then according to the Hohenberg-Kohn theorem this changed density corresponds uniquely to an external potential $v + \delta v$. For the system with external potential $v + \delta v$ we have the corresponding Kohn-Sham system and the coupling-constant integrated pair-correlation factor $\bar{g} + \delta\bar{g}$. So the derivative of $\bar{g}$ occurring in the response potential may be regarded as the response of $\bar{g}$ to density changes $\delta\rho$ caused by potential changes δv. The response potential does not affect the energy directly, but it does so indirectly since it determines, as part of the Kohn-Sham potential, the SCF density in the Kohn-Sham calculation. As we will see below (cf. also ref. [7]), the response part of the potential has less pronounced features than the hole potential but is certainly required to obtain accurate KS orbitals and densities.

It is of course also possible to use eq.(19) when deriving v_{xc}. It is convenient to use for the exact kinetic energy $\langle T \rangle = T[\rho]$ and for the kinetic energy of the noninteracting electrons $T_s[\rho]$ the expressions derived in ref. [10]:

$$
\begin{aligned}
T[\rho] &= T_W + \int \rho(\vec{r})v_{kin}(\vec{r})d\vec{r} \\
T_s[\rho] &= T_W + \int \rho(\vec{r})v_{s,kin}(\vec{r})d\vec{r}
\end{aligned}
\qquad (27)
$$

Here T_W is the Weiszäcker kinetic energy for a density ρ, which is just N times the kinetic energy of the normalized density amplitude ("density orbital") $\sqrt{\rho/N}$,

$$T_W[\rho] = N \int \sqrt{\frac{\rho}{N}}(-\frac{1}{2}\nabla^2)\sqrt{\frac{\rho}{N}}d\vec{r} \qquad (28)$$

The kinetic potential v_{kin} can be related to the electron correlation by expressing it in terms of the conditional amplitude $\Psi/\sqrt{\rho}$ or in terms of the derivative of the one-electron density matrix [10]. Taking the functional derivative of E_{XC} of eq.(19) to obtain v_{xc} leads of course to kinetic potentials and their responses:

$$E_{XC} = \frac{1}{2}\int \frac{\rho(1)(g(1,2)-1)\rho(2)}{r_{12}}d2d1 + \int \rho(1)(v_{kin}(1) - v_{s,kin}(1))d1 \qquad (29)$$

$$v_{xc} = \frac{\delta E_{XC}}{\delta\rho} = v_{scr} + v_{scr}^{response} + (v_{kin} - v_{s,kin}) + (v_{kin}^{resp} - v_{s,kin}^{resp}) \qquad (30)$$

The kinetic potential plays an important role in the typical molecular left-right correlation effect [10]. This particular form of correlation leads to a peak in v_{kin} at the bond midpoint (a ridge around the bond midplane) in molecules. Other features of these potentials, such as step structure in v_{kin}^{resp} in atoms, are discussed in ref. [7]. In the present paper however we will concentrate on atoms, where exchange is the dominating factor in the correlation between electron motions. We will for this case restrict ourselves to the exchange-only density-functional theory as made operational in the optimized potential method [27, 28]. This will enable us to highlight the relative importance of the screening and the screening-response parts of the Kohn-Sham potential and to discuss the accuracy of the LDA and GGA approximations for these potentials.

4   Screening and response potentials in exchange-only density functional theory

Given an external potential, one may ask for the local potential that generates the orbitals that minimize the energy of a single-determinantal wavefunction. This problem, nowadays often referred to as the exchange-only DFT, has been addressed by Sharp and Horton [29] and Talman and Shadwick [27]. The local potential obtained within this "optimized potential model" (OPM) is plotted in fig. 2 for Cd.

The correlation hole of the single-determinantal wavefunction is just the Fermi hole and the screening part of the effective potential is just the potential due to this hole:

$$\rho^{Xhole}(\vec{r}_2\sigma|\vec{r}_1\sigma\prime) \;=\; \delta_{\sigma\sigma\prime}\frac{\Gamma_X(\vec{r}_1\sigma,\vec{r}_2\sigma)}{\rho(\vec{r}_1\sigma)} \tag{31}$$

$$= \; -\frac{\sum_{i,j=1}^{N_\sigma} \phi_{i\sigma}(\vec{r}_1)\phi_{i\sigma}^*(\vec{r}_2)\phi_{j\sigma}^*(\vec{r}_1)\phi_{j\sigma}(\vec{r}_2)}{\rho_\sigma(\vec{r}_1)} \tag{32}$$

$$v_{Xscr\sigma}(\vec{r}_1) = \int \frac{\rho^{Xhole}(\vec{r}_2\sigma|\vec{r}_1\sigma)}{r_{12}}d\vec{r}_2 \tag{33}$$

This potential is often referred to as the Slater potential, since it is has been analyzed thoroughly by Slater [31, 32]. The Slater potential $v_{Xscr}(\vec{r}_1)$ ($V_{OPM,scr}$ in the figure) does reflect the shell structure, having different slopes within different shells, but it exhibits less structure than the full exchange potential, in that the characteristic intershell maxima of the latter are missing. It has been emphasized that these "bumps" are needed to lower the total energy [33]. Interestingly, as shown by the plot of the difference $v_X^{OPM} - v_{scr}^{OPM} = v_{resp}^{OPM}$, the maxima originate by the addition of a very simply structured response potential to the Slater potential: v_{resp}^{OPM} is flat over the shell regions and exhibits steps in going from one shell to the next. (This intriguing step structure of v_{resp}^{OPM} is reminiscent of the step structure found for the response part of v_{kin}, cf. ref. [7].)

The origin of the step structure in the screening-response potential is related to special properties of the exchange hole. It has been observed by Luken and coworkers [34, 35] that the shape of the exchange hole is almost independent of the position $\vec{r}_1$ of the reference electron as long as this position stays within an atomic shell, or, in molecules, within the region covered by one localized orbital. The hole density is then very similar to just the shell density or the localized orbital density. When the reference position crosses a shell boundary the hole undergoes rapid change to the shape that is characteristic for the new shell. This can be illustrated by considering the hole density as the square of a hole amplitude and expanding the amplitude in the orbitals from which the determinantal wavefunction is build. When the reference position is within a certain shell, only the orbitals that belong to that shell acquire significant coefficients, cf. ref. [16]. The phenomenon is evident in the plot of the pair correlation factor $g_\sigma(\vec{r}_1, \vec{r}_2)$ for the OPM determinantal wavefunction of Be in fig. 3 (the hole density is just the plotted $(g_\sigma(\vec{r}_1, \vec{r}_2) - 1)$ times $\rho(\vec{r}_2)$).

The boundary between the $1s$ and $2s$ shells is at ca. 1 bohr. Note that as long as the reference position r_1 is inside the $1s$ shell, the hole as function of r_2 is approximately $-\rho(r_2)$ within the $1s$ region, and negligible outside that region. However, when r_1 is in the $2s$ region, the hole is negligible in the $1s$ region and equal to $-\rho(r_2)$ in the $2s$ region. In order to derive that these properties of the hole lead to the observed step structure in the screening-response potential, we have to consider more closely eq.25 (the bars can be omitted since we are dealing with the exchange only case where the averaging process implied by the coupling-constant integration does not occur). Using for the pair correlation factor in the exchange-only case the well known expression

$$g_{s\sigma}(\vec{r}_2, \vec{r}_3) = 1 - \frac{\sum_{i,j}^{N_\sigma} \phi_i(\vec{r}_2)\phi_i^*(\vec{r}_3)\phi_j(\vec{r}_3)\phi_j^*(\vec{r}_2)}{\rho_\sigma(\vec{r}_2)\rho_\sigma(\vec{r}_3)}, \tag{34}$$

it is clear that taking the derivative with respect to $\rho_\sigma(\vec{r}_1)$ requires knowledge of

$$\frac{\delta\phi_i(\vec{r}_2)}{\delta\rho_\sigma(\vec{r}_1)} = \int \frac{\delta\phi_i(\vec{r}_2)}{\delta v_{s\sigma}(\vec{r}_3)}\frac{\delta v_{s\sigma}(\vec{r}_3)}{\delta\rho_\sigma(\vec{r}_1)}d\vec{r}_3 \tag{35}$$

The derivative of ϕ_i with respect to the Kohn-Sham potential is known in terms of the Kohn-Sham orbitals and one-electron energies:

$$\frac{\delta\phi_i(\vec{r}_2)}{\delta v_{s\sigma}(\vec{r}_3)} = -G_{i\sigma}(\vec{r}_2, \vec{r}_3)\phi_{i\sigma}(\vec{r}_3) \tag{36}$$

where $G_{i\sigma}$ is the following Green's function:

$$G_{i\sigma}(\vec{r}_2, \vec{r}_3) = \sum_{j\neq i} \frac{\phi_j(\vec{r}_2)\phi_j^*(\vec{r}_3)}{\epsilon_{j\sigma} - \epsilon_{i\sigma}} \tag{37}$$

The functional derivative of the Kohn-Sham potential with respect to the density is the inverse density response function:

$$\frac{\delta v_{s\sigma}(\vec{r}_3)}{\delta\rho_\sigma(\vec{r}_1)} = \chi_{s\sigma}^{-1}(\vec{r}_3, \vec{r}_1) \tag{38}$$

It is possible to derive a simple expression for the screening-response potential by application of an "effective energy denominator" type of approximation for the Green's function, as has been done before by Sharp and Horton [29] and Talman and Shadwick [27]. We refer to ref. [8] for details of the derivation. The resulting expression for the screening-response potential becomes

$$v_{scr,\sigma}^{resp}(\vec{r}_1) = \sum_{i}^{N_\sigma-1} w_{i\sigma}\frac{|\phi(\vec{r}_1)|^2}{\rho_\sigma(\vec{r}_1)} \tag{39}$$

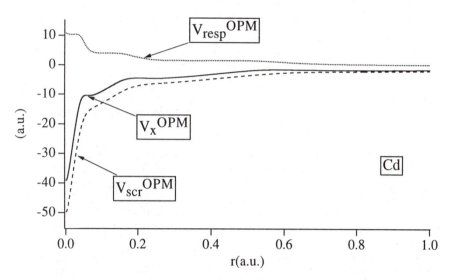

Figure 2. Exchange potential of the optimized potential model and its components for the cadmium atom.

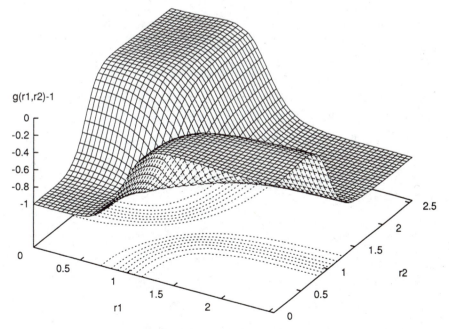

Figure 3. The pair-correlation factor as a function of the radial coordinates of electrons for the beryllium atom.

where the constant $w_{i\sigma}$ is defined in terms of the expectation value of the local exchange potential $v_{x\sigma}$ over the orbital $\phi_{i\sigma}$ minus a sum over two-electron integrals of exchange type:

$$w_{i\sigma} = \langle \phi_i | v_X | \phi_i \rangle - \sum_k^{N_\sigma} \langle \phi_{k\sigma} \phi_{i\sigma} | \frac{1}{r_{12}} | \phi_{i\sigma} \phi_{k\sigma} \rangle \tag{40}$$

This is precisely the second term (now to be identified with the response part of the potential) of the approximate expression that had been obtained earlier by Krieger, Li and Iafrate [28] for the local KS potential in the exchange-only case. KLI derived their approximation using the same approximation to the Green's function. It has been demonstrated by Krieger et al. that their approximation is very accurate, while providing at the same time a computationally more tractable way to calculate the potential than the Talman-Shadwick procedure. From the form of the expression 39 we may understand its step character, the height of the steps being governed by the constants $w_{i\sigma}$. We note that the summation in eq. 39 does not include the highest occupied orbital. The step height in the region of the outermost orbital is zero, the potential is therefore short-ranged and does not destroy the $-1/r$ asymptotic behaviour of the hole potential $v_{X\,scr}$.

It is a computational disadvantage that still the exchange-type two-electron integrals over the KS orbitals have to be calculated to obtain the step heights w_i. It is possible to simplify the representation of the response part $v_{X\,scr}^{resp}$ even further by the *Ansatz* that w_i can be written as a function of the orbital energies ϵ_i [14]. The form of the function can be chosen such that various requirements are met:

    a) Gauge invariance. Addition of a constant to the eigenvalues ϵ_i, by adding a constant to the potential, should not affect the steps. To satisfy this requirement, we choose w_i to be a function of the difference $(\mu - \epsilon_i)$

$$w_i = f(\mu - \epsilon_i), \tag{41}$$

where μ is the Fermi level of a given system, which is equal to the one-electron energy of the highest occupied orbital, $\mu = \epsilon_N$.

    b) Proper scaling. The exchange potential and its components v_{scr} and v_{scr}^{resp} have the following scaling property

$$v_x([\rho_\lambda]; \vec{r}) = \lambda v_x([\rho]; \lambda \vec{r}) \tag{42}$$

where

$$\rho_\lambda(\vec{r}) = \lambda^3 \rho(\lambda \vec{r}) \tag{43}$$

while ϵ_i has the scaling property

$$\epsilon_i[\rho_\lambda] = \lambda^2 \epsilon_i[\rho(\vec{r})] \tag{44}$$

To provide (42), the function f from eq. 41 should scale as follows

$$f(\lambda^2(\mu - \epsilon_i)) = \lambda f(\mu - \epsilon_i) \tag{45}$$

and so we find the square root of $(\mu - \epsilon_i)$ to be the properly scaling function f

$$w_i = f(\mu - \epsilon_i) = K[\rho]\sqrt{\mu - \epsilon_i} \tag{46}$$

c) Short range. By definition, (46) satisfies the condition $w_N = 0$. Owing to this, the highest occupied orbital ϕ_N does not contribute to the numerator of (39), thus providing the short-range behaviour of our proposed model response potential v_{resp}^{mod}

$$v_{resp}^{mod}(\vec{r}) = K[\rho] \sum_{i=1}^{N} \sqrt{\mu - \epsilon_i} \frac{|\phi_i(\vec{r})|^2}{\rho(\vec{r})} \tag{47}$$

d) Correct electron gas limit. It is desirable that the model response potential be correct in the limit of a homogeneous electron gas. $K[\rho]$ in (47) is a numerical coefficient, which can be determined from this requirement. For the gas of density ρ the exact $v_{X scr}^{resp}$ has the form

$$v_{X scr}^{resp} = \frac{k_F}{2\pi} \tag{48}$$

where k_F is the Fermi wavevector

$$k_F = (3\pi^2 \rho)^{\frac{1}{3}} \tag{49}$$

Putting v_{resp}^{mod} of eq. (47) to be equal to (48), one can calculate $K_g[\rho]$. For the homogeneous electron gas the Kohn-Sham orbitals and eigenvalues are given by

$$\phi_{\vec{k}}(\vec{r}) = \frac{1}{\sqrt{V}} e^{i\vec{k}\vec{r}} \tag{50}$$

where V is the volume of the system and

$$\epsilon_{\vec{k}} = \frac{k^2}{2} + v_x[\rho] + v_c[\rho] \tag{51}$$

The Fermi level is given by

$$\mu = \frac{k_F^2}{2} + v_x[\rho] + v_c[\rho] \tag{52}$$

Inserting the above expression in (47), we obtain

$$v_{resp}^{mod} = \frac{K_g[\rho]}{\sqrt{2}\rho V} \sum_{|\vec{k}| < k_F} \sqrt{k_F^2 - k^2} \tag{53}$$

A replacement of the sum in (53) by an integral yields

$$v_{resp}^{mod}(\vec{r}) = \frac{K_g[\rho]}{\sqrt{2}(2\pi)^3 \rho} \int_0^{k_F} \sqrt{k_F^2 - k^2} 4\pi k^2 dk =$$

$$= \frac{K_g[\rho]k_F^4}{2\sqrt{2}(2\pi)^2 \rho} \int_0^1 \sqrt{1 - x^2} x^2 dx = \frac{3\pi K_g}{16\sqrt{2}} k_F \tag{54}$$

From (48) and (54) the $K_g[\rho]$ value is defined by

$$K_g[\rho] = K_g = \frac{8\sqrt{2}}{3\pi^2} \approx 0.382 \tag{55}$$

which is valid for a homogeneous electron gas of arbitrary density, i.e. in this case $K_g[\rho]$ does not depend on ρ.

It is also possible to determine the constant $K[\rho]$ during the SCF calculations in a self-consistent manner. We refer to ref. [14] for details. The self-consistently determined values K_{sc} differ little from the homogeneous electron gas value.

We now investigate the performance of the GGA potentials in the exchange-only case, both for the screening and for the response parts. We use as GGA exchange energy density $\epsilon_x^{GGA}[\rho](\vec{r})$ the LDA plus the gradient correction introduced by Becke [5] (Becke88). This defines the screening and screening response potentials according to the following equations:

$$E_X^{GGA} = \int \rho(\vec{r})\epsilon_x^{GGA}[\rho](\vec{r})d\vec{r} = \frac{1}{2}\int \rho(\vec{r})v_{Xscr}^{GGA}(\vec{r})d\vec{r} \qquad (56)$$

and

$$v_{Xscr}^{GGA} = 2\epsilon_X^{GGA} \qquad (57)$$

$$v_{Xscr}^{resp,GGA} = \int \rho(\vec{r}')\frac{\delta\epsilon_x^{GGA}(\vec{r}')}{\delta\rho(\vec{r})}d\vec{r}' - \epsilon_x^{GGA}(\vec{r}). \qquad (58)$$

Full OPM calculations [27] provide us with a benchmark accurate v_X, the accurate screening potential v_{Xscr} can be evaluated from the orbitals using the defining equation for the Fermi hole potential (or Slater potential)

$$\begin{aligned} v_S(\vec{r}_1) &= \int \frac{\rho(\vec{r}_2)[g_x([\rho];\vec{r}_1,\vec{r}_2)-1]}{|\vec{r}_1-\vec{r}_2|}d\vec{r}_2 \\ &= -\frac{1}{2}\sum_{i=1}^{N}\sum_{j=1}^{N}\int \frac{\phi_i(\vec{r}_1)\phi_i^*(\vec{r}_2)\phi_j^*(\vec{r}_1)\phi_j(\vec{r}_2)}{\rho(\vec{r}_1)}d\vec{r}_2 \qquad (59) \end{aligned}$$

The response potential is then simply obtained as the difference between the full v_X and the screening part v_S. In fig. 4 we compare the accurate OPM Slater potential, calculated with the OPM orbitals, with the GGA approximation $v_S^{GGA} = 2\epsilon_X^{GGA}$. Here v_S^{GGA} has been calculated from the self-consistent GGA density but use of the OPM density would have given essentially the same curve: the differences in the shapes of v_S^{OPM} and v_S^{GGA} are almost completely due the different expressions, eq. 59 resp. eq. 57.

It is known that the GGA (Becke88) exchange energies are quite accurate, so v_S^{GGA} times ρ integrates to the correct number, but the figure shows that also locally the GGA energy density is reasonably close to the OPM one. This also holds for heavier atoms (cf. Ca, Kr, Cd in ref. [36]. There is of course still room for improvement, in particular the behaviour close to the nucleus is different and the GGA energy density is not deep enough in the $1s$ region. However, the GGA response potentials, which are compared in fig. 5, show much more significant deviations from OPM. It is clear that $v_{Xscr}^{resp,GGA}$, denoted as v_{resp}^{GGA} in the figure, does not have the required short range. It is responsible for the wrong asymptotic behaviour of the GGA exchange potential, which causes for instance the wrong orbital energies (see above). In the second place we note the singular behaviour of $v_{Xscr}^{resp,GGA}$ at the nucleus. It is indeed easily derived that the GGA exchange energy density leads to a Coulombic singularity in the response potential [6]. Finally, the striking step like behaviour of the response potential, which

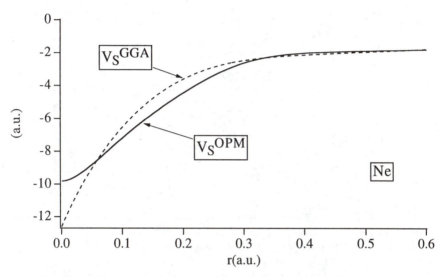

Figure 4. Comparison of the OPM Slater potential with its GGA approximation for the neon atom.

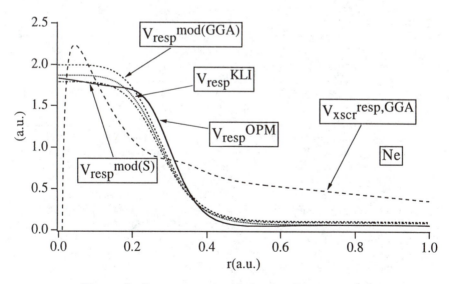

Figure 5. Response potentials for the neon atom.

Table   II. Comparison of differences (in mHartrees) of the total energies calculated self-consistently with the exact functional v_S and various approximations to v_{resp}, with the OPM total energies E_{tot}^{OPM} [37] (the latter are given in Hartrees and are always more negative).

ATOM	$-E_{tot}^{OPM}$	v_{KLI}^{resp}	$v_{mod}^{resp}(K_{sc})$	$v_{mod}^{resp}(K_g)$	neglect of v_{resp}
Be	14.572	0	0	0	11
Ne	128.545	0	1	0	44
Mg	199.612	1	2	2	79
Ar	526.812	2	3	4	109
Ca	676.752	2	4	4	146
Zn	1777.834	4	6	6	258
Kr	2752.043	4	5	5	288
Sr	3131.533	4	7	7	324
Cd	5465.114	6	6	6	419
Xe	7232.121	7	12	11	450

is clearly visible in v_{resp}^{OPM}, is not represented well by the GGA approximation, although it clearly does make an attempt to build the required repulsive shape of the response potential. Fig. 5 also shows the response potential as given by the KLI expression, eq. 39, and two cases where our model expression has been used. In one case, denoted $v_{resp}^{mod(GGA)}$ the orbitals have been used from a self-consistent calculation with our v_{resp}^{mod} plus the GGA screening potential v_S^{GGA}, in the other case, denoted $v_{resp}^{mod(S)}$ our model response potential has been used in conjunction with the full Slater screening potential (eq.59). All these potentials are reasonably accurate, $v_{resp}^{mod(GGA)}$ giving a somewhat too high step.

In order to show both the importance of the response part of the potential and the quality of the KLI approximation as well as our modelling, we give in table II the deviation from the total OPM energy when the response part of the potential is omitted or when one of the approximations is used. For the screening part in all cases the Slater potential has been used. Both the electron gas value for K and the self-consistently determined K have been used in our model response potential.

The total energy is always determined from the self-consistently obtained orbitals in the same way, i.e. using $(1/2)v_S$ as energy density. Therefore, the OPM total energy is always the lowest one, since the OPM local potential precisely determines the orbitals having this property. Note that the response part of the potential is not related directly to the total energy, it only serves to determine the orbitals and the density so that the energy becomes minimal. The results in the table show that the various model response potentials are doing pretty well. This does not mean that any response potential would do. The results obtained by omitting the response part of the potential altogether show that this distorts the orbitals and density sufficiently to give large deviations in the total energy, in spite of its variational stability.

It is also interesting to see if the use of appropriate response potentials improves the poor GGA results for the highest orbital energy.

In table III we show the deviations of the ϵ_{HOMO} with respect to the OPM result for the same models as used in the previous table. The deviations for the KLI response

Table   III. Comparison of differences between the OPM values for the highest occupied orbital energy ϵ_N^{OPM} (in mHartrees) and those calculated self-consistently with the exact functional v_S and various approximations to v_{resp}

ATOM	$-\epsilon_N^{OPM}$	v_{KLI}^{resp}	$v_{mod}^{resp}(K_{sc})$	$v_{mod}^{resp}(K_g)$	neglect of v_{resp}
Be	309	0	1	6	-17
Ne	851	2	21	30	-61
Mg	253	1	5	5	-31
Ar	591	2	18	21	-47
Ca	196	1	10	10	-29
Zn	293	1	-14	-14	-64
Kr	523	1	20	20	-44
Sr	179	1	12	10	-30
Cd	265	0	2	1	-65
Xe	456	1	23	22	-40

potential are at the mHartree level, our model response potential gives a deviation about an order of magnitude larger. This is still an order of magnitude better than the pure GGA potential.

5   Conclusions

We have investigated in this paper the relationship between the one-electron local effective Kohn-Sham potential of density functional theory and the physics of electron correlation as embodied in the behaviour of exchange and correlation holes. It has proven useful to break up the exchange-correlation potential in a part that is just the potential of the hole charge density (the screening potential) and a part that is related to the response of the hole to density changes (the response potential). The first is equivalent to the energy density per particle and is therefore directly related to the density functional for the energy. The response potential is needed in order to generate accurate densities from the solutions of the KS equations.

We wish to emphasize that the KS orbitals are not just mathematical constructs, without physical meaning, that are only useful for the construction of the density. The connection that we have discussed in this paper between the KS potential and the physics of electron correlation (described here in terms of Fermi and Coulomb holes), implies that the field that is described by the KS potential is no less "physical" than is the field of other one-electron models, such as Hartree-Fock. There is therefore no reason why one should not use the KS orbitals in qualitative MO considerations, that are so useful in rationalizing a large body of chemistry [38, 39]. As a matter of fact, there are certain advantages to the use of Kohn-Sham orbitals. They possess some properties that are implicitly assumed in perturbative MO considerations. In the first place the highest occupied orbital energy is exactly equal to the first ionization energy of the molecule. This means that the orbital energy levels are positioned correctly with respect to each other for PMO arguments to be used ("orbital control"). In particular, the virtual orbitals of the KS calculations do not exhibit the undesirable upward shift that is characteristic of Hartree-Fock. The virtual KS orbitals are solutions in exactly the same potential as the occupied orbitals. Since this potential contains a full hole charge potential (see above) it is "self-interaction free", also for the virtual orbitals. In practice this means that the occupied and virtual orbitals have both realistic one-electron energies and spatial extent (no unphysically diffuse virtual orbitals) and possess the typical bonding and antibonding characteristics that one expects on the basis of elementary MO considerations. Therefore the results of Kohn-Sham calculations, apart from providing the well-known highly accurate results for total energies, may be easily interpreted in the familiar qualitative MO language. An additional advantage is the property that the KS orbitals yield the exact density. This is a very desirable feature when calculating electrostatic interaction contributions ("charge control"). Semi-quantitative energy decomposition methods, in which the electrostatic interaction energy as well as frontier orbital interaction terms play a role, have for instance been advocated by Morokuma [40, 41] and others [42, 43, 44]. Since the pioneering work of Ziegler [45, 46, 47] in the application of energy decomposition methods within the framework of Kohn-Sham theory (initially in the context of exchange-only LDA or $X\alpha$), numerous applications have appeared, see refs. [48, 49, 50, 51, 52] and references therein.

References

[1] P. Hohenberg and W. Kohn, *Phys. Rev.* 136B:864, 1965

[2] W. Kohn, L.J. Sham, Phys. Rev. 140A:1133, 1965

[3] J.P. Perdew, Phys. Rev. B33:8822, 1986

[4] J.P. Perdew and Wang Yue, *Phys. Rev.* B33:8800, 1986

[5] A.D. Becke, *Phys. Rev.* A38:3098, 1988

[6] R. van Leeuwen and E.J. Baerends, Phys. Rev. A49:2421, 1994

[7] O. Gritsenko, R. van Leeuwen and E.J. Baerends, *J. Chem. Phys.* 101:8955, 1994

[8] R. van Leeuwen, O. Gritsenko and E.J. Baerends, *Z. Physik D* 33:229, 1995

[9] O. Gritsenko, R. van Leeuwen, E. van Lenthe and E.J. Baerends, *Phys. Rev.* A51:1944, 1995

[10] M. A. Buijse, E.J. Baerends and J.G. Snijders, *Phys. Rev.* A40:4190, 1989

[11] R. van Leeuwen and E.J. Baerends, *Int. J. Quantum Chem.* 52:711, 1994

[12] Wang Yue and R.G. Parr, *Phys. Rev.* A47:R1591, 1993

[13] Qingsheng Zhao, R.C. Morrison and R.G. Parr, *Phys. Rev.* A50:2138, 1994

[14] O. Gritsenko, R. van Leeuwen and E.J. Baerends, *Phys. Rev.* A52:1870, 1995

[15] R. van Leeuwen and E.J. Baerends, *Phys. Rev.* A51:170, 1995

[16] M.A. Buijse and E.J. Baerends, in: *Density Functional Theory of Molecules, Clusters and Solids*, ed. D.E. Ellis, , p.1, Kluwer, 1995

[17] A.V. Bunge and R.O. Esquivel, *Phys. Rev.* a33:853, 1986

[18] S.H. Vosko, L. Wilk and M. Nusair, *Can. J. Phys.* 58:1200, 1980

[19] C.O. Almbladh and U. von Barth, *Phys. Rev.* B31:3231, 1985

[20] C. Filippi, C.J. Umrigar and M. Taut *J. Chem. Phys.* 100:1290, 1994

[21] J. Harris and R.O. Jones, *J. Phys. F* 4:1170, 1974

[22] O. Gunnarsson and B.I. Lundqvist, *Phys. Rev.* B13:4274, 1976

[23] D.C. Langreth and J. P. Perdew, *Sol. State Comm.* 31:567, 1975

[24] R.M.Dreizler and E.K.U.Gross, *Density Functional Theory: An Approach to the Quantum Many-Body Problem*, Springer-Verlag, Berlin, 1990.

[25] B.G.Johnson, P.M.W.Gill, and J.A.Pople. *J. Chem. Phys.*, 98:5612, 1992.

[26] A.D.Becke. *J. Chem. Phys.*, 96:2155, 1992.

[27] J.D. Talman and W.F. Shadwick, *Phys. Rev.* A14:36, 1976

[28] J. B. Krieger, Y. Li and G. J. Iafrate, *Phys. Rev.* A45:101, 1992

[29] R.T. Sharp and G. K. Horton, *Phys. Rev.* 90:317, 1953

[30] A.D. Becke, *J. Chem. Phys.*, 97:9173, 1993

[31] J.C. Slater, *Phys. Rev.* 81:385, 1951

[32] J.C. Slater, *Quantum Theory of Atomic Structure, Vol. 2*, MacGraw-Hill, 1960

[33] Y. Wang, J.P. Perdew, J.A. Chevary, L.D. Macdonald and S. H. Vosko, *Phys. Rev.* A41:78, 1990

[34] W.L. Luken and D. N. Beratan, *Theoret. Chim. Acta* 61:265, 1982

[35] W.L. Luken, *Int. J. Quantum Chem.* 22:889, 1982

[36] O. Gritsenko, R. van Leeuwen and E.J. Baerends, *Int. J. Quantum Chem.* to be published

[37] Y. Li, J. B. Krieger and G.J. Iafrate, *Phys. Rev.*A47:165, 1993

[38] T.A. Albright, J.K. Burdett and M.H. Whangbo, *Orbital Interactions in Chemistry*, Wiley, 1985

[39] A. Rauk, *Orbital Interaction Theory of Organic Chemistry*, Wiley, 1994

[40] K. Morokuma, *J. Chem. Phys.* 55:1236, 1971

[41] K. Kitaura and K. Morokuma, *Int. J. Quantum Chem.* 10:325, 1976

[42] S. Wolfe, D.J. Mitchell and M.H. Whangbo, *J. Am. Chem. Soc.* 110:1936, 1978

[43] F. Bernardi, A. Bottoni, A. Mangini and G. Tonachini, *J. Molec. Struct. (Theochem)* 86:163, 1986

[44] A.J. Stone and R.W. Erskine, *J. Am. Chem. Soc.* 102:7185, 1980

[45] T. Ziegler and A. Rauk, *Theor. Chim. Acta* 46:1, 1977

[46] T. Ziegler and A. Rauk, *Inorg. Chem.* 18:1558, 1979

[47] T. Ziegler and A. Rauk, *Inorg. Chem.* 18:1755, 1979

[48] H. Jakobson and T. Ziegler, *J. Am. Chem. Soc.* 116:3667, 1994

[49] E.J. Baerends and A. Rozendaal, in: *Quantum Chemistry: The Challenge of Transition Metal Complexes*, p. 159, Reidel, 1986

[50] A. Rosa and E.J. Baerends, *New. J. Chem.* 15:815, 1991

[51] A. Rosa and E.J. Baerends, *Inorg. Chem.* 33:584, 1993

[52] F.M. Bickelhaupt, N.N. Nibbering, E.M. van Wezenbeek and E.J. Baerends, *J. Phys. Chem.* 96:4864, 1992

Chapter 3

Conventional Quantum Chemical Correlation Energy Versus Density-Functional Correlation Energy

E. K. U. Gross, M. Petersilka, and T. Grabo

Institut für Theoretische Physik, Universität Würzburg, Am Hubland, D–97074 Würzburg, Germany

We examine the difference between the correlation energy as defined within the conventional quantum chemistry framework and its namesake in density-functional theory. Both correlation energies are rigorously defined concepts and satisfy the inequality $E_c^{QC} \geq E_c^{DFT}$. We give numerical and analytical arguments suggesting that the numerical difference between the two rigorous quantities is small. Finally, approximate density functional correlation energies resulting from some popular correlation energy functionals are compared with the conventional quantum chemistry values.

In quantum chemistry (QC), the exact correlation energy is traditionally defined as the difference between the exact total energy and the total selfconsistent Hartree-Fock (HF) energy:

$$E_{c,\text{exact}}^{QC} := E_{\text{tot,exact}} - E_{\text{tot}}^{HF} \quad . \tag{1}$$

Within the framework of density-functional theory (DFT) [1, 2], on the other hand, the correlation energy is a functional of the density $E_c^{DFT}[\rho]$. The exact DFT correlation energy is then obtained by inserting the exact ground-state density of the system considered into the functional $E_c^{DFT}[\rho]$, i. e.

$$E_{c,\text{exact}}^{DFT} = E_c^{DFT}[\rho_{\text{exact}}] \tag{2}$$

0097–6156/96/0629–0042$15.00/0

In practice, of course, neither the quantum chemical correlation energy (1) nor the DFT correlation energy (2) are known exactly. Nevertheless, both quantities are rigorously defined concepts.

The aim of the following section is to give a coherent overview of how the correlation energy is defined in the DFT literature [3–14] and how this quantity is related to the conventional QC correlation energy. The two exact correlation energies $E_{c,\text{exact}}^{\text{QC}}$ and $E_{c,\text{exact}}^{\text{DFT}}$ are generally not identical. They satisfy the inequality $E_{c,\text{exact}}^{\text{QC}} \geq E_{c,\text{exact}}^{\text{DFT}}$. Furthermore we will give an analytical argument indicating that the difference between the two exact quantities is small.

In the last section we compare the numerical values of *approximate* conventional QC correlation energies with *approximate* DFT correlation energies resulting from some popular DFT correlation energy functionals. It turns out that the difference between DFT correlation energies and QC correlation energies is smallest for the correlation energy functional of Colle and Salvetti [15, 16] further indicating [17] that the results obtained with this functional are closest to the exact ones.

Basic Formalism

We are concerned with Coulomb systems described by the Hamiltonian

$$\hat{H} = \hat{T} + \hat{W}_{\text{Clb}} + \hat{V} \tag{3}$$

where (atomic units are used throughout)

$$\hat{T} = \sum_{i=1}^{N} \left(-\frac{1}{2} \nabla_i^2 \right) \tag{4}$$

$$\hat{W}_{\text{Clb}} = \frac{1}{2} \sum_{\substack{i,j=1 \\ i \neq j}}^{N} \frac{1}{|\mathbf{r}_i - \mathbf{r}_j|} \tag{5}$$

$$\hat{V} = \sum_{i=1}^{N} v(\mathbf{r}_j) \quad . \tag{6}$$

To keep the following derivation as simple as possible, we choose to work with the traditional Hohenberg-Kohn [18] formulation rather than the constrained-search representation [4, 19, 20] of DFT. In particular, all ground-state wavefunctions (interacting as well as non-interacting) are assumed to be non-degenerate. By virtue of the Hohenberg-Kohn theorem [18] the ground-state density ρ uniquely determines the external potential $v = v[\rho]$ and the ground-state wave function $\Psi[\rho]$. If $v_0(\mathbf{r})$ is a given external potential characterizing a particular physical system, the Hohenberg-Kohn total-energy functional is defined as

$$E_{v_0}[\rho] = \langle \Psi[\rho]| \hat{T} + \hat{W}_{\text{Clb}} + \hat{V}_0 |\Psi[\rho]\rangle \quad . \tag{7}$$

As an immediate consequence of the Rayleigh-Ritz principle, the total-energy functional (7) is minimized by the exact ground-state density ρ_{exact} corresponding to the potential v_0, the minimum value being the exact ground-state energy, i. e.

$$E_{\text{tot,exact}} = E_{v_0} [\rho_{\text{exact}}] \quad . \tag{8}$$

In the context of the Kohn-Sham (KS) scheme [21] the total-energy functional is usually written as

$$E_{v_0}[\rho] = T_s[\rho] + \int \rho(\mathbf{r}) v_0(\mathbf{r}) \, d^3r + \frac{1}{2} \int \int \frac{\rho(\mathbf{r})\rho(\mathbf{r}')}{|\mathbf{r} - \mathbf{r}'|} \, d^3r \, d^3r' + E_{\text{xc}}[\rho] \tag{9}$$

where $T_s[\rho]$ is the kinetic-energy functional of non-interacting particles. By virtue of the Hohenberg-Kohn theorem, applied to non-interacting systems, the density ρ uniquely determines the single-particle potential $v_s[\rho]$ and the ground-state Slater-determinant

$$\Phi^{\text{KS}}[\rho] = \frac{1}{\sqrt{N!}} \det \left\{ \varphi_{j\sigma}^{\text{KS}}[\rho] \right\} \tag{10}$$

and hence $T_s[\rho]$ is given by

$$\begin{aligned} T_s[\rho] &= \langle \Phi^{\text{KS}}[\rho] | \hat{T} | \Phi^{\text{KS}}[\rho] \rangle \\ &= \sum_{\sigma=\uparrow,\downarrow} \sum_{j=1}^{N_\sigma} \int \varphi_{j\sigma}^{\text{KS}}[\rho](\mathbf{r})^* \left(-\frac{1}{2}\nabla^2 \right) \varphi_{j\sigma}^{\text{KS}}[\rho](\mathbf{r}) \, d^3r \quad . \end{aligned} \tag{11}$$

We mention in passing that the Hohenberg-Kohn theorem can also be formulated for a "Hartree-Fock world" [22], implying that the HF density uniquely determines the external potential. Consequently the HF ground-state determinant is a functional of the density as well:

$$\Phi^{\text{HF}}[\rho] = \frac{1}{\sqrt{N!}} \det \left\{ \varphi_{j\sigma}^{\text{HF}}[\rho] \right\} \quad . \tag{12}$$

The resulting kinetic-energy functional

$$\begin{aligned} T^{\text{HF}}[\rho] &= \langle \Phi^{\text{HF}}[\rho] | \hat{T} | \Phi^{\text{HF}}[\rho] \rangle \\ &= \sum_{\sigma=\uparrow,\downarrow} \sum_{j=1}^{N_\sigma} \int \varphi_{j\sigma}^{\text{HF}}[\rho](\mathbf{r})^* \left(-\frac{1}{2}\nabla^2 \right) \varphi_{j\sigma}^{\text{HF}}[\rho](\mathbf{r}) \, d^3r \end{aligned} \tag{13}$$

is different from $T_s[\rho]$ because the orbitals in (11) come from a *local* single-particle potential $v_s[\rho]$ while the orbitals in (13) come from the *nonlocal* HF potential $v^{\text{HF}}[\rho]$. However, the numerical difference between $T^{\text{HF}}[\rho]$ and $T_s[\rho]$ has been found to be rather small [14].

The remaining term, $E_{\text{xc}}[\rho]$, on the right hand side of equation (9) is termed the exchange-correlation (xc) energy. Comparison of equation (9) with equation (7) shows that the xc-energy functional is formally given by

$$E_{\text{xc}}[\rho] = \langle \Psi[\rho] | \hat{T} + \hat{W}_{\text{Clb}} | \Psi[\rho] \rangle - T_s[\rho] - \frac{1}{2} \int \int \frac{\rho(\mathbf{r})\rho(\mathbf{r}')}{|\mathbf{r} - \mathbf{r}'|} \, d^3r \, d^3r' \quad . \tag{14}$$

In density-functional theory the exact exchange-energy functional is defined by

$$E_{\mathrm{x}}^{\mathrm{DFT}}[\rho] := \langle \Phi^{\mathrm{KS}}[\rho] | \hat{W}_{\mathrm{Clb}} | \Phi^{\mathrm{KS}}[\rho] \rangle - \frac{1}{2} \int \int \frac{\rho(\mathbf{r})\rho(\mathbf{r}')}{|\mathbf{r} - \mathbf{r}'|} \, d^3r \, d^3r' \quad . \tag{15}$$

This is identical with the ordinary Fock functional

$$E_{\mathrm{x}}^{\mathrm{HF}}[\varphi_{j\sigma}] = -\frac{1}{2} \sum_{\sigma=\uparrow,\downarrow} \sum_{j,k=1}^{N_\sigma} \int \int d^3r \, d^3r' \, \frac{\varphi_{j\sigma}^*(\mathbf{r})\varphi_{k\sigma}^*(\mathbf{r}')\varphi_{k\sigma}(\mathbf{r})\varphi_{j\sigma}(\mathbf{r}')}{|\mathbf{r} - \mathbf{r}'|} \tag{16}$$

evaluated, however, with the KS Orbitals, i. e.

$$E_{\mathrm{x}}^{\mathrm{DFT}}[\rho] = E_{\mathrm{x}}^{\mathrm{HF}}\left[\varphi_{j\sigma}^{\mathrm{KS}}[\rho]\right] \quad . \tag{17}$$

The DFT correlation-energy functional is then given by

$$E_{\mathrm{c}}^{\mathrm{DFT}}[\rho] = E_{\mathrm{xc}}[\rho] - E_{\mathrm{x}}^{\mathrm{DFT}}[\rho] \quad . \tag{18}$$

Inserting the respective definitions (14) and (17) of $E_{\mathrm{xc}}[\rho]$ and $E_{\mathrm{x}}^{\mathrm{DFT}}[\rho]$ we find

$$E_{\mathrm{c}}^{\mathrm{DFT}}[\rho] = \langle \Psi[\rho] | \hat{T} + \hat{W}_{\mathrm{Clb}} | \Psi[\rho] \rangle - T_{\mathrm{s}}[\rho] - \frac{1}{2} \int \int \frac{\rho(\mathbf{r})\rho(\mathbf{r}')}{|\mathbf{r} - \mathbf{r}'|} \, d^3r \, d^3r' - E_{\mathrm{x}}^{\mathrm{HF}}\left[\varphi_{j\sigma}^{\mathrm{KS}}[\rho]\right] \quad . \tag{19}$$

In terms of the Hartree-Fock total-energy functional

$$
\begin{aligned}
E_{v_0}^{\mathrm{HF}}[\varphi_{j\sigma}] = {} & \sum_{\sigma=\uparrow,\downarrow} \sum_{j=1}^{N_\sigma} \int \varphi_{j\sigma}(\mathbf{r})^* \left(-\frac{1}{2}\nabla^2\right) \varphi_{j\sigma}(\mathbf{r}) \, d^3r + \int \rho(\mathbf{r}) \, v_0(\mathbf{r}) \, d^3r \\
& + \frac{1}{2} \int \int \frac{\rho(\mathbf{r})\rho(\mathbf{r}')}{|\mathbf{r} - \mathbf{r}'|} \, d^3r \, d^3r' + E_{\mathrm{x}}^{\mathrm{HF}}[\varphi_{j\sigma}]
\end{aligned}
\tag{20}
$$

and the total-energy functional (7) the DFT correlation energy (19) is readily expressed as

$$E_{\mathrm{c}}^{\mathrm{DFT}}[\rho] = E_{v_0}[\rho] - E_{v_0}^{\mathrm{HF}}\left[\varphi_{j\sigma}^{\mathrm{KS}}[\rho]\right] \quad . \tag{21}$$

By equation (2), the exact DFT correlation energy is then obtained by inserting the exact ground-state density ρ_{exact} (corresponding to the external potential v_0) into the functional (21). By virtue of equation (8) one obtains

$$E_{\mathrm{c,exact}}^{\mathrm{DFT}} = E_{\mathrm{tot,exact}} - E_{v_0}^{\mathrm{HF}}\left[\varphi_{j\sigma}^{\mathrm{KS}}[\rho_{\mathrm{exact}}]\right] \quad . \tag{22}$$

The conventional quantum chemical correlation energy, on the other hand, is given by

$$E_{\mathrm{c,exact}}^{\mathrm{QC}} = E_{\mathrm{tot,exact}} - E_{v_0}^{\mathrm{HF}}\left[\varphi_{j\sigma}^{\mathrm{HF}}[\rho_{\mathrm{HF}}]\right] \tag{23}$$

where $\varphi_{j\sigma}^{\mathrm{HF}}[\rho_{\mathrm{HF}}]$ are the usual selfconsistent HF orbitals corresponding to the external potential v_0, i. e. ρ_{HF} is that very HF density which uniquely corresponds to the external potential v_0. Of course, ρ_{HF} and ρ_{exact} are generally not identical. Comparison of (22) with (23) shows that

$$E_{\mathrm{c,exact}}^{\mathrm{DFT}} = E_{\mathrm{c,exact}}^{\mathrm{QC}} + \left(E_{v_0}^{\mathrm{HF}}\left[\varphi_{j\sigma}^{\mathrm{HF}}[\rho_{\mathrm{HF}}]\right] - E_{v_0}^{\mathrm{HF}}\left[\varphi_{j\sigma}^{\mathrm{KS}}[\rho_{\mathrm{exact}}]\right] \right) \quad . \tag{24}$$

This is the central equation relating the DFT correlation energy to the QC correlation energy. Since the HF orbitals $\varphi_{j\sigma}^{\mathrm{HF}}[\rho_{\mathrm{HF}}]$ are the ones that minimize the HF total-energy functional (20), the inequality

$$E_{v_0}^{\mathrm{HF}}\left[\varphi_{j\sigma}^{\mathrm{HF}}[\rho_{\mathrm{HF}}]\right] \le E_{v_0}^{\mathrm{HF}}\left[\varphi_{j\sigma}^{\mathrm{KS}}[\rho_{\mathrm{exact}}]\right] \tag{25}$$

must be satisfied and it follows from equation (24) that

$$E_{c,\mathrm{exact}}^{\mathrm{QC}} \ge E_{c,\mathrm{exact}}^{\mathrm{DFT}} \tag{26}$$

This was first recognized by Sahni and Levy [3]. Equation (24) tells us that, as a matter of principle, selfconsistent DFT results for the correlation energy should not be compared directly with the conventional quantum chemical correlation energy but rather with the right-hand side of equation (24). In practice, of course, quantum-chemical correlation energies and ground-state densities are known only approximately, e. g. , from configuration-interaction (CI) calculations. Hence,

$$E_{\mathrm{tot,CI}} - E_{\mathrm{tot}}^{\mathrm{HF}}\left[\varphi_{j\sigma}^{\mathrm{KS}}[\rho_{\mathrm{CI}}]\right] \tag{27}$$

is the quantity the selfconsistent DFT correlation energy should in principle be compared with. The second term of (27) is readily computed by employing one of the standard techniques [13, 23, 24, 25] of calculating the KS potential and its orbitals from a given CI density. In the following we shall argue, however, that the difference between $E_{c,\mathrm{exact}}^{\mathrm{DFT}}$ and $E_{c,\mathrm{exact}}^{\mathrm{QC}}$ can be expected to be small. To see this we rewrite equation (24) as

$$
\begin{aligned}
E_{c,\mathrm{exact}}^{\mathrm{DFT}} - E_{c,\mathrm{exact}}^{\mathrm{QC}} = & \left(E_{v_0}^{\mathrm{HF}}\left[\varphi_{j\sigma}^{\mathrm{HF}}[\rho_{\mathrm{HF}}]\right] - E_{v_0}^{\mathrm{HF}}\left[\varphi_{j\sigma}^{\mathrm{KS}}[\rho_{\mathrm{x-only}}]\right]\right) \\
& + \left(E_{v_0}^{\mathrm{HF}}\left[\varphi_{j\sigma}^{\mathrm{KS}}[\rho_{\mathrm{x-only}}]\right] - E_{v_0}^{\mathrm{HF}}\left[\varphi_{j\sigma}^{\mathrm{KS}}[\rho_{\mathrm{exact}}]\right]\right) .
\end{aligned} \tag{28}
$$

where $\rho_{\mathrm{x-only}}$ is the ground-state density of an exact exchange-only DFT calculation [26, 27] and $\varphi_{j\sigma}^{\mathrm{KS}}[\rho_{\mathrm{x-only}}]$ are the corresponding KS orbitals. The first difference on the right-hand side of equation (28) is known to be small [26, 27]. The second difference, on the other hand, is easily seen to be of *second* order in $(\rho_{\mathrm{x-only}} - \rho_{\mathrm{exact}})$ and is therefore expected to be small as well:

$$
\begin{aligned}
& E_{v_0}^{\mathrm{HF}}\left[\varphi_{j\sigma}^{\mathrm{KS}}[\rho_{\mathrm{x-only}}]\right] - E_{v_0}^{\mathrm{HF}}\left[\varphi_{j\sigma}^{\mathrm{KS}}[\rho_{\mathrm{exact}}]\right] \\
& = \int d^3r \left.\frac{\delta E_{v_0}^{\mathrm{HF}}\left[\varphi_{j\sigma}^{\mathrm{KS}}[\rho]\right]}{\delta\rho(\mathbf{r})}\right|_{\rho_{\mathrm{x-only}}} \cdot (\rho_{\mathrm{x-only}}(\mathbf{r}) - \rho_{\mathrm{exact}}(\mathbf{r})) + O(\rho_{\mathrm{x-only}} - \rho_{\mathrm{exact}})^2 \\
& = \int d^3r\, \mu \cdot (\rho_{\mathrm{x-only}}(\mathbf{r}) - \rho_{\mathrm{exact}}(\mathbf{r})) + O(\rho_{\mathrm{x-only}} - \rho_{\mathrm{exact}})^2 \\
& = 0 + O(\rho_{\mathrm{x-only}} - \rho_{\mathrm{exact}})^2
\end{aligned}
$$

The second equality follows from the fact that $\rho_{\mathrm{x-only}}$ minimizes the density functional $E_{v_0}^{\mathrm{HF}}\left[\varphi_{j\sigma}^{\mathrm{KS}}[\rho]\right]$. Hence we conclude that $E_{c,\mathrm{exact}}^{\mathrm{DFT}} - E_{c,\mathrm{exact}}^{\mathrm{QC}}$ should be small. This estimate is confirmed by results of accurate variational and quantum Monte

Carlo calculations on H^-, He, Be^{+2}, Ne^{+8} [13] and Be and Ne [28] as can be seen from Table 1. There, the conventional quantum chemical correlation energies of these systems are compared with the "exact" DFT correlation energies calculated from equation (22). For all elements and ions shown, the relation (26) is confirmed, as expected. The difference between the DFT and the conventional QC correlation energies is found to be small compared with the total correlation energies. However, the absolute differences, being sometimes as high as a few mHartrees, are of the same order of magnitude as the deviations between experimental total energies and total energies calculated with *approximate* state-of-the art density functionals [17].

Table 1: *Comparison of exact DFT correlation energies with conventional quantum chemical correlation energies (QC) [29].* Δ *denotes the difference between the QC and the DFT correlation energy (in Hartree units).* $\Delta\%$ *denotes the value of* $|E_{c,\text{exact}}^{\text{QC}} - E_{c,\text{exact}}^{\text{DFT}}|/|E_{c,\text{exact}}^{\text{DFT}}|$ *in percent.*

	DFT	QC	Δ	$\Delta\%$
H^-	$-0.041\,995$	$-0.039\,821$	$+0.002\,174$	5.2
He	$-0.042\,107$	$-0.042\,044$	$+0.000\,063$	0.2
Be^{+2}	$-0.044\,274$	$-0.044\,267$	$+0.000\,007$	0.02
Ne^{+8}	$-0.045\,694$	$-0.045\,693$	$+0.000\,001$	0.002
Be	$-0.096\,2$	$-0.094\,3$	$+0.001\,9$	2.0
Ne	-0.394	-0.390	$+0.004$	1.0

To conclude this section, we mention that there exists yet another possibility of defining a density functional for the correlation energy [4–11,13]:

$$\tilde{E}_c[\rho] = E_{v_0}[\rho] - E_{v_0}^{\text{HF}}\left[\varphi_{j\sigma}^{\text{HF}}[\rho]\right] \tag{29}$$

where $\varphi_{j\sigma}^{\text{HF}}[\rho]$ are the HF orbitals corresponding to the density ρ (see equation (12)). If the exact density ρ_{exact} is inserted in (29) $\varphi_{j\sigma}^{\text{HF}}[\rho_{\text{exact}}]$ are the HF orbitals corresponding to some unknown external potential $\tilde{v}_0$ whose HF density is ρ_{exact}. The decomposition

$$\tilde{v}_0(\mathbf{r}) =: v_0(\mathbf{r}) + \tilde{v}_c(\mathbf{r}) \tag{30}$$

makes clear that on the single-particle level the definition (29) leads to a hybrid scheme featuring the ordinary *non-local* HF exchange potential combined with the *local* correlation potential $\tilde{v}_c(\mathbf{r})$. In the present paper, this hybrid scheme will not be further investigated. We only mention that, with arguments similar to the one leading to (26) $\tilde{E}_c$ satisfies the inequalities:

$$\tilde{E}_c[\rho_{\text{exact}}] \leq E_{c,\text{exact}}^{\text{QC}} \leq \tilde{E}_c[\rho_{\text{HF}}] \tag{31}$$

as was first pointed out by Savin, Stoll and Preuss [8].

Correlation Energies from Various DFT Approximations

For further analysis, we compare in Tables 2, 3 and 4 the DFT correlation energies resulting from various approximations to $E_c^{\mathrm{DFT}}[\rho]$. LYP denotes the correlation-energy functional by Lee, Yang and Parr [30], PW91 the generalized gradient approximation by Perdew and Wang [31], and LDA the conventional local density approximation in the parametrisation of E_c by Vosko, Wilk and Nusair [32]. The first column, denoted by CS and KLI-CS, respectively, shows the results of a recently developed scheme which employs an optimized effective potential (OEP) including correlation effects [17]. In this scheme the full integral equation of the optimized effective potential method [33, 34],

$$\sum_{i=1}^{N_\sigma} \int d^3r' \left(V_{\mathrm{xc}\sigma}^{\mathrm{OEP}}(\mathbf{r}') - u_{\mathrm{xci}\sigma}(\mathbf{r}') \right) \left(\sum_{\substack{k=1 \\ k\neq i}}^{\infty} \frac{\varphi_{k\sigma}^*(\mathbf{r})\varphi_{k\sigma}(\mathbf{r}')}{\varepsilon_{k\sigma} - \varepsilon_{i\sigma}} \right) \varphi_{i\sigma}(\mathbf{r})\varphi_{i\sigma}^*(\mathbf{r}') + c.c. = 0 \tag{32}$$

with

$$u_{\mathrm{xci}\sigma}(\mathbf{r}) := \frac{1}{\varphi_{i\sigma}^*(\mathbf{r})} \frac{\delta E_{\mathrm{xc}}[\varphi_{j\sigma}]}{\delta\varphi_{i\sigma}(\mathbf{r})} \tag{33}$$

is solved semi-analytically by an approved method due to Krieger, Li and Iafrate [35, 36, 37]:

$$V_{\mathrm{xc}\sigma}^{\mathrm{OEP}}(\mathbf{r}) \approx V_{\mathrm{xc}\sigma}^{\mathrm{KLI}}(\mathbf{r}) = \frac{1}{\rho_\sigma(\mathbf{r})} \sum_{i=1}^{N_\sigma} \rho_{i\sigma}(\mathbf{r}) \left[u_{\mathrm{xci}\sigma}(\mathbf{r}) + \left(\bar{V}_{\mathrm{xci}\sigma}^{\mathrm{KLI}} - \bar{u}_{\mathrm{xci}\sigma} \right) \right] \tag{34}$$

where the constants $\left(\bar{V}_{\mathrm{xci}\sigma}^{\mathrm{KLI}} - \bar{u}_{\mathrm{xci}\sigma} \right)$ are the solutions of the set of linear equations

$$\sum_{i=1}^{N_\sigma-1} \left(\delta_{ji} - M_{ji\sigma} \right) \left(\bar{V}_{\mathrm{xci}\sigma}^{\mathrm{KLI}} - \bar{u}_{\mathrm{xci}\sigma} \right) = \bar{V}_{\mathrm{xcj}\sigma}^{\mathrm{S}} - \bar{u}_{\mathrm{xcj}\sigma} \qquad j = 1, \ldots, N_\sigma \tag{35}$$

with

$$M_{ji\sigma} := \int d^3r \frac{\rho_{j\sigma}(\mathbf{r})\rho_{i\sigma}(\mathbf{r})}{\rho_\sigma(\mathbf{r})}, \tag{36}$$

$$V_{\mathrm{xc}\sigma}^{\mathrm{S}}(\mathbf{r}) := \sum_{i=1}^{N} \frac{\rho_{i\sigma}(\mathbf{r})}{\rho_\sigma(\mathbf{r})} u_{\mathrm{xci}\sigma}(\mathbf{r}). \tag{37}$$

Here, $\bar{u}_{\mathrm{xcj}\sigma}$ denotes the average value of $u_{\mathrm{xcj}\sigma}(\mathbf{r})$ taken over the density of the $j\sigma$ orbital, i. e.

$$\bar{u}_{\mathrm{xcj}\sigma} = \int \rho_{j\sigma}(\mathbf{r})u_{\mathrm{xcj}\sigma}(\mathbf{r})d^3r \tag{38}$$

and similarly for $\bar{V}_{\mathrm{xcj}\sigma}^{\mathrm{S}}$. Like in the conventional Kohn-Sham method, the xc-potential resulting from equation (34) leads to a single-particle Schrödinger equation with a *local* effective potential

$$\left(-\frac{\nabla^2}{2} + v_0(\mathbf{r}) + \int \frac{\rho(\mathbf{r}')}{|\mathbf{r} - \mathbf{r}'|} d^3r' + V_{\mathrm{xc}\sigma}^{\mathrm{OEP}}(\mathbf{r}) \right) \varphi_{j\sigma}(\mathbf{r}) = \varepsilon_{j\sigma}\varphi_{j\sigma}(\mathbf{r}) \tag{39}$$

$$(j = 1, \ldots, N_\sigma \quad \sigma = \uparrow, \downarrow).$$

The selfconsistent solutions $\varphi_{j\sigma}(\mathbf{r})$ of equation (39) with lowest single-particle energies $\varepsilon_{j\sigma}$ minimize the total-energy functional

$$
\begin{aligned}
E_{v_0}^{\mathrm{OEP}}[\varphi_{j\sigma}] = &\sum_{\sigma=\uparrow,\downarrow}\sum_{i=1}^{N_\sigma}\int \varphi_{i\sigma}^*(\mathbf{r})\left(-\frac{1}{2}\nabla^2\right)\varphi_{i\sigma}(\mathbf{r})\,d^3r \\
&+\int \rho(\mathbf{r})\,v_0(\mathbf{r})\,d^3r \\
&+\frac{1}{2}\int\int \frac{\rho(\mathbf{r})\rho(\mathbf{r}')}{|\mathbf{r}-\mathbf{r}'|}\,d^3r\,d^3r' \\
&-\frac{1}{2}\sum_{\sigma=\uparrow,\downarrow}\sum_{j,k=1}^{N_\sigma}\int\int d^3r\,d^3r'\,\frac{\varphi_{j\sigma}^*(\mathbf{r})\varphi_{k\sigma}^*(\mathbf{r}')\varphi_{k\sigma}(\mathbf{r})\varphi_{j\sigma}(\mathbf{r}')}{|\mathbf{r}-\mathbf{r}'|} \\
&+E_{\mathrm{c}}^{\mathrm{CS}}[\{\varphi_{j\sigma}\}].
\end{aligned}
\tag{40}
$$

In the above equation, $E_{\mathrm{c}}^{\mathrm{CS}}$ denotes the Colle-Salvetti functional [15, 16] for the correlation-energy given by

$$
\begin{aligned}
E_{\mathrm{c}}^{\mathrm{CS}} = &-ab\int \gamma(\mathbf{r})\xi(\mathbf{r})\left[\sum_\sigma \rho_\sigma(\mathbf{r})\sum_i |\nabla\varphi_{i\sigma}(\mathbf{r})|^2 - \frac{1}{4}|\nabla\rho(\mathbf{r})|^2\right. \\
&\left. -\frac{1}{4}\sum_\sigma \rho_\sigma(\mathbf{r})\triangle\rho_\sigma(\mathbf{r}) + \frac{1}{4}\rho(\mathbf{r})\triangle\rho(\mathbf{r})\right]d^3r \\
&-a\int \gamma(\mathbf{r})\frac{\rho(\mathbf{r})}{\eta(\mathbf{r})}\,d^3r,
\end{aligned}
\tag{41}
$$

where

$$
\gamma(\mathbf{r}) = 4\,\frac{\rho_\uparrow(\mathbf{r})\rho_\downarrow(\mathbf{r})}{\rho(\mathbf{r})^2},
\tag{42}
$$

$$
\eta(\mathbf{r}) = 1 + d\rho(\mathbf{r})^{-\frac{1}{3}},
\tag{43}
$$

$$
\xi(\mathbf{r}) = \frac{\rho(\mathbf{r})^{-\frac{5}{3}}e^{-c\rho(\mathbf{r})^{-\frac{1}{3}}}}{\eta(\mathbf{r})}.
\tag{44}
$$

The constants a, b, c and d are given by

$$
\begin{aligned}
a &= 0.04918, &\quad b &= 0.132, \\
c &= 0.2533, &\quad d &= 0.349.
\end{aligned}
$$

In Table 2, the four *approximate* DFT correlation energy functionals are evaluated at the exact densities [13, 28] of H^-, He, Be^{+2}, Ne^{+8}, Be, Ne and compared with the *exact* DFT correlation energies given by equation (22). On average, the KLI-CS values are superior.

In Table 3 selfconsistent DFT correlation energies are compared with QC values taken from [38]. In these selfconsistent calculations the *approximate* correlation-energy functionals $E_{\mathrm{c}}^{\mathrm{LYP}}$, $E_{\mathrm{c}}^{\mathrm{PW91}}$, $E_{\mathrm{c}}^{\mathrm{LDA}}$ are complemented with the *approximate* exchange-energy functionals $E_{\mathrm{x}}^{\mathrm{B88}}$ [39], $E_{\mathrm{x}}^{\mathrm{PW91}}$ [31] and $E_{\mathrm{x}}^{\mathrm{LDA}}$, respectively. In the KLI-CS case, the DFT exchange-energy functional (17) is of course treated exactly. The numerical data show three main features:

Table 2: *Non-relativistic absolute correlation energies resulting from various approximate DFT correlation energy functionals, evaluated at the exact ground-state densities [13, 28] of the respective atoms (in Hartree units). Exact values are from [13, 38].* $|\Delta|\%$ *denotes the mean value of* $|E_c - E^{\text{DFT}}_{c,\text{exact}}|/|E^{\text{DFT}}_{c,\text{exact}}|$ *in percent.*

	CS	LYP	PW91	LDA	EXACT
H^-	0.0297	0.0299	0.0320	0.0718	0.0420
He	0.0416	0.0438	0.0457	0.1128	0.0421
Be^{+2}	0.0442	0.0491	0.0535	0.1512	0.0443
Ne^{+8}	0.0406	0.0502	0.0617	0.2030	0.0457
Be	0.0936	0.0955	0.0950	0.2259	0.0962
Ne	0.375	0.383	0.381	0.745	0.394
$|\Delta|\%$	8.2	9.5	15.4	175	

Table 3: *Non-relativistic absolute correlation energies of first and second row atoms from selfconsistent calculations with various DFT approximations. QC denotes the conventional quantum chemistry value [38].* $|\Delta|\%$ *denotes the mean value of* $\left|(E^{\text{DFT}}_c - E^{\text{QC}}_c)/E^{\text{QC}}_c\right|$ *in percent. All other numbers in Hartree units.*

	KLI-CS	BLYP	PW91	LDA	QC
He	0.0416	0.0437	0.0450	0.1115	0.0420
Li	0.0509	0.0541	0.0571	0.1508	0.0453
Be	0.0934	0.0954	0.0942	0.2244	0.0943
B	0.1289	0.1287	0.1270	0.2906	0.1249
C	0.1608	0.1614	0.1614	0.3587	0.1564
N	0.1879	0.1925	0.1968	0.4280	0.1883
O	0.2605	0.2640	0.2587	0.5363	0.2579
F	0.3218	0.3256	0.3193	0.6409	0.3245
Ne	0.3757	0.3831	0.3784	0.7434	0.3905
Na	0.4005	0.4097	0.4040	0.8041	0.3956
Mg	0.4523	0.4611	0.4486	0.8914	0.4383
Al	0.4905	0.4979	0.4891	0.9661	0.4696
Si	0.5265	0.5334	0.5322	1.0418	0.5050
P	0.5594	0.5676	0.5762	1.1181	0.5403
S	0.6287	0.6358	0.6413	1.2259	0.6048
Cl	0.6890	0.6955	0.7055	1.3289	0.6660
Ar	0.7435	0.7515	0.7687	1.4296	0.7223
$|\Delta|\%$	3.13	4.52	5.10	120	

Table 4: *Non-relativistic absolute correlation energies of atoms from selfconsistent calculations with various DFT approximations. All numbers in Hartree units.*

	KLI-CS	BLYP	PW91		KLI-CS	BLYP	PW91
K	0.8030	0.7821	0.7994	Rb	1.7688	1.7832	1.9509
Ca	0.8269	0.8329	0.8467	Sr	1.8222	1.8355	2.0056
Sc	0.8832	0.8855	0.9033	Y	1.8763	1.8863	2.0671
Ti	0.9371	0.9374	0.9613	Zr	1.9281	1.9363	2.1307
V	0.9882	0.9882	1.0198	Nb	1.9475	1.9558	2.1899
Cr	1.0073	1.0086	1.0736	Mo	1.9905	2.0003	2.2551
Mn	1.0812	1.0861	1.1375	Tc	2.0796	2.0874	2.3412
Fe	1.1597	1.1620	1.2158	Ru	2.1571	2.1637	2.4254
Co	1.2324	1.2331	1.2933	Rh	2.2278	2.2340	2.5081
Ni	1.3009	1.3010	1.3700	Pd	2.3123	2.3154	2.6074
Cu	1.3693	1.3694	1.4562	Ag	2.3561	2.3649	2.6705
Zn	1.4273	1.4303	1.5212	Cd	2.4146	2.4247	2.7373
Ga	1.4704	1.4753	1.5768	In	2.4600	2.4704	2.7964
Ge	1.5101	1.5174	1.6343	Sn	2.5024	2.5135	2.8577
As	1.5465	1.5570	1.6917	Sb	2.5419	2.5544	2.9193
Se	1.6177	1.6288	1.7662	Te	2.6134	2.6252	2.9965
Br	1.6795	1.6912	1.8393	I	2.6763	2.6876	3.0726
Kr	1.7355	1.7493	1.9112	Xe	2.7338	2.7456	3.1475

1. For most atoms, the absolute value of E_c^{QC} is smaller than the absolute correlation energy obtained with any DFT method, as it should be according to the relation (26).

2. The values of E_c^{KLI-CS}, E_c^{LYP}, E_c^{PW91} and E_c^{QC} agree quite closely with each other while the absolute value of E_c^{LDA} is too large roughly by a factor of two. We mention that due to the well known error cancellation between E_x^{LDA} and E_c^{LDA}, the resulting LDA values for total xc energies are much better.

3. The difference between E_c^{DFT} and E_c^{QC} is smallest for the E_c^{KLI-CS} values, larger for E_c^{LYP} and largest for E_c^{PW91}. The difference between E_c^{QC} and E_c^{DFT} has three sources:

   (a) The values of E_c^{QC} are only approximate, i. e. not identical with $E_{c,exact}^{QC}$.

   (b) The values of E_c^{DFT} are only approximate, i. e. not identical with $E_{c,exact}^{DFT}$.

   (c) As shown in the last section, the exact values $E_{c,exact}^{QC}$ and $E_{c,exact}^{DFT}$ are not identical.

Currently it is not known with certainty which effect gives the largest contribution. However, with the arguments given in the last section, we expect the contribution of $\left(\text{c}\right)$ to be small. Assuming that the quoted values of E_c^{QC} are very close to $E_{c,\text{exact}}^{QC}$ we conclude that E_c^{KLI-CS} is closest to $E_{c,\text{exact}}^{DFT}$.

Table 4 shows correlation energies of atoms K through Xe obtained with the various selfconsistent DFT approaches. In almost all cases, the absolute KLI-CS values for E_c are smallest and the ones from PW91 are largest, while the LYP values lie in between. In most cases, E_c^{KLI-CS} and E_c^{BLYP} agree within less than 1 % while $|E_c^{PW91}|$ is larger (by up to 10 %) as the atomic number Z increases. We emphasize that reliable values for E_c^{QC} do not exist for these atoms.

Acknowledgments

We thank C. Umrigar for providing us with the exact densities and KS potentials for H^-, He, Be^{+2}, Ne^{+8}, Be and Ne. We gratefully appreciate the help of Dr. E. Engel especially for providing us with a Kohn-Sham computer code and for some helpful discussions. We would also like to thank Professor J. Perdew for providing us with the PW91 xc subroutine. This work was supported in part by the Deutsche Forschungsgemeinschaft.

Literature cited

[1] R. M. Dreizler, E.K.U. Gross. *Density Functional Theory;* Springer-Verlag: Berlin Heidelberg, 1990
[2] R.G. Parr, W. Yang. *Density-Functional Theory of Atoms and Molecules;* Oxford University Press: New York, 1989
[3] V. Sahni, M. Levy, Phys. Rev. B **33**, 3869 (1986)
[4] M. Levy, Proc. Natl. Acad. Sci. USA **76**, 6062 (1979)
[5] S. Baroni, E. Tuncel, J. Chem. Phys. **79**, 6140 (1983)
[6] M. Levy, J. P. Perdew, V. Sahni, Phys. Rev. A **30**, 2745 (1984)
[7] H. Stoll, A. Savin. In *Density Functional Methods in Physics*; R. M. Dreizler, J. da Providencia, Eds.: NATO ASI Series B123; Plenum: New York London, 1985; p 177.
[8] A. Savin, H. Stoll, H. Preuss, Theor. Chim. Acta **70**, 407 (1986)
[9] M. Levy, R. K. Pathak, J. P. Perdew, S. Wei, Phys. Rev. A **36**, 2491 (1987)
[10] M. Levy, Phys. Rev. A **43**, 4637 (1991)
[11] A. Görling, M. Levy, Phys. Rev. A **45**, 1509 (1992)
[12] A. Görling, M. Levy, Phys. Rev. A **47**, 13105 (1993)
[13] C.J. Umrigar, X. Gonze, Phys. Rev. A **50**, 3827 (1994)
[14] A. Görling, M. Ernzerhof, Phys. Rev. A, in press (1995)
[15] R. Colle, D. Salvetti, Theor. Chim. Acta **37**, 329 (1975)
[16] R. Colle, D. Salvetti, Theor. Chim. Acta **53**, 55 (1979)
[17] T. Grabo, E.K.U. Gross, Chem. Phys. Lett., **240**, 141 (1995); Erratum: ibid. **241**, 635 (1995)

[18] P. Hohenberg, W. Kohn, Phys. Rev. **136**, B864 (1964)

[19] E.H. Lieb. In *Physics as Natural Philosophy;* A. Shimony, H. Feshbach, Eds.; MIT Press: Cambridge, 1982; p. 111; a revised version appeared in Int. J. Quant. Chem. **24**, 243 (1983)

[20] M. Levy, Phys. Rev. A **26**, 1200 (1982)

[21] W. Kohn, L.J. Sham, Phys. Rev. **140**, A1133 (1965)

[22] P.W. Payne, J. Chem. Phys. **71**, 490 (1979)

[23] A. Görling, Phys. Rev. A **46**, 3753 (1992)

[24] R. van Leeuwen, E.J. Baerends, Phys. Rev. A **49**, 2412 (1994)

[25] Q. Zhao, R. C. Morrison, R. G. Parr, Phys. Rev. A **50**, 2138 (1994)

[26] E. Engel, J.A. Chevary, L.D. Macdonald, S.H. Vosko, Z. Phys. D **23**, 7 (1992)

[27] E. Engel, S.H. Vosko, Phys. Rev. A **47**, 2800 (1993)

[28] C. J. Umrigar, X. Gonze, unpublished; a preliminary version of the Ne data was published in *High Performance Computing and its Application to the Physical Sciences*, proceedings of the Mardi Gras '93 Conference, D. A. Browne *et al.* Eds.; World Scientific, Singapore: 1993

[29] The QC values for Be and Ne are taken from [38]; the QC values for the two electron systems are the differences between the total energies taken from [13] and HF total energies obtained with our program.

[30] C. Lee, W. Yang, R.G. Parr, Phys. Rev. B **37**, 785 (1988)

[31] J.P. Perdew. In *Electronic structure of solids '91*, P. Ziesche and H. Eschrig Eds.; Akademie Verlag, Berlin: 1991; and J.P. Perdew and Y. Wang, Tulane University, unpublished.

[32] S.J. Vosko, L. Wilk, M. Nusair, Can. J. Phys. **58**, 1200 (1980)

[33] R.T. Sharp, G.K. Horton, Phys. Rev. **90**, 317, (1953)

[34] J.D. Talman, W.F. Shadwick, Phys. Rev. A **14**, 36 (1976)

[35] J.B. Krieger, Y. Li, G.J. Iafrate, Phys. Rev. A **45**, 101 (1992)

[36] J.B. Krieger, Y. Li, G.J. Iafrate, Phys. Rev. A **46**, 5453 (1992)

[37] J.B. Krieger, Y. Li, G.J. Iafrate, Phys. Rev. A **47**, 165 (1993)

[38] S.J. Chakravorty, S.R. Gwaltney, E.R. Davidson, F.A. Parpia, C. Froese Fischer, Phys. Rev. A **47**, 3649 (1993)

[39] A.D. Becke, Phys. Rev. A **38**, 3098 (1988)

Chapter 4

Ab Initio Molecular Dynamics with the Projector Augmented Wave Method

Peter E. Blöchl[1], Peter Margl[2], and Karlheinz Schwarz[3]

[1]IBM Research Division, Zurich Research Laboratory, Säumerstrasse 4,
CH−8803 Rüschlikon, Switzerland
[2]Department of Chemistry, University of Calgary,
2500 University Drive Northwest, Calgary, Alberta T2N 1N4, Canada
[3]Technical University Vienna, Getreidemarkt 9/158,
A−1060 Vienna, Austria

An introduction to the ab-initio molecular dynamics approach of Car and Parrinello and to the projector augmented wave method is given. The projector augmented wave method is an all-electron electronic structure method that allows ab-initio molecular dynamics simulations to be performed accurately and efficiently even for first-row and transition metal elements. We describe the supercell approach and how it can be extended to isolated charged or polar molecules. Applications to organometallic compounds, including ferrocene and the fluxional molecule beryllocene, demonstrate the capabilities of this methodology.

Ten years ago, Car and Parrinello (1) invented the ab-initio molecular dynamics approach, which allows one to simulate the motion of the atoms from first principles. This tool was promising for applications in chemistry, because it combines the accuracy of density functional calculations with the ability to study the finite temperature dynamics of molecules. A new electronic structure method, the projector augmented wave (PAW) method (2), overcomes the limitations of the pseudo-potential approach, which has been employed in most ab-initio molecular dynamics simulations. The PAW method makes the full all-electron wave functions accessible and allows first-row and transition-metal elements to be studied with moderate computational effort. This method has proven useful in a number of recent applications to physics (3, 4), chemistry (5–9), and biochemistry (10). The expectations that ab-initio molecular dynamics allows direct simulation of chemical processes on a picosecond time scale and a nanometer length scale are now being rapidly realized.

0097–6156/96/0629–0054$15.00/0
© 1996 American Chemical Society

In this article we sketch the basic ideas underlying the current methodology and describe some recent applications. We make an attempt to show the general picture at the expense of details that can be found in more specialized publications. Furthermore this paper is restricted to our own work, and thus the reader should be aware that there are other interesting developments (*11–13*) in this field that are not mentioned here.

Ab-initio Molecular Dynamics Approach

Electronic structure calculations are usually performed for a static atomic structure. The dynamical behavior can be studied with molecular dynamics techniques that describe the interaction between atoms by parameterized inter-atomic potentials with parameters adjusted to experiment. Numerous such potentials have been devised and tuned to different environments. Their computational simplicity allows very large systems to be studied for long time scales. However, parameterized inter-atomic potentials have difficulties describing chemical reactions, because of the complexity of the interactions during bond breaking and formation as well as the lack of experimental information required to specify the parameters of the essential part of the potential energy surface.

Ab-initio molecular dynamics is one way to combine the virtues of these two distinct approaches. Here the dynamics of the system is simulated, but in each time step the forces are obtained directly from an accurate electronic structure calculation. Even though some simulations have been performed with the Hartree–Fock method (*14–16*), most ab-initio molecular dynamics calculations are based on density functional theory (*17, 18*). Within the ab-initio molecular dynamics approach, it is currently possible to simulate systems having about a hundred atoms for a few to a few tens of picoseconds. Let us list here the time scales of some common processes to give a feeling for their magnitude:

- Electronic transitions corresponding to visible light occur on a time scale of less than 2.5 fsec, so they can be regarded as instantaneous because the simulation is discretized in steps of about 0.25 fsec.

- A period of a carbon–hydrogen stretch frequency takes about 10 fsec–0.01 psec. Other molecular stretch vibrations and hydrogen bond-bending modes take about twice as long, i.e. 20 fsec or longer.

- Chemical reactions have two distinct time scales: one is the waiting time discussed above, and the other is the reaction event. If we talk about chemical reactions we can estimate the time it takes to overcome a barrier E_A, i.e. the waiting time, as $1/\nu \exp(+E_A/k_B T)$, where ν is a typical molecular frequency. If we use $\nu = 1/20$ fsec and at room temperature we find a waiting time of 1.4 psec for a barrier of a 10 kJ/mol corresponding to the breaking of a hydrogen bond, but 10^7 psec for a barrier of 50 kJ/mol such as in low-barrier chemical reactions.

This implies that at room temperature we can easily study molecular vibrations and low barrier processes such as hydrogen bond rearrangement or the fluxional motion of organometallic compounds. Note, however, that ab-initio molecular dynamics does not describe nuclear tunnel processes. Chemical reactions involving larger barriers, however, can be studied directly by simulations only at elevated temperatures. Above 1000 K we can access reactions involving barriers up to about 50 kJ/mol on a picosecond time scale. Whereas the waiting time is often substantially larger than the affordable simulation time, the reaction event typically occurs on the time scale of molecular vibrations, i.e. a few tenths or hundredths of a picosecond, which is well within the reach of ab-initio molecular dynamics simulations. Thus for systems with large barriers we first must locate the transition states using standard methods, and then we can study the dynamics of a particular reaction event by direct molecular dynamics simulations.

What is the underlying idea of ab-initio molecular dynamics? As in classical molecular dynamics, the nuclear trajectories can be obtained from Newton's law

$$M_i \ddot{R}_i = -\nabla_{R_i} E \quad , \tag{1}$$

where R_i is a coordinate of atom i, M_i its mass and E the total energy expression. This equation can be discretized using the Verlet algorithm, which replaces the acceleration by

$$\ddot{R}_i(t) = \frac{R_i(t + \Delta) - 2R_i(t) + R_i(t - \Delta)}{\Delta^2} \quad , \tag{2}$$

where Δ is the time step. The Verlet algorithm appears to be superior to other, more complex algorithms if large time steps are chosen (*19*). The stability limit of the Verlet algorithm lies at $\Delta = T_{min}/3$, where T_{min} is the oscillation period of the fastest vibration of the system. Hence the time step for a system with C–H bonds must clearly be shorter than 3 fsec. In practice, we use a time step of 0.25 fsec.

If we use the total energy of density functional theory to derive the forces, we must know the self-consistent wave functions at each time step. Considering that a typical simulation requires several thousand to ten thousand time steps, it is hardly possible to perform an independent self-consistent calculation for each time step. This problem has been solved by Car and Parrinello with an ingeniously simple trick: They introduced an additional classical equation of motion for the quantum mechanical wave functions describing the electrons

$$m_\Psi |\ddot{\Psi}_n\rangle = -H|\Psi_n\rangle + \sum_m |\Psi_m\rangle \Lambda_{mn} \quad , \tag{3}$$

where H is the Hamiltonian and Λ_{mn} the matrix of Lagrange multipliers, which ensure that the wave functions remain orthogonal to each other; m_Ψ is a fictitious mass for the wave functions, a free parameter that ideally should be chosen very small. A typical value for m_Ψ is 1000 a.u.

It is not immediately obvious that this coupled system of equations yields meaningful results. The main requirement is that the electrons remain close to the electronic ground state, i.e. at the Born–Oppenheimer surface. This requires the wave function coefficients to be "cold". At the same time, however, the nuclei may be "hot", corresponding to a high temperature. As with any thermally coupled system the temperatures will equilibrate, ultimately rendering the forces acting on the nuclei meaningless because the density functional theory requires the wave functions to be in their ground state. The underlying reason why ab-initio molecular dynamics simulations are still feasible is the adiabatic principle. It states that the thermal coupling of two subsystems with well separated vibrational spectra is small. This is indeed the case, at least for systems with a finite HOMO–LUMO gap, and hence the heat transfer from hot nuclei to cold wave functions is minute, often not even noticeable on a picosecond time scale. In practice the simulation is made stationary using thermostats for the electronic and the atomic subsystem absorbing the remaining heat transfer (20, 21).

The adiabatic principle can easily be demonstrated for a simple classical model: It consists of a light particle with a mass m and a position r and a heavy particle with a mass M and a coordinate R. The two particles are coupled by a spring with a given force constant c. The light particle is analogous to the wave functions, whereas the heavy particle is analogous to the nuclei. This model reproduces many of the typical features of the coupled system of equations in ab-initio molecular dynamics such as slow heat transfer and increased effective mass of the heavy particles (or the nuclei). We obtain a coupled system of equations

$$M\ddot{R} = F(t) - c(R - r)$$
$$m\ddot{r} = -c(r - R) \quad , \tag{4}$$

where $F(t)$ is a time-dependent force acting on the nuclei. This system is transformed into a center-of-mass coordinate $X = (MR + mr)/(M + m)$ and a relative coordinate $x = r - R$. If we fix the position of the heavy particle and relax the light particle, the relative coordinate vanishes. Hence $x = 0$ characterizes the instantaneous ground state or the Born–Oppenheimer surface. Now our system of equations reads

$$(M + m)\ddot{X} = F(t)$$
$$m\ddot{x} = -c(1 + \frac{m}{M})x - \frac{m}{M}F(t) \quad . \tag{5}$$

For the adiabatic principle to work, the frequency $\omega = \sqrt{c/m}$ should be much higher than the frequencies contributing to $F(t)$. This frequency can be controlled by choosing the mass of the light particle, which is analogous to the fictitious mass of the wave functions m_Ψ, sufficiently small. That this is indeed the case in ab-initio molecular dynamics, at least for closed-shell systems, has been demonstrated previously (22). In this limit we can determine the relative coordinate analytically

$$x(t) = -\frac{1}{\omega^2}\frac{F(t)}{m + M} + A\cos\left(\sqrt{\left(1 + \frac{m}{M}\right)}\omega t + \phi\right) \quad , \tag{6}$$

where A and ϕ are arbitrary constants.

We can now summarize the main features that also occur in ab-initio molecular dynamics.

- If we start on the Born–Oppenheimer surface, for example $x = \dot{x} = F = \dot{F} = 0$, the coefficient A remains zero for all times. The remaining deviation $x(t) = (1/\omega^2)(F(t)/m + M)$ from the instantaneous ground state remains small because of the small factor $1/\omega^2$.

- The light particle can also embark on a free oscillation described by the second term in equation 6. This is an uncontrolled deviation from the Born–Oppenheimer surface. If such oscillations are large, the forces acting on the nuclei in an ab-initio molecular dynamics simulation become meaningless. However, in the adiabatic limit, where ω goes to infinity, the particles decouple and there is no heat transfer. Therefore the systems remains on the Born–Oppenheimer surface, if that is where we started. In practice, if the adiabatic principle is only fulfilled approximately, the amplitude of the free oscillation will increase, but only very slowly. Thus it can easily be controlled by a small constant friction or a thermostat acting on the small particle (21).

- The center of mass variable follows the forces like a particle with a slightly heavier mass $M + m$. This implies that the nuclear mass needs to be renormalized, so that the effective mass of the electronic wave functions, which is fictitious, is accounted for. Ways to estimate this renormalization have been described previously (2).

- The position of the heavy particle is, however, not identical to the center of mass variable. The difference is proportional to the deviation of the light particle from the ground state and multiplied by a small factor $m/(M + m)$. In many simulations a close look reveals a tiny jitter in the atomic trajectories, which is reminiscent of the remaining small free oscillation of the wave functions. However, these oscillations are not random and cancel exactly as long as the electrons remain close to the Born–Oppenheimer surface.

It should be noted that such difficulties as the necessity of renormalizing the masses are so small that they remained undetected for a long time. However, as they surfaced, remedies have been developed.

Projector Augmented Wave Method

Before one can perform an electronic structure calculation, one must choose a basis set. Here, a tradeoff between the size of the basis set needed to obtain converged results and the effort to evaluate the total energy, the Hamiltonian, and the overlap-matrix elements, must be found. The smallest basis sets are probably obtained with augmented wave methods such as the linear muffin tin orbital (LMTO) method (23), which uses sophisticatedly composed basis sets. At the other extreme are the pure plane wave methods. Simple plane waves require enormous basis set sizes, so in practice pseudopotentials (24, 25) must be employed. On the other hand, the integrations are trivial because the relevant operators can be cast in a diagonal or separable (26, 27) form using Fast Fourier transforms (FFT), making the computational effort per basis function minute. Plane wave methods are promising, in particular for density functional theory, because here the electron density needs to be represented on a real space grid. For plane waves this is achieved by just another FFT, whereas the same operation easily becomes costly for more complex basis sets. A further advantage of plane wave methods is that an enormous know-how has been accumulated with them for ab-initio molecular dynamics.

However, even when pseudopotentials are employed, the basis set size required for describing first-row and transition-metal elements can be daunting. Further-more, pseudopotentials obscure the electron density near the nucleus. This puts information that depends on the local environment close to the nucleus, such as electric field gradients or hyperfine parameters, out of reach except through indi-rect reconstruction techniques (28). These are clear disadvantages for most typical applications in chemistry.

The design goal for the projector augmented wave (PAW) method was to com-bine the advantages of the plane wave pseudopotential method with that of the all-electron augmented wave methods such as the linear augmented plane wave method (23).

The PAW method borrows the idea of augmentation from the existing all-electron augmented wave methods and uses composed basis sets. In this way augmented wave methods can accommodate the various shapes of the wave func-tions in different regions. Close to the nucleus the kinetic energy is large, ruling out the use of a plane wave expansion. However, in this so-called augmentation region the potential is almost spherically symmetric, suggesting an expansion of the wave function into functions similar to atomic orbitals, which are separated according to their angular momenta and treated on a radial grid. Here the potential is domi-nated by the nucleus and the tightly bound core electrons, which are affected only slightly by the chemical environment of the atom. Thus, a small number of basis functions is sufficient to describe the wave functions in this region. On the other hand, far from the nucleus the potential is shallow and the kinetic energy of the electrons is small. This is the region where the chemical bonds are located, so the wave function varies drastically from one molecular environment to another. Hence plane waves, which form a complete basis set, seem ideally suited to this region.

The PAW method describes the all-electron valence wave function as follows: far from the nucleus, i.e. farther than the covalent radius, the wave function is described by a plane wave expansion. Near the nuclei, however, we subtract and add partial waves, which are similar to atomic orbitals, to the plane wave part in order to incorporate the proper nodal structure of the true wave functions. Thus, the plane wave part near the nucleus can be a smooth continuation of the wave function outside the atomic regions. The partial wave expansions are sufficiently localized within the augmentation regions that the overlap of partial waves centered on different nuclei need not be considered.

In the PAW method an all-electron valence wave function $|\Psi\rangle$ is therefore a superposition of three parts

$$|\Psi\rangle = |\tilde{\Psi}\rangle + \sum_i |\phi_i\rangle\langle\tilde{p}_i|\tilde{\Psi}\rangle - \sum_i |\tilde{\phi}_i\rangle\langle\tilde{p}_i|\tilde{\Psi}\rangle \quad . \tag{7}$$

Equation 7 is the definition of the basis set used in the PAW method. It defines the variational degree of freedom for the all-electron wave functions. The plane wave expansion coefficients of $|\tilde{\Psi}\rangle$ correspond to the orbital coefficients. Equation 7 above can alternatively be read as a linear transformation from a fictitious pseudo-wave function to the corresponding all-electron wave function. We briefly comment on each of the three individual terms.

$|\tilde{\Psi}\rangle$: A smooth function extending over all space, which can be expanded in plane waves. This is called a pseudo-wave function and is denoted $|\tilde{\Psi}\rangle$. One can think of the tilde as an operator that turns the all-electron wave function into the corresponding pseudo-wave function.

$\sum_i |\phi_i\rangle\langle\tilde{p}_i|\tilde{\Psi}\rangle$: A partial wave expansion of the all-electron wave function near an atom. Partial waves are solutions of the Schrödinger equation for a given energy and the atomic all-electron potential (note: they need not be bound states!). They are denoted by the symbol $|\phi_i\rangle$. In practice they are calculated in a spherical harmonics expansion on a radial logarithmic grid. The index i refers to a particular nuclear site, the angular momentum quantum numbers, and an additional index that simply counts the partial waves for the same site and angular momenta. In practice we use one or two partial waves per site and angular momentum. The highest main angular momentum of the partial waves is typically equal to or one higher than that of the occupied valence shells.

$\sum_i |\tilde{\phi}_i\rangle\langle\tilde{p}_i|\tilde{\Psi}\rangle$: A partial wave expansion of the pseudo-wave function in terms of pseudo-partial waves, denoted $|\tilde{\phi}_i\rangle$. Again, the tilde denotes that this is the pseudo-version of an all-electron partial wave. For every all-electron partial wave, a corresponding pseudo-partial wave is defined such that the pseudo-partial wave is identical to the all-electron partial wave beyond the covalent radius. The shape of the pseudo-partial waves will determine the general shape of the plane wave part, and therefore they are chosen as smooth functions.

Both partial wave expansions become identical beyond the atomic regions. The coefficients are obtained as scalar products of the pseudo-wave function with projector functions $\langle\tilde{p}_i|$ that are constructed such that each scalar product picks out the weight of the corresponding pseudo-partial wave from the pseudo-wave function. The projector functions are entirely localized within the atomic regions and must fulfill the condition $\langle\tilde{p}_i|\tilde{\phi}_i\rangle = \delta_{i,j}$. Once partial waves and projector functions are defined, they are fixed during the self-consistency or molecular dynamics simulation.

As is common practice in most augmented wave methods, we employ the frozen core approximation. Thus the atomic core states are calculated accurately for an isolated atom and subsequently transferred to the molecular or crystalline environment. This approximation has proven accurate for total energies and forces, as long as the core states are calculated accurately for the isolated atom. Semi-core states, which may be affected by different chemical bonding, can be treated explicitly as valence states, should it be necessary.

How do the projector functions pick out the weight of a partial wave? We assume that the pseudo-partial waves form a complete set within an augmentation region. This means that any function $\tilde{\Psi}(r)$, which may be defined in all space, can be written – within the augmentation region – in the form $\tilde{\Psi}(r) = \sum_i \tilde{\phi}(r) c_i$ if the coefficients c_i are chosen accordingly. We now want to determine the weight of a partial wave $\tilde{\phi}_j$ in an arbitrary pseudo-wave function $|\tilde{\Psi}\rangle$ by building the corresponding scalar product:

$$\langle \tilde{p}_j | \tilde{\Psi} \rangle = \sum_i \langle \tilde{p}_j | \tilde{\phi} \rangle c_i = c_j \quad . \tag{8}$$

The sum simply drops out because of the orthogonality condition between projector functions and pseudo-partial waves mentioned above.

It turns out that the expectation values and the total energy expression can be decomposed like the wave function into one part that involves only smooth functions and two contributions from functions on a radial grid multiplied by spherical harmonics. The total energy expression of the PAW method is identical to the density functional total energy in the frozen core approximation if the plane wave basis is complete and if the partial wave expansions are complete in the corresponding augmentation regions. In reality these expansions have to be truncated. It turns out that convergence is achieved rapidly and that the accuracy can be controlled by increasing the number of terms in the expansion. Forces, Hamiltonian, and overlap matrix can be evaluated analytically.

Even though the PAW method is an all-electron method, there is a close connection between it and the pseudopotential approach. Because of the similarity of the plane wave expressions of PAW and the pseudopotential method, which are also the computationally most demanding operations, the numerical techniques for this part are related. Furthermore, the pseudopotential approach can be *derived* from the PAW method by a well-defined approximation. If the total energy contributions related to the difference between plane wave part and the full all-electron wave function are expanded only to linear order in the deviation from the atomic density, working pseudopotentials are obtained. This also shows that nonlinear terms from exchange and correlation and from the on-site Coulomb repulsion, which are particularly important for transition metal elements, may cause transferability problems in the pseudopotential approach.

Supercells, Isolated Molecules, and Clusters

If plane waves are used, periodic images are created automatically. This implies that a plane-wave-based method having discretized reciprocal space vectors always describes a periodic crystal. If nonperiodic systems are investigated, these artificial periodic unit cells are called supercells. They can be made sufficiently large so that wave functions of periodic images no longer overlap. This requires a vacuum region approximately 6 Å thick between the periodic images. The supercell approach is very convenient, because it allows crystals, surfaces, interfaces, point defects, and isolated molecules to be studied with the same tools.

A reverse approach is often taken in quantum chemistry, namely that the local electronic structure of extended systems is studied using isolated fragments. For molecules, this is the natural approach. For crystals, on the other hand, such fragments contain several hundreds to several thousands of atoms in order to create an intact region of more than 6 Å about a local center to be investigated. This is a task which is rarely feasible and scarcely done.

However, the supercell approach encounters a serious problem if isolated molecules are studied that carry a charge or an electrostatic dipole. The electrostatic interaction is long-ranged, and does not vanish within the vacuum region separating periodic images. Hence this artificial electrostatic interaction between the periodic images has to be subtracted. We have devised one solution to this problem (29). The idea is to construct a point charge model of the molecular charge distribution that reproduces the electrostatic potential outside the molecule. Once that has been obtained, the electrostatic interaction between the periodic images of the point charge model, which is obtained by Ewald sums, is identical to that of the original charge distribution. The underlying point charge model is obtained directly from an intermediate model of atom-centered spherical Gaussians. This Gaussian density is fitted to the true density with a bias function that enhances the weight of the small Fourier components of the true charge density, which give rise to the long-range electrostatic potential while suppressing large Fourier components. The electrostatic potential is obtained as the analytical derivative of the total energy, including the subtraction of the interaction energy with respect to the charge density.

These point charges can be used in future applications. Most empirical interatomic force fields employed in classical molecular dynamics are based on atom-centered point charges. Our calculated point charges may provide a reasonable starting point to parameterize force fields. They may also allow a quantum mechanical calculation of a reacting group of molecules to be coupled to an environment such as a solvent that is described by the simpler classical molecular dynamics approach (30).

Applications

The methodology described in the previous sections has been applied to bulk crystals, surfaces, and molecules. Here we summarize the results of some recent calculations on organometallic compounds.

Molecular Ground-State Properties. In this section we demonstrate that the PAW method is capable of describing chemical bonding among elements throughout the periodic table. We focus our attention on molecules containing metal atoms and first-row atoms because they pose particular demands on quantum-chemical methods and therefore serve as convenient examples to show the potential of the PAW technique. First, electron correlation effects, which are difficult to include beyond Hartree–Fock calculations, play an important role in transition metal com-

Table I. Parameters and results of ground-state calculations using PAW. Bond lengths are given in angstroms. Calculations were performed in cubic or orthorhombic cells between 8 and 11 Å in size. Invariance of results with respect to the box geometry was tested in each case. PAW calculations of Fe_2 and FeCO were performed with a plane wave cutoff of 60 Ryd, calculations on Cp and CpLi with 50 Ryd, numbers in parentheses were obtained with a cutoff of 30 Ryd. Stretch frequencies (^{a}) are given in cm^{-1}

Compound	Variable	PAW	Other Calc.	Expt.
Fe_2 ($^7\Delta_u$)	r_{Fe-Fe}	1.995(1.990)	1.96^{31}	2.02^{32}
			1.97^{33}	1.87^{34}
	$\omega_{Fe-Fe}{}^a$	407	412^{31}	299.6^{35}
			409^{33}	
$C_5H_5^-$($^2B_2$)	$r_{C=C}$	1.367(1.369)	1.351^{36}	-
	r_{C-C}	1.466(1.466)	1.483^{36}	-
	r_{C-C}	1.401(1.402)	1.411^{36}	-
C_5H_5Li($^1A_1$)	r_{C-C}	1.423(1.423)	1.422^{37}	-
	r_{C-H}	1.096(1.101)	1.070^{37}	-
	r_{Li-C}	2.102(2.102)	-	-
FeCO($^5\Sigma^-$)	r_{Fe-C}	1.864(1.859)	1.891^{38}	-
	r_{C-O}	1.176(1.176)	1.192^{38}	-
FeCO($^3\Sigma^-$)	r_{Fe-C}	1.687(1.685)	1.717^{38}	-
	r_{C-O}	1.186(1.186)	1.209^{38}	-
$FeCp_2$(D_{5h})	r_{Fe-Cp}	(1.607)	1.585^{39}	1.65^{40}
	r_{C-C}	(1.426)	1.421^{39}	1.440^{40}
	r_{C-H}	(1.096)	1.093^{39}	1.104^{40}
$Fe(CO)_5$(D_{3h})	$r_{Fe-C(ax)}$	(1.781)	1.768^{39}	1.807^{41}
	$r_{Fe-C(eq)}$	(1.778)	1.769^{39}	1.827^{41}
	$r_{C-O(ax)}$	(1.165)	1.145^{39}	1.152^{41}
	$r_{C-O(eq)}$	(1.168)	1.147^{39}	1.152^{41}
$Ru(Cp)_2$(D_{5h})	r_{Ru-C}	(2.168)	-	2.196^{40}
	r_{C-C}	(1.438)	-	1.440^{40}
	r_{C-H}	(1.098)	-	1.130^{40}

pounds and, second, the above-mentioned atoms have rapidly oscillating valence wave functions near the nucleus, even if pseudopotentials are employed, which make them demanding when dealt with using plane-wave-based techniques.

Ground-state calculations have been performed for Fe_2 in the $^7\Delta_u$ state, the cyclopentadiene radical (Cp) in the $^2B_2$ ground state, lithium cyclopentadienide (LiCp) in the $^1A_1$ ground state and the iron monocarbonyl (FeCO) molecule in the $^5\Sigma^-$ and the $^3\Sigma^-$ states. Further examples include ferrocene, iron pentacarbonyl, and ruthenocene. We employed the local density approximation; the local spin density approximation was used for those molecules that have a nonsinglet electronic ground state, for which we used the parameterization by Perdew and Zunger (*42*). The results of these tests are given in Table I. Further technical details can be found in Refs. (*6, 7*).

Our results are in good agreement with experimental and other theoretical results, establishing that equilibrium geometries and vibrational frequencies can be computed reliably by the PAW method, applying the supercell approach. The basis set size to achieve this goal is modest, which means that calculations can easily be carried out on a workstation.

Vibrational Properties: Ferrocene. A first-principles molecular dynamics trajectory contains a wealth of information about molecular motions far transcending the analytic power of "static" techniques. The accuracy of the calculated

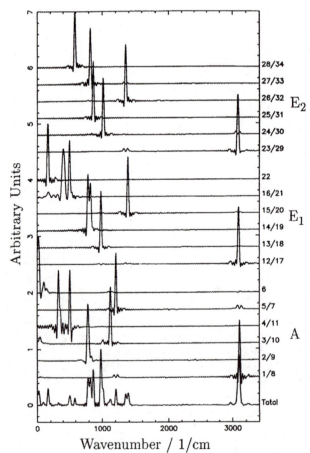

Figure 1: Vibrational spectrum of ferrocene divided according to irreducible representations and eigenvectors as calculated by PAW. Spectra are arranged in groups of six belonging to different irreducible representations, from bottom to top: Total (no symmetry and no eigenmode projection) spectrum; in groups of six spectra: spectra belonging to the $A(=A_1 + A_2)$ irreducible representations, spectra for E_1, spectra for the E_2 irreducible representation. Reproduced with permission from (*7*). Copyright 1994 American Institute of Physics.

molecular motions can be tested rigorously by comparing them to the experimentally accessible vibrational spectra. A method developed by Kohanoff et al. (*43*) and improved by the authors (*7*) allows the vibrational frequencies and eigenvectors to be extracted directly from a dynamical trajectory. This method has been applied to the ferrocene molecule (*7*), for which a vibrational analysis has long been missing due to experimental difficulties. Frequency shifts for deuterated ferrocene and the corresponding eigenvectors are obtained by diagonalizing the dynamical matrix after renormalizing it according to the changed masses. The accuracy of the procedure was tested and confirmed on the benzene molecule, for which the vibrational assignments are well established.

All 57 vibrations of ferrocene have been determined. At the bottom of Figure 1, the total Fourier transform (FT) vibrational spectrum of ferrocene is shown. This spectrum is too complicated for a complete vibrational assignment because only about 15 of the 34 fundamentals are discernible. However, by projecting them onto the vibrational modes, the individual contributions can be clearly separated (spectra 1–34 in Figure 1). Another independent vibrational analysis of ferrocene within density functional theory, albeit using second derivatives of the total energy (*44*), agrees well with our results. Based on these data, we partially revised the previously accepted assignments (*45, 46*) for four modes of ferrocene as shown in Table II.

Table II. New assignments proposed for the eigenmodes of ferrocene based on the present first-principles molecular dynamics investigation. The leftmost column denotes the mode number using the nomenclature introduced by Lippincott and Nelson (*45*). The numbers are the peak locations found for the modes in the PAW simulation or by experiment. Correlations are drawn between the PAW frequency and the experimental frequency by using the same superscripts for related frequencies. Superscripts C and E are unmatched in the other column, indicating that a peak in the region of 823/810 was probably not resolved in experimental spectra and the experimental peak at 1255/1250 is probably not due to a fundamental vibration.

Mode	PAW	Expt.
$\nu_{24/30}$	$1020/1011^A$	$1191/1189^D$
$\nu_{25/31}$	$868/863^B$	$1058/1055^A$
$\nu_{27/33}$	$823/810^C$	$897/885^B$
$\nu_{5/7}$	$1205/1205^D$	$1255/1250^E$

Fluxional Dynamics of Organometallics: Beryllocene. Organometallics often exhibit interesting dynamical phenomena such as fluxional structural rearrangements. Such phenomena can ideally be studied by direct simulation, because the waiting time between fluxional events mentioned above is often extremely short (of the order of picoseconds) and on the other hand because the anharmonicities of such complex total energy surfaces are fully taken into account. We have studied beryllocene, an extreme case, for which fluxionality completely blurs even low-temperature experiments so that its structure has remained unclear until recently. The fluxionality is due to the much weaker interaction of Be with the Cp rings than with Fe in Ferrocene, due to the lack of d-electrons.

Our geometry optimizations yield the energetic ordering η^1-η^5 < η^2-η^5 < η^3-η^3 < η^5-η^5-D_{5h} < η^5-η^5-D_{5d} of the different isomers. All isomers are energetically separated by less than 11 kJ/mol. We have performed several molecular dynamics simulations totaling $\approx$15 ps. During these simulations, we observed intramolecular transformations, which can be divided into two classes as sketched in Figure 2.

- In the "gear-wheel" mechanism, the bond between Be and the η^1 ring migrates from one carbon atom to the next on the same ring. The initial and final states are identical except that the η^1 ring is rotated 72° about its axis. The transition state is η^2-η^5-coordinated and has an activation energy of 5 kJ/mol.

- The "molecular-inversion" mechanism interchanges the roles of the η^1 and the η^5 rings by a motion of the Be atom parallel to the ring planes from the centrally bonded position of one ring to that of the other ring [Figure 2 (bottom)]. The transition state for this mechanism is an η^3-η^3 configuration of C_{2h} symmetry with an activation barrier of 8 kJ/mol.

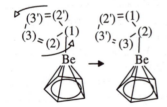

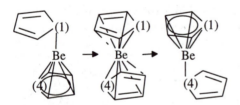

Figure 2: Schematic drawing of the intramolecular rearrangement pathways of beryllocene. Top: gear-wheel mechanism. Bottom: molecular-inversion mechanism. Reproduced with permission from (9). Copyright 1995 American Institute of Physics.

During a simulation of 8 psec at 400 K, we observed ten successful gear-wheel events and two molecular inversions. Taking into account all our accumulated molecular dynamics data, the gear-wheel and the molecular-inversion mechanisms have rates of 1–4 psec^{-1} and 0.3–1.5 psec^{-1}, respectively. Whereas NMR measurements of the spin relaxation times of the ^{9}Be nucleus (47) suggested that the two mechanisms occur simultaneously, we find that they are mostly well separated and occur on different time scales. The measured rates for molecular rearrangement are $10^{10\pm1}$s^{-1}. These measurements should be compared only to the molecular inversion process, because the spin-lattice relaxation of the ^{9}Be nucleus should be caused primarily by the inversion of the molecular dipole. Given the lower temperature of the experiment (≈ 300 K) and the viscosity of the solvent, the agreement between experiment and the PAW simulations is quite satisfactory.

The crucial parameter for rearrangements is apparently the tilt angle between the Cp rings. At 400 K we find on average large tilt angles $\approx 20°$ compared to the ground-state tilt of $8°$. This is caused by the larger entropy of the η^1-η^5 configuration. Molecular inversions and gear-wheel processes occur primarily for small tilt angles, where the Be atom exhibits large amplitude oscillations.

Conclusion and Outlook

We hope to have shown that the PAW method in combination with the ab-initio molecular dynamics approach is an interesting method for applications both in physics and chemistry. Currently, a number of new calculations are in progress. Coordination complexes of up to one hundred atoms are being studied to unravel the reaction mechanisms of enantioselective catalysts. The adsorption properties of methanol in a zeolite have been investigated using dynamical simulations (*48*). It has proven possible to study this process for a complete sodalite crystal with periodic boundary conditions and with no symmetry constraints for about 10 psec. Other areas involve adsorption properties of organic molecules on semiconductor surfaces, molecular crystals (*49*), superionic conductors, and many more (*50*).

Acknowledgments

This work has been supported in part by the Austrian Science Foundation (FWF) projects P9213, P10842, and J01099, as well as by grants from the Swiss Federal Office for Education and Science. It is part of the IBM joint research project 80601511.

References

[1] Car, R.; Parrinello, M. *Phys. Rev. Lett.* **1985**, *55*, 2471.

[2] Blöchl, P. E. *Phys. Rev. B* **1994**, 50, 17953.

[3] Fisher, A. J.; Blöchl, P. E. *Phys. Rev. Lett.* **1993**, *70*, 3263.

[4] Fisher, A. J.; Blöchl, P. E. In *Computations for the Nano-Scale* Blöchl, P. E., Joachim, C., Fisher, A. J., Eds.; Series E: Applied Sciences; Kluwer: Dordrecht, 1993, Vol. 240, pp 185-197.

[5] Nesper, R.; Vogel, K.; Blöchl, P. E. *Angew. Chem. Int'l. Ed. Engl.* **1993**, *32*, 768.

[6] Margl, P.; Schwarz, K.; Blöchl, P. E. In *Computations for the Nano-Scale* Blöchl, P. E.; Joachim, C.; Fisher, A. J., Eds.; Series E: Applied Sciences; Kluwer: Dordrecht, 1993, Vol. 240, pp. 153-162.

[7] Margl, P.; Schwarz, K.; Blöchl, P. E. *J. Chem. Phys.* **1994**, *100*, 8194.

[8] Margl, P.; Schwarz, K.; Blöchl, P. E. *J. Amer. Chem. Soc.* **1994**, *116*, 11177.

[9] Margl, P.; Schwarz, K.; Blöchl, P. E. *J. Chem. Phys.* **1995**, *103*, 683.
[10] Carloni, P.; Blöchl, P. E.; Parrinello, M. *J. Phys. Chem.* **1995**, *99*, 1339.
[11] Galli, G.; Parrinello, M. In *Computer Simulation in Materials Science*, Meyer, M.; Pontikis, V., Eds.; Series E: Applied Sciences; Kluwer: Dordrecht, 1991, Vol. 205, pp. 283-304.
[12] Payne, M. C.; Teter, M. P.; Allan, D. C.; Arias, T. A.; Joannopoulos, J. D. *Rev. Mod. Phys.* **1992**, *64*, 1045.
[13] Singh, D. J. *Planewaves, Pseudopotentials and the LAPW Method*, Kluwer: Dordrecht, 1994.
[14] Hartke, B.; Carter, E. A. *J. Chem. Phys.* **1992**, *97*, 6569.
[15] Gibson, D. A.; Carter, E. A. *J. Phys. Chem.* **1993**, *97*, 13429.
[16] Hartke, B.; Gibson, D. A.; Carter, E. A. *Int. J. Quant. Chem.* **1993**, *45*, 59.
[17] Hohenberg, P.; Kohn, W. *Phys. Rev. B* **1964**, *136*, 664.
[18] Kohn, W.; Sham, L. J. *Phys. Rev. B* **1965**, *140*, 1133.
[19] Berendsen, H. J. C.; van Gunsteren, W. F. In *Molecular-Dynamics Simulation of Statistical-Mechanical Systems*, Ciccotti, G.; Hoover, W. G., Eds.; Varenna Notes; North-Holland: Amsterdam, 1983, Vol. 97, p. 43.
[20] Nosé, S. *Mol. Phys.* **1984**, *52*, 255.
[21] Blöchl, P. E.; Parrinello, M. *Phys. Rev. B* **1992**, *45*, 9413.
[22] Pastore, G.; Smargiassi, E.; Buda, F. *Phys. Rev. A* **1991**, *44*, 6334.
[23] Andersen, O. K. *Phys. Rev. B* **1975**, *12*, 3060.
[24] Hamann, D. R.; Schlüter, M.; Chiang, C. *Phys. Rev. Lett.* **1979**, *43*, 1494.
[25] Vanderbilt, D. *Phys. Rev. B* **1990**, *41*, 7892.
[26] Kleinman, L.; Bylander, D. M. *Phys. Rev. Lett.* **1982**, *48*, 1425.
[27] Blöchl, P. E. *Phys. Rev. B* **1990**, *41*, 5414.
[28] Van de Walle, C. G.; Blöchl, P. E. *Phys. Rev. B* **1993**, *47*, 4244.
[29] Blöchl, P. E. *J. Chem. Phys.*, **1995**, in press.
[30] Field, M. J. In *Computer Simulations of Biomolecular Systems* Gunsteren, W. F.; Weiner, P. K.; Wilkinson, A. J., Eds.; Escom: Leiden, 1993, Vol. 2, pp. 82-123.
[31] Dhar, S.; Kestner, N. R. *Phys. Rev. A* **1988**, *38*, 1111.
[32] Purdum, H.; Montano, P. A.; Shenoy, G. K.; Morrison, T. *Phys. Rev. B* **1982**, *25*, 4412.
[33] Sarnthein, J. unpublished LAPW result in a 6.21×6.21×6.35 Å unit cell, **1992**.
[34] Montano, P. A.; Shenoy, G. K. *Solid State Commun.* **1980**, *35*, 53.
[35] Moskovits, M.; DiLella, D. P. *J. Chem. Phys.* **1980**, *73*, 4917.
[36] McKee, M. L. *J. Phys. Chem.* **1992**, *96*, 1683.
[37] Waterman, K. C.; Streitwieser Jr., A. *J. Am. Chem. Soc.* **1984**, *106*, 3138.
[38] Barnes, L. A.; Rosi, M.; Bauschlicher, C. W. *J. Chem. Phys.* **1991**, *94*, 2031.
[39] Fan, L.; Ziegler, T. *J. Chem. Phys.* **1991**, *95*, 7401.
[40] Haaland, A.; Nilsson, J. E. *Acta Chem. Scand.* **1968**, *22*, 2653.
[41] Beagley, B.; Schmidling, D. G. *J. Mol. Struct.* **1974**, *22*, 466.
[42] Perdew, J. P.; Zunger, A.; *Phys. Rev. B* **1981**, *23*, 5048.
[43] Kohanoff, J. *Comp. Mater. Sci.* **1994**, *2*, 221.
[44] Berces, A.; Ziegler, T.; Fan, L. *J. Phys. Chem.* **1994**, *98*, 1584.
[45] Lippincott, E. R.; Nelson, R. D. *Spectrochim. Acta* **1958**, *10*, 307.

[46] Bodenheimer, J. S.; Nelson, R. D. *Spectrochim. Acta A* **1973**, *29*, 1733.
[47] Nugent, K. W.; Beattie, J. K.; Field, L. D. *J. Phys. Chem.* **1989**, *93*, 5371.
[48] Nusterer, E.; Blöchl, P. E.; Schwarz, K. *Angew. Chem.* **1996**, in press.
[49] Hoerner, C.; Blöchl, P. E.; Margl, P.; Koenig, C. submitted to *Phys. Rev. B.*
[50] Margl, P.; Ziegler, T.; Blöchl, P. E. *J. Amer. Chem. Soc.* **1996**, in press.

Chapter 5

A Gaussian Implementation of Yang's Divide-and-Conquer Density-Functional Theory Approach

Alain St-Amant

Department of Chemistry, University of Ottawa, 10 Marie Curie, Ottawa, Ontario K1N 6N5, Canada

A gaussian implementation of Yang's divide-and-conquer approach to density functional theory has been created. Divide-and-conquer approaches to the fits of the density and the exchange-correlation potentials have been developed. The concept of extended buffer space is introduced. It extends the spatial extent of buffer space while keeping the number of basis functions under control. Tests on dipeptide, tripeptide, and tetrapeptide analogues are presented. The results suggest that we will have to go to much larger systems before our gaussian divide-and-conquer approach outperforms the conventional gaussian density functional approach.

Computational chemists are increasingly turning towards density functional theory (DFT) [1] to perform caclulations previously done by conventional *ab initio* methods [2]. DFT allows us to treat systems much larger than those currently feasible within any correlated post-Hartree-Fock approach. Nevertheless, truly large systems are beyond conventional DFT methods. In theory, we could work directly with the electronic density, a tremendous computational simplification. However, in practise, we must make use of molecular orbitals, and DFT schemes thus scale between N^2 and N^3, where N is related to system size.

Yang has proposed a divide-and-conquer (DAC) approach offering linear scaling [3]. Eliminating the need for molecular orbitals that span the entire molecule, it brings us closer to an approach dealing directly with the density. The DAC approach has been implemented [3] within a program similar to DMol [4]. We wish to implement a DAC scheme within the linear combination of gaussian type orbitals (LCGTO)-DFT formalism, originally developed by Dunlap, Connolly, and Sabin [5], popularized by such programs as DGauss [6] and DeFT [7]. The fitting procedures that make LCGTO-DFT so efficient are not found within a DMol-type scheme. We must address this point within our gaussian DAC approach.

0097–6156/96/0629–0070$15.00/0

For the systems of interest to us, we demand a higher level of accuracy than that found in previous DAC schemes. We would like to see our DAC scheme introduce errors, versus conventional gaussian DFT, no greater than 0.2 kcal mol^{-1} in relative conformational energies of small peptides. It is important to note that this 0.2 kcal mol^{-1} threshold is not the error relative to experiment, but rather to conventional calculations where the only difference lies in the fact that the DAC approach is not used. We wish to establish a well-defined DAC protocol that will consistently give us this kind of precision so that we can be assured that any major errors are a result of the DFT method and not the DAC implementation.

The Kohn-Sham Equations

In the Kohn-Sham (KS) approach [8] to DFT, the density, $\rho(\mathbf{r})$, is expressed as the sum of the square moduli of N_{occ} doubly-occupied (we could readily generalize this to the spin-polarized open shell case) KS molecular orbitals,

$$\rho(\mathbf{r}) = 2 \sum_i^{N_{occ}} |\psi_i(\mathbf{r})|^2 . \tag{1}$$

Within the KS approach, the electronic energy is partitioned as follows,

$$E\left[\rho(\mathbf{r})\right] = T\left[\rho(\mathbf{r})\right] + U\left[\rho(\mathbf{r})\right] + E_{xc}\left[\rho(\mathbf{r})\right] . \tag{2}$$

$T\left[\rho(\mathbf{r})\right]$ is the kinetic energy of non-interacting electrons,

$$T\left[\rho(\mathbf{r})\right] = 2 \sum_i^{N_{occ}} \int \psi_i(\mathbf{r}) \frac{-\nabla^2}{2} \psi_i(\mathbf{r}) d\mathbf{r} . \tag{3}$$

$U\left[\rho(\mathbf{r})\right]$ is simply the classical coulomb energy,

$$U\left[\rho(\mathbf{r})\right] = \sum_A^{nuclei} \int \frac{\rho(\mathbf{r}) Z_A}{|\mathbf{r} - \mathbf{R_A}|} d\mathbf{r} + \frac{1}{2} \int \int \frac{\rho(\mathbf{r})\rho(\mathbf{r'})}{|\mathbf{r} - \mathbf{r'}|} d\mathbf{r} d\mathbf{r'} , \tag{4}$$

where $\{Z_A\}$ and $\{\mathbf{R_A}\}$ are the nuclear charges and coordinates. $E_{xc}\left[\rho(\mathbf{r})\right]$ is made to contain all the remaining effects of exchange and correlation (XC).

The KS molecular orbitals, $\{\psi_i(\mathbf{r})\}$, are obtained by solving the KS equations,

$$\hat{H}\psi_i(\mathbf{r}) = \varepsilon_i \psi_i(\mathbf{r}), \tag{5}$$

where $\hat{H}$ is the KS operator. It is given by

$$\hat{H} = \frac{-\nabla^2}{2} + \sum_A^{nuclei} \frac{Z_A}{|\mathbf{r} - \mathbf{R_A}|} + \int \frac{\rho(\mathbf{r'})}{|\mathbf{r} - \mathbf{r'}|} d\mathbf{r'} + v_{xc}(\mathbf{r}) \tag{6}$$

where $v_{xc}(\mathbf{r})$ is the XC potential,

$$v_{xc}(\mathbf{r}) = \frac{\delta E_{xc}\left[\rho(\mathbf{r})\right]}{\delta\rho(\mathbf{r})} . \tag{7}$$

The total energy can be expressed in terms of the KS eigenvalues, $\{\varepsilon_i\}$,

$$
E\left[\rho(\mathbf{r})\right] = 2 \sum_{i}^{N_{occ}} \varepsilon_i - \frac{1}{2} \int \int \frac{\rho(\mathbf{r})\rho(\mathbf{r}')}{|\mathbf{r}-\mathbf{r}'|} d\mathbf{r} d\mathbf{r}'
$$

$$
- \int \rho(\mathbf{r}) v_{xc}(\mathbf{r}) d\mathbf{r} + E_{xc}\left[\rho(\mathbf{r})\right] + \sum_{A}^{nuclei} \sum_{B<A} \frac{Z_A Z_B}{R_{AB}}. \tag{8}
$$

In the above expression, the sum of the KS eigenvalues is first corrected for the double-counting of coulomb repulsion. It is then corrected for the fact that $\hat{H}$ contains $v_{xc}(\mathbf{r})$. Its contribution is subtracted out and replaced by $E_{xc}\left[\rho(\mathbf{r})\right]$. The final term is the trivial internuclear repulsion term.

The Conventional Gaussian Density Functional Approach

In the conventional LCGTO-DFT approach [5, 6], the KS orbitals are expressed as a linear combination of N contracted gaussian basis functions,

$$
\psi_i(\mathbf{r}) = \sum_{\mu}^{N} C_{\mu i} \chi_\mu(\mathbf{r}). \tag{9}
$$

The coefficient matrix, $\mathbf{C}$, is obtained by solving

$$
\mathbf{HC} = \mathbf{SC}\varepsilon, \tag{10}
$$

where the matrix elements $H_{\mu\nu}$ and $S_{\mu\nu}$ are given by

$$
H_{\mu\nu} = \int \chi_\mu(\mathbf{r}) \left\{ \frac{-\nabla^2}{2} + \sum_{A}^{nuclei} \frac{Z_A}{|\mathbf{r}-\mathbf{R_A}|} + \int \frac{\rho(\mathbf{r}')}{|\mathbf{r}-\mathbf{r}'|} d\mathbf{r}' + v_{xc}(\mathbf{r}) \right\} \chi_\nu(\mathbf{r}) d\mathbf{r} \tag{11}
$$

and

$$
S_{\mu\nu} = \int \chi_\mu(\mathbf{r}) \chi_\nu(\mathbf{r}) d\mathbf{r}. \tag{12}
$$

ε is the vector of KS eigenvalues. Combining equations (1) and (9), $\rho(\mathbf{r})$ is now

$$
\rho(\mathbf{r}) = 2 \sum_{i}^{N_{occ}} \left| \sum_{\mu}^{N} C_{\mu i} \chi_\mu(\mathbf{r}) \right|^2 = 2 \sum_{\mu}^{N} \sum_{\nu}^{N} \sum_{i}^{N_{occ}} C_{\mu i} C_{\nu i} \chi_\mu(\mathbf{r}) \chi_\nu(\mathbf{r}). \tag{13}
$$

Barring any simplification, $H_{\mu\nu}$ will require the evaluation of four-center two-electron integrals, leading to a formal N^4 scaling. However, $\rho(\mathbf{r})$ is fit by an auxiliary set of M uncontracted gaussians, $\{\varphi_k(\mathbf{r})\}$,

$$
\rho(\mathbf{r}) \approx \tilde{\rho}(\mathbf{r}) = \sum_{k}^{M} a_k \varphi_k(\mathbf{r}). \tag{14}
$$

Vector $\mathbf{a}$ is obtained by a variational, analytical, procedure [5] that minimizes

$$
\int \int \frac{[\rho(\mathbf{r}) - \tilde{\rho}(\mathbf{r})][\rho(\mathbf{r}') - \tilde{\rho}(\mathbf{r}')]}{|\mathbf{r}-\mathbf{r}'|} d\mathbf{r} d\mathbf{r}' \tag{15}
$$

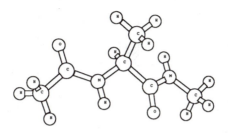

Figure I. Alanine Dipeptide.

subject to the constraint that $\tilde{\rho}(\mathbf{r})$ remains normalized to the total number of electrons in the system. Inserting $\tilde{\rho}(\mathbf{r})$ into equation (11), $H_{\mu\nu}$ now requires only the evaluation of three-center two-electron integrals. Since M and N are of the same order [9], this step now formally scales as only N^3.

The XC potential is fit by a second set of M auxiliary gaussians [5], $\{\phi_l(\mathbf{r})\}$,

$$v_{xc}(\mathbf{r}) \approx \tilde{v}_{xc}(\mathbf{r}) = \sum_{l}^{M} b_l \phi_l(\mathbf{r}). \tag{16}$$

Unlike the fit to $\rho(\mathbf{r})$, this fit can neither be analytical nor variational. Vector b, is obtained by least squares fitting $v_{xc}(\mathbf{r})$ on a set of grid points, $\{R_I\}$,

$$\sum_{I}^{points} [v_{xc}(\mathbf{R_I}) - \tilde{v}_{xc}(\mathbf{R_I})]^2 \, W_I \tag{17}$$

with associated weights, $\{W_I\}$. To evaluate $v_{xc}(\mathbf{R_I})$, we must first synthesize $\rho(\mathbf{R_I})$, via equation (13). This scales as N^2. Since the number of grid points scales linearly with system size, as does N, this procedure over the entire set of grid points also scales as N^3. Inserting $\tilde{v}_{xc}(\mathbf{r})$ into equation (11), $H_{\mu\nu}$ requires the evaluation of three-center overlap integrals, yet another N^3 step.

Divide-and-Conquer Density Functional Theory

It is apparent from equation (3) that the kinetic energy component of the total energy requires the use of molecular orbitals delocalized over the entire extent of a molecule. Yang has recently proposed a DAC approach [3] that brings us closer to true density functional approaches where we work directly with $\rho(\mathbf{r})$ and eliminate the need for molecular orbitals. A rigorous theoretical justification of the method is given elsewhere [3]. Here we will focus on describing the approach within a linear combination of atomic orbitals (LCAO) context.

A molecule may be divided into a set of N_{sub} chemically intuitive subsystems. For example, the alanine dipeptide analogue, Figure I, may be partitioned into three CH_3 fragments, two NH fragments, two CO fragments, and a CH fragment.

The electronic density can be arbitrarily partitioned in the following fashion,

$$\rho(\mathbf{r}) = \sum_{\alpha}^{N_{sub}} \rho^{\alpha}(\mathbf{r}) \tag{18}$$

where

$$\rho^{\alpha}(\mathbf{r}) = \sum_{\alpha}^{N_{sub}} p^{\alpha}(\mathbf{r})\rho(\mathbf{r}) = 2 \sum_{\alpha}^{N_{sub}} p^{\alpha}(\mathbf{r}) \sum_{i}^{N_{occ}} |\psi_i(\mathbf{r})|^2 , \tag{19}$$

provided that $p^{\alpha}(\mathbf{r})$ is a function that obeys, at each point in space, the constraint

$$\sum_{\alpha}^{N_{sub}} p^{\alpha}(\mathbf{r}) = 1. \tag{20}$$

We can now change the summation over doubly-occupied orbitals in equation (19) to one over the N_{occ} doubly-occupied orbitals and the N_{vir} empty virtual orbitals,

$$\rho(\mathbf{r}) = 2 \sum_{\alpha}^{N_{sub}} p^{\alpha}(\mathbf{r}) \sum_{i}^{N_{occ}+N_{vir}} \eta(\varepsilon_F - \varepsilon_i) |\psi_i(\mathbf{r})|^2 . \tag{21}$$

In the above expression, ε_F is the Fermi energy and η is a Heavyside function equal to one when $\varepsilon_i \leq \varepsilon_F$ and zero when $\varepsilon_i > \varepsilon_F$.

From equation (21), we see that if we make $p^{\alpha}(\mathbf{r})$ go to zero as we move away from the atoms in subsystem α, we need only know $\psi_i(\mathbf{r})$ in the vicinity of subsystem α. This can be accomplished with the following definition for $p^{\alpha}(\mathbf{r})$,

$$p^{\alpha}(\mathbf{r}) = \frac{\sum_{A\in\alpha}^{atoms} \left[\rho_{atom}^A(|\mathbf{r} - \mathbf{R_A}|)\right]^2}{\sum_{B}^{atoms} \left[\rho_{atom}^B(|\mathbf{r} - \mathbf{R_B}|)\right]^2} \tag{22}$$

where $\rho_{atom}^A(|\mathbf{r} - \mathbf{R_A}|)$ is the spherical atomic density of atom A, centered at $\mathbf{R_A}$, and the summations in the numerator and the denominator run over all the atoms in subsystem α and over all the atoms in the molecule, respectively.

Having built $p^{\alpha}(\mathbf{r})$ to localize the contributions of $\{\psi_i(\mathbf{r})\}$, we begin to implement a DAC approach and approximate the true $\{\psi_i(\mathbf{r})\}$ by subsystem orbitals, $\{\psi_i^{\alpha}(\mathbf{r})\}$. A modified version of equation (10) is solved for each subsystem α,

$$\mathbf{H}^{\alpha}\mathbf{C}^{\alpha} = \mathbf{S}^{\alpha}\mathbf{C}^{\alpha}\varepsilon^{\alpha}, \tag{23}$$

where $\mathbf{H}^{\alpha}$ and $\mathbf{S}^{\alpha}$ contain the elements of $\mathbf{H}$ and $\mathbf{S}$ where both basis functions are part of subsystem α. With $\mathbf{C}^{\alpha}$, each $\psi_i^{\alpha}(\mathbf{r})$ is expanded, via equation (9), within subsystem α's N^{α} basis functions.

Plugging the subsystem orbitals into equation (21), $\rho(\mathbf{r})$ is approximated by

$$\rho(\mathbf{r}) \approx \tilde{\rho}(\mathbf{r}) = 2 \sum_{\alpha}^{N_{sub}} p^{\alpha}(\mathbf{r}) \sum_{i}^{N^{\alpha}} \eta(\varepsilon_F - \varepsilon_i^{\alpha}) |\psi_i^{\alpha}(\mathbf{r})|^2 . \tag{24}$$

To avoid oscillations in the course of a calculation and achieve self-consistency, it is necessary to work with a unique value of ε_F. We must thus further approximate $\rho(\mathbf{r})$ by replacing the Heavyside function with a Fermi function

$$\tilde{\rho}(\mathbf{r}) = 2 \sum_{\alpha}^{N_{sub}} p^{\alpha}(\mathbf{r}) \sum_{i}^{N^{\alpha}} \frac{1}{1 + e^{-\beta(\varepsilon_F - \varepsilon_i^{\alpha})}} |\psi_i^{\alpha}(\mathbf{r})|^2 \tag{25}$$

where β is an adjustable parameter that makes the Fermi function approach the Heavyside function as it is increased. Fortunately, the final results are fairly insensitive to its precise value and we have simply used the previously suggested value [3]. Finally, ε_F is chosen such that the approximate DAC density, $\tilde{\rho}(\mathbf{r})$, is normalized to the number of electrons in the molecule, N_e,

$$\int \tilde{\rho}(\mathbf{r})d\mathbf{r} = 2 \sum_{\alpha}^{N_{sub}} \sum_{i}^{N^{\alpha}} \frac{1}{1 + e^{-\beta(\varepsilon_F - \varepsilon_i^{\alpha})}} \int p^{\alpha}(\mathbf{r}) |\psi_i^{\alpha}(\mathbf{r})|^2 \, d\mathbf{r} = N_e. \tag{26}$$

The divide-and-conquer total energy,

$$E[\tilde{\rho}(\mathbf{r})] = 2 \sum_{\alpha}^{N_{sub}} \sum_{i}^{N^{\alpha}} \frac{1}{1 + e^{-\beta(\varepsilon_F - \varepsilon_i^{\alpha})}} \varepsilon_i^{\alpha} \int p^{\alpha}(\mathbf{r}) |\psi_i^{\alpha}(\mathbf{r})|^2 \, d\mathbf{r}$$
$$- \frac{1}{2} \int \int \frac{\tilde{\rho}(\mathbf{r})\tilde{\rho}(\mathbf{r}')}{|\mathbf{r} - \mathbf{r}'|} d\mathbf{r}d\mathbf{r}' - \int \tilde{\rho}(\mathbf{r})v_{xc}(\mathbf{r})d\mathbf{r} + E_{xc}[\tilde{\rho}(\mathbf{r})]. \tag{27}$$

is simply the appropriately modified version of equation (8).

In practise, it is found that expanding $\{\psi_i^{\alpha}(\mathbf{r})\}$ within the basis of only those atoms actually in the subsystem is too severe an approximation [3]. Therefore, equation (23) is solved within the basis of the subsystem's atoms and its associated buffer atoms. Buffer atoms "neighbor" the subsystem. As the size of the buffer space increases, we project $\hat{H}$ onto a larger portion of the molecule's total orbital basis. The $\{\psi_i^{\alpha}(\mathbf{r})\}$ in turn approach $\{\psi_i(\mathbf{r})\}$ and the approximation made in equation (24) improves. Defining the buffer space involves a compromise between accuracy (adding more buffer atoms) and computational efficiency (adding less buffer atoms). It is important to note that even though N^{α} increases as we add buffer atoms, $\{p^{\alpha}(\mathbf{r})\}$ remains unaffected. Contributions to $\rho(\mathbf{r})$ are still localized about only the true subsystem atoms. Contributions to $\rho(\mathbf{r})$ in the vicinity of the buffer atoms will be generated by their own subsystem calculations.

The computational advantages DAC DFT are twofold. First, it is inherently parallelizable. The bulk of the CPU time is spent in the independent construction and diagonalization of subsystem KS matrices H^{α}. Second, for very large systems, it will achieve linear scaling. For small molecules, buffer space may extend over a significant fraction of the molecule, and it would be computationally inefficient to work on a collection of subsystems that are almost as large as the original system. However, as the molecule's size increases and buffer space spans but a small fraction of the molecule, the CPU requirements of any subsystem calculation will plateau at a certain value provided a fast multipole method is used to handle the long range electrostatics [10]. At this point, the total CPU time will scale linearly with the number of subsystems, which in turn scales linearly with system size. Linear scaling is thus achieved. DAC DFT is a truly promising approach for tackling very large systems with today's massively parallel supercomputers.

A Gaussian Implementation of Divide-and-Conquer DFT

The auxiliary basis sets of conventional gaussian DFT present a new problem within DAC DFT. Unless we adopt a DAC approach for the fits of $\rho(\mathbf{r})$ and $v_{xc}(\mathbf{r})$, there is no point in adopting a gaussian DAC approach at all.

In our DAC approach to fitting $\rho(\mathbf{r})$, we would like to end up with a final expression that closely resembles that found in conventional LCGTO-DFT,

$$\tilde{\tilde{\rho}}(\mathbf{r}) = \sum_k^M a_k \varphi_k(\mathbf{r}),$$ (28)

where the M auxiliary basis functions, $\{\varphi_k(\mathbf{r})\}$, still span the entire molecule. The double tilde notation is used to indicate that we are approximating the true LCGTO-DFT $\rho(\mathbf{r})$ in two ways: we are adopting a DAC approach to get an approximate density which we in turn fit.

Our scheme is the following. Within each subsystem, $\rho^\alpha(\mathbf{r})$ is constructed,

$$\rho^\alpha(\mathbf{r}) = p^\alpha(\mathbf{r}) \sum_i^{N^\alpha} \frac{1}{1 + e^{-\beta(\varepsilon_F - \varepsilon_i^\alpha)}} \left| \sum_\mu^{N^\alpha} C_{\mu i} \chi_\mu(\mathbf{r}) \right|^2.$$ (29)

$\rho^\alpha(\mathbf{r})$ is fit within the auxiliary bases of both the subsystem and buffer atoms,

$$\rho^\alpha(\mathbf{r}) \approx \tilde{\rho}^\alpha(\mathbf{r}) = \sum_k^{M^\alpha} a_k^\alpha \varphi_k(\mathbf{r}).$$ (30)

M^α is the number of auxiliary functions in the subsystem and its associated buffer. Unfortunately, the fitting procedure can no longer be variational, nor analytical. The complexity of the expression for $\rho^\alpha(\mathbf{r})$ does not permit an analytical approach. We must go to a numerical approach where the $|\mathbf{r} - \mathbf{r}'|^{-1}$ factor present in equation (15) is no longer feasible. We thus lose the variational aspect of the fitting procedure. We must perform a straightforward numerical fit of $\rho^\alpha(\mathbf{r})$, minimizing

$$\sum_I^{points} [\rho^\alpha(\mathbf{R_I}) - \tilde{\rho}^\alpha(\mathbf{R_I})]^2 W_I$$ (31)

on a grid spanning the entire molecule. However, $p^\alpha(\mathbf{r})$ localizes $\rho^\alpha(\mathbf{r})$ to a small region of space, thus eliminating a large number of points.

Once all the subsystems fit, we simply sum the subsystem fit coefficient vectors. Their sum, $\mathbf{a^T}$, is then used to construct the fit to the molecule's total density,

$$\tilde{\tilde{\rho}}(\mathbf{r}) = \sum_\alpha^{N_{sub}} \tilde{\rho}^\alpha(\mathbf{r}) = \sum_\alpha^{N_{sub}} \sum_k^{M^\alpha} a_k^\alpha \varphi_k(\mathbf{r}) = \sum_k^M \left(\sum_\alpha^{N_{sub}} a_k^\alpha \right) \varphi_k(\mathbf{r}) = \sum_k^M a_k^T \varphi_k(\mathbf{r}).$$ (32)

We now have a fit of $\rho(\mathbf{r})$ without ever having to deal with the entire molecule at any point other than the trivial addition of the subsystem fit coefficients.

To fit $v_{xc}(\mathbf{r})$, we must take a slightly different approach. At any point in space, we must know $\rho(\mathbf{r})$ at that point before we can assign it a value of $v_{xc}(\mathbf{r})$. Fortunately, in the course of constructing $\tilde{\rho}(\mathbf{r})$, we had to evaluate $\rho^\alpha(\mathbf{r})$ at each point where $p^\alpha(\mathbf{r})$ was non-negligible. At each point, we continually sum up the contribution of the $\rho^\alpha(\mathbf{r})$'s. Once we have the total unfitted DAC density, $\tilde{\rho}(\mathbf{r})$, at each point, we evaluate $\tilde{v}_{xc}(\mathbf{r})$ at each point. If we were to use a gradient-corrected functional, we would have to construct the derivatives of $\tilde{\rho}(\mathbf{r})$ as well. This adds an extra cost, but it can be done using the same approach used to construct $\tilde{\rho}(\mathbf{r})$. In this paper, we will limit ourselves to the local density approximation (LDA).

We may now localize the contribution of each subsystem to $\tilde{v}_{xc}(\mathbf{r})$. This is achieved by using the same partition function as before

$$\tilde{v}_{xc}^\alpha(\mathbf{r}) = p^\alpha(\mathbf{r})\tilde{v}_{xc}(\mathbf{r}). \tag{33}$$

The subsystem contributions, $\tilde{v}_{xc}^\alpha(\mathbf{r})$, are then fit, minimizing

$$\sum_I^{points} \left[\tilde{v}_{xc}^\alpha(\mathbf{R_I}) - \tilde{v}_{xc}(\mathbf{R_I})\right]^2 W_I \tag{34}$$

over the grid points where $p^\alpha(\mathbf{r})$ is non-negligible. This is done within the M^α XC auxiliary basis functions of the subsystem and its associated buffer, yielding

$$\tilde{v}_{xc}^\alpha(\mathbf{r}) \approx \tilde{\tilde{v}}_{xc}^\alpha(\mathbf{r}) = \sum_l^{M^\alpha} b_l^\alpha \phi_l(\mathbf{r}). \tag{35}$$

We have again adopted a double tilde notation for the same reasons as above. As with $\tilde{\rho}(\mathbf{r})$, the subsystem fit coefficients are summed together to yield $\tilde{v}_{xc}(\mathbf{r})$,

$$\tilde{v}_{xc}(\mathbf{r}) = \sum_\alpha^{N_{sub}} \tilde{\tilde{v}}_{xc}^\alpha(\mathbf{r}) = \sum_\alpha^{N_{sub}} \sum_l^{M^\alpha} b_l^\alpha \phi_l(\mathbf{r}) = \sum_l^M \left(\sum_\alpha^{N_{sub}} b_l^\alpha\right) \phi_l(\mathbf{r}) = \sum_l^M b_l^T \phi_l(\mathbf{r}). \tag{36}$$

Once $\tilde{\rho}(\mathbf{r})$ and $\tilde{v}_{xc}(\mathbf{r})$ are known, the matrix elements of $\mathbf{H}^\alpha$ for any subsystem can be evaluated analytically, as in the conventional LCGTO-DFT approach.

It should now be noted that as system size increases, both the number of subsystems and the number of auxiliary basis functions grow linearly. The method thus formally scales as N^2 overall. However, the XC integrals are three-center overlap integrals that vanish rapidly as the auxiliary basis function is moved away from the subsystem containing the two orbital basis functions. The same cannot be said for the three-center two-electron integrals. However, as we move away from a subsystem, we can replace $\tilde{\rho}(\mathbf{r})$ with point charges. The CPU costs associated with these point charges should be negligible and we should recover linear scaling.

Extended Buffer Space

Applications of DAC DFT have to date used buffer atoms in an "all or nothing" fashion. If an atom is judged to be sufficiently close to the subsystem, its entire basis is added to the subsystem's buffer. Otherwise, its basis is completely ignored. We report here the first use of what we call extended buffer space.

To use an extended buffer, we must first define various levels of basis set sophistication. For the orbital bases, we can have, for example, double-ζ with polarization (DZP), double-ζ (DZ), and single-ζ (SZ) bases. For the auxiliary bases, conventional LCGTO-DFT usually has a set of s functions augmented by a second set of s, p, and d functions constrained to having the same exponent. We will call this the L2 basis. We will create a lower level basis, L1. It is simply the L2 basis with the p and d functions dropped from its second set. In our calculations, the L2 basis for a heavy atom has four s functions augmented by three sets of s, p, and d functions, for 34 functions in all. Hydrogen atoms have three and one, respectively, for 13 functions in all [9].

With various levels of basis set sophistication, we can phase out buffer space gradually. For example, if we are performing a DZ+P/L2 calculation, then subsystem atoms and very near neighbor atoms, true buffer atoms, would carry DZ+P orbital bases and L2 auxiliary bases. More distant neighboring atoms would be in the extended buffer and would carry DZ/L1 bases. Atoms even further away would still be in the extended buffer, but they would see their orbital bases dropped to SZ. At a certain distance, we would truncate the buffer space. The goal of this sample DAC calculation is to reproduce the conventional DZ+P/L2 results. An atom will carry a DZ+P/L2 basis within its own subsystem calculation. For other subsystem calculations, the quality of this atom's basis will be determined by its proximity to the subsystem under study, going from DZ+P/L2 to DZ/L1 to SZ/L1 to nothing as the subsystems get further and further away. The extended buffer approach succeeds in greatly extending the spatial extent of buffer space while keeping N^{α} and M^{α} under control. The value of these last two numbers ultimately determines the cost of a subsystem calculation.

Results and Discussion

In an attempt to establish a reliable protocol for gaussian DAC DFT, we have carried out a series of tests on a dipeptide (Figure I), a tripeptide (Figure II), and a tetrapeptide (Figure III). The dipeptide geometries were optimized at the LDA level while those of the tripeptide and tetrapeptide were subsets of the many local minima on the PM3 potential energy surfaces of these systems. We have divided each molecule into three subsystems for testing purposes. To reduce CPU time, all tests have been performed within the LDA. For such systems, the LDA has been shown to be entirely inadequate [11]. However, a protocol proven reliable within the LDA will no doubt remain reliable once gradient-corrections are introduced.

Our first tests were performed on the alanine dipeptide analogue (ADA), Figure I. ADA is divided into three subsystems: the two terminal methyl groups and everything else remaining in the interior. Both methyl subsystems are in the buffer of the central subsystem. The central subsystem is in the buffer of each methyl subsystem. Each methyl subsystem is in the extended buffer of the other. The distance between any atom in one methyl subsystem and any atom in the other is at least 5.3 Å in all four ADA conformers studied. Table I lists the results

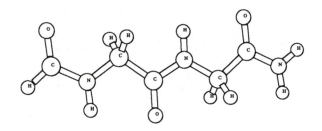

Figure II. FOR-GLY-GLY-NHE Tripeptide.

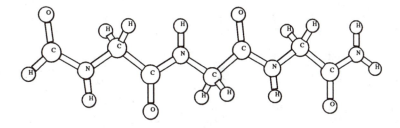

Figure III. FOR-GLY-GLY-GLY-NHE Tetrapeptide.

of tests of various protocols. The orbital and auxiliary bases assigned to atoms in the subsystem and its associated buffer are in the second and third columns. The orbital and auxiliary bases assigned to extended buffer atoms are in the fourth and fifth columns. With four conformers, there are six relative conformationl energies. The root mean square (rms) errors in the relative conformational energies in our DAC calculations are in the sixth column, under $rms\ \Delta\Delta E$. The rms errors in the four total energies are in the last column, under $rms\ \Delta E$. Errors are calculated against conventional DFT calculations which use the bases listed in the second and third columns and the approach outlined in the section on our implementation of gaussian DAC DFT. In a sense, our reference numbers are results of DAC calculations on molecules with only one subsystem. Since our test DAC DFT calculations have the central subsystem's buffer extending over the entire molecule, the DAC DFT approach is clearly not meant to be more computationally efficient for ADA than the conventional DFT approach. Our goal is to test various DAC DFT protocols on relatively small systems.

Protocols I to III make no use of extended buffer space. The orbital and auxiliary bases of the opposing methyl subsystem are not included in each methyl subsystem calculation. This is not the same as setting KS matrix elements to zero in conventional DFT: the central subsystem sees the entire KS matrix, with no matrix elements arbitrarily set to zero. Clearly, $rms\ \Delta\Delta E$ is unacceptable. Buffer space cannot be truncated at 5.3 Å. The $rms\ \Delta E$ values are of the same

Table I. Test of DAC Protocols on the Alanine Dipeptide Analogue

protocol	subsystem/buffer basis orbital	subsystem/buffer basis auxiliary	extended buffer basis orbital	extended buffer basis auxiliary	rms $\Delta\Delta E^a$ (kcal mol^{-1})	rms ΔE^b (kcal mol^{-1})
I	DZP	L2	none	none	2.28	1.68
II	DZ	L2	none	none	2.23	1.59
III	SZ	L2	none	none	2.07	2.58
IV	DZP	L2	DZP	none	1.64	2.19
V	DZ	L2	DZ	none	1.55	2.09
VI	SZ	L2	SZ	none	2.06	2.59
VII	DZP	L2	none	L2	0.34	0.89
VIII	DZ	L2	none	L2	0.37	0.93
IX	SZ	L2	none	L2	0.03	0.02
X	DZP	L2	SZ	L2	0.02	0.02
XI	DZ	L2	SZ	L2	0.02	0.02
XII	SZ	L2	SZ	L2	0.00	0.00
XIII	DZP	L2	DZ	L2	0.01	0.00
XIV	DZ	L2	DZ	L2	0.00	0.00
XV	DZP	L2	none	L1	0.14	0.90
XVI	DZ	L2	none	L1	0.22	0.95
XVII	SZ	L2	none	L1	0.12	0.09
XVIII	DZP	L2	SZ	L1	0.18	0.13
XIX	DZ	L2	SZ	L1	0.15	0.11
XX	SZ	L2	SZ	L1	0.17	0.12
XXI	DZP	L2	DZ	L1	0.19	0.12
XXII	DZ	L2	DZ	L1	0.16	0.11

^{a} Root mean square error in relative conformational energies.
^{b} Root mean square error in total energies.

order. Unfortunately, errors in the total energy are sometimes positive, sometimes negative. This accounts for the larger rms $\Delta\Delta E$ values. As will be seen, this will be a recurring theme: accurate relative energies require accurate total energies.

Protocols IV to VI introduce orbital bases to the extended buffer. Protocols VII to IX do the same with auxiliary bases. The quality of the bases in the extended buffer is the same as in the buffer. In other words, we are extending the buffer space over the entire molecule for either the orbital or auxiliary basis. Adding only orbital bases to the extended buffer does little to improve the situation. Relative energies are slightly better; total energies are slightly worse. Adding only auxiliary bases to the extended buffer has a profound effect. Relative and total energies are greatly improved. We now have a clear distinction between SZ and DZ or DZP bases. Protocol IX, using a SZ orbital basis, is clearly acceptable, with an rms $\Delta\Delta E$ of 0.03 kcal mol^{-1} and an rms ΔE of 0.02 kcal mol^{-1}. The ability to use smaller buffers with SZ bases has been observed in previous

DAC studies [3]. Unfortunately, SZ-quality results are not acceptable for the vast majority of applications. Protocols VII and VIII with DZ and DZP bases are not quite accurate enough, with rms $\Delta\Delta E$ values of over 0.3 kcal mol^{-1}.

Keeping L2 auxiliary bases in the extended buffer, we investigate the effects of adding lower level orbital bases to the extended buffer. Protocols X to XII use a SZ/L2 extended buffer. Protocols XIII and XIV use a DZ/L2 extended buffer. We do not list results for SZ/L2 calculations with a DZ/L2 extended buffer as it would be pointless to have a protocol where extended buffer atoms have better bases than subsystem atoms. The simple addition of SZ bases to the extended buffer reduces errors dramatically. For protocols X and XI, the rms $\Delta\Delta E$ and rms ΔE values are only 0.02 kcal mol^{-1}. The errors for protocol XII are zero since we are performing DAC calculations using the full orbital and auxiliary bases in each subsystem. Raising the quality of the orbital basis in the extended buffer to DZ eliminates virtually all errors. This type of accuracy is far more than required. We have clearly established the usefulness of extended buffer space, for all DAC approaches, not only our gaussian implementation. Taking DZP calculations as an example, protocol VII clearly establishes the need to somehow include a terminal methyl's orbital basis in the opposing methyl's buffer space. Protocol X shows, however, that there is no need to add the full DZP basis (30 basis functions) to the opposing methyl's buffer. The SZ basis (8 basis functions) suffices. We have managed to eliminate 22 orbital basis functions from each methyl subsystem calculation, with little loss of precision.

Protocols XV to XXII explore the possibility of using a lower quality, L1, auxiliary basis in the extended buffer. Protocols XV to XVII do not add orbital bases to the extended buffer. Their rms $\Delta\Delta E$ values are fortuitously better than those of protocols VII to IX, where the superior L2 basis was used in the extended buffer. The rms ΔE values are relatively unaffected upon going from L2 to L1. Focussing only upon the SZ results, errors of about 0.1 kcal mol^{-1} are introduced as we go from an L2 to an L1 auxiliary basis in the extended buffer. Protocols XVIII to XXII are just protocols X to XIV with the auxiliary basis in the extended buffer dropped from L2 to L1. Since protocols X to XIV were, for all intents and purposes, exact, protocols XVIII to XXII are good tests of the effect of going down to an L1 basis on buffer atoms beyond 5.3 Å. We see that acceptable rms $\Delta\Delta E$ and rms ΔE values of 0.2 and 0.1 kcal mol^{-1} are consistently introduced. Errors in relative energies are greater than those in total energies as DAC calculations neither systematically overestimate, nor underestimate, total energies. The extended buffer approach allows us to reduce the number of auxiliary basis functions on the opposing methyl from 73 to 19, almost a factor of four, while maintaining acceptable errors. To see the effect of not having any auxiliary basis functions on the opposing methyl, one need only look back at the totally unacceptable errors obtained within protocols IV to VI.

Table II lists results of DAC calculations on the FOR-GLY-GLY-NHE (two glycine residues capped by formyl and amide groups) tripeptide. It was partitioned into three subsystems, in one of two ways. In scheme A, the terminal peptide groups (CONH$_2$) form subsystems. The remaining interior atoms form

Table II. Test of DAC Protocols on FOR-GLY-GLY-NHE

partitioning	extended buffer		errors in total energy			
	orbital	auxiliary	(kcal mol^{-1})			
A	none	L2	−1.14	−3.28	−1.68	+1.62
B			−0.83	−0.86	−1.20	+0.94
A	SZ	L2	−0.06	−0.28	−0.11	−0.37
B			0.00	−0.01	−0.05	−0.47
A	DZ	L2	0.00	−0.01	−0.01	−0.08
B			+0.01	0.00	0.00	−0.09
A	none	L1	−1.49	−5.67	−1.43	+2.19
B			−0.77	−0.89	−1.16	+1.79
A	SZ	L1	−0.18	−2.94	+0.17	+0.82
B			+0.06	−0.01	+0.01	+0.90
A	DZ	L1	−0.11	−2.65	+0.29	+1.09
B			+0.06	−0.01	+0.06	+1.26

Minimum distances between terminal subsystem atoms, partitioning A: 6.2, 3.8, 3.9 and 2.2 Å; partitioning B: 8.7, 5.6, 5.7 and 2.2 Å. Subsystem atoms and buffer atoms have DZP/L2 bases.

the third. In scheme B, the terminal formyl and amide groups form subsystems, the remaining atoms again forming the third. In schemes A and B, both terminal subsystems are in the buffer of the central subsystem and the central subsystem is in the buffer of each terminal subsystem. The terminal subsystems are in each other's extended buffer. Four conformers were studied. The minimum distances between two atoms in separate terminal subsystems are 6.2, 3.8, 3.9, and 2.2 Å within scheme A, and 8.7, 5.6, 5.7, and 2.2 Å within scheme B. Since DZP/L2-quality will be required in future applications, these tests always use DZP/L2 bases on subsystem and buffer atoms. Table II lists the individual errors in total energies. The ADA results indicate that we must limit these errors to 0.1 kcal mol^{-1} if we are to achieve 0.2 kcal mol^{-1} accuracy in relative energies.

With no orbital bases in the extended buffer, unacceptable errors arise, even at distances of 8.7 Å (-0.77 and -0.83 kcal mol^{-1} with the L1 and L2 auxiliary bases). As in ADA, an SZ/L1 extended buffer gives good results when the extended buffer begins 5.6 or 5.7 Å beyond the subsystem. A very bad error, -2.94 kcal mol^{-1}, arises if we phase in an SZ/L1 extended buffer at a distance of only 3.8 Å. Similar conclusions are drawn for the DZ/L1 extended buffer. An SZ/L2 extended buffer is almost acceptable at a distance of 3.8 or 3.9 Å. The error within scheme A for the second conformer, -0.28 kcal mol^{-1}, is perhaps a bit too high to tolerate. A DZ/L2 extended buffer works well at distances as small as 2.2 Å. Results with a DZ/L2 buffer phased in beyond 3.8 Å are essentially exact.

Table III lists results of DAC calculations on the FOR-GLY-GLY-GLY-NHE tetrapeptide. It is partitioned into three subsystems and its buffer and extended buffer is assigned in the same fashion as the tripeptide. Four conformers were

Table III. Test of DAC Protocols on FOR-GLY-GLY-GLY-NHE

partitioning	extended buffer		errors in total energy (kcal mol^{-1})			
	orbital	auxiliary				
A	none	L2	−0.22	+4.91	−0.14	−1.02
B			−0.15	−1.09	−0.19	−0.65
A	SZ	L2	−0.01	−0.14	−0.01	−0.04
B			0.00	−0.06	−0.01	−0.01
A	DZ	L2	0.00	+0.23	0.00	−0.01
B			0.00	+0.01	0.00	0.00
A	none	L1	−0.26	+2.69	−0.03	+0.48
B			−0.18	−0.56	−0.16	−0.17
A	SZ	L1	−0.05	−1.64	+0.09	+1.40
B			−0.03	+0.48	+0.03	+0.46
A	DZ	L1	−0.04	−1.27	−0.11	+1.43
B			−0.04	+0.54	+0.03	+0.46

Minimum distances between terminal subsystem atoms, partitioning A: 9.7, 2.3, 7.9 and 4.5 Å; partitioning B: 12.3, 2.3, 8.9, and 4.5 Å. Subsystem atoms and buffer atoms have DZP/L2 bases.

studied. The minimum distances between atoms in separate terminal subsystems are 9.7, 2.3, 7.9, and 4.5 Å in scheme A, and 12.3, 2.3, 8.9, and 4.5 Å in scheme B. With no orbital bases in the extended buffer, 0.2 kcal mol^{-1} errors still arise at distances as large as 9.7 or 12.3 Å. SZ/L2 and DZ/L2 extended buffers essentially provide exact results if phased in at 4.5 Å. At 2.3 Å, the DZ/L2 extended buffer generates a +0.23 kcal mol^{-1} error in the second conformer within scheme A, suggesting we were somewhat lucky with the DZ/L2 buffer at 2.2 Å in the tripeptide. Results with the SZ/L1 and DZ/L1 extended buffers strongly discourage using an L1 basis in the extended buffer if phased in as close as 4.5 Å. The DZ/L2 errors are essentially non-existent in the fourth conformer, but with the DZ/L1 extended buffer, errors of +1.43 and +0.46 kcal mol^{-1} are observed.

Our tests suggest that DZ orbital bases can be safely introduced to the extended buffer at distances just under 3 Å. The orbital bases in the extended buffer can then be dropped to SZ at distances just over 4 Å. The distance at which we can eliminate orbital bases from the extended buffer is still unclear. Tests show that doing this at 10 – 12 Å almost meets our accuracy criterion. For the auxiliary bases, we can go to an L1 basis in the extended buffer at 5 – 6 Å. Doing so under 5 Å proved disasterous in certain instances. The ADA results showed that our results are far more sensitive to dropping auxiliary, rather than orbital, bases from the extended buffer. We can expect to go to at least 10 – 12 Å with the L1 bases in the extended buffer. The distance at which we can stop adding auxiliary bases to the extended buffer is still very much in question.

Conclusion

The spatial extent of buffer space is disappointingly large. Clearly, the DAC approach will not be more efficient for systems as small as our test peptides. Fortunately, the use of an extended buffer space helps us dramatically reduce the number of basis functions in a subsystem calculation. The extended buffer approach should prove to be very useful in any future DAC DFT program. The sensitivity of our results to dropping auxiliary bases from the extended buffer suggests that the DAC approach is better suited to DFT formalisms that do not fit $\rho(\mathbf{r})$ and $v_{xc}(\mathbf{r})$ [4]. Future work may alleviate this problem. We must first establish whether this sensitivity can be attributed to either the fit of $\rho(\mathbf{r})$ or $v_{xc}(\mathbf{r})$. If the fit of $\rho(\mathbf{r})$ is the source of our problems, readjustments of the positions and weights of grid points used to fit $\rho^{\alpha}(\mathbf{r})$, equation (31), might help. If the fit of $v_{xc}(\mathbf{r})$ is at fault, redefining the partitioning function, $p^{\alpha}(\mathbf{r})$, in equation (33), further localizing a subsystem's contribution to $v_{xc}(\mathbf{r})$, will help.

Acknowledgements

We wish to thank the Natural Sciences and Engineering Research Council of Canada and the University of Ottawa for financial support.

References

[1] Parr, R. G.; Yang, W. *Density-Functional Theory of Atoms and Molecules*; Oxford University Press: New York, 1989.

[2] *Density Functional Methods in Chemistry*; Labanowski, J. K., Andzelm, J. W., Eds.; Springer-Verlag: New York, 1991.

[3] Yang, W. *J. Mol. Struct. (THEOCHEM)* **1992**, *225*, 461.

[4] Delley, B. *J. Chem. Phys.* **1990**, *92*, 508.

[5] Dunlap, B. I.; Connolly, J. W. D.; Sabin, J. R. *J. Chem. Phys.* **1979**, *71*, 3396.

[6] Andzelm, J.; Wimmer, E. *J. Chem. Phys.* **1992**, *96*, 1280.

[7] DeFT and related documentation is distributed free of charge from the Ohio Supercomputing Center archives (http://www.osc.edu/ccl/cca.html).

[8] Kohn, W.; Sham, L. J. *Phys. Rev. A* **1965**, *140*, 1133.

[9] Godbout, N.; Salahub, D. R.; Andzelm, J.; Wimmer, E. *Can. J. Chem.* **1992**, *70*, 560.

[10] White, C. A.; Johnson, B. G.; Gill, P. M. W.; Head-Gordon, M. *Chem. Phys. Lett.* **1994**, *230*, 8.

[11] St-Amant, A., in *Reviews in Computational Chemistry: Volume 7*; Lipkowitz, K. B., Boyd, D. B., Eds.; VCH Publishers: New York, 1995.

Chapter 6

Direct Ab Initio Dynamics Methods for Calculating Thermal Rates of Polyatomic Reactions

Thanh N. Truong, Wendell T. Duncan, and Robert L. Bell

Department of Chemistry, University of Utah, Salt Lake City, UT 84112

We present a direct *ab initio* dynamics methodology for calculating thermal rate constants from density functional theory (DFT). Dynamical theory is based on a variational transition state theory plus multi-dimensional semi-classical tunneling approximations. Potential energy surface information is calculated from a combined DFT/*ab initio* Molecular Orbital theory approach. We also present applications of this method to predicting detailed dynamics of a hydrogen abstraction reaction and proton transfer in a model biological system to illustrate its versility, accuracy and prospects for molecular modeling of reactive dynamics of polyatomic chemical reactions.

I. INTRODUCTION

The prediction of reaction rates from first principles allows one to make direct comparisons between theory and experiment and hence to deduce reaction mechanisms on the molecular level. For this reason, it has been a major goal of theoretical chemistry. However, it also has been a challenge particularly for polyatomic reactions for the following reasons. The conventional approach of reactive dynamical calculations using either the full quantal dynamics, classical or semiclassical trajectory method, or variational transition state theory (VTST) requires the availability of an accurate analytical potential energy function (PEF).[1-3] Developing such a potential energy function is not a trivial task and is a major obstacle for the

0097–6156/96/0629–0085$15.00/0

dynamical study of a new reaction despite the steady improvement in computer speed. This is because; i) the explicit functional form for a potential energy function is somewhat arbitrary and mostly depends on the investigator's intuition, ii) fitting this functional form to a set of *ab initio* energy points and any available experimental data is tedious and yet does not guarantee convergence or correct global topology, iii) the number of energy points needed grows geometrically with the number of geometrical internal coordinates. As the system size increases, this task becomes much more complex if it can still be accomplished at all. Thus, developments of new methodologies for studying dynamics, kinetics and mechanisms of large polyatomic reactions are of great interest.

Direct dynamics methods,[3-41] including those being developed in our lab[5,12,34-37] offer a viable alternative for studying chemical reactions of complex systems. In the direct dynamics approach, all required energies and forces for each geometry that is important for evaluating dynamical properties are obtained directly from electronic structure calculations rather than from empirical analytical force fields. Our earlier contributions to this area include the development of two different methodologies for calculating thermal rate constants and related properties. One approach is to estimate thermal rate constants and tunneling contributions using the interpolated Variational Transition State Theory (VTST) which has proven useful when available accurate *ab initio* electronic structure information is limited.[16] The other approach is to use a semiempirical molecular orbital Hamiltonian at the Neglect of Diatomic Differential Overlap (NDDO) level as a fitting function in which parameters have been readjusted to accurately represent activation barriers.[17] Full VTST calculations with multidimensional semiclassical tunneling approximations then can be carried out using this NDDO Hamiltonian with specific reaction parameters. Both of these approaches have been successfully applied to various chemical reactions.[16,17,23,24,26,32,39,42,43] However, many difficulties persist. For instance, in the former interpolated VTST approach, it is difficult to correlate vibrational modes in the transition state region to reactant and product asymptotes when mode crossings occur as they often do. In the later, it may prove to be difficult to adjust the original NDDO parameters to accurately describe the transition state region if the original NDDO potential energy surface differs significantly from the reference accurate *ab initio* surface. Recent development in combining both approaches[22,41] has some promise.

In this chapter, we will focus only on our recent contributions to the development of direct *ab initio* dynamics methods in which no experimental data other than physical constants were used for calculating thermal rate constants of gas-phase polyatomic reactions. The dynamical method is based on full VTST theory plus multi-dimensional semiclassical tunneling corrections. The main difference with our previous work, however, is in the way the potential energy surface information is obtained. In our new approach, desired quantities are obtained directly from *ab initio* electronic structure calculations, thus no fitting is involved. For quantitative predictions of kinetic properties, the potential energy surface must be adequately accurate. If such information is to be calculated from a sufficiently accurate level of *ab initio* molecular orbital theory, the computational demand can be substantial. In this case, these methods are only useful for small systems, and thus they stop short of our goal. To alleviate this difficulty, we have introduced two new methodologies which can be used in combination. One is a focusing technique or an adaptive grid method in which more computational resources are spent on regions that are most sensitive to the dynamics and less resources elsewhere. This allows one to obtain an optimal accuracy with a minimum computational cost at a given level of theory. The other is the use of a computationally less demanding electronic structure method, density functional theory (DFT), for the computationally most expensive step in obtaining the potential energy information required for rate calculations. The later raises an interesting and important question. Would DFT methods be sufficiently accurate for this purpose?

Rapid developments in new functionals have significantly improved the accuracy of DFT methods in the past few years. Previously, most applications of DFT were for predicting properties of stable equilibrium structures.[44-46] Recently, more studies[47-53] on the accuracy of DFT methods for transition state properties have been reported. A general conclusion is that for transition state properties the non-local DFT and the hybrid DFT methods in which a portion of Hartree-Fock exchange is included yield results of comparable accuracy to the second-order Møller-Plesset (MP2) method but at a much cheaper computational cost, particulary for large systems. Thus, the computational advantage of DFT would allow application of the direct *ab initio* dynamics method to studying reactions involving larger polyatomic molecules.

To illustrate the applicability, accuracy and versatility of this direct *ab initio* dynamics approach, we present two different applications. One is the hydrogen abstraction $CH_4 + H <\longrightarrow CH_3 + H_2$ reaction. This reaction has served as a prototype reaction involving polyatomic molecules and has played an important role in the theoretical and experimental developments of chemical kinetics. In addition, it has an intrinsic importance to combustion kinetics and is of fundamental interest to organic reaction mechanisms. For this reason, ample experimental rate data is available for comparison. Also this reaction is small enough so that accurate *ab initio* MO calculations can also be performed to test the accuracy of DFT methods. The second example is the proton transfer in formamidine-water complex which has important implications in biological processes. Due to the limited space, we can only focus on the accuracy of the methodology and not so much on the chemistry of these reactions. We refer readers to our original papers[5,34,35] for such discussion.

II. THEORY

A. Variational transition state theory

Variational transition state theory and multidimensional semiclassical tunneling methods have been described in detail elsewhere.[54-60] In this chapter, we only capture the essence of the theory and the approximations involved in the applications presented here. VTST is based on the idea that by varying the dividing surface along the minimum energy path (MEP) to minimize the rate, one can minimize the error due to "recrossing" trajectories. The MEP is defined as the steepest descent path from the saddle point to both the reactant and product directions in the mass-weighted cartesian coordinate system. The reaction coordinate s is then defined as the distance along the MEP with the origin located at the saddle point and is positive on the product side and negative on the reactant side. For a canonical ensemble at a given temperature T, the canonical variational theory (CVT) rate constant for a bimolecular reaction is given by

$$k^{CVT}(T) = \min_{s} k^{GT}(T,s) \tag{1}$$

where

$$k^{GT}(T,s) = \frac{\sigma}{\beta h} \frac{Q^{GT}(T,s)}{\Phi^{R}(T)} e^{-\beta V_{MEP}(s)}. \tag{2}$$

In these equations, $k^{GT}(T,s)$ is the generalized transition state theory rate constant at the dividing surface which intersects the MEP at s and is orthogonal to the MEP at the intersection point. σ is the symmetry factor accounting for the possibility of more than one symmetry-related reaction path and can be calculated as the ratio of the product of the reactant rotational symmetry numbers to the transition state one. For example, the rotational symmetry numbers for CH_4 (T_d), CH_3 (D_{3h}), H_2 ($D_{\infty h}$) and the $H_3C..H..H$ generalized transition state (C_{3v}) are 12, 6, 2 and 3, respectively. Consequently, σ equals 4 for both the forward and reverse directions of the

$CH_4 + H \leftrightarrow CH_3 + H_2$ reaction. β is $(k_B T)^{-1}$ where k_B is Boltzman's constant and h is Planck's constant. $\Phi^R(T)$ is the reactant partition function (per unit volume for bimolecular reactions). $V_{MEP}(s)$ is the classical potential energy (also called the Born-Oppenheimer potential) along the MEP with its zero of energy at the reactants, and $Q^{GT}(T,s)$ is the internal partition function of the generalized transition state at s with the local zero of energy at $V_{MEP}(s)$. Both $\Phi^R(T)$ and $Q^{GT}(T,s)$ partition functions are approximated as products of electronic, vibrational and rotational partition functions. For the electronic partition function, the generalized transition state electronic excitation energies and degeneracies are assumed to be the same as at the transition state. For rotations, since the rotational energy levels are generally closely spaced, replacing the quantal rotational partition functions by the classical ones yields very little loss in accuracy. For vibrations, in the present study, the partition functions are calculated quantum mechanically within the framework of the harmonic approximation. Note that anharmonicity sometimes can have a noticeable effect on reaction rates particularly at higher temperatures. However, it is not included in the applications presented below.

The canonical variational transition state theory described above yields the hybrid (i.e. classical reaction path motion with vibrational degrees of freedom quantized) rate constants. Furthermore, if the generalized transition state is located at the saddle point ($s=0$), eq. (2) reduces to conventional transition state theory.

To include quantal effects for motion along the reaction coordinate, we multiply the CVT rate constants by a transmission coefficient, κ (T). Thus, the final quantized rate constant is

$$k(T) = \kappa^{CVT/G}(T) k^{CVT}(T) \qquad (3)$$

B. Multidimensional Semiclassical Tunneling Methods

To calculate the transmission coefficient, we first approximate the effective potential for tunneling to be the vibrationally adiabatic ground-state potential curve defined by

$$V_a^G(s) = V_{MEP}(s) + \varepsilon_{int}^G(s) \qquad (4)$$

where $\varepsilon_{int}^G(s)$ denotes the zero-point energy in vibrational modes tranverse to the MEP. The ground state transmission coefficient, $\kappa^{CVT/G}(T)$, is then approximated as the ratio of the thermally averaged multidimentional semiclassical ground-state transmission probability, $P^G(E)$, for reaction in the ground state to the thermally averaged classical transmission probability for one-dimensional scattering by the ground-state effective potential $V_a^G(s)$.[24,56-65] If we denote the CVT transition state for temperature T as $s_*^{CVT}(T)$, the value of $V_a^G\{s_*^{CVT}(T)\}$, denoted as $E_*(T)$, is the quasiclassical ground-state threshold energy, then the transmission coefficient $\kappa^{CVT/G}(T)$ can be expressed as

$$\kappa^{CVT/G}(T) = \frac{\displaystyle\int_0^\infty P^G(E)e^{-E/k_bT}dE}{\displaystyle\int_{E_*(T)}^\infty e^{-E/k_bT}dE}. \tag{5}$$

The semiclassical transmission probability $P^G(E)$ accounts for both nonclassical reflection at energies above the quasiclassical threshold and also nonclassical transmission, i.e., tunneling, at energies below that threshold. However, the Boltzmann factor in Eq. (5) makes tunneling the far more important contribution.

Several approximations for the semiclassical transmission probability $P^G(E)$ are available, however, only two, namely, the zero-curvature[58] and the centrifugal-dominant small-curvature semiclassical adiabatic ground-state[65] approximations used in the present study are presented here. For convenience, we label them as ZCT and SCT for the zero-curvature tunneling and small-curvature tunneling cases, respectively. Since the ZCT approximation is a special case of the SCT approximation, we present only the formalism for the SCT below.

The SCT used here is a generalization of the Marcus-Coltrin approximation in which the tunneling path is distorted from the MEP out to the concave-side vibrational turning point in the direction of the internal centrifugal force. This phenomenon is commonly refered to as "corner cutting". Instead of defining the tunneling path explicitly, the centrifugal effect is included by replacing the reduced mass by an effective reduced mass, $\mu_{eff}(s)$, which is used to evaluate imaginary action integrals and thereby tunneling probabilities. Note that in the mass-weighted cartesian coordinate system, the reduced mass μ is set equal to 1 amu. The ground-state transmission probability at energy E is

$$P^G(E) = \frac{1}{\left\{ 1 + e^{-2\theta(E)} \right\}}$$

(6)

where $\theta(E)$ is the imaginary action integral evaluated along the tunneling path,

$$\theta(E) = \frac{2\pi}{h} \int_{s_l}^{s_r} \sqrt{2\mu_{eff}(s)\left| E - V_a^G(s) \right|} \, ds$$

(7)

and where the integration limits, s_l and s_r, are the reaction-coordinate turning points defined by

$$V_a^G\left[s_l(E) \right] = V_a^G\left[s_r(E) \right] = E$$

(8)

Note that the ZCT results can be obtained by setting $\mu_{eff}(s)$ equal to μ in Eq. (7). The effect of the reaction-path curvature is included in the effective reduced mass $\mu_{eff}(s)$ which is given by

$$\mu_{eff}(s) = \mu \times \min \left\{ \begin{matrix} \exp\left\{ -2\bar{a}(s) - [\bar{a}(s)]^2 + \left(d\bar{t}/ds \right)^2 \right\} \\ 1 \end{matrix} \right.$$

(9)

where

$$\bar{a}(s) = \left| \kappa(s)\bar{t}(s) \right|$$

(10)

The magnitude of the reaction-path curvature $\kappa(s)$ is given by

$$\kappa(s) = \left(\sum_{m=1}^{3N-7} \left[\kappa_m(s) \right]^2 \right)^{1/2}$$

(11)

where the summation is over all 3N-7 generalized normal modes and $\kappa_m(s)$ is the reaction-path curvature component along mode m given by[66]

$$\kappa_m(s) = -\mathbf{L}_m^T \, \mathbf{F} \frac{\nabla V}{|\nabla V|^2}$$

(12)

and where $\mathbf{L}_m^T$ is the transpose of the generalized normal mode eigenvector of mode m, $\mathbf{F}$ is the force constant matrix (Hessian matrix),

∇V is the gradient. Finally, $\bar{t}(s)$ is the maximum concave-side vibrational displacement along the curvature direction for which there is no tunneling in the vibrational coordinates. Within the harmonic approximation, $\bar{t}(s)$ is given by

$$\bar{t}(s) = \left(\frac{\kappa \hbar}{\mu} \right)^{1/2} \left\{ \sum_{m}^{F-1} \left[\kappa_m(s) \right]^2 w_m^2(s) \right\}^{-1/4} \tag{13}$$

where w_m is the generalized vibrational frequency of mode m.

As described above, in order to carry out full VTST calculations and multi-dimensional semiclassical zero- or small-curvature tunneling corrections, geometries, energies, gradients and Hessians are needed at the stationary points and along the MEP. Hessian calculations along the MEP are computationally the most expensive step. Below we describe two different methodologies for minimizing the computational cost of this step.

C. Focusing technique

The focusing technique was developed to assure the convergence of the calculated rate constants with a minimal number of Hessians required. This method involves two separate steps. First, a preliminary rate calculation with a coarse Hessian grid is carried out to estimate regions containing the temperature-dependent canonical transition states or those having large curvature where the "corner cutting" effect would be large. A finer Hessian grid is then used for these regions to improve the accuracy of the calculated CVT rate constants and the SCT tunneling probability. The technique will be illustrated in more detail below.

D. Employing DFT methods

Non-local DFT methods such as the combination of Becke's Half-and-Half[67] (BH&H) or three parameter[68] (B3) hybrid exchange with Lee-Yang-Parr[69] (LYP) correlation functionals can be used to calculate the geometries and Hessians along the MEP. The hybrid BH&H functional as implemented in the G92/DFT program[70] consists of 50% Hartree-Fock and 50% Slater exchange contribution. The DFT energies, however, are not always sufficiently accurate. In this case, to obtain more a accurate potential energy along the MEP, one can either perform a series of single point calculations at a more accurate level of theory with a larger basis set for selected points along the MEP or scale $V_{MEP}(s)$ by a constant factor to match a more accurately calculated classical barrier and reaction energy.

From our experience, we have found that the B3LYP method is slightly more accurate than the BH&HLYP method for calculating equilibrium structures. For finding transition state structures, the BH&HLYP method has shown to be more reliable for open-shell sytems. For closed-shell systems, our preliminary results indicate that both B3LYP and BH&HLYP are adequate. In practice, we often carry out accurate *ab initio* MO calculations at the stationary points to check the accuracy of DFT methods prior to their use in rate calculations.

III. APPLICATIONS

A. $CH_4 + H <\longrightarrow> CH_3 + H_2$ reaction

We used this reaction as a test case where the reaction valley (geometries, energies, gradients and Hessians along the MEP) was calculated with both the Quadratic Configuration Interaction including all Single and Double excitations (QCISD) and the hybrid BH&HLYP methods. Both methods use the same 6-311G(d,p) basis set. In particular, we have performed a new benchmark converged direct *ab initio* dynamics rate calculations for this reaction. In the new calculations, the reaction path was calucated at the QCISD/6-311G(d,p) with the step size of 0.01 amu$^{1/2}$bohr which is an order of magnitude smaller than in our previous study.[34] Hessian grids are also much finer. Furthermore, instead of scaling the potential energy along the MEP as in the previous study, single point PMP4/6-311+G(2df,2pd) calculations were performed at the Hessian grids.

Reaction energies and barrier heights are listed in Table 1. The BH&H-LYP classical barriers were found to be too low by 2.7 and 0.6 kcal/mol for the forward and reverse reactions, respectively, as compared to the CCSD(T)/cc-pVQZ results. Single-point spin projected fourth-order Møller-Plesset perturbation theory (PMP4) calculations with the larger 6-311+G(2df,2pd) basis set at the BH&H-LYP/6-311G(d,p) geometries bring the differences in the classical barriers to less than 0.7 kcal/mol and also yield the reaction enthalpy at 0 K to be -0.36 kcal/mol as compared to the CCSD(T)/cc-pVQZ value of -0.3 kcal/mol and the experimental value from JANAF tables[74] of -0.02 kcal/mol.

Table 1: Heat of reaction and barrier heights[a] (kcal/mol) for the CH_4 +H <--> CH_3 +H_2 reaction

Level	ΔE	ΔH_0^0	$\Delta V_f^{\ddagger}$	$\Delta V_r^{\ddagger}$
BH&HLYP/6-311G(d,dp)	1.4	-1.9	12.6 (e11.1)[a]	11.2 (13.0)[a]
PMP4/6-311+G(2df,2pd) //BH&HLYP/6-311G(d,p)	2.9	-0.4	14.6 (13.1)	11.7 (13.5)
QCISD/6-311G(d,p)	2.5	-0.7	16.3 (14.8)	13.8 (15.5)
PMP4/6-311+G(2df,2pd) //QCISD/6-311G(d,p)	3.6	0.3	14.6 (13.0)	10.9 (12.7)
CCSD(T)/cc-pVQZ //CCSD(T)/cc-VQZ[b]	3.5	-0.3	15.3 (13.1)	11.8 (13.3)
J3[c]	2.8	-0.02	12.9 (11.8)	10.1 (11.9)
Expt.[d]	2.6[d]	-1.3[d] -0.02[e]	(13.3 ± 0.5)[d]	(14.6 ± 0.4)[d]

[a] Zero-point energy corrected barriers are given in the parentheses.

[b] From Ref. 71.

[c] From Ref. 72.

[d] From Ref. 73.

[e] From Ref. 74.

Similar results were obtained if the QCISD geometries were used in the PMP4 calculations except the calculated reaction enthalpy is slightly positive. Using the BH&H-LYP zero-point energy correction, the PMP4//BH&H-LYP zero-point energy corrected barriers for both forward and reverse reactions are within 0.2 kcal/mol of the CCSD(T)/cc-pVQZ values and are also in good agreement with experimental data.[73]

The geometries along the minimum energy path calculated by both the BH&H-LYP and QCISD levels are shown in Figure 1. We found that the BH&H-LYP method yields the active C-H_a and H_a-H bond lengths and H-C-H_a angle as functions of the reaction coordinate in excellent agreement with the QCISD results. More specifically, Figure 1 shows an unnoticeable difference in the active bond lengths and a difference of less than 1 degree in the H-C-H_a angle between the two methods.

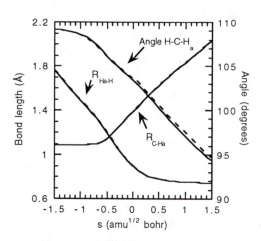

Figure 1: BH&HLYP (solid lines) and QCISD (dased lines) geometrical parameters along the MEP of the H + $CH_4 \rightarrow H_2$ + CH_3 reaction vs the reaction coordinate s.

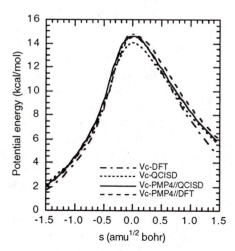

Figure 2: Classical potential energy curves { PMP4//QCISD (solid), PMP4//BH&HLYP (dashed), scaled BH&HLYP (dashed-dotted), scaled QCISD (dotted)}.

The good agreement between PMP4//QCISD and PMP4//BH&HLYP potential curves as shown in Fig. 2 indicates that the accuracy of the potential energy along the MEP can be improved by carrying out single point PMP4/6-311+G(2df,2pd) calculations at selected points along the DFT MEP. When such single point calculations on the Hessian grids are computationally expensive, it is possible to scale the potential energy along the MEP by a constant factor to adjust the barrier height to more accurate calculations. As shown in Fig. 2, we scaled the BH&HLYP and QCISD potential curves to best reproduce both the forward

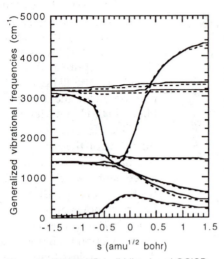

Figure 3: BH&HLYP (solid lines) and QCISD (dashed lines) harmonic frequencies along the reaction coordinate s.

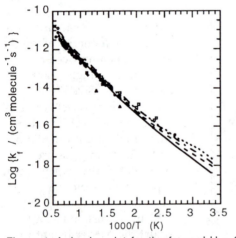

Figure 4: Arrhenius plot for the forward H + CH$_4$ reaction. Symbols are experimental data. Lines are the CVT/SCT results {PMP4//QCISD (solid), PMP4//BH&HLYP (dashed), scaled BH&HLYP (dotted), scaled QCISD (dashed-dotted)}.

and reverse classical barriers calculated at the CCSD(T)/cc-pVQZ level of theory by Kraka et al.[71] Note that both the scaled DFT and QCISD classical potential curves have about the same width compared to the PMP4//QCISD curves, though the barrier heights differ by about 1 kcal/mol. In general, scaling the reaction profile by a constant to obtain more accurate barrier heights does not guarantee to improve the shape as well as the asymtotic regions of the reaction profile. It is possible, however, to add a small number of single point calculations as discussed above and to interpolate the energy corrections along the MEP. Such a procedure is now being tested in our lab.

Generalized frequencies calculated at the BH&H-LYP/6-311G(d,p) level versus the reaction coordinate are plotted in Figure 3 along with the previous QCISD/6-311G(d,p) results. Note that excellent agreement was found between the BH&H-LYP and QCISD results, though the former are slightly larger by about 3%.

The Ahrrenius plots of the calculated and experimental forward and reverse rate constants are

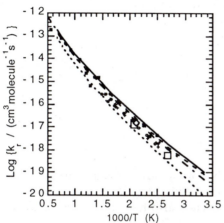

Figure 5: Arrhenius plot of the $CH_3 + H_2$ reaction. Symbols are experimental data. Lines are the CVT/SCT results {PMP4//QCISD (solid), PMP4//BH&HLYP(dashed), scaled BH&HLYP (dotted), scaled QCISD (dashed-dotted)}.

shown in Figures 4 and 5, respectively. Note that our CVT/SCT-PMP4//BH&H-LYP results for the forward rate constants are in excellent agreement with the experimental data for the temperature range from 300-1500 K with the largest deviation factor of 1.5 compared to the recent recommended experimental values.[75] Similar results were found for the PMP4//QCISD calculations. The deviation factor is slightly larger for the reverse rate constants from both the PMP4//QCISD and PMP4//BH&HLYP calculations, particularly ranging from 2.5-3.0 for the temperature range from 300-1500 K. Rate constants calculated from the scaled BH&HLYP and QCISD potential are also in reasonably good agreement with the experimental data.

The above results show that we can use computationally less demanding DFT methods to provide geometries and Hessians along the MEP. DFT energies may not be sufficient for rate calculations, however, potential energies along the MEP can be improved by scaling by a factor to reproduce classical forward and/or reverse barriers from more accurate calculations or performing a series of single point calculations at a more accurate *ab initio* MO level.

B. Proton transfer in formamidine-water complex

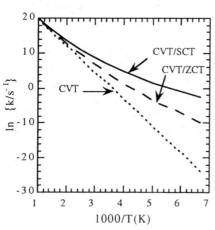

We used proton transfer in the formamidine-water complex as a basic model for studying proton transfer in hydrogen bonded systems, though formamidine and its amidine class also have their own biological and pharmaceutical importance.

Previous studies[5,76] found that adding a water molecule to bridge the proton donor and acceptor sites stabilizes the transition state. This lowers the barrier by 27 kcal/mol, and as a consequence, significantly enhances the transfer rate. Due to the small proton mass, we also found that the tunneling contribution is significant, particularly at low temperatures, by comparing the CVT and CVT/SCT results. Furthermore, the "corner cutting" effect included in the CVT/SCT calculations on the multidimensional surface greatly enhances the tunneling probability for proton transfer in the formamidine-water complex (see Figure 6). This is illustrated by the large increase in the CVT/SCT rate constants when compared to the CVT/ZCT rate. Note that in ZCT calculations, tunneling is restricted to be along the MEP. The reaction valley in this case was calculated at the MP2 level. The potential energy along the MEP was further scaled by a factor of 1.123 to match the

Figure 6: Arrhenius plot of calculated CVT, CVT/ZCT and CVT/SCT thermal rate constants. (Adapted from Ref. 5).

CCSD(T)//MP2 classical barrier heights. In all *ab initio* MO and DFT calculations for this system, the 6-31G(d,p) basis set was used.

The large "corner cutting" tunneling effect requires much more potential energy surface information than just in the vicinity of the saddle point. However, due to the size of this system, one would like to minimize the number of MP2 Hessian calculations without a significant loss to the accuracy. Thus, it is a good example to illustrate the accuracy of our focusing technique described below.

For this discussion, the MEP was calculated at the MP2 level with a maximum of 28 Hessian points evenly distributed between s values of 0 and 1.2 amu$^{1/2}$bohr. Due to the symmetry of the MEP, this is equivalent to a total of 59 Hessian points, including three stationary points, for the entire MEP. Using the CVT/SCT thermal rate constants calculated from these 28 calculated Hessian points as the reference point, we have calculated CVT/SCT rate constants with the number of Hessian points less than the full 28, which were chosen by the focusing technique, and plotted in Figure 7 with the percent difference in the rate constant at 300 K versus the number of Hessians used. We found that a minimum of 10 Hessian points is required for the convergence of the rate constant to within 10%. Furthermore, using a low-pass filtering technique to remove noise in the calculated effective reduced mass slightly improves this convergence. Note that even with 5 Hessian points, the calculated rate constant at 300 K converges to within a factor of two in the 28-point case. For reactions with a smaller tunneling contribution than this case, a smaller number of Hessians may be sufficient. In addition, not including the "corner cutting" effect from the bending modes, i.e. vibrational modes with frequency less than 1800 cm^{-1}, only introduces an error of less than 20%.

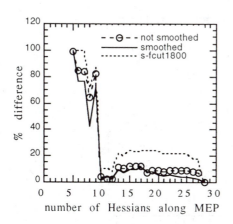

Figure 7: Convergence of the calculated rate constant at 300 K as functions of the number of Hessians along the MEP. (Adapted from Ref. 5)

Finally, using the MP2 rate results for the tautomerization in the **formamidine-water complex**

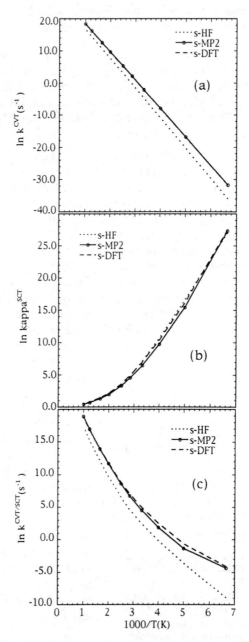

Figure 8: Arrhenius plots of the individual contributions, (a) CVT rate, (b) SCT transmission coefficients, and (c) the total CVT/SCT rate constants vs 1000/T (K).

(Adapted from Ref. 5.)

as a reference point, we have investigated the accuracy using the non-local BH&H-LYP DFT method for calculating the potential energy information. In this case, we also used the same barrier scaling procedure described above. In addition, we also investigated another computationally less demanding approach. That is to use HF theory, but in addition to scaling the classical barrier to the CCSD(T)//MP2 value, the HF frequencies were also scaled by a factor of 0.9. In both HF and DFT cases, the MEP's were calculated with the same step size of 0.1 $amu^{1/2}$bohr as used in the MP2 calculations. The Arrhenius plot for the MP2, HF and DFT CVT and CVT/SCT rate constants are also shown in Figure 8. We found the HF-CVT rate constants are noticeably smaller than the MP2 and DFT rates. As a result, the final HF CVT/SCT rate constants are also smaller and the differences increase as the temperature decreases. The excellent agreement between the MP2 and DFT rate constants further supports the above conclusion on the use of non-local DFT methods for direct *ab initio* dynamics calculations. For instance at 300 K, the HF rate constant is smaller than the MP2 value by a factor of 8.0 while the DFT rate constant is larger by a factor of 1.6.

We have further developed these direct *ab initio* dynamics methods to calculate vibrational-state selected rates of polyatomic reactions.[12,37] The results so far are encouraging. Particularly, it allows one to correlate features on the potential surface to state specific chemistry. Thermal and vibrational-state selected rate constants of polyatomic reactions now can be routinely calculated from *ab initio* MO and/or DFT methods by using our TheRate (Theoretical Rate) program.[77] We are in the process of developing new methodologies for including anharmonicity and large curvature tunneling contributions within our direct *ab initio* dynamics approach.

IV. CONCLUSION

The direct *ab initio* dynamics method we described above with the use of a non-local DFT method to provide potential energy information and a focusing technique to minimize the number of Hessians required offers a promising alternative for studying kinetics, dynamics and mechanisms of large polyatomic reactions. Within the same methodology, one can further extend the dynamical theory to treat chemical reactions on crystal surfaces as well as in solutions. Such steps are now being taken in our lab.

Acknowledgement

This work was supported in part by the University of Utah and by the National Science Foundation through an NSF Young Investigator Award to T.N.T..

References

[1] G. C. Schatz, *Rev. Mod. Phys.* **61,** 669 (1989).

[2] *Potential Energy Surfaces and Dynamics Calculations*, Vol. , edited by D. G. Truhlar (Plenum, New York, 1981).

[3] D. G. Truhlar, M. S. Gordon, and R. Steckler, *Chem. Rev.* **87,** 217 (1987).

[4] K. K. Baldridge, M. S. Gordon, D. G. Truhlar, and R. Steckler, *J. Phys. Chem.* **93,** 5107 (1989).

[5] R. Bell and T. N. Truong, *J. Chem. Phys.* **101,** 10442 (1994).

[6] B. Calef and A. Redondo, *Chem. Phys. Letters* **223,** 1 (1994).

[7] R. Car and M. Parrinello, *Phys. Rev. Lett.* **55,** 2471 (1985).

[8] R. Car and M. Parrinello, in *Simple Molecular Systems at Very High Density*, edited by A. Polian, P. Loubeyre, and N. Boccara (Plenum, New York, 1989), pp. 455.

[9] W. Chen, W. L. Hase, and H. B. Schlegel, *Chem. Phys. Letters* **228,** 436 (1994).

[10] S. M. Colwell and N. C. Handy, *J. Chem. Phys.* **82,** 1281 (1985).

[11] C. J. Doubleday, J. W. J. McIver, and M. Page, *J. Phys. Chem.* **92,** 4367 (1988).

[12] W. T. Duncan and T. N. Truong, *J. Chem. Phys.* **103,** 9642 (1995).

[13] M. J. Field, *Chem. Phys. Lett.* **172,** 83 (1990).

[14] B. C. Garrett, M. L. Koszykowski, C. F. Melius, and M. Page, *J. Phys. Chem.* **94,** 7096 (1990).

[15] B. C. Garrett and C. F. Melius, in *Theoretical and Computational Models for Organic Chemistry*, edited by S. J. Formosinho, I. G. Csizmadia, and L. G. Arnaut (Kluwer, Dordrecht, 1991), pp. 25-54.

[16] A. Gonzalez-Lafont, T. N. Truong, and D. G. Truhlar, *J. Chem. Phys.* **95,** 8875 (1991).

[17] A. Gonzalez-Lafont, T. N. Truong, and D. G. Truhlar, *J. Phys. Chem.* **95,** 4618 (1991).

[18] S. K. Gray, W. H. Miller, Y. Yamagychi, and H. F. Schaefer, *J. Am. Chem. Soc.* **103,** 1900 (1981).

[19] J. C. Greer, R. Ahlrichs, and I. V. Hertel, *Z. Phys. D - Atoms, Molecules and Clusters* **18,** 413 (1991).

[20] B. Hartke and E. A. Carter, *J. Chem. Phys.* **97,** 6569 (1992).

[21] T. Helgaker, E. Uggerud, and H. J. A. Jensen, *Chem. Phys. Lett.* **173,** 145 (1990).

[22] W.-P. Hu, Y.-P. Liu, and D. G. Truhlar, *J. Chem. Soc. Faraday Trans.* **90,** 1715 (1994).

[23] Y.-P. Liu, G. C. Lynch, T. N. Truong, D.-h. Lu, and D. G. Truhlar, *J. Am. Chem. Soc.* **115,** 2408 (1993).

[24] Y.-P. Liu, D.-h. Lu, A. Gonzalez-Lafont, D. G. Truhlar, and B. C. Garrett, *J. Am. Chem. Soc.* **115,** 7806 (1993).

[25] D. Malcome-Lawes, *J. Amer. Chem. Soc. Faraday Trans. 2* **71,** 1183 (1975).

[26] V. S. Melissas and D. G. Truhlar, *J. Phys. Chem.* **98,** 875 (1994).

[27]K. Moroduma and S. Kato, in *Potential Energy Surfaces and Dynamics Calculations*, edited by D. G. Truhlar (Plenum, New York, 1981), pp. 243-264.

[28]D. K. Remler and P. A. Madden, *Mol. Phys* **70**, 921 (1990).

[29]A. Tachibana, I. Okazaki, M. Koizumi, K. Hori, and T. Yamabe, *J. Am. Chem. Soc.* **107**, 1190 (1985).

[30]A. Tachibana, H. Fueno, and T. Yamabe, *J. Am. Chem. Soc.* **108**, 4346 (1986).

[31]D. G. Truhlar and M. S. Gordon, *Science* **249**, 491 (1990).

[32]T. N. Truong and D. G. Truhlar, *J. Chem. Phys.* **93**, 1761 (1990).

[33]T. N. Truong and J. A. McCammon, *J. Am. Chem. Soc.* **113**, 7504 (1991).

[34]T. N. Truong, *J. Chem. Phys.* **100**, 8014 (1994).

[35]T. N. Truong and W. T. Duncan, *J. Chem. Phys.* **101**, 7408 (1994).

[36]T. N. Truong and T. J. Evans, *J. Phys. Chem.* **98**, 9558 (1994).

[37]T. N. Truong, *J. Chem. Phys.* **102**, 5335 (1995).

[38]E. Uggerud and T. Helgaker, *J. Am. Chem. Soc.* **114**, 4265 (1992).

[39]A. A. Viggiano, J. Paschekewitz, R. A. Morris, J. F. Paulson, A. Gonzalez-Lafont, and D. G. Truhlar, *J. Am. Chem. Soc.* **113**, 9404 (1991).

[40]I. Wang and M. J. Karplus, *J. Am. Chem. Soc.* **95**, 8160 (1973).

[41]D. G. Truhlar, in *The Reaction Path in Chemistry: Current Approaches and Perspectives*, edited by D. Heidrich (Kluwer Academic, 1995), pp. 229.

[42]V. S. Melissas and D. G. Truhlar, *J. Chem. Phys.* **99**, 3542 (1993).

[43]V. S. Melissas and D. G. Truhlar, *J. Chem. Phys.* **99**, 1013 (1993).

[44]J. Andzelm and E. Wimmer, *J. Chem. Phys.* **96**, 1280 (1992).

[45]B. G. Johnson, P. M. W. Gill, and J. A. Pople, *J. Chem. Phys.* **98**, 5612 (1993).

[46]*Density Functional Methods in Chemistry*, Vol. , edited by J. K. Labanowski and J. W. Andzelm (Springer-Verlag, New York, 1991).

[47]Y. Abashkin and N. Russo, *J. Chem. Phys.* **100**, 4477 (1994).

[48]R. V. Stanton and K. M. Merz, Jr., *J. Chem. Phys.* **100**, 434 (1994).

[49]Q. Zhang, R. Bell, and T. N. Truong, *J. Phys. Chem.* **99**, 592 (1995).

[50]J. Baker, J. Andzelm, M. Muir, and P. R. Taylor, *Chem. Phys. Letters* **237**, 53 (1995).

[51]T. Ziegler, *Chem. Rev.* **91**, 651 (1991).

[52]L. Fan and T. Ziegler, *J. Chem. Phys.* **92**, 3645 (1990).

[53]L. Y. Fan and T. Ziegler, *J. Am. Chem. Soc.* **114**, 3823 (1992).

[54]B. C. Garrett and D. G. Truhlar, *J. Chem. Phys.* **70**, 1593 (1979).

[55]D. G. Truhlar and B. C. Garrett, *Accounts Chem. Res.* **13**, 440 (1980).

[56]D. G. Truhlar, A. D. Isaacson, R. T. Skodje, and B. C. Garrett, *J. Phys. Chem.* **86**, 2252 (1982).

[57]D. G. Truhlar and B. C. Garrett, *Annu. Rev. Phys. Chem.* **35**, 159 (1984).

[58]D. G. Truhlar, A. D. Isaacson, and B. C. Garrett, in *Theory of Chemical Reaction Dynamics*, Vol. 4, edited by M. Baer (CRC Press: Boca Raton, Florida, 1985), pp. 65-137.

[59]D. G. Truhlar and B. C. Garrett, *J. Chim. Phys.* **84**, 365 (1987).

[60]S. C. Tucker and D. G. Truhlar, in *New Theoretical Concepts for Understanding Organic Reactions*, edited by J. Bertran and I. G. Csizmadia (Kluwer, Dordrecht, Netherlands, 1989), pp. 291-346.

[61]B. C. Garrett, N. Abushalbi, D. J. Kouri, and D. G. Truhlar, *J. Chem. Phys.* **83**, 2252 (1985).

[62]M. M. Kreevoy, D. Ostovic, D. G. Truhlar, and B. C. Garrett, *J. Phys. Chem.* **90**, 3766 (1986).

[63]A. D. Isaacson, B. C. Garrett, G. C. Hancock, S. N. Rai, M. J. Redmon, R. Steckler, and D. G. Truhlar, *Computer Phys. Comm.* **47**, 91 (1987).

[64]B. C. Garrett, T. Joseph, T. N. Truong, and D. G. Truhlar, *Chem. Phys.* **136**, 271 (1989).

[65]D.-h. Lu, T. N. Truong, V. S. Melissas, G. C. Lynch, Y. P. Liu, B. C. Garrett, R. Steckler, A. D. Isaacson, S. N. Rai, G. C. Hancock, J. G. Lauderdale, T. Joseph, and D. G. Truhlar, *Computer Phys. Comm.* **71**, 235 (1992).

[66]M. Page and J. W. McIver Jr., *J. Chem. Phys.* **88**, 922 (1988).

[67]A. D. Becke, *J. Chem. Phys.* **98**, 1372 (1993).

[68]A. D. Becke, *J. Chem. Phys.* **98**, 5648 (1993).

[69]C. Lee, W. Yang, and R. G. Parr, *Phys. Rev. B* **37**, 785 (1988).

[70]M. J. Frisch, G. W. Trucks, H. B. Schlegel, P. M. W. Gill, B. G. Johnson, M. W. Wong, J. B. Foresman, M. A. Robb, M. Head-Gordon, E. S. Replogle, R. Gomperts, J. L. Andres, K. Raghavachari, J. S. Binkley, C. Gonzalez, R. L. Martin, D. J. Fox, D. J. Defrees, J. Baker, J. J. P. Stewart, and J. A. Pople, G92/DFT, *Revision* G.3, (Gaussian, Inc., Pittsburgh, 1993).

[71]E. Kraka, J. Gauss, and D. Cremer, *J. Chem. Phys.* **99**, 5306 (1993).

[72]T. Joseph, R. Steckler, and D. G. Truhlar, *J. Chem. Phys.* **87**, 7036 (1987).

[73]H. Furue and P. D. Pacey, *J. Phys. Chem.* **94**, 1419 (1990).

[74]*JANAF Thermochemical Tables*, edited by M. W. Chase, Jr., C. A. Davies, J. R. Downey, Jr., D. J. Frurip, R. A. McDonald, and A. N. Syverud (J. Phys. Chem. Ref. Data, Vol. 14, 1985).

[75]D. L. Baulch, C. J. Cobos, R. A. Cox, C. Esser, P. Frank, T. Just, J. A. Kerr, M. J. Pilling, J. Troe, R. W. Walker, and J. Warnatz, *J. Phys. Chem. Ref. Data* **21**, 441 (1992).

[76]K. A. Nguyen, M. S. Gordon, and D. G. Truhlar, *J.Am. Chem.Soc.* **113**, 1596 (1991).

[77]T. N. Truong and W. T. Duncan, TheRate program (Univeristy of Utah, Salt Lake City, 1993). More information on this program can be obtained from the web page, http://www.chem.utah.edu/mercury/TheRate/TheRate.html, or send an email to TheRate@mail.chem.utah.edu.

Chapter 7

Comparison of Local, Nonlocal, and Hybrid Density Functionals Using Vibrational Absorption and Circular Dichroism Spectroscopy

P. J. Stephens[1], F. J. Devlin[1], C. S. Ashvar[1], K. L. Bak[2], P. R. Taylor[3], and M. J. Frisch[4]

[1]Department of Chemistry, University of Southern California, Los Angeles, CA 90089–0482
[2]UNI-C, Olof Palmes Allé 38, DK–8200 Aarhus N, Denmark
[3]San Diego Supercomputer Center, P.O. Box 85608, San Diego, CA 92186–9784
[4]Lorentzian, Inc., 140 Washington Avenue, North Haven, CT 06473

Ab initio calculations of vibrational unpolarized absorption and circular dichroism spectra of 6,8-dioxabicyclo[3.2.1] octane are reported. The harmonic force field is calculated via Density Functional Theory using three density functionals: LSDA, BLYP, B3LYP. The basis set is 6-31G*. Spectra calculated using the hybrid B3LYP functional give the best agreement with experimental spectra, demonstrating that this functional is superior in accuracy to the BLYP and LSDA functionals.

Density Functional Theory (DFT) (*1*) is increasingly the methodology of choice in ab initio calculations. At the same time, the number, variety, and sophistication of density functionals is also increasing. The choice of DFT is thus accompanied by the problem of selecting the optimum functional.

In this paper, we demonstrate the utility of vibrational unpolarized absorption spectra and vibrational circular dichroism spectra in assessing the accuracies of density functionals. Specifically, we report calculations of the vibrational absorption and circular dichroism spectra of 6,8-dioxabicyclo[3.2.1]octane, **1**, using three density functionals and evaluate the relative accuracies of these functionals by comparison of the predicted spectra to experiment.

Broadly speaking, density functionals in active use today can be classed as: (i) local; (ii) non-local; or (iii) hybrid. Local functionals

0097–6156/96/0629–0105$15.00/0

are the simplest and were the first to be used. Non-local (or "gradient") corrections were then added, creating non-local functionals. Very recently, even more sophisticated functionals have been introduced, based on the Adiabatic Connection Method of Becke (2), which are referred to as hybrid functionals. In this work we utilize one functional from each class: (i) the Local Spin Density Approximation (LSDA) functional; (ii) the non-local Becke-Lee-Yang-Parr (BLYP) functional; and (iii) the Becke 3-Lee-Yang-Parr (B3LYP) hybrid functional.

Since the development and implementation of analytical derivative methods for DFT energy gradients, vibrational frequencies have been used extensively in evaluating the accuracies of density functionals (3). The recent development and implementation of analytical derivative methods for DFT energy second derivatives (Hessians) (4) has greatly increased the efficiency of calculations of DFT harmonic frequencies. In contrast, vibrational intensities have not been substantially utilized. In this work we demonstrate the advantages of incorporating vibrational intensities in the evaluation of density functionals. We further demonstrate the additional advantages of the use of both unpolarized absorption and circular dichroism intensities. Unpolarized vibrational absorption spectroscopy is widely utilized and well-understood. Vibrational Circular Dichroism (VCD) spectroscopy (5) is not yet as widely utilized. Our work will illustrate its substantial potential.

Methods

DFT harmonic force fields and atomic polar tensors (APTs) were calculated using GAUSSIAN 92/DFT and the three density functionals:
1) LSDA (local spin density approximation): this uses the standard local exchange functional (6) and the local correlation functional of Vosko, Wilk, and Nusair (VWN) (7).
2) BLYP: this combines the standard local exchange functional with the gradient correction of Becke (6) and uses the Lee-Yang-Parr correlation functional (8) (which also includes density gradient terms).
3) Becke3LYP: this functional is a hybrid of exact (Hartree-Fock) exchange with local and gradient-corrected exchange and correlation terms, as first suggested by Becke (2). The exchange-correlation functional proposed and tested by Becke was

$$E_{xc} = (1 - a_0)E_x^{LSDA} + a_0 E_x^{HF} + a_x \Delta E_x^{B88} + E_c^{LSDA} + a_c \Delta E_c^{PW91} \qquad (1)$$

Here ΔE_x^{B88} is Becke's gradient correction to the exchange functional, and ΔE_c^{PW91} is the Perdew-Wang gradient correction to the correlation functional (9). Becke suggested coefficients $a_0 = 0.2$, $a_x = 0.72$, and

$a_c = 0.81$ based on fitting to heats of formation of small molecules. Only single-point energies were involved in the fit; no molecular geometries or frequencies were used. The Becke3LYP functional in Gaussian 92/DFT uses the values of a_0, a_x, and a_c suggested by Becke but uses LYP for the correlation functional. Since LYP does not have an easily separable local component, the VWN local correlation expression has been used to provide the different coefficients of local and gradient corrected correlation functionals:

$$E_{xc}^{B3LYP} = (1 - a_0)E_x^{LSDA} + a_0 E_x^{HF} + a_x \Delta E_x^{B88} + a_c E_c^{LYP} + (1 - a_c)E_c^{VWN} \quad (2)$$

The standard fine grid in Gaussian 92/DFT (*10*) was used in all DFT calculations. This grid was produced from a basic grid having 75 radial shells and 302 angular points per radial shell for each atom and by reducing the number of angular points for different ranges of radial shells, leaving about 7000 points per atom while retaining similar accuracy to the original (75,302) grid. Becke's numerical integration techniques (*11*) were employed.

Atomic axial tensors (AATs) (*12*) were calculated using the distributed origin gauge (*12,13*), in which the AAT of nucleus λ, $(M_{\alpha\beta}^{\lambda})^0$, with respect to origin O is given by

$$(M_{\alpha\beta}^{\lambda})^0 = (I_{\alpha\beta}^{\lambda})^{\lambda} + \frac{i}{4\hbar c} \sum_{\gamma\delta} \epsilon_{\beta\gamma\delta} R_{\lambda\gamma}^0 P_{\alpha\delta}^{\lambda} \quad (3)$$

where $\bar{R}_{\lambda}^0$ is the equilibrium position of nucleus λ relative to origin O, $P_{\alpha\beta}^{\lambda}$ is the APT of nucleus λ, and $(I_{\alpha\beta}^{\lambda})^{\lambda}$ is the electronic AAT of nucleus λ calculated with the origin at $\bar{R}_{\lambda}^0$. "Distributed" AATs, $(I_{\alpha\beta}^{\lambda})^{\lambda}$, were calculated at the SCF level using Gauge-Invariant Atomic Orbitals (GIAOs) (*14*) <u>via</u> the SIRIUS/ABACUS program suite (*15*). (At this time, DFT code for AATs is not available.)

Harmonic force fields, APTs and AATs were calculated using the 6-31G* basis set. Vibrational frequencies, dipole strengths and rotational strengths were calculated thence, and in turn used to synthesize unpolarized absorption and circular dichroism spectra.

Results

Experimental unpolarized absorption and circular dichroism spectra of **1** have been reported by Wieser and coworkers (*16*). (Note that ref. 16 also diagrams the structure of **1**.) The spectra over the range 800-1500 cm^{-1} are reproduced in Figures 1 and 2. Calculated spectra, obtained from calculated frequencies, dipole strengths and rotational strengths using Lorentzian band shapes (*17*) and an arbitrarily chosen

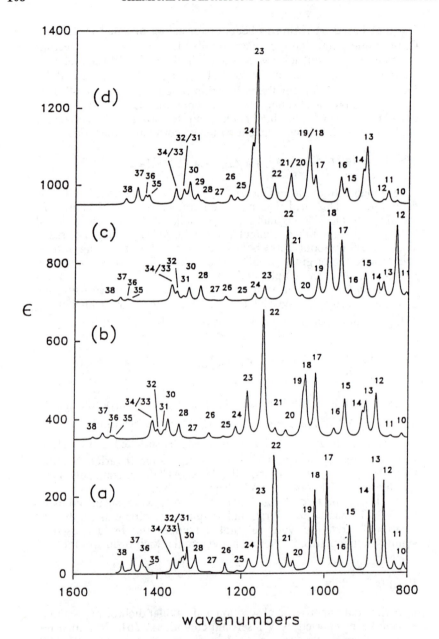

Figure 1: Vibrational unpolarized absorption spectra of **1**: a) experimental spectrum (*16*); b)-d) calculated spectra, using the b) B3LYP, c) BLYP and d) LSDA functionals. Bandshapes in b)-d) are Lorentzian; γ=4.0 cm[-1] for all bands. Fundamentals are numbered.

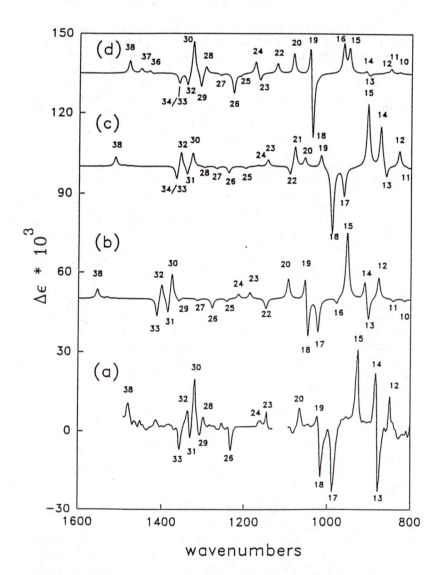

Figure 2: Vibrational circular dichroism spectra of **1**: a) experimental spectrum (*16*); b)-d) calculated spectra, using the b) B3LYP, c) BLYP and d) LSDA functionals. Bandshapes in b)-d) are Lorentzian; $\gamma=4.0$ cm^{-1} for all bands. Fundamentals are numbered.

band width (half-width at half-height, $\gamma=4$ cm[-1]), are also displayed in Figures 1 and 2.

The absorption and VCD spectra obtained using the B3LYP functional are in excellent qualitative agreement with the experimental spectra, allowing for an overall shift to higher frequency. Together, the absorption and VCD spectra permit assignment of the fundamentals 10-38 of **1**. Fundamentals 10-28 and 35-38 are resolved and assignable from the absorption spectrum. Fundamentals 12-15, 17-20, 23, 24, 26, 28-33, and 38 are resolved and assignable from the VCD spectrum. Fundamentals 33 and 34 are predicted to be 3 cm[-1] apart and of comparable absorption intensity; they can both be assigned to the absorption band just above 1360 cm[-1]. The VCD intensity of fundamental 34 is predicted to be weak and only fundamental 33 contributes to the VCD spectrum therefore. Note that the assignment of modes 30-32 on the basis of the absorption spectrum alone is uncertain, due to the presence of overtone/combination bands in this region. The VCD spectrum removes the uncertainty. Wieser and coworkers reported experimental frequencies, dipole strengths and rotational strengths for **1**. Their values for the bands assigned above as fundamentals are given in Table 1, together with the calculated results.

The absorption and VCD spectra predicted using the BLYP functional are quite similar to those obtained from the B3LYP functional for fundamentals 23-38, allowing for an overall shift to lower frequency. In contrast, the spectra for fundamentals 12-22 are substantially different. The changes in relative absorption and VCD intensities of fundamentals 21 and 22, of fundamentals 18 and 19, and of fundamentals 12 and 13/14 are especially noticeable. The BLYP spectra for fundamentals 12-22 are overall in worse agreement with experiment.

The absorption and VCD spectra predicted using the LSDA functional are **very** different from those obtained from the B3LYP and BLYP functionals. They are also in **very** much worse agreement with experiment.

On the basis of these results, we conclude that the three functionals can be ranked in relative accuracy:

B3LYP > BLYP >> LSDA

This conclusion is consistent with that arrived at in our earlier study of 4-methyl-2-oxetanone (*18*) and in a forthcoming study of a set of 10, small chiral molecules (*19*). It is also consistent with a comparison of calculated and experimental harmonic frequencies for 11 small molecules (*20*): at the 6-31G* basis set level, the mean absolute percentage deviations of calculated frequencies were found to be 2.0%, 3.3% and 3.5% for the B3LYP, BLYP and LSDA functionals respectively.

It is important to emphasize that a comparison of calculated and observed frequencies alone does **not** lead to the same conclusion. Observed frequencies are lower than harmonic frequencies due to

**TABLE 1: Frequencies, Dipole Strengths and
Rotational Strengths of 1[a]**

fund.	calculation [b]			experiment [c]		
	$\bar{\upsilon}$	D	R	$\bar{\upsilon}$	D	R
38	1555	4.1	6.7	1484	13.1	19.4
37	1532	12.0	1.0	1458	20.7	
36	1513	5.4	0.4	1438	11.9	
35	1506	4.8	-0.2	1435/1430	5.8/6.7	
34	1415	17.2	1.6	1365	11.1	
33	1412	25.0	-15.5	1362	14.9	-18.8
32	1401	14.6	12.6	1343	21.8	24.9
31	1386	11.8	-14.1	1338	11.4	-29.8
30	1376	39.4	21.3	1331	37.6	52.3
29	1359	0.8	-2.8	1315	5.8	-15.2
28	1349	30.9	0.4	1310	28.1	11.6
27	1314	1.7	-1.7	1276	0.8	
26	1278	12.8	-8.1	1240	11.5	-16.4
25	1243	4.2	-2.5	1211		
24	1214	25.9	3.9	1182	34.4	
23	1187	112.6	5.5	1157	126.3	13.6
22	1149	328.7	-9.6	1124/1119	171.5/205.9	
21	1120	20.4	0.1	1089	33.4	-9.6
20	1095	18.6	19.8	1076	17.0	19.3
19	1053	78.0	32.5	1033	77.4	20.3
18	1047	149.9	-47.3	1022	190.8	-62.3
17	1023	180.7	-34.9	993	248.5	-80.0
16	979	25.0	-6.5	962	32.6	6.4
15	953	117.3	78.4	939	120.2	88.0
14	911	64.5	24.2	893	120.1	63.3
13	903	103.2	-30.1	882	216.0	-74.0
12	878	145.3	26.5	857	174.6	32.5
11	842	8.9	-3.3	832	26.1	
10	815	16.7	-3.0	809	21.2	

Footnotes:

a. Frequencies in cm^{-1}, dipole strengths in 10^{-40} esu^2 cm^2, rotational
   strengths in 10^{-44} esu^2 cm^2. Rotational strengths are for (1R, 5S)-**1**.
b. B3LYP/DFT force field and APTs; GIAO/SCF Distributed AATs;
   6-31G* basis set.
c. From reference 16.

anharmonicity. For the 11 small molecules referred to above, mean absolute percentage deviations of calculated and observed frequencies were 4.5%, 2.6% and 2.5% for B3LYP, BLYP and LSDA respectively. Using observed frequencies would thus lead to the erroneous conclusion that the BLYP functional is more accurate than the B3LYP functional. In the case of 1, comparison of calculated and observed frequencies would likewise indicate BLYP to be superior to B3LYP. Only if the anharmonicity corrections are included can the relative accuracies of the functionals be reliably assessed using frequencies.

Conclusion

Our calculations for 1 convincingly demonstrate that the hybrid density functional, B3LYP, provides a harmonic force field—and, thence, normal coordinates and unpolarized absorption and VCD intensities—of accuracy greater than obtained using the BLYP and LSDA functionals. It follows that, in future predictions of vibrational spectra using DFT, hybrid functionals such as B3LYP are the best choice of the currently available options.

Our calculations also demonstrate the utility of vibrational spectra in defining the relative accuracies of density functionals. It is likely that functionals more accurate than the current generation of hybrid functionals will be developed in the future. Vibrational spectra should be useful in assessing their accuracy. In contrast, with the exception of molecules where anharmonicity-corrected vibrational frequencies are available, the comparison of frequencies will not be useful since the deviations between calculation and experiment will be dominated by anharmonicity.

Lastly, we have illustrated again the added benefits of using both unpolarized absorption and VCD spectra in comparing theoretical and experimental spectra.

Literature Cited

(1)   Ziegler, T. *Chem. Revs.* 1991, **91**, 651.
(2)   Becke, A.D. *J. Chem. Phys.* 1993, **98**, 1372, 5648.
(3)   See for example: Johnson, B.G.; Gill, P.M.W.; Pople, J.A. *J. Chem. Phys.* 1993, **98**, 5612.
(4)   Johnson, B.G.; Frisch, M.J. *Chem. Phys. Lett.* 1993, **216**, 133; Johnson, B.G.; Frisch, M.J. *J. Chem. Phys.* 1994, **100**, 7429; Komornicki, A.; Fitzgerald, G. *J. Chem. Phys.* 1993, **98**, 1398.
(5)   Stephens, P.J.; Lowe, M.A.; *Ann. Rev. Phys. Chem.* 1985, **36**, 213.
(6)   Becke, A.D. In *The Challenge of d and f Electrons: Theory and Computation*; Salahub, D.R., Zerner, M.C., Eds.; American Chemical Society: Washington, D.C., 1989, Chapter 12, pp 165-179.

(7) Vosko, S.H.; Wilk, L.; Nusair, M. *Can. J. Phys.* 1980, **58**, 1200.

(8) Lee, C.; Yang, W.; Parr, R.G. *Phys. Rev. B.* 1988, **37**, 785.

(9) Perdew, J.P. In *Electronic Structure of Solids*; Ziesche, P., Eschrig, H., Eds.; Akademie Verlag: Berlin, 1991.

(10) Trucks, G.W.; Frisch, M.J. To be published.

(11) Becke, A.D. *J. Chem. Phys.* 1988, **88**, 2547.

(12) Stephens, P.J.; Jalkanen, K.J.; Amos, R.D.; Lazzeretti, P.; Zanasi, R. *J. Phys. Chem.* 1990, **94**, 1811.

(13) Stephens, P.J. *J. Phys. Chem.* 1987, **91**, 1712.

(14) Bak, K.L.; Jørgensen, P.; Helgaker, T.; Ruud, K.; Jensen, H.J. Aa *J. Chem. Phys.* 1993, **98**, 8873; Bak, K.L.; Jørgensen, P.; Helgaker, T.; Ruud, K.; Jensen, H.J. Aa *J. Chem. Phys.* 1994, **100**, 6620; Bak, K.L.; Jørgensen, P.; Helgaker, T.; Ruud, K. *Faraday Disc.* 1994, **99**, 0000.

(15) SIRIUS, Jensen, H.J. Aa., Aagren, H. and Olsen, J.; ABACUS, Helgaker, T., Bak, K.L., Jensen, H.J. Aa., Jørgenson, P., Kobayashi, R., Koch, H., Mikkelson, K., Olsen, J., Ruud, K., Taylor, P.R., and Vahtras, O.

(16 Eggimann, T.; Shaw, R.A.; Wieser, H.J. *J. Phys. Chem.* 1991, **95**, 591; Eggimann, T.; Ibrahim, N.; Shaw, R.A.; Wieser, H. *Can. J. Chem.* 1993, **71**, 578.

(17) Kawiecki, R.W.; Devlin, F.J.; Stephens, P.J.; Amos, R.D.; Handy, N.C. *Chem. Phys. Lett.* 1988, **145**, 411; Kawiecki, R.W.; Devlin, F.J.; Stephens, P.J.; Amos, R.D. *J. Phys.Chem.* 1991, **95**, 9817.

(18) Stephens, P.J.; Devlin, F.J.; Chabalowski, C.F.; Frisch, M.J. *J. Phys. Chem.* 1994, **98**, 11623.

(19) Devlin, F.J.; Finley, J.W.; Stephens, P.J.; Frisch, M.J. *J. Phys. Chem.*, in press.

(20) Finley, J.W.; Stephens, P.J. *J. Mol. Struct.* (*Theochem*), in press.

Chapter 8

Polymers and Muffin-Tin Orbitals

Michael Springborg, Catia Arcangeli[1], Karla Schmidt, and Heiko Meider

**Department of Chemistry, University of Konstanz,
D–78434 Konstanz, Germany**

A method for calculating electronic properties of polymeric materials is reviewed. It is based on the density-functional formalism of Hohenberg and Kohn. The eigenfunctions to the Kohn-Sham equations are expanded in a basis set of linearized muffin-tin orbitals, but no shape approximation for the potential is made. The materials of interest are assumed to be infinite, periodic, noninteracting, helical chains, and the symmetry is explicitly used in constructing helical Bloch waves. As examples of the calculation of structural properties we discuss sulphur helices and the dimerization in trans polyacetylene. A method that allows for calculating energies of defect-induced orbitals is demonstrated on polarons in polythiophene and solitons in hydrogen fluoride. As an intermediate between insulating and semiconducting polymers we discuss polybutadiene. Momentum densities for different periodic $C_2H_2Cl_2$ chains are presented, and the importance of including spin-orbit couplings when studying optical properties is demonstrated for bismuth chains. Finally, a method for studying doped chains is presented and preliminary results for the dimerization of a linear doped carbon chain are reported.

During the last few years electronic-structure calculations within the Hohenberg-Kohn formalism (1) have become important for the studies of properties of molecular systems, as is impressively demonstrated by other contributions of this volume. One has now reached a state where many materials can be treated almost routinely with density-functional methods [see, e.g., (2)]. The systems that can be treated most easily are either finite in all three dimensions or infinite and periodic in the three dimensions. When the systems are neither finite nor periodic in all three dimensions, it is considerably more difficult to treat them with density-functional methods.

[1]Current address: Max-Planck-Institut für Festkörperforschung, D–70569 Stuttgart, Germany

0097–6156/96/0629–0114$15.00/0

Polymeric materials, the subject of the present contribution, are a representative example. We shall here classify a material consisting of very long chains for which the intrachain interactions are much stronger than the interchain interactions as a polymer. Thus, this class contains also many systems that usually not are given the label polymer. Assuming that certain idealizations to be described below are reasonable, we have developed a density-functional method (*3,4*) for studying these materials. This method shall briefly be reviewed in the next section.

Subsequently we discuss various recent applications of the method as well as some extensions. We start out with discussing the total energy of perfect, defect-free helical polymers as a function of structure for two periodic, infinite polymers, i.e. a sulphur helix and trans polyacetylene. However, in many cases structural defects, and in particular localized orbitals induced by these defects, are important for the properties of the materials, and we therefore discuss how calculations on the periodic structures can give information on these states. This discussion is examplified through applications on polarons in polythiophene and solitons in hydrogen fluoride. In a further study of the electronic orbitals, but this time for a periodic system, we consider an intermediate system between insulating and semiconducting polymers, i.e. polybutadiene. In order to demonstrate that the density-functional calculations also can give useful information on other properties than the total energy and the single-particle levels, we continue by studying momentum space densities, and show that these can give information that is not directly accessible by position-space studies. In particular we demonstrate this for three periodic $(C_2H_2Cl_2)_x$ polymers. A recent extension of our computational scheme is devoted to spin-orbit couplings, and we show that these are important when discussing optical properties of polymeric chains containing heavier atoms. This discussion is examplified through results for bismuth chains. In the last application we present results for calculations on doped chains, in particular on the dimerization of a simple conjugated polymer.

Helical Polymers and the LMTOs

Neglecting interchain interactions we shall assume that the system of our interest is periodic, infinite, isolated, helical, and with a straight helical axis. A hypothetic example of such a chain is shown schematically in Figure 1. It has two atoms per unit cell and it is obvious that any pair of atoms is equivalent to any other pair.

The primitive symmetry operation that maps the polymer onto itself is a combined rotation of v and translation of h. Since the chain is assumed to be isolated, v does not need to be commensurate with 2π.

The helical symmetry is explicitly used by defining local atom-centered right-handed coordinate systems that have the z axes parallel to the helical axis and the x axes pointing away therefrom. Basis functions, electron densities, potentials, etc. will then, when described in the local coordinate systems, be equivalent for different unit cells. Moreover, helical Bloch waves generated from the atom-centered basis functions can be used in creating symmetry-adapted basis functions in a manner completely analogous to that used for the more conventional translational symmetry. Finally, zigzag and translational symmetries are special cases of the helical symmetry corresponding to $v = \pi$ and $v = 2\pi$, respectively.

In our approach, the eigenfunctions to the Kohn-Sham equations

$$\left[-\frac{\hbar^2}{2m}\nabla^2 + V(\vec{r})\right]\psi_i(\vec{r}) = \epsilon_i\psi_i(\vec{r}) \tag{1}$$

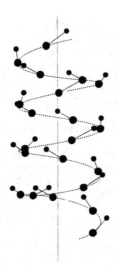

Figure 1. A hypothetic helical polymer with two atoms per unit cell.
The vertical dashed line represents
the helical axis and the dashed curve
symbolizes the screw axis symmetry.

are expanded in a basis of linearized muffin-tin orbitals (LMTOs). These are
defined as follows.

We separate space into atom-centered non-overlapping (muffin-tin) spheres
and the interstitial region. Inside any sphere (at $\vec{R}$, e.g.) we expand the potential
in angular components,

$$V(\vec{r}) = \sum_L V_{\vec{R},L}(|\vec{r} - \vec{R}|)Y_L(\widehat{r - R}) \tag{2}$$

where $L \equiv (l, m)$, and $\widehat{r - R}$ is a unit vector along the direction of $\vec{r} - \vec{R}$.
Subsequently, we solve numerically the one-dimensional Kohn-Sham equations

$$[-\frac{\hbar^2}{2m}\nabla^2 + V_{\vec{R},(0,0)}\frac{1}{\sqrt{4\pi}}]\phi_{\vec{R},L,\kappa}(\vec{r}) = \epsilon_{\nu,\vec{R},L,\kappa}\phi_{\vec{R},L,\kappa}(\vec{r}) \tag{3}$$

inside the sphere for reasonable energies $\epsilon_{\nu,\vec{R},L,\kappa}$. We also define

$$\dot{\phi}_{\vec{R},L,\kappa}(\vec{r}) = \frac{\partial}{\partial\epsilon_{\nu,\vec{R},L,\kappa}}\phi_{\vec{R},L,\kappa}(\vec{r}). \tag{4}$$

An LMTO centered on an atom (at $\vec{R}_2$) is then defined as a Hankel function
in the interstitial region,

$$c_{\vec{R}_2,L_2,\kappa} \cdot K_{L_2}(\kappa, \vec{r} - \vec{R}_2) \equiv c_{\vec{R}_2,L_2,\kappa} \cdot \frac{i\kappa^{l_2+1}}{(2l_2 - 1)!!}h_{l_2}^{(1)}(\kappa|\vec{r} - \vec{R}_2|)Y_L(\widehat{r - R_2}) \tag{5}$$

augmented continuously and differentiably inside another sphere (e.g., at $\vec{R}_1$)
with ϕ and $\dot{\phi}$ functions

$$\sum_{L_1}[\Pi_{L_2L_1}(\kappa, \vec{R}_2 - \vec{R}_1)\phi_{\vec{R}_1,L_1,\kappa}(\vec{r}) + \Omega_{L_2L_1}(\kappa, \vec{R}_2 - \vec{R}_1)\dot{\phi}_{\vec{R}_1,L_1,\kappa}(\vec{r})]. \tag{6}$$

Thereby the constants c, Π, and Ω are defined [for details, see (3)]. We notice that for $\vec{R}_1 = \vec{R}_2$ the sum in equation 6 reduces to one term $(L_2 = L_1)$ for which $\Pi = 1$ per definition.

As an illustration of our approach we show in Figure 2 the potential inside an oxygen sphere for a CO_2 molecule as well as an LMTO centered on the same atom.

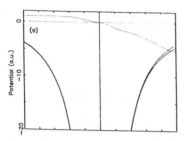 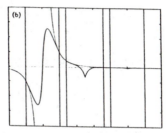

Figure 2. Results for a linear CO_2 molecule with the C atom placed at the origin and the O atoms at $z = \pm 2.5$ a.u. The sphere sizes were chosen equal to 1.0 a.u. (a) shows the full potential (solid curve) and its s component (long-dashed curve) inside the sphere at the O atom at $z = -2.5$ a.u., along the z axis, whereas the short-dashed curve shows the difference scaled by a factor of 10. (b) shows a p_z LMTO centered on the same atom. The dashed curve is the Hankel function of equation 5, and the vertical dashed lines mark the boarders of the muffin-tin spheres.

The LMTOs are per construction good approximations to the exact solutions to equation 1 since the potential is close to spherically symmetric (cf. Figure 2a). Moreover, they adopt themselves to the potential (cf. Figure 2b) as this changes during the iterative process of solving the Kohn-Sham equations. We stress, however, that the muffin-tin potential (spherical symmetry inside the spheres; constant in the interstitial region) is used only in defining the basis functions: in calculating matrix elements etc. we include the full potential. Gaussian and Slater-type orbitals can also be considered constructed from a spherically symmetric potential!

Our method may appear very different from more well-known methods for electronic-structure calculations. This is, however, not the case as, e.g., can be seen by considering the electron density. Inside any sphere (at $\vec{R}$) this becomes

$$\rho(\vec{r}) = \sum_{i=1}^{occ.} \sum_{L_1,L_2} \sum_{\kappa_1,\kappa_2} \left[a^i_{\vec{R}_k,L_1,\kappa_1} \cdot \phi_{\vec{R}_k,L_1,\kappa_1}(\vec{r}-\vec{R}_k) + b^i_{\vec{R}_k,L_1,\kappa_1} \cdot \dot{\phi}_{\vec{R}_k,L_1,\kappa_1}(\vec{r}-\vec{R}_k) \right]^*$$

$$\times \left[a^i_{\vec{R}_k,L_2,\kappa_2} \cdot \phi_{\vec{R}_k,L_2,\kappa_2}(\vec{r}-\vec{R}_k) + b^i_{\vec{R}_k,L_2,\kappa_2} \cdot \dot{\phi}_{\vec{R}_k,L_2,\kappa_2}(\vec{r}-\vec{R}_k) \right]$$

$$\equiv \sum_L \rho_{\vec{R},L}(|\vec{r}-\vec{R}|)Y_L(\widehat{r-R}) \tag{7}$$

which is closely related to the form used in various density-functional methods for crystalline materials. Using the form (7) the potential can be evaluated relatively easily and written in the form (2). We mention that the constants a and b are obtained from the solutions to the Kohn-Sham equations.

In the interstitial region we make, equivalent to the approach of many other density-functional methods for molecular systems, a least-squares fit,

$$\rho(\vec{r}) = \sum_{i=1}^{occ.} \sum_{L_1,L_2} \sum_{\kappa_1,\kappa_2} \sum_{\vec{R}_1,\vec{R}_2} \left[d^i_{\vec{R}_1,L_1,\kappa_1} \cdot K_{L_1}(\kappa_1, \vec{r}-\vec{R}_1) \right]^* \left[d^i_{\vec{R}_2,L_2,\kappa_2} \cdot K_{L_2}(\kappa_2, \vec{r}-\vec{R}_2) \right]$$

$$\simeq \sum_{\lambda} \sum_{L} \sum_{\vec{R}} r_{\vec{R},L,\lambda} \cdot K_L(\lambda, \vec{r}-\vec{R}) \tag{8}$$

and the potential is then expressed as

$$V(\vec{r}) \simeq \sum_{\lambda} \sum_{L} \sum_{\vec{R}} v_{\vec{R},L,\lambda} \cdot K_L(\lambda, \vec{r}-\vec{R}). \tag{9}$$

Details about this and other aspects of the calculations can be found in (3,4). One important aspect shall, however, be mentioned. Due to the expressions (2), (7)–(9) all densities and potentials appear as sums of single-atom components. This offers the possibility to construct good first approximations by superimposing those for individual atoms, that, moreover, can be obtained from calculations on small model systems where the atoms are embedded in the proper chemical environment.

Finally, in the studies whose results will be reported below we use the local approximation of von Barth and Hedin (5) in describing exchange and correlation unless stated otherwise.

Applications

Sulphur Helices. Density-functional methods are first of all methods for calculating the total energy as a function of structure. Our first example shall therefore concentrate on calculated structural properties of a specific helix, i.e., the sulphur helix. This has only one atom per unit cell, and we need accordingly three parameters for specifying the structure. As those we choose the nearest-neighbour bond length r, the bond angle α, and the dihedral angle γ. Table I gives our optimized values of those (6) together with the experimental values (7). Except for the bond length, our values agree well with the experimental values. Moreover, as Figure 3 shows, the total-energy surface is very flat around the minimum which makes the optimized values very sensitive to small inaccuracies in the calculations, and due to the relatively strong dependence of r on α the discrepancy in r may be explained therefrom.

Table I. Our optimized values of the structural parameters of a sulphur helix together with experimental values

	r (Å)	α (deg.)	γ (deg.)
Present	4.22	109	86.5
Experiment	3.90	106	85.3

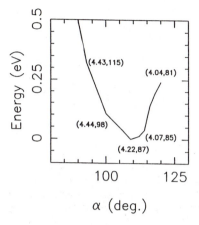

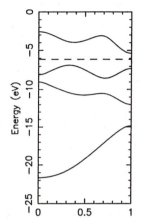

Figure 3. Total energy (eV/atom) for a sulphur helix as a function of α. The two other structural parameters (r, γ) have been optimized for each value of α and are also shown.

Figure 4. Band structures for the optimized sulphur helix. $k = 0$ and $k = 1$ is the zone center and zone edge, respectively, and the dashed line represents the Fermi level.

Several years ago it was discussed whether the sulphur helix was a 7/2 (8) or a 10/3 (9) helix. These values correspond to $v = 103°$ and $v = 108°$, respectively. Our optimized value $v = 107°$ agrees well with those.

The sulphur helix is semiconducting as the band structures in Figure 4 show. An optimized zigzag or linear sulphur chain is on the other hand metallic (6) and we suggest accordingly that the fact that the helical structure is the one with the lowest total energy may be explained as being due to a helical Peierls' transition.

Trans Polyacetylene. The structure of trans polyacetylene $(CH)_x$ is shown in Figure 5, where we also show the structural parameters that have been optimized using the present method (4).

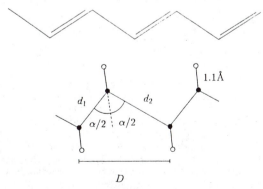

Figure 5. Structure of trans polyacetylene together with the definition of the various structural coordinates. Some of the structural degrees of freedom were kept frozen in the calculations as indicated. Dark and light circles represent carbon and hydrogen atoms, respectively.

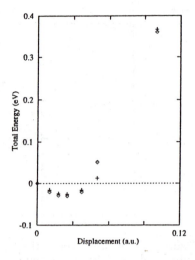

Figure 6. Total energy in eV per C_2H_2 unit relative to that of the undimerized structure as a function of dimerization for LDA (squares) and GGA (crosses).

The covalent bonds of trans polyacetylene are formed by energetically lowlying σ bonds between carbon sp^2 hybrids and hydrogen $1s$ functions, plus π bonds between the C atoms. The π orbitals are those appearing closest to the Fermi level. With effectively one (π) electron per (CH) site, the system experiences a Peierls' distortion (dimerization) giving alternating shorter and longer C–C bonds. As shown in Table II this is found experimentally (10,11) and theoretically both with our method (4) and with Hartree-Fock calculations (12). In addition, we observe in Table II a good agreement between all three sets of parameters.

Table II. Structural parameters (cf. Figure 5) for trans polyacetylene from experiment, our calculations, and Hartree-Fock calculations

	d_1 (Å)	d_2 (Å)	α (deg.)	D (Å)
Present	1.36	1.46	128	2.53
Experiment	1.36	1.44	123	2.46
Hartree-Fock	1.33	1.48	124	2.48

Trans polyacetylene is the prototype for a larger class of conjugated polymers that contain backbones with alternating C–C single and double bonds. These have a number of interesting properties due to π electrons, that have attracted a large research activity, both from basic science and for industrial applications [see, e.g., (13)]. Many of these properties are related conceptually to the interchange of the double and single bonds, i.e., to the relatively high mobility or polarizability of the π electrons. For trans polyacetylene the interchange of the double and single bonds leads to a new structure energetically degenerate with the original one, but for many other conjugated polymers the two structures differ in energy. For trans polyacetylene, interfaces, so-called solitons, separating regions with different bond-length alternations may exist. For all conjugated polymers polarons may exist. Polarons are confined structural defects containing the energetically unfavourable bond-length alternation embedded in the low-energy structure. These defects are highly mobile and may act as carriers of charge, spin, or energy.

One of the main ingredients in understanding these defects is that of understanding the dimerization, i.e., the size of the bond-length alternation as well as the energy gain upon this distortion. For trans polyacetylene in particular, it has been suggested that the Peierls' mechanism is not responsible for the experimentally observed dimerization, and that an accurate description of correlation effects, beyond that of the local-density approximation (LDA), is required in order to correctly describe the dimerized ground state of this system (14). In order to test the role of correlation in the dimerization process we have studied the total energy as a function of bond-length alternation using both the LDA of von Barth and Hedin expression (5), and the more accurate gradient corrected exchange-correlation density functional proposed by Perdew and Wang (15,16), known as the generalized gradient approximation (GGA). We fixed all bond angles at 120° as well as the unit-cell length at 2.41 Å. By displacing the CH units rigidly parallel to the polymer axis alternatingly in one or the other direction we obtained the results shown in Figure 6.

It is seen that the results depend only weakly on the applied functional. Moreover, the energy gain upon dimerization is very small (about 0.03 eV per C_2H_2 unit), explaining the high mobility of the solitons as well as their stability. The results of Figure 6 may subsequently form a part of the basis for further theoretical studies of the solitons. We consider, however, this to be beyond the scope of the present contribution and refer the interested reader to (4).

Polythiophene. Polythiophene, Figure 7, is another of the conjugated polymers mentioned above. We shall for this discuss further how the structural defects can be studied.

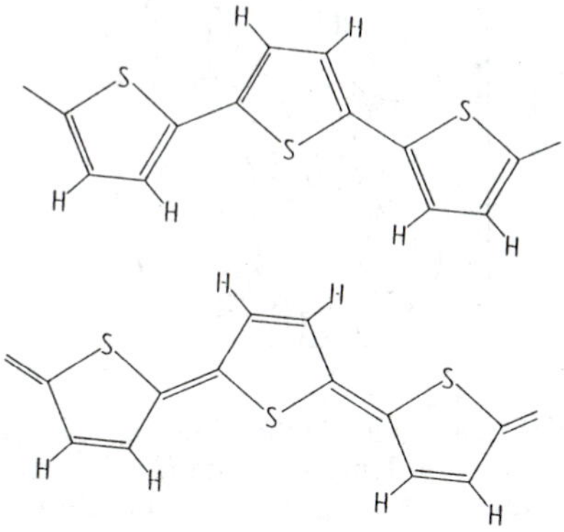

Figure 7. (upper part) The aromatic and (lower part) the quinoid structure of polythiophene.

It is known that the lowest energy is found for the aromatic structure, whereas the quinoid structure corresponds to a metastable structure. A polaron is a finite region where the structure changes from the aromatic towards the quinoid structure and back again. The importance of this structural defect comes from the fact that it induces states in the gap separating occupied and empty bands. Thereby it can accommodate charge that moreover can be transported through the chain as the defect moves. It is of importance to be able to identify these defects, since they play a crucial role in many transport processes, and optical spectroscopy devoted to detecting the gap states offers one method. Theory is then needed in interpreting the experimental results.

The simplest single-particle models predict that two polaron-induced gap levels appear exactly symmetrically in the gap [see, e.g., (17)], and deviations from this are often interpreted as indicating important correlation effects. This is, e.g., the case for two recent experimental studies of derivatives of polythiophene (18,19).

It would be highly desirable to obtain more accurate theoretical estimates on the positions of the gap states. However, the presence of the defects destroys the periodicity of the polymer, making density-functional studies very difficult. Mapping the results of such studies on periodic chains onto a model Hamiltonian that subsequently can be used to study the defects requires very detailed density-functional studies and is therefore also a far from trivial task.

In order to circumvent these problems we have developed a scheme that allows for accurate estimates of the positions of the defect-induced gap states in quasi-one-dimensional materials (20,21). It is not restricted to the present LMTO method or to density-functional methods, but is general for single-particle models. We shall illustrate it here for the polarons in polythiophene.

In Figure 8a we show the π bands of a realistic structure for polythiophene, and in Figures 8b and 8c the electron densities of the two orbitals closest to the Fermi level (22). When passing from the aromatic to the quinoid structure, the energy of the HOMO will move upwards as is evident from the contour plot of Figure 8b. Similarly, the energy of the LUMO will go down, but not as much since this orbital has dominating components on the sulphur atoms which hardly experience the structural changes. In total we estimate the changes of the frontier orbitals to be as shown schematically in Figure 9. This picture is actually confirmed by more detailed density-functional calculations.

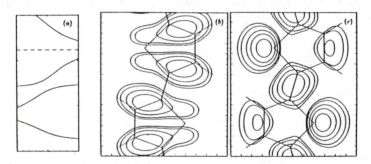

Figure 8. (a) π band structures of polythiophene. The dashed line represents the Fermi level. (b) and (c) contour plots of the density of the (b) HOMO and (c) LUMO in a plane parallel to but above that of the polymer backbone.

In order to estimate the positions of the polaron-induced gap states we consider the maximal aromatic-to-quinoid distortion within the polaron. For this geometry we consider the band structures of the equivalent periodic structure (see Figure 9). The band edges of this are then very good approximations to the positions of the gap states. As is seen in Figure 9, we can thereby immediately explain the asymmetry of the gap states as observed experimentally without having to assume strong correlation effects. We stress that this apparently simple prescription is more exact than may immediately be anticipated.

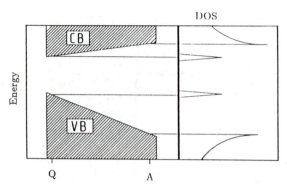

Figure 9. (left part) Schematic representation of the changes of the bands around the Fermi level for polythiophene when passing from the aromatic (A) to the quinoid (Q) structure. VB and CB denotes valence band and conduction band, respectively. Also shown (right part) is the estimated density of states (DOS) for a polaron-containing chain.

Hydrogen Fluoride. Hydrogen-bonded systems represent another class of materials for which defects are important. For those, solitons are believed to be responsible for proton transport [see, e.g., (23)]. Therefore a mechanism like that of Figure 10 is assumed relevant. In Figure 10 we see that two types of defect can be introduced, where one involves charged units (e.g., HMH^+ and M^- units in Figure 10) and is called ionic defects, whereas the other type involves only neutral units (i.e., MH units in Figure 10) and is called Bjerrum defects.

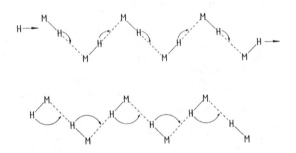

Figure 10. Proton transport in a hydrogen-bonded chain via solitons. M represents a smaller or larger group of atoms. The upper part shows ionic defects, since the incoming H atom may be charged, whereas the lower part shows Bjerrum defects. The propagation proceeds from left to right in the upper part and from right to left in the lower.

Almost exclusively, experimental and theoretical studies devoted to demonstrating the existence of these defects have focused on structural and vibrational properties. We wanted, however, to explore whether these also might have characteristic electronic responses. We thereby applied the same scheme as above for polarons in polythiophene and studied accordingly various structures resembling periodic solitons. As the system of interest we chose hydrogen fluoride $(HF)_x$, corresponding to M in Figure 10 being a single fluorine atom. We examined chains with four HF units per unit cell (24).

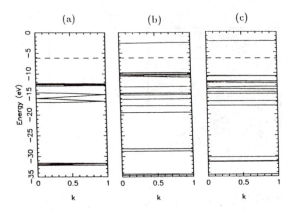

Figure 11. Band structures of (a) periodic undistorted $(HF)_x$, (b) periodically repeated ionic defects, and (c) periodically repeated Bjerrum defects. The calculations were performed for chains with four HF units per unit cell. $k = 0$ and $k = 1$ represent the center and boundary of the Brillouin zone, and the dashed line is the Fermi level.

Some representative results are shown in Figure 11. It is clearly seen that the introduction of defects leads to significant modifications of the band structures and in particular to the occurrence of bands in regions where the unperturbed system has none. For the experimentally more realistic low concentrations of solitons these extra bands may show up as extra features in the optical spectra. They may be so strong that they can be detected and thus give more information on the properties and existence of the solitons.

Polybutadiene — a Copolymer between Polyethylene and Polyacetylene. Polybutadiene, $(CH_2CH_2CHCH)_x$, has recently attracted some interest [see, e.g., (25) and references therein] as a possible conducting polymer similar to polyacetylene and polythiophene discussed above. On the other hand, considering the polymer as a copolymer between polyethylene, $(CH_2)_x$, and polyacetylene, $(CH)_x$, one would expect that the π electrons are localized to the CH segments. This will in turn give small π band widths and a low conductivity. In order to explore this in further detail we have studied the electronic properties of two isomers of polybutadiene and compared them with those of polyethylene and polyacetylene (26).

In Figure 12 we show the band structures for the four systems. For the cis isomer (Figure 12c) both bands closest to the Fermi level are of π symmetry, whereas for the trans isomer (Figure 12b) an additional steeper σ band defines the top of the valence bands. Thus, we see that the π bands of polybutadiene are flat supporting the above interpretation that the π electrons are localized to the CH units. This is further supported by analyzing the electronic orbitals (26). Moreover, the gap is fairly large and resembles more that found for the insulating polyethylene than that found for the semiconducting polyacetylene. Finally, only the lowest valence bands (between -21 and -16 eV) are broad and easily recognized for all the polymers. These correspond to C–C σ bonds.

Momentum Densities in $(C_2H_2Cl_2)_x$ Polymers. Within the common formulation of density-functional formalism, the methods are constructed for pro-

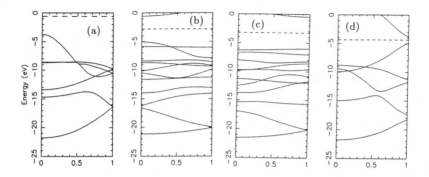

Figure 12. Band structures of (a) polyethylene, (b) trans and (c) cis polybutadiene, and (d) polyacetylene. The dashed lines mark the Fermi levels, and $k = 0$ and $k = 1$ are the center and the edge of the Brillouin zone, respectively.

ducing correct total energies and electron densities in position space. Moreover, the calculated single-particle energies are usually good approximations to electronic excitation energies. But also electron densities in momentum space can in principle be calculated with density-functional methods. In the exact formulation this is not only that obtained from the Fourier-transformed Kohn-Sham orbitals $\psi_i(\vec{r})$, but the so-called Lam-Platzman correction should be added (27). As a first — and often good — approximation one may, however, neglect this correction, and so we shall do here.

Momentum densities are accessible with various experimental techniques [see, e.g., (28)] and they provide useful information that only with difficulty can be obtained by other methods. It turns out, however, that instead of analyzing the momentum density $\rho(\vec{p})$ directly, it is more convenient to study its Fourier transform, the so-called reciprocal form factor $B(\vec{s})$ (29). Within a single-particle picture this obeys

$$B(\vec{s}) = \int \rho(\vec{r})e^{i\vec{p}\cdot\vec{s}}d\vec{p} = \sum_{i=1}^{\text{occ.}} \int \psi_i(\vec{r})\psi_i^*(\vec{r} + \vec{s})d\vec{r}. \tag{10}$$

We shall here discuss and compare it for three stoichiometrically identical but structurally different polymers. The polymers consist of a zigzag backbone of carbon atoms. Two additional atoms, H and/or Cl, are attached to each carbon atom. The chains consist of repeated units each with two C atoms. The three systems we shall study are (I) one with alternating two H atoms and two Cl atoms attached to the C atoms, and (II and III) chains with one H and one Cl atom attached to each C atom either on the same (II) or on alternating (III) sides of the plane of the C atoms (30).

In Figure 13 we show the results. We see that the z components possess a characteristic minimum around 7 a.u. for all systems as well as a node at the value of the lattice constant (~ 4.5 a.u.). More interesting are, however, the differences in the x and y components. As equation 10 shows, $B(\vec{s})$ is the sum of the overlaps of the eigenfunctions with themselves but displaced $\vec{s}$. Thus, the fact that $B(\vec{s})$ is most rapidly decaying for the x and y components in Figure 13a indicates that the orbitals of this material are those most localized in those directions. On the other hand, they are the least decaying in Figure 13c for which

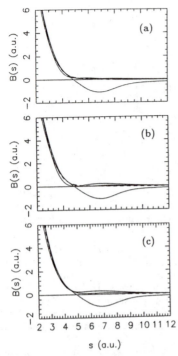

Figure 13. The reciprocal form factor $B(\vec{s})$ for the three $C_2H_2Cl_2$ polymers with the C backbone placed in the (x, z) plane and the z axis parallel to the polymer axis. The three full curves for $s = 6$ a.u. from below are for $\vec{s}$ parallel to the z, x, and y axis, respectively. The dashed curve represents the spherical average.

we thus conclude the orbitals to be most delocalized. As discussed elsewhere (30) some extra features can be related to resonances between the orbitals centered at atoms of neighbouring CX_2 units. We stress that these differences are hardly observable in direct studies of the electron density in position space, and momentum-space studies thus offer a valuable alternative to obtain information about the electronic orbitals.

Spin-Orbit Couplings and Bismuth Chains. Recently Romanov succeeded in synthesizing thin Bi wires inside the channels of a mordenite crystal (31). Subsequent x-ray scattering experiments provided some information on the structure and additional information was obtained through optical experiments. However, for the latter he was completely lacking reference data and compared therefore with free-atom data. In order to study this approximation in more detail and to get additional information on the system we have performed calculations on various Bi chains (32). But since Bi is a heavy atom, the inclusion of relativistic effects, including spin-orbit couplings, becomes important when discussing optical properties.

The general spin-orbit Hamilton operator is given through

$$\hat{H}_{so} = \frac{\hbar}{2m^2c^2}\hat{\vec{s}} \cdot \left[(\vec{\nabla}V(\vec{r})) \times \hat{\vec{p}}\right]. \tag{11}$$

The dependence of $\hat{H}_{so}$ on $\vec{\nabla}V$ makes it a good approximation to consider only that part of space where this term is large, i.e. the region close to the nuclei. Moreover, since the dominating part of the potential here is spherically symmetric we shall only include that part (cf. Figure 2a). We accordingly end up with the following approximate expression

$$\hat{H}_{so} \simeq \frac{1}{2m^2c^2r} \frac{\partial V_{\vec{R},(0,0)}(r)}{\partial r} \frac{1}{\sqrt{4\pi}} (\hat{\vec{s}} \cdot \hat{\vec{l}}), \qquad (12)$$

which is considered non-zero only inside the muffin-tin spheres.

As a first approximation we have included the Hamilton operator of equation (11) only perturbatively. I.e., we have only added it in calculating the band structures once the Kohn-Sham equations have been solved self-consistently.

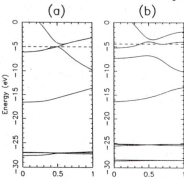

Figure 14. Band structure (a) without and (b) with the inclusion of spin-orbit couplings for a linear Bi chain with Bi–Bi bond lengths of 5.6 a.u.

Figure 14 shows the band structures for a linear chain of Bi atoms with a Bi–Bi interatomic distance set equal to a typical value of 5.6 a.u. It is seen that the bands split up into three groups, of which the lowest one (around -25 – -30 eV) is due to mainly $5d$ functions, the next one (at -15 eV) to $6s$ functions, and last group is due to $6p$ functions. We see that the $6s$ functions are not affected by the spin-orbit couplings, as is well known, whereas the $5d$ bands split as much as 3–4 eV. The $5d$ bands are moreover very flat, indicating that these orbitals are well localized to the atoms, so that they are expected to have large matrix elements for $\hat{H}_{so}$. However, the $6p$ functions are also split by the spin-orbit couplings, although their larger band widths indicate considerably smaller localization. Finally, the large width of these bands show that it is not a good approximation to compare results of optical experiments with electronic levels of single atoms. This is even more pronounced when considering other quasi-one-dimensional forms of Bi chains, for which the atoms have a higher coordination (*32*).

Doped Chains. In a number of cases the chains are doped. This is the case for some of the so-called MX chains that are a part of a crystal and that are surrounded by oppositely charged counterions [see, e.g., (*13*)]. Also for some of the systems already discussed in this contribution charged states are relevant, e.g., some of the defect-containing structures of the conjugated polymers and the hydrogen-bonded chains. It is accordingly desirable to be able to study doped chains, too.

The main problem when studying these theoretically is that when considering periodic structures, a periodically repeated non-vanishing charge will lead to a diverging electrostatic potential. Thus, the charge has to be compensated somehow, if periodicity is retained. This could, e.g., be done by placing extra point charges at certain points but it leaves the arbitrariness of choosing their positions. For three-dimensional crystalline materials a constant background density forms a useful approach but for the quasi-one-dimensional chains the volume of the three-dimensional unit cell is infinite leading to a vanishing background density.

As an alternative we have chosen to add constant background densities inside the muffin-tin spheres only (33). Their values are determined so that the total background density of each sphere is the same for all spheres. These extra charge densities can be easily taken into account in the calculation with a procedure almost equivalent to the one used for the frozen-core density and the nuclei.

In preliminary studies we have applied this approach on a linear carbon chain (33). The neutral linear carbon chain with all C–C bond lengths equal has a half-filled, doubly degenerate π band around the Fermi level. A dimerization (bond-length alternation) lowers the total energy and opens up a gap at the Fermi energy. This Peierls' mechanism is completely analogous to that observed for trans polyacetylene. For the doped chain it may, however, be (partly) suppressed since the dimerization does not lead to the occurrence of a gap at the Fermi level. We studied this by considering linear chains with two C atoms per unit cell and with a unit-cell length equal to 2.64 Å. Moreover, we monitored the relative total energy as a function of the size of an alternating displacement of the C atoms along the polymer axis. The results are shown in Figure 15.

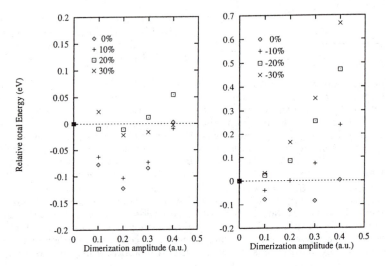

Figure 15. Relative total energy per C_2 unit for various linear doped C chains as a function of the displacement of the atoms relative to the undimerized structure. The labels give the number of electrons (in fractions of one per C_2 unit) added to (left panel) and removed from (right panel) the chains.

Figure 15 shows that the dimerization reduces as charge is added to the chains. However, although no gap appears at the Fermi level for any of the doped chains independent of the dimerization, the dimerization does not disappear abruptly when charging the system. For the neutral system, the dimerization leads to a stabilization of occupied orbitals and a destabilization of unoccupied orbitals, but from a total-energy point of view the latter is unimportant. When adding electrons, the destabilization, however, becomes important, but it can only partly compensate the stabilization, so that the dimerization is not completely suppressed. An equivalent picture also holds when removing electrons, although Figure 15 clearly shows that there is an asymmetry between positively and negatively doped chains. In total, this explains why the dimerization is reduced but not removed for the doped chains.

Conclusions

In the present contribution we have briefly reviewed a density-functional method for the calculation of electronic properties of helical polymers and have given some examples of applications. The materials of interest are both finite (in two dimensions) and infinite (in one dimension), which leads to considerable complications in the computational scheme. Moreover, for the general helical polymer it is necessary (and a simplification, too) to work with functions defined in local atom-centered coordinate systems.

The examples were mainly chosen from our most recent work, thereby illustrating the 'state of the art'. For a more detailed, but less recent, review, the reader is referred to (*34*).

Some of the important issues at the moment are to be able to calculate forces, which will make geometry optimizations orders of magnitude simpler. Furthermore, applying constrained density-functional calculations in order to estimate parameters for (extended) Hubbard models is highly relevant, as was most clearly demonstrated in our discussion of polythiophene. And, finally, further studies of the effects of distortions (for instance due to structural defects or extra charges) are very desirably.

Acknowledgments. Parts of this work were supported by the state of Baden-Württemberg through the Zentrum II at the University of Konstanz and by the Deutsche Forschungsgemeinschaft through project no. II C1-Sp 439/2-1. Other parts were supported by the European Union within the project 'Coherent non-linear dynamics of complex physical and biological systems' (project no. SC1*-CT91-0705), and by Nato through a travel grant, no. CRG 890533. Finally, the participation at the symposium was made possible by a grant from the Deutsche Forschungsgemeinschaft.

Literature Cited

[1] Hohenberg, P.; Kohn, W. *Phys. Rev.* **1964**, *136*, B864.
[2] Kohn, W.; Sham, L. J. *Phys. Rev.* **1965**, *140*, A1133.
[3] Springborg, M.; Andersen, O. K. *J. Chem. Phys.* **1987**, *87*, 7125.
[4] Springborg, M.; Calais, J.-L.; Goscinski, O.; Eriksson, L. A. *Phys. Rev. B* **1991**, *44*, 12713.
[5] von Barth, U.; Hedin, L. *J. Phys. C* **1972**, *5*, 1629.
[6] Springborg, M.; Jones, R. O. *Phys. Rev. Lett.* **1986**, *57*, 1145.
[7] Donohue, J. *The Structures of the Elements*, J. Wiley and Sons: New York, 1974.
[8] Pauling, L. *Proc. Natl. Acad. Sci. U.S.A.* **1949**, *35*, 495.
[9] Prins, J. A.; Schenk, J.; Hospel, P. A. M. *Physica* **1956**, *22*, 770.
[10] Yannoni, C. S.; Clarke, T. C. *Phys. Rev. Lett.* **1983** *51*, 1191.
[11] Begin, D.; Saldi, F.; Lelaurain, M.; Billaud, D. *Solid State Commun.* **1990**, *76*, 591.
[12] Karpfen, A.; Petkov, J. *Solid State Commun.* **1979**, *29*, 251.
[13] Proceedings of International Conference on Science and Technology of Synthetic Metals 1992, *Synth. Met.* **1993**, *55–57*.
[14] Ashkenazi, J.; Pickett, W. E.; Krakauer, H.; Wang, C. S.; Klein, B. M.; Chubb, S. R. *Phys. Rev. Lett.* **1989**, *62*, 2016.
[15] Perdew, J. P.; Wang, Y. *Phys. Rev. B* **1992**, *45*, 13244.

[16] Perdew, J. P. In *Electronic Structure of Solids '91*; Ziesche, P.; Eshrig, H., Eds.; Akademie Verlag: Berlin, Germany, 1991.

[17] Heeger, A. J.; Kivelson, S.; Schrieffer, J. R.; Su, W.-P. *Rev. Mod. Phys.* **1988**, *60*, 781.

[18] Ziemelis, K. E.; Hussain, A. T.; Bradley, D. D. C.; Rühe, J.; Wegner, G. *Phys. Rev. Lett.* **1991**, *66*, 2231.

[19] Gustafsson, J. C.; Pei, Q.; Inganäs, O. *Solid State Commun.* **1993**, *87*, 265.

[20] Springborg, M. *Solid State Commun.* **1994**, *89*, 665.

[21] Springborg, M. *Z. Phys. B.* **1994**, *95*, 363.

[22] Springborg, M. *J. Phys. Condens. Matt.* **1992**, *4*, 101.

[23] *Proton Transfer in Hydrogen-Bonded Systems*; Bountis, T., Ed.; Plenum Press, New York, 1992.

[24] Springborg, M. *Chem. Phys.*, in press.

[25] Dai, L.; Mau, A. W. H.; Griesser, H. J.; Winkler, D. A. *Macromol.* **1994**, *27*, 6728.

[26] Meider, H.; Springborg, M., University of Konstanz, unpublished results.

[27] Lam, L.; Platzman, P. M. *Phys. Rev. B* **1974**, *9*, 5122.

[28] Proceedings of Sagamore X Conference on Charge, Spin and Momentum Densities, *Z. Naturforsch. A* **1993**, *48*.

[29] Pattison, P.; Weyrich, W.; Williams, B. *Solid State Commun.* **1977**, *21*, 967.

[30] Springborg, M. *Chem. Phys.*, in press.

[31] Romanov, S. *J. Phys. Condens. Matt.* **1993**, *5*, 1081.

[32] Schmidt, K.; Springborg, M., University of Konstanz, unpublished data.

[33] Arcangeli, C.; Springborg, M.; Albers, R. C., University of Konstanz and Los Alamos National Laboratory, unpublished data.

[34] Springborg, M. *Int. Rev. Phys. Chem.* **1993**, *12*, 241.

Chapter 9

Structures and Interaction Energies of Mixed Dimers of NH_3, H_2O, and HF by Hartree–Fock, Møller–Plesset, and Density-Functional Methodologies

Carlos P. Sosa[1], John E. Carpenter[1], and Juan J. Novoa[2]

[1]Cray Research, Inc., 655 E. Long Oak Drive, Eagan, MN 55121
[2]Department de Quimica Fisica, Facultat de Quimica,
Universitat de Barcelona, Av. Diagonal 647, Barcelona 08028, Spain

The intermolecular interaction energies for hydrogen-bonded complexes have been investigated using Hartree-Fock (HF), Møller-Plesset (MP2) and density functional theory (DFT). Interaction energies were computed for the following complexes: NH_3--NH_3, NH_3--H_2O, NH_3--HF, H_2O--H_2O, H_2O--HF, and HF--HF. Pople's split valence (6-31++G(2d,2p)) basis sets were employed throughout these calculations. The predicted interaction energies for these complexes using the B3-LYP hybrid method (without ZPE and BSSE) are: -4.02, -7.15, -14.26, -4.97, -9.30, and -4.73 kcal/mol, respectively. Charge transfer computed using the natural bond order (NBO) scheme suggests that the local density approximation tends to overestimate the amount of charge transferred between donor and acceptor when compared to MP2 results.

Density functional theory (DFT) started to be successfully applied to molecular problems in mid 1980's. Slater (*1*) pioneered the idea of the electron-gas approximation as a simplification of the Hartree-Fock equations. Methods such as the so-called Xα method have been widely used in band structure calculations (*2*). However, it was not until the mid 1960's when Kohn, Hohenberg, and Sham (*3-4*) provided the foundation of what is now widely accepted as density functional theory. They showed that density functional theory is a many-body theory that can stand in the same footing as more conventional theories such as Hartree-Fock and post-Hartree-Fock methods.

   Density functional theory (*5-7*), offers a promising tool that may be applied to large systems. It includes correlation effects in a form in which the cost is only N^3; where N is the number of basis functions. In practice, the scaling can be lower (*8*). Also, because of recent developments of analytic gradient techniques, progress in functionals for gradient corrections to the local spin density approximation (LSD) (*3-*

4), and the availability of versatile software, density functional theory is becoming a popular tool in chemistry (*9*). Calculation of thermodynamic data for polyatomic molecules (*10-15*) shows that the gradient-corrected functionals provide energetics typically better than the HF method. The gradient-corrected results are closer to correlated methods (e.g., MP2), or better (*16*). Also, in recent papers (*13-15,17-18*), we presented systematic studies using density functional methods for bond dissociation, rotational barriers and transition structures. However, prediction of barrier heights in some cases has turned out to be more problematic. For example, the predicted DFT barrier heights for ethylene + butadiene (Diels-Alder) is much too low (with and without gradient corrections) suggesting that charge transfer may play an important role in strongly interacting systems (*18*).

In a further test of this hypothesis, we have carried out a series of calculations on dimers of three hydrogen bonding molecules (*17*). In the present study we want to extend our studies to include dimers selected to cover interaction energies between NH_3 and NH_3, H_2O, and HF, H_2O and H_2O and HF, and finally HF interacting with HF. The goal of this work has been (a) to establish how well the S-VWN, B-LYP, and B3-LYP reproduce the MP2 interaction energy using the 6-31++G(2d,2p) basis sets, (b) how the optimum energy structures computed using DFT compares with the predicted structures at the MP2 level and (c) how charge transfer in DFT compares to MP2.

Computational Section.

First principles calculations were carried out using the Gaussian 92/DFT program (*19*) on a Cray C90. The Kohn-Sham equations were solved as described by Pople and co-workers (*8*). The Coulomb and exact exchange integrals were evaluated analytically. All the integrals for the approximate exchange and correlation contributions to the energy were carried out over a numerical grid (*20*).

In the present study we have tested within the local spin density (LSD) approximation the combinations: Slater-Dirac (S) exchange (*1, 21*) and Vosko, Wilk, and Nusair (VWN) correlation functional (*22*) (S-VWN). Gradient corrections were introduced using Becke (*23*) exchange with the Lee, Yang and Parr (*24*) correlation (B-LYP). The LYP correlation is the density functional energy formulation of the Colle-Salvetti correlation energy (*25*). We have also tested Becke's three parameter hybrid functional with gradient corrections provided by the LYP functional (B3-LYP) (*26-27*) .

All the calculations for the NH_3, H_2O, and HF dimers have been performed with Pople's split-valence 6-31++G(2d,2p) (these basis sets include diffuse functions on heavy and hydrogen atoms, the polarization function space consists of two d's on heavy atoms and two p's on hydrogen) (*28-30*) basis sets. All equilibrium geometries were fully optimized with analytical gradient methods at the HF, MP2 , and DFT levels. The interaction energies were computed as the difference between the complex and the monomers. Harmonic vibrational frequencies were carried out for all the complexes. Only the zero-point energy (ZPE) corrections are presented in this study.

The method of natural bond orbital (NBO) analysis was employed to study charge transfer. As previously pointed out, the NBO technique uses information from diagonalizing the atomic sub-blocks of the density matrix (*31*). This procedure involves the orthogonalization of the atomic orbital basis to form the natural atomic orbitals, and a bond orbital transformation from the natural atomic orbitals to the final natural bond orbitals.

Results and Discussion.

The geometries for these complexes were predicted using density functional methods. All the optimized geometries were compared with conventional *ab initio* methods. Hartree-Fock and MP2 have been extensively employed in many chemical studies (*32*). They provide a very reliable way to calibrate density functional methods applied to chemical reactions and specially to weakly-bound complexes (*33-37*).

The calculated geometries for this series of hydrogen bond complexes are presented in Table I. In this section we only summarize our results using DFT. Figure 2 illustrates all the parameters used throughout these calculations for the NH_3, H_2O, and HF mixed dimers. Table II shows all the interaction energies, charge transfer and dipole moments.

Geometries.

The prediction of the structure of the ammonia dimer has proven to be a difficult task (*38-47*). Three structures have mainly been suggested as the equilibrium geometry of this dimer. The first structure corresponds to the classical structure as defined by Nelson et al. (*41*) (Figure 1a). In this approach one of the N-H bonds is collinear with the C3 axis of the other NH_3 molecule. One with Cs symmetry (*47*) (Figure 1b), and the other structure corresponds to a cyclic structure with C2h symmetry(*40*) (Figure 1c). Recently, Tao and Klemperer (*43*) have investigated the equilibrium geometry of the NH_3--NH_3 dimer using 6-311+G(3d,2p) and [7s5p3d,4s1p] extended with bond functions at the Hartree-Fock and MPn levels. They concluded that when the basis sets are enlarged with bond functions the C2h structure becomes more stable. On the other hand, basis sets without bond functions predict the Cs structure to be more stable. Recently, Muguet and Robinson (*45-46*) have reported a new method to correct the basis sets superposition error. In their study they carry out an extensive analysis of the BSSE for the NH_3--NH_3 dimer and presented a section discussing Tao and Klemperer's (*43*) work. The objective of the present study is mainly to asses the difference in charge transfer between DFT and MP2 calculations for this series of mixed dimers .

Three different type of dimers containing ammonia are presented in Table I. The predicted r (See Figure 2) distance at the Hartree-Fock level (not shown in Table I) for NH_3--NH_3, NH_3--H_2O, and NH_3--HF are 3.484, 3.092, and 1.809 Å, respectively. The effect of correlation corrections at the MP2 level is to shorten the r distance by about 0.1 to 0.2 Å. On the other hand, density functional theory using S-

Table I.   Optimized geometries for all the complexes[1]

Level		NH_3-NH_3	NH_3-H_2O	NH_3-HF	H_2O-H_2O	H_2O-HF	HF-HF
MP2	r	3.240	2.927	1.674	2.901	1.702	2.726
	r(HaX)	1.016	0.974	0.957	0.968	0.938	0.923
	r(HbX)	1.013	1.013	1.013			
	r(HcX)	1.013	1.013		0.961	0.961	0.926
	θa	125.7	117.9	111.9	109.0	113.0	
	θb	19.4	6.4	0.0	5.3	0.0	6.3
	θc	73.8	99.6	111.8			110.0
	θd	107.0	104.9		104.6		
S-VWN	r	2.985	2.739	1.523	2.712	1.538	2.548
	r(HaX)	1.037	1.002	0.996	0.989	0.969	0.948
	r(HbX)	1.024	1.024	1.023			
	r(HcX)	1.023	1.023		0.972	0.973	0.937
	θa	123.9	117.0	111.0	103.8	110.2	
	θb	15.1	6.6	0.0	6.7	0.0	8.0
	θc	80.3	99.1	111.0			
	θd	107.8	106.0				
B-LYP	r	3.325	2.944	1.654	2.942	1.696	2.758
	r(HaX)	1.030	0.991	0.981	0.982	0.958	0.942
	r(HbX)	1.025	1.024	1.024			
	r(HcX)	1.025	1.025		0.974	0.974	0.937
	θa	120.0	105.7	111.1	107.8	111.9	
	θb	8.6	5.1	0.0	5.0	0.0	6.6
	θc	95.0	123.6	112.1			109.4
	θd	106.8	104.8		104.7		
B3-LYP	r	3.295	2.933	1.660	2.916	1.690	2.725
	r(HaX)	1.021	0.978	0.964	0.976	0.944	0.931
	r(HbX)	1.016	1.015	1.016			0.926
	r(HcX)	1.016	1.016		0.963	0.964	
	θa	121.0	105.4	111.7	108.2	113.3	
	θb	10.2	5.0	0.0	4.8	0.0	6.8
	θc	90.7	123.4	111.2			109.8
	θd	107.4	105.5		105.2		

[1]Bond length in Å; angles in degrees.

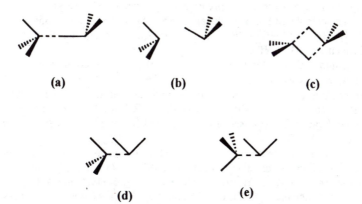

Figure 1. (a) Classical Structure, (b) Cs Symmetry, (c) Cyclic Structure, (d) Eclipsed Structure, (e) Staggered.

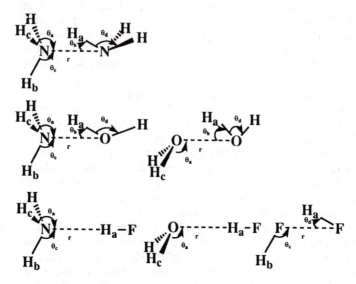

Figure 2. Optimized parameters for this series of complexes.

VWN (local approximation) tends to underestimate the r distance when compared to Hartree-Fock or MP2 calculations. S-VWN underestimates the r distance between 0.2 to about 0.3 Å for these three ammonia complexes. Gradient corrections play an important role. The B-LYP functional that contains gradient corrections for the exchange and correlation terms predicts an r distance in close agreement to MP2 results. The calculated B-LYP r distances are: 3.325, 2.944, and 1.654 Å for NH_3--NH_3, NH_3--H_2O, and NH_3--HF, respectively. Similar results are obtained using the hybrid functional (B3-LYP).

In general, all the DFT parameters for these three complexes are in good agreement with MP2 results. In the case of NH_3--H_2O, we have identified two conformers, one conformer has the Ha-O bond eclipsed with respect to Hb-N bond (Figure 1d) and a second structure has these two bonds staggered (Figure 1e). Hartree-Fock favors the eclipsed structure while MP2 and DFT predict the staggered structure to be the most stable.

Microwave and radio frequency spectra for NH_3--H_2O have reported a vibrationally averaged distance (N--O) of 2.983 Å (*48*). Also, previous *ab initio* calculations (*49-52*) at the Hartree-Fock level using 6-31G* and 4-31G basis sets have predicted distances (N--O) of 3.05 and 2.95 Å, respectively. These results are in agreement with our Hartree-Fock results (not shown in Table I). On the other hand, MP2, B-LYP, and B3-LYP tend to shorten this distance and predict values within experimental uncertainty.

Tables I also summarizes all the optimized geometries (See Figure 2) for H_2O--H_2O, H_2O--HF, and HF--HF dimers. All these structures were optimized using the 6-31++G(2d,2p) basis sets. This basis set were selected based on previous experience with the HF--HF dimer (17). Calculated interaction energies and BSSE computed with aug-cc-pVTZ and 6-31++G(2d,2p) have been found in good agreement (*17*). We have recently carried out a comparison between *ab initio* conventional methods and DFT for H_2O and HF dimers (*17*). In this section we only summarize some of the trends as a function of mixed dimers.

Recently, Feller (*53*) and Kim and Jordan (*54*) have also extensively studied the H_2O--H_2O dimer using Hartree-Fock and post-Hartree-Fock methods as well as DFT. Kim and Jordan (*54*) have reported good agreement between the predicted B3-LYP geometries and the MP2 and experimental geometries. Examination of the results in Table I indicates considerable changes in the r distance at the DFT level as different functionals are compared (*17*). Adding gradient corrections to the local approximation functional tend to elongate the r distance. In all the cases S-VWN underestimates the r distance when compared to MP2 values.

All the optimized geometries for the HF--HF dimer are collected in Table I. All our results obtained using MP2 are in good agreement with previous calculations at similar levels of theory (*17*). In the case of the HF--HF dimer (See Figure 2), DFT calculations tend to have a marked basis set effect. This is particularly true for the r distance (*17*). In general, S-VWN underestimates the r distance and B-LYP and B3-LYP brings the predicted DFT geometries in good agreement with MP2 results.

Similar to the ammonia dimers, replacing H_2O in the water dimer with a stronger donor (HF) tends to decrease the r distance. Experimentally, the r(F--O) distance has been reported to be 2.66 Å (*52*). The predicted S-VWN value of 2.506 Å clearly underestimates the experimental results. On the other hand, MP2, B-LYP, and B3-LYP predict $r(F--O)$ distances that are ± 0.03 Å within experimental result.

Energetics.

Previously, we reported basis set effects and interaction energies with and without counterpoise corrections for HF--HF, H_2O--H_2O, NH_3--NH_3, $C_2H_2--H_2O$, and CH_4--H_2O (*17*). All the interaction energies computed with the 6-31++G(2d,2p) basis sets were found to be in good agreement with energies obtained using aug-cc-pVTZ basis sets. The BSSE error for these systems with large basis sets is less than 1 kcal/mol. Therefore, all the interaction energies presented in this work do not include BSSE. On the other hand, smaller basis sets without diffuse functions do not describe these interactions properly (*17*). It was also found that DFT tends to be more susceptible to the basis sets truncation than conventional methods such as MPn (*17*). In contrast to our previous study, here we report MP2 and DFT calculations for NH_3--X, H_2O-- X, and HF--X where X corresponds to NH_3, H_2O, and HF. This provides a systematic study of interaction energies and charge transfer as a function of different substituents for NH_3, H2O, and HF dimers. Figure 3 illustrates a qualitative matrix of interactions for the set of dimers presented in this study. For this series of mixed dimers, we have looked at the effect of charge transfer using DFT and MP2 methods for weak, moderate, and strong interactions.

Interaction energies (ΔE), charge transfer (nCT), and dipole moments (μ) for all the dimers are collected in Table II. The first series of dimers correspond to NH_3-- NH_3, NH_3--H_2O, and NH_3--HF. The ordering of the interaction energies for these ammonia complexes increases as the dipole moment of the substituent increases, that is, HF > H_2O > NH_3. Inspecting Table II reveals that S-VWN (local approximation) overestimates the interaction energy when compared to MP2 results. On the other hand, B-LYP and B3-LYP give results that are consistent with MP2. B-LYP and B3-LYP predict interaction energies that are only within 1 kcal/mol error compared to MP2.

In addition to the NH_3 series of dimers, Table II also provides energies for H_2O--H_2O and H_2O--HF. These two dimers may be analyzed in a similar fashion as the NH_3 dimers. As it would be expected from the previous series, H_2O--HF shows a stronger interaction than H_2O--H_2O. Similar trends may be observed for these dimers. S-VWN clearly overestimates the interaction energy between 4 to 6 kcal/mol when compared to all the other methods presented in Table II. The interaction energy of the H_2O--H_2O dimer has been deduced from experimental data to be 5.4±0.7 (*55*) and 5.4±0.2 kcal/mol (*56*) (vibrational and temperature effects have been subtracted). Recent calculations for this interaction energy at the B3-LYP/aug-cc-pVTZ has been reported to be -4.57 kcal/mol (*54*). MP2 calculations using large basis sets have provided estimates of the H_2O--H_2O interaction energy in the range of -4.7 to -5.1 kcal/mol (*17, 47, 53, 54*). The predicted B-LYP and B3-

Table II. Calculated Interaction Energies, Charge Transfer, and Dipole Moments[1-3].

Level		NH_3-NH_3	NH_3-H_2O	NH_3-HF	H_2O-H_2O	H_2O-HF	HF-HF
MP2	ΔE	-4.48	-7.44	-13.93	-5.40	-9.31	-4.83
	nCT	0.00671	0.03076	0.07484	0.01935	0.03963	0.0754
	μ	1.86	3.61	4.72	2.75	4.24	3.51
	ΔZPE	1.38	2.25	1.69	2.26	2.94	1.92
S-VWN	ΔE	-6.54	-12.03	-21.03	-9.49	-15.23	-8.76
	nCT	0.031410	0.06753	0.12355	0.04588	0.0768	0.0420
	μ	2.38	3.96	5.22	2.68	4.55	3.44
	ΔZPE	1.60	2.31	3.01	2.45	2.94	2.07
B-LYP	ΔE	-3.59	-6.81	-14.07	-4.58	-8.92	-4.39
	nCT	0.01358	0.03792	0.09063	0.02184	0.04849	0.0203
	μ	2.63	3.69	4.86	2.64	4.27	3.37
	ΔZPE	1.36	2.31	2.96	2.08	2.78	1.78
B3-LYP	ΔE	-4.02	-7.15	-14.26	-4.97	-9.30	-4.73
	nCT	0.01219	0.03350	0.08189	0.01967	0.04324	0.0187
	μ	2.49	3.67	4.81	2.60	4.32	3.39
	ΔZPE	1.38	2.28	3.03	2.18	2.83	1.84

[1]Interaction and zero-point energies in kcal/mol.
[2]Charge transfer corresponds to the number of electrons transfered from donor to acceptor.
[3]Dipole moment in Debye.

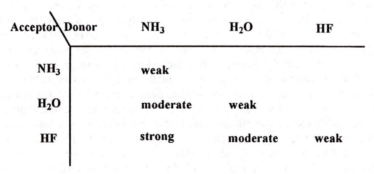

Figure 3. Qualitative interaction matrix.

LYP interaction energies are: -4.58 and -4.97 kcal/mol, respectively. These results are consistent with previous DFT studies on the H_2O--H_2O dimer (57).

In the case of H_2O--HF, the estimated experimental interaction energy (vibrational and temperature effects have been subtracted) corresponds to 7.17 ± 1.6 kcal/mol (58). Our predicted MP2, B-LYP, and B3-LYP are within the upper limit of the estimated experimental value.

In addition to the two previous series of complexes, Table II also summarizes interaction energies, charge transfer, and dipole moments for the HF--HF dimer. In this case, Pine and Howard (59) have estimated the interaction to be -4.56 (+0.29, -0.27) kcal/mol. Michael et al. (60) have reported a theoretical interaction energy of -4.55 kcal/mol. It is not surprising, S-VWN overestimates this interaction as well. B-LYP and B3-LYP interaction energies are: -4.39 and -4.73 kcal/mol, respectively. Both in good agreement when compared to previous theoretical results and our MP2.

A manifestation of the behavior of the interaction energies may be observed in the amount of charge transfer as a function of the method. In the cases of NH_3 and H_2O dimers, the decrease of the r distance presumably reflects cooperative efforts as the dipole of the substituent increases. For example, NH_3 is a better electron donor than H_2O or HF, for NH_3--H_2O, NH_3--HF, H_2O--H_2O, and H_2O--HF, it is expected that the dimers containing NH_3 will show a stronger hydrogen bond. This type of trend may also be observed for H_2O--HF and HF--HF, where H_2O is a better donor than HF, and a stronger hydrogen bond for the H_2O--HF case is expected as well.

In general, DFT appears to qualitatively reproduce these observations, however, a more careful analysis of the charge transfer indicates that DFT tends to overestimate the amount of charge that has been transferred between donor and acceptor. This is particularly true for the local approximation. Examination of Table II indicates that S-VWN tends to overestimate the charge transfer in all the complexes studied here. In other words, it favors more charge transfer and more repulsion can be overcome. This effect may be responsible for the underestimation of the r distance using the local approximation (S-VWN) for all these dimers.

The effect of gradient corrections is to bring the value of nCT in closer agreement with MP2 results. B-LYP tends to correct the overestimation of the charge transfer (nCT) at the local density approximation, in some cases by as much as a factor of two. B3-LYP predicts charge transfer in even closer agreement to MP2 results. The largest discrepancies correspond to the NH_3--NH_3 and HF--HF, where B-LYP and B3-LYP still overestimate the amount of charge transfer when compared to MP2.

Acknowledgments

We thank the management of Cray Corporate Network for his generous allocation of computer time in the Cray C-90 and Y-MP machines. JJN also thanks the "Fundació Catalana per a la Recerca" (Catalunya, Spain) for making possible his visit to Cray Research and the Spanish DGICYT for its continuous support under Project PB92-0655-C02-02.

Literature Cited.

1. Slater, J. C. *Phys. Rev.* **1951**, *81*, 385.
2. Slater, J. C. *The Self-Consistent Field for Molecules and Solids, Quantum Theory of Molecules and Solids* , McGraw-Hill: New York, NY, 1974, Vol. 4.
3. Hohenberg, P.; Kohn, W. *Phys. Rev.* **1964**, *B136*, 864.
4. Kohn, W.; Sham, L. J. *Phys. Rev.* **1965**, *A140*, 1133.
5. Parr, R. G.; Yang, W. *Density-Functional Theory of Atoms and Molecules*, Oxford University Press: New York, NY, 1989.
6. Levy, M.; *Proc. Natl. Acad. Sci. (USA)* **1979**, *76*, 6062.
7. Jones, R. O.; Gunnarsson, O. *Reviews of Modern Physics* **1990**, *61*, 689.
8. Johnson, B. G.; Gill, P. M. W. Pople, J. A. *J. Chem. Phys.* **1993**, *98*, 5612.
9. Ziegler, T. *Chemical Review* **1991**, *91*, 651.
10. Becke, A. D. *J. Chem. Phys.* **1986**, *84*, 4524.
11. Becke, A. D. *The Challenge of d and f Electrons: Theory and Computation*; ACS Symposium Series No 394; ACS: Washington D. C., N.H., 1989.
12. Becke, A. D. *J. Chem. Phys.* **1988**, *88*, 2547.
13. Andzelm, J.; Sosa, C.; Eades, R. A. *J. Phys. Chem.* **1993**, *97*, 4664.
14. Sosa, C.; Andzelm, J.; Lee, C.; Blake, J. F.; Chenard, B. L.; Butler, T. W. *Int. J. Quantum Chem.* **1994**, *49*, 511.
15. Sosa, C.; Lee, C. *J. Chem. Phys.* **1993**, *98*, 8004.
16. Wimmer, E.; Andzelm, J. *J. Chem. Phys.* **1992**, *96*, 1280.
17. Novoa, J. J.; Sosa, C. *J. Phys. Chem.* **1995**, *99*, 15837.
18. Carpenter, J. E.; Sosa, C. *J. Mol. Structure (THEOCHEM)* **1994**, *311*, 325.
19. *Gaussian 92/DFT, Revision G.3*, Frisch, M. J.; Trucks, G. W.; Schlegel, H. B.; Gill, P. M. W.; Johnson, B. G.; Wong, M. W.; Joresman, J. B.; Robb, M. A.; Head-Gordaon, M.; Replogle, E. S.; Gomperts, R.; Andres, J. L.; Raghavachari, K.; Binkley, J. S.; Gonzalez, C.; Martin, R. L.; Fox, D. J.; DeFrees, D. J.; Baker, J.; Stewart, J. J. P.; Pople, J. A. Gaussian, Inc: Pittsburgh PA, 1993 .
20. Johnson, B. G.; Frisch, M. J. *Chem. Phys. Lett.* **1993**, *216*, 133.
21. Dirac, P. A. M. *Proc. Cambridge Phil. Soc.* **1930**, *26*, 133.
22. Vosko, S. H.; Wilk, L.; Nusair, M. *Can. J. Phys.* **1980**, *58*, 1200.
23. Becke, A. D. *Phys. Rev.* **1988**, *A38*, 3098.
24. Lee, C; Yang, W.; Parr, R. G. *Phys. Rev.* **1993**, *B37*, 785.
25. Colle, R.; Salvetti, D. *Theor. Chim. Acta* **1975**, *37*, 329.
26. Frisch, M. J.; Frisch, AE.; Foresman, J. B. *Gaussian 92/DFT User's Reference Release Notes*, Gaussian, Inc: Pittsburgh PA, 1993.
27. Becke, A. D. *J. Chem. Phys.* **1993**, *98*, 1372; **1993**, *98*, 5648.
28. Hehre, W. J.; Ditchfield, R.; Pople, J. A. *J. Chem. Phys.* **1972**, *56*, 2257.
29. Hariharan, P. C.; Pople, J. A. *Mol. Phys.* **1974**, *27*, 209.
30. Clark, T.; Chandrasekhar, J.; Spitznagel, G. W.; Schleyer, P. v. R. *J. Comp. Chem.* **1983**, *4*, 294.
31. Reed, A. E.; Weinstock, R. B.; Weinhold, F. *J. Chem. Phys.* **1985**, *83*, 2257.
32. Hehre, W. J.; Radom, L.; Schleyer, P. v. R.; Pople, J. A. *Ab Initio Molecular Orbital Theory;* John Wiley & Sons: New York, NY, 1986.

33. Buckingham, A. D.; Fowler, P. W.; Hutson, J. M.. *Chem. Rev.* **1988**, *88*, 963 and references therein.
34. Hobza, P.; Zahradnik, R. *Intermolecular Complexes*; Elsevier: Amsterdam, The Netherlands, 1988.
35. Chalasinski, G.; Gutowski, M. *Chem. Rev.* **1988**, *88*, 943.
36. Liu, B; McLean, A. D. *J. Chem. Phys.* **1989**, *91*, 2348 and references therein.
37. Scheiner, S. *Reviews in Computational Chemistrsy;* VCH: New York, NY, 1991, Vol. 2, pp. 165.
38. Sagarik, K. D.; Ahlrichs, R.; Brode, S. *Mol. Phys.* **1986**, *57*, 1247.
39. Hassett, D. M..; Marsden, C. J.; Smith, B. J. *Chem. Phys. Lett.* **1991**, *183*, 449.
40. Latajka, Z.; Scheiner, S. *J. Chem. Phys.* **1986**, *84*, 341.
41. Nelson, D. D., Jr.; Fraser, G. T.; Klemperer, W. *J. Chem. Phys.* **1985**, *83*, 6201.
42 Nelson, D. D., Jr.; Klemperer, W.; Fraser, G. T.; Lovas, F. J.; Suenram, D. *J. Chem. Phys.* **1987**, *87*, 6364.
43. Tao, F-. M.; Klemperer, W. *J. Chem. Phys.* **1993**, *99*, 5976.
44. Olfhof, E. H. T., Van der Avoird, A.; Wormer, P. E. S. *J. Chem. Phys.* **1994**, *101*, 8430.
45. Muguet, F. F.; Robinson, G. W. *J. Chem. Phys.* **1995**, *102*, 3648.
46 Muguet, F. F.; Robinson, G. W.; Basses-Muguet, M. P. *J. Chem. Phys.* **1995**, *102*, 3655.
47. Frisch, M. J.; Del Bene, J. E.; Binkley, J. S.; Schaefer III, H. F. *J. Chem. Phys.* **1986**, *84*, 2279.
48. Herbine, P.; Dyke, T. R. *J. Chem. Phys.* **1985**, *83*, 3768.
49. Dill, J. D.; Allen, L. C.; Tropp, W. C.; Pople, J. A. *J. Am. Chem. Soc.* **1975**, *97*, 7220.
50. Kollman, P. A.; Allen, L. C. *J. Am. Chem. Soc.* **1971**, *93*, 4991.
51. Lathan, W. A.; Curtiss, L. A.; Hehre, W. J.; Lisle, J. B.; Pople, J. A. *Prog. Phys. Org. Chem.* **1974**, *11*, 175.
52. Bevan, J. W.; Kisiel, Z.; Legon, A. C.; Miller, D. J.; Rogers, S. C. *Proc. R. Soc. London Ser.* **1980**, *A372*, 441.
53. Feller, D. *J. Chem. Phys.* **1992**, *96*, 6104.
54. Kim, K.; Jordan, K. D. *J. Phys. Chem.* **1994, **98*, 10089.
55. Curtis, L. A.; Frurip, D. J.; Blander, M *J. Chem. Phys.* **1979**, *71*, 2703.
56. Raimers, J.; Watts, R.; Kein, M. *Chem. Phys.* **1982**, *64*, 92.
57. Xantheas, S. S. *J. Chem. Phys.* **1995**, *102*, 4505.
58. Thomas, R. K. *Proc. R. Soc. London Ser.* **1975**, *A344*, 579.
59. Pine, A. S.; Howard, B. J. *J. Chem. Phys.* **1986**, *84*, 590.
60. Michael, D. W.; Dykstra, C. E.; Lisy, J. M. *J. Chem. Phys.* **1984**, *81*, 5998.

Chapter 10

Free Energy Perturbation Calculations Within Quantum Mechanical Methodologies

Robert V. Stanton, Steven L. Dixon, and Kenneth M. Merz, Jr.[1]

Department of Chemistry, 152 Davey Laboratory, Pennsylvania State University, University Park, PA 16802

Newly developed techniques for use in quantum free energy and potential of mean force calculations are presented. These techniques are the first to provide a direct means of performing arbitrary perturbations in single-topology quantum mechanical systems. The theoretical and practical considerations for carrying out the associated molecular dynamics simulations are discussed. Simulations are conducted for systems that are purely quantum mechanical, and for systems that involve a combination of quantum mechanical and molecular mechanical atoms. Preliminary results demonstrate these procedures to constitute a powerful tool in free energy calculations, with the potential to significantly increase the accuracy of simulations on both large-scale and small-scale systems.

Free energy perturbation (FEP)(1-3) and potential of mean force (PMF)(4, 5) calculations within molecular mechanical (MM) force fields have proven to be two of the most powerful computational methodologies available. These techniques, which have been developed and refined over the last two decades, allow access to important thermodynamic quantities, such as the relative binding free energies of ions within an ionophore(6), or inhibitors to an enzyme.(7) While other techniques, such as linear response theory(8), have been developed to calculate these same quantities they may require additional parameterization. FEP and PMF calculations, in contrast, are done using statistical averages of the energy differences between the various intermediate states considered in the calculation and thus may be applied to any system for which a potential energy surface is known. Many elegant applications of these theoretical techniques within molecular dynamics (MD) calculations currently exist within the literature.(4, 5, 7, 9-14)

Recently, quantum mechanical (QM) MD(15-18), as well as coupled potential MD(19-22) calculations have become computationally practical. In such calculations a system is modeled using a QM Hamiltonian, within the Born-Oppenheimer approximation, and is propagated classically through time using Newtonian mechanics. Coupled potentials (CP), in addition to the QM portion of

[1]Corresponding author

the system, also include atoms which are modeled molecular mechanically. The division of a large system into QM and MM portions allows for a portion of the system to be modeled using the more accurate QM potential function while still including environmental effects from the MM portion of the system. Typically, the QM calculation involves a region of particular interest such as the solute within a solvent system or the active site within an enzyme. Ideally the entire system would be modeled using quantum mechanics, however due to the scaling of the computational expense of QM calculations with the size of the system (between n^3 and n^7, where n is the number of basis functions) fully QM representations of large systems remain impractical.

With the increased use of quantum mechanics for MD simulations, a natural extension of these simulations is their use with FEP and PMF calculations. Unfortunately, though, the extrapolation of the FEP and PMF methods used within MM calculations to QM potentials is not as straightforward as might be hoped. In a MM FEP calculation the relative free energy difference between two states is calculated by slowly perturbing the parameters appropriate for one state into those of the other. For example, if a ketone group were being perturbed into an alcohol, the bond, angle, torsional, and van der Waals (vdW) parameters as well as the atomic charges, would be coupled to a perturbation parameter λ such that $\lambda = 1$ corresponds to the ketone and $\lambda = 0$ the alcohol. Two approaches are available for the calculation of the intermediate states. These are referred to as the single and dual topology methods. In the dual topology method the forces and energy of the fractional lambda state are generated by weighted averages of these values from independent calculations at $\lambda = 1$ and $\lambda = 0$. The single topology method, alternatively, uses weighted averages of the parameter values to calculate forces and energy. A study contrasting the single and dual topology methods showed the single topology method to converge more quickly than the dual in many cases.(23)

If single topology methods are to be used within QM systems techniques must be developed which allow for the calculation of systems corresponding to fractional values of the perturbation parameter. Four possible system perturbations can be defined within QM-FEP calculations: 1) changing an atom type.; 2) changing the number of atoms; 3) changing the number of electrons; and 4) changing the number of atomic orbital basis functions. Perturbations between systems can, and usually do, incorporate many of these changes simultaneously. Recently, within our lab we have developed methods allowing for each of these four types of perturbations. We will outline these methods and the results of test calculations in the remainder of this paper.

An alternative method for the calculation of FEPs within QM and CP systems has been demonstrated by Warshel and Wesolowski.(24) In these calculations they determine the free energy for changing the QM starting and end point structures into their MM equivalent. Then by calculating the MM FEP they are able to complete a free energy cycle and obtain the relative free energy difference between the two QM systems. The method presented here involves a direct QM conversion and requires no MM intermediate.

Theoretical Approach

In the next four subsections we will discuss the individual techniques necessary to change the number of electrons, atoms and orbitals within a system as well as the identity of certain atoms. All of the derivations and tests shown here are done within the semiempirical PM3 Hamiltonian(25, 26), although they could easily be extended for use with other QM potential functions.

Number of Electrons. Here we formulate the fractional electron method to allow for non-integer numbers of electrons in a QM system. The approach relies on the use of a pseudo-closed-shell expression for the electronic energy, where fractional occupation numbers for the MO's is assumed at the outset. If M is the number of atomic orbitals (AO), n is the total number of electrons in the system and n_k is the number of electrons in the kth molecular orbital (MO) then it is possible to define the electronic energy of a system using the pseudo-closed shell formula:

$$E_{elec} = \sum_{\mu=1}^{M}\sum_{v=1}^{M}\left(H_{\mu v} + G_{\mu v}\right)P_{\mu v} \tag{1}$$

where,

$$H_{\mu v} = \int \chi_{\mu}^{\dagger}(1)\left[-\frac{1}{2}\nabla_1^2 - \sum_A V_A(r_1)\right]\chi_v(1)\delta\,\tau_1 \tag{2}$$

$$G_{\mu v} = \sum_{\lambda=1}^{M}\sum_{\sigma=1}^{M}\left[(\mu v|\lambda\sigma) - \frac{1}{2}(\mu\sigma|\lambda v)\right]P_{\lambda\sigma} \tag{3}$$

$$P_{\mu v} = \sum_{k=1}^{M} n_k \mathbf{c}_{\mu k}^{\dagger}\mathbf{c}_{vk} \tag{4}$$

$$(\mu v|\lambda\sigma) = \int\int \chi_{\mu}^{\dagger}(1)\chi_v(1)\frac{1}{r_{12}}\chi_{\lambda}^{\dagger}(2)\chi_{\sigma}(2)d\,\tau_1 d\,\tau_2 \tag{5}$$

Note that in this derivation the number, n_k, of electrons in a MO may take on a non-integer value ($0 \le n_k \le 2$). While not representative of a physical system non-integer electrons have been used previously in, for example, the so-called half-electron method.(27)

Energy gradients and thus atomic forces can be computed analytically only if the electronic energy is variationally optimized with respect to the MO's. Further, the energy expression (1) implicitly assumes that the MO's are orthonormal. Thus we require,

$$\frac{\delta E}{\delta \mathbf{c}_k} = 0 \tag{6}$$

$$\mathbf{c}_i^{\dagger}\mathbf{c}_j = \delta_{ij} \tag{7}$$

$$\mathbf{c}_i^{\dagger}\mathbf{S}\mathbf{c}_j = \delta_{ij} \tag{8}$$

Equation 8 is the standard *ab initio* Hartree-Fock formulation, while equation 7 is employed with semiempirical treatments. To keep the discussion general we will focus on the Hartree-Fock procedure and not the semiempirical methods. Extension to semiempirical approaches is straightforward.

The quantity which must be minimized is the Lagrangian L,

$$L = E_{elec} - 2\sum_i^{occ}\sum_j^{occ}\varepsilon_{ij}\left(\mathbf{c}_i^{\dagger}\mathbf{S}\mathbf{c}_j - \delta_{ij}\right) \tag{9}$$

where ε_{ij} are Lagrange multipliers. The minimization is most conveniently carried out with respect to $c_k^{\dagger}$ while treating c_k as formally independent. Thus we have,

$$0 = \frac{\delta L}{\delta c_k^{\dagger}} = \frac{\delta}{\delta c_k^{\dagger}} \left\{ \sum_{\mu=1}^{M} \sum_{\nu=1}^{M} [H_{\mu\nu} + G_{\mu\nu}] P_{\mu\nu} - 2 \sum_{i}^{occ} \sum_{j}^{occ} \varepsilon_{ij} \left(c_k^{\dagger} S c_j - \delta_{ij} \right) \right\} \quad (10)$$

Formal differentiation, followed by a series of algebraic manipulations yields,

$$0 = \sum_{\mu=1}^{M} \sum_{\nu=1}^{M} \left(H_{\mu\nu} + G_{\mu\nu} \right) \frac{\delta P_{\mu\nu}}{\delta c_k^{\dagger}} + \sum_{\lambda=1}^{M} \sum_{\sigma=1}^{M} \frac{\delta P_{\lambda\sigma}}{\delta c_k^{\dagger}} [G_{\sigma\lambda}]^{\dagger} - 2 \sum_{j}^{occ} \varepsilon_{kj} S c_j \quad (11)$$

Using the Hermitian nature of G the subscripts of G can be interchanged and the expression can be rewritten as,

$$0 = \sum_{\mu=1}^{M} \sum_{\nu=1}^{M} \left(H_{\mu\nu} + G_{\mu\nu} \right) \frac{\delta P_{\mu\nu}}{\delta c_k^{\dagger}} + \sum_{\lambda=1}^{M} \sum_{\sigma=1}^{M} \frac{\delta P_{\lambda\sigma}}{\delta c_k^{\dagger}} G_{\lambda\sigma} - 2 \sum_{j}^{occ} \varepsilon_{kj} S c_j \quad (12)$$

Rearranging terms gives,

$$0 = \sum_{\mu=1}^{M} \sum_{\nu=1}^{M} \left(H_{\mu\nu} + 2G_{\mu\nu} \right) \frac{\delta P_{\mu\nu}}{\delta c_k^{\dagger}} - 2 \sum_{j}^{occ} \varepsilon_{kj} S c_j \quad (13)$$

$$0 = \sum_{\mu=1}^{M} \sum_{\nu=1}^{M} \left(H_{\mu\nu} + 2G_{\mu\nu} \right) n_k c_{\nu k} e_{\mu} - 2 \sum_{j}^{occ} \varepsilon_{kj} S c_j \quad (14)$$

$$0 = n_k (\mathbf{H} + 2\mathbf{G}) c_k - 2 \sum_{j}^{occ} \varepsilon_{kj} S c_j \quad (15)$$

And finally,

$$n_k \mathbf{F} c_k = 2 \sum_{j}^{occ} \varepsilon_{kj} S c_j \quad (16)$$

where the Fock matrix $\mathbf{F}$ is defined as $(\mathbf{H}+2\mathbf{G})$.
For a closed-shell system , the familiar equation

$$\mathbf{F} c_k = \sum_{j}^{occ} \varepsilon_{kj} S c_j \quad (17)$$

would result because $n_k=2$ for all occupied MO's. In the case of the fractional electron method, however, at least one occupation number will be less than 2 and equation 17 is retained.

At this point it is customary to define a unitary transformation $\mathbf{U}$ of the occupied MO's that would diagonalize the matrix of Lagrange multipliers ε_{kj}, so that the problem to solve has the general form,

$$\mathbf{F} c_k = \varepsilon_k S c_k \quad (18)$$

This is easily justified in the case of a closed-shell system because the wavefunction and all the molecular properties calculated from it are unaffected by a unitary transformation of the occupied MO's. Unfortunately, there may be no actual wavefunction that corresponds to the fractional electron system, and the energy will be affected by any unitary transformation that mixes fully occupied MO's with partially occupied MO's. So the unitary transformation argument cannot be used to convert equation 17 to equation 18. However, as long as equation 17 is satisfied for some set of Lagrange multipliers that preserve the orthonormality of the MO's, the corresponding vectors c_k will be nontrivial solutions that coincide with an extremum of E_{elec}. This is all that is required in the fractional electron method, thus a diagonal matrix of Lagrange multipliers may be assumed, and the MO's are solutions of.

$$n_k \mathbf{F} \mathbf{c_k} = 2\varepsilon_k \mathbf{S} \mathbf{c_k} \tag{19}$$

Note that equation 20 can be related to equation 19 by defining an F'

$$\mathbf{F}'_k \mathbf{c}_k = \varepsilon'_k \mathbf{S} \mathbf{c}_k \tag{20}$$

$$\mathbf{F}'_k = \frac{n_k}{2} \mathbf{F} \tag{21}$$

The eigenvalues of equation 20 are related to those of equation 19 by

$$\varepsilon'_k = \frac{n_k}{2} \varepsilon_k \tag{22}$$

Thus by making n_k a function of λ we can alter the number of electrons in a system. The states corresponding to fractional λ values will be not be physical, however since the transformation between states is smooth and the free energy a state function only the free energy difference between the initial and final states can be determined.

Number of Atomic Orbitals. Changing the number of atomic orbitals can be done by scaling the Fock matrix elements corresponding to these orbitals with the perturbation parameter λ. This requires that the perturbations be done in the direction in which the number of atomic orbitals decreases. As λ approaches zero, the energy levels of the disappearing atomic orbitals becomes very high so that they no longer contribute to the occupied molecular orbitals, and thus do not affect the energy for the system. This simple method allows the deletion of atomic orbitals without the generation of a new independent Fock matrix as would be required for a dual topology QM approach.

Identity of an Atom. If the number of valence electrons and the locations of the atomic orbitals on two molecular species are identical, the energies for these two systems will exhibit the same mathematical dependence on the atomic parameters, independent of the identity of the element. A brief derivation of the PM3 approximation(25, 26, 28), which is largely based on the AM1(29) and MNDO(30) method, is given here as the dependency of the parameters is essential. The total heat of formation of a molecule, within PM3, is given as a combination of electronic and nuclear repulsion energies E_{elect} and E_{nuc} along with the experimental heat of formation of the atoms, and the calculated electronic energy for the gaseous atom E_{el}^A.

$$\Delta H_f = E_{elect} + E_{nuc} - \sum_A E_{el}^A + \sum_A \Delta H_f^A \tag{23}$$

E_{el}^A is taken to be a function of (1) the ground state atomic orbital (AO) population of i, (given by the density matrix element P_{ii}), (2) the one-electron energies for AO i of the ion resulting from removal of the valence electrons, U_{ii}, and (3) the two-electron one-center integrals, $<iiljj>$ and $<ijlij>$.

$$E_{el}^A = f\left(P_{ii}, U_{ii}, (ii|jj), (ij|ij)\right) \tag{24}$$

The two electron one center integrals are given as $<ss|ss>=G_{ss}$, $<ss|pp>=G_{sp}$, $<pp|pp>=G_{pp}$, $<pp|p'p'>=G_{p2}$ and $<sp|sp>=H_{sp}$, while U_{ii} is taken to be U_{ss} or U_{pp}. For G_{p2} the p and p' indices represent different Cartesian p orbitals on the same center. The nuclear energy is expressed as the summation over the individual nuclear-nuclear interactions.

$$E_{nuc} = \sum_{i<j} E_n(i,j) \tag{25}$$

Within PM3 $E_n(i,j)$ is given as:

$$E_n(i,j) = Z_i Z_j \left(S_i S_i | S_j S_j\right) \bullet (1 + e^{(-\alpha_i R_{ij})} + e^{(-\alpha_j R_{ij})}) + \frac{Z_i Z_j}{R_{ij}} \bullet$$
$$\left(\sum_k a_{ki} \exp\left[-b_{ki}(R_{ij} - c_{ki})^2\right] + \sum_k a_{kj} \exp\left[-b_{kj}(R_{ij} - c_{kj})^2\right] \right) \tag{26}$$

The simple exponential forms in equation 26 were used in the original MNDO formulation(30, 31), while the Gaussian terms were added within AM1(29) and PM3(25, 26, 28) to facilitate the representation of hydrogen bonded systems by reducing the long-range repulsion of the core-core term used in MNDO. The adjustable semiempirical parameters associated with the exponential and Gaussian terms in equation 4 are α_i, a_{ki}, b_{ki} and c_{ki}, while Z_i is the number of valence electrons on atom i and $<s_i s_i | s_j s_j>$ is a two center two electron repulsion integral.

The electronic energy (see equation 23) can be expressed as

$$E_{elect} = \frac{1}{2} \sum_{i=1} \sum_{j=1} P_{ij}\left(H_{ij} + F_{ij}\right) \tag{27}$$

Where P_{ij} is the density matrix, H_{ij} is the one-electron matrix and F_{ij} is the Fock matrix. The diagonal Fock matrix elements are given by,

$$F_{\mu\mu} = U_{\mu\mu} + \sum_{B\neq A} V_{\mu\mu,B} + \sum_v^A P_{vv}\left[(\mu\mu|vv) - \frac{1}{2}(\mu v|\mu v)\right] + \sum_{B\neq A} \sum_\lambda^B \sum_\sigma^B P_{\lambda\sigma}(\mu\mu|\lambda\sigma) \tag{28}$$

where μ is centered on atom A. If μ and v are both on atom A the off diagonal elements are,

$$F_{\mu v} = \sum_{B\neq A} V_{\mu v,B} + \frac{1}{2} P_{\mu v}\left[3(\mu v|\mu v) - (\mu\mu|vv)\right] + \sum_{B\neq A} \sum_\lambda^B \sum_\sigma^B P_{\lambda\sigma}(\mu v|\lambda\sigma) \tag{29}$$

While if μ is on A and v is on B the off diagonal elements take the form,

$$\mathbf{F}_{\mu\nu} = \frac{1}{2}\left(\beta_\mu + \beta_\nu\right)S_{\mu\nu} - \frac{1}{2}\sum_\lambda^A \sum_\sigma^B \left(\mu\lambda|\nu\sigma\right) \tag{30}$$

In equations 28 and 29 β_ν and β_μ are semiempirical parameters, $<\mu\nu|\lambda\sigma>$ represents the two electron two center integrals and $S_{\mu\nu}$ is the overlap of Slater type orbitals of the form,

$$\phi_\mu = Nr^{n-1}e^{-r\xi}Y_l^m(\theta,\phi) \tag{31}$$

Here ξ is an AO exponent which can be labeled as ξ_s or ξ_p, depending on whether it corresponds to a s-type or p-type AO.

In PM3 18 parameters are defined for each heavy atom (*i.e.* C, O, N, *etc.*) while hydrogen has only 11 because the p AO parameters are unnecessary. The parameters for the heavy atoms are U_{ss}, U_{pp}, β_s, β_p, ξ_s, ξ_p, α_i, G_{ss}, G_{sp}, G_{pp}, G_{p2}, H_{sp}, a_{1i}, b_{1i}, c_{1i}, a_{2i}, b_{2i}, and c_{2i}. If the parameters are linearly scaled from the values appropriate for one element to those of another an interconversion of element types can be achieved (*i.e.* $V(\lambda) = \lambda V_A + (1-\lambda)V_B$, where V_A and V_B are the parameter type appropriate to system A and B).

One additional complication does arise if overlap integrals are required for an atom that is undergoing a change in principal quantum number (e.g., Zn^{+2} -> Cd^{+2}). Since the Slater orbitals depend on the principal quantum number the overlap integral must reflect the change in this parameter. This may be done rather abruptly using integer steps in discrete windows of the simulation, or it can be done gradually by calculating the overlap integral for each of the principal quantum numbers and then assuming a linear interpolation between the two values. An alternative method which would alleviate this problem altogether would be to imply a numerical quadrature scheme to estimate overlap integrals between pseudo-slater-type oribtals with non-integer quantum numbers. The linear interpolation is used here, however, because it is straightforward to apply within the existing framework of MOPAC.

Number of Atoms. Changing the number of atoms in a system is a special case of changing atomic identities. Atoms being deleted are perturbed into dummy particles for which all parameter values are zero. Typically, this technique would be combined with one or more of the methods described above as the number of orbitals and electrons is quite likely to also change when an atom is deleted.

Except for numerical instabilities, the non-physical nature of states arising from such a treatment is not a concern because the free energy which is being calculated in the simulation is a state function and, hence, it is independent of the path between the initial and final states. As long as the initial and final states remain unchanged the path between them will have no effect on the overall result.

Computational Procedure

The simulations in this study were done using the semiempirical MO method PM3(25, 26, 28) within the context of the QM/MM coupled potential recently implemented within our lab.(14) A modified version of MOPAC 5.0(32, 33), was used for the QM calculations and the driver MD program was a customized version of AMBER.(34) The PM3 Hamiltonian was selected because of its superior correlation with experimental heats of formation for the compounds studied herein.(25) From a methodological perspective, AM1,(29) or MNDO,(30) or any other LCAO based method could have been used.

The gas phase fractional electron method simulations were done with 0.5 ps of initial equilibration and 9.0 ps for the perturbation using a 0.1 fs time step. A very short timestep was employed to ensure an accurate integration of the equations of motions during the perturbation from one QM state to another. The simulations were done using a slow growth(35) methodology in which the perturbation parameter lambda (λ) controlled the number of electrons present in the system. Perturbations between element types and to dummy atoms were controlled through the direct scaling of the semiemperical parameters.(36) Each system was tightly coupled to a temperature bath which was maintained at 300K using the Berendsen method.(37) In addition, a holonomic constraint technique was employed to help maintain the geometry of the compounds as atoms were deleted.(38) Bonds which changed lengths between the starting state ($\lambda=0$) and the final state ($\lambda=1$) were coupled to λ and constrained to change smoothly been the appropriate gas-phase values during the simulation.

In the calculation of the solvation free energies the MM solute molecules where placed within an initial solvent box of dimension 21 x 21 x 21Å with approximately 275 TIP3P(39) water molecules. The simulations used constant pressure (1 atm.) and temperature (300K) conditions.(37) The time step used for the two systems studied here were different due to the presence of high frequency motions within the ammonium ion and methane which are not present in chloride or fluoride. A time step of 1.5 fs seconds was used in the perturbation of chloride to fluoride while a time step of 0.5 fs was used for the perturbation of ammonium ion to methane. In both cases SHAKE was used to constrain the bond lengths of the solvent molecules at their equilibrium values.(38) The resulting systems where then equilibrated for 30 ps and then free energy data was collected for an additional 60 ps using double wide sampling. To examine the hysteresis of the FEP simulation a second run was conducted using the end point of the initial 30 ps of equilibration to begin a second 30 ps of equilibration and then an additional 60 ps of double wide sampling. During the sampling phase for the chloride to fluoride ion interconversion eleven simulations were used to obtain ten free energy windows to complete the perturbation from $\lambda=0$ to $\lambda=1$. For methane to the ammonium ion twenty-one windows were used to affect the interconversion.

Another aspect of the calculations done in solvent is the treatment of long range interactions, which is qualitatively handled through the use of a Born correction.(40, 41) It is important to note that while the Born corrections are equivalent for two identically charged atoms (which are taken to be point charges), this will not be the case when the charge of the system is changed along with the atomic identity. Where this occurs the Born correction must be added to our calculated free energies. In the case of a point charge using a 9.0Å cutoff the Born correction is 18.2 kcal/mole. The value of the Born correction for a non-point ion is more difficult to evaluate. For example, when considering the ammonium ion the actual sphere in which solvent is "seen" by the solute is ~10Å in radius because of the 1Å N-H bond. For this size sphere a Born correction of 16.4 kcal/mole is calculated. However, the correct value to use is not obvious as pointed out in a similar study by Boudon and Wipff(42) where they used a point charge-derived Born correction for the ammonium ion.

The thermodynamic cycle used for the calculation of solvation free energies is given as Figure 1. Since the Gibb's free energy is a state function this cycle gives,

$$\Delta\Delta G_{gas}^{x\to y} + \Delta G_{solv}^{y} - \Delta\Delta G_{solv}^{x\to y} - \Delta G_{solv}^{x} = 0 \qquad (32)$$

Upon rearrangement $\Delta\Delta G_{solv}^{x\to y}$ can be calculated as:

$$\Delta\Delta G_{solv}^{x\to y} = \Delta G_{solv}^{y} - \Delta G_{solv}^{x} + \Delta\Delta G_{gas}^{x\to y} \qquad (33)$$

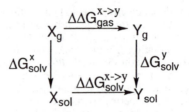

Figure 1: Thermodynamic cycle

As mentioned above when the charge of the system changes a Born correction must be added to $\Delta\Delta G_{solv}^{x\to y}$ to account for long range interactions. Consideration of this gives:

$$\Delta\Delta G_{solv}^{x\to y} = \Delta G_{solv}^{y} - \Delta G_{solv}^{x} + \Delta\Delta G_{gas}^{x\to y} - \Delta\Delta G_{born} \qquad (34)$$

Results

The results of the first series of calculations done to study perturbations between atoms types are summarized in Table 1. The first involved the perturbation of a chloride ion into a fluoride ion in the presence of MM water. The $\Delta\Delta G_{solv}^{Cl^-\to F^-}$ was calculated to be -4.8±0.6 kcal/mole through the perturbation of the QM atoms within our methodology. Equation 32 can be used to compare this calculated number with an estimated experimental value. The free energy of solvation for chloride has been reported experimentally to be -75(43) -75.8(44) and -82.9 kcal/mole(45) while the value for fluoride has been reported to be -103.8(44), -105(43) and -112.8 kcal/mole(45). The value of $\Delta\Delta G_{gas}^{Cl^-\to F^-}$ is the energetic difference in the gas phase between the two atomic species or simply the difference between their heats of formation (*i.e.* $\Delta H_f^{Cl^-} = $ -51.2 kcal/mole, $\Delta H_f^{F^-} = $ -31.2 kcal/mole and $\Delta\Delta G_{gas}^{Cl^-\to F^-} = $ 20.0 kcal/mol in PM3). This gives a $\Delta\Delta G_{solv}^{Cl^-\to F^-}$ (from equation 13) of between -8.0 and -10.0 kcal/mole (average value of -9.3 kcal/mol) depending on the set of experimental solvation free energies chosen. In this case we have only evaluated $\Delta\Delta G_{solv}^{Cl^-\to F^-}$ using experimental solvation free energies obtained from the same source.(43-45) If the solvation free energy values obtained from the various sources (using the same standard state) were mixed a larger range of values for $\Delta\Delta G_{solv}^{Cl^-\to F^-}$ would result. The experimental value for $\Delta\Delta G_{solv}^{Cl^-\to F^-}$ results is in reasonable accord with our calculated value of -4.8±0.6 kcal/mole. The difference between the estimated average experimental value and the calculated $\Delta\Delta G_{solv}^{Cl^-\to F^-}$ can be improved by deriving Lennard-Jones parameters that are suitable for coupled potential simulations of chloride and fluoride.(46)

Table 1: Free Energies.[a]

quantity	Cl⁻ to F⁻	NH₄⁺ to CH₄
$\Delta\Delta G_{solv}^{x\to y}(calc)$	-4.8±0.6	-94.6±1.5
$\Delta\Delta G_{solv}^{x\to y}(exp)$[b]	-8.0- -10.0	-102.8- -111
$\Delta\Delta G_{gas}^{x\to y}$	20.0	-166.4

a) All energies are given in kcal/mole.
b) Estimated from equations 13 and 14.

The second system studied involved the perturbation of the ammonium ion into methane in aqueous solution, and the results for these simulations are summarized in Table 1. The number of valence electrons in this perturbation is maintained even though the overall charge changes. The calculated $\Delta\Delta G_{solv}^{NH_4^+ \to CH_4}$ was -94.6±1.5 kcal/mole. For comparison an estimate of the experimental value for $\Delta\Delta G_{solv}^{NH_4^+ \to CH_4}$ can be calculated using equation 33. In this equation $\Delta\Delta G_{gas}^{NH_4^+ \to CH_4}$ can again be taken from the difference of the heats of formation for the two compounds in the gas phase ($\Delta H_f^{NH_4^+}$=153.4 kcal/mole and $\Delta H_f^{CH_4}$=-13.0 kcal/mole) as evaluated by PM3.(26) This results in a $\Delta\Delta G_{gas}^{NH_4^+ \to CH_4}$ value of -166.4 kcal/mole, which was additionally verified through a gas phase free energy simulation as a check of our methodology. The experimental $\Delta G_{solv}^{CH_4}$ is 1.93(47) kcal/mole while $\Delta G_{solv}^{NH_4^+}$ has been reported to be -69.8(45), -74.8(48), and -78 kcal/mole(49). Using equation 33 and the values for $\Delta G_{solv}^{CH_4}$, $\Delta G_{solv}^{NH_4^+}$, $\Delta\Delta G_{gas}^{NH_4^+ \to CH_4}$ and a Born correction of 16.4 kcal/mole we estimate that $\Delta\Delta G_{solv}^{NH_4^+ \to CH_4}$ is between -102.8 to -111 kcal/mole. Our calculated value (-94.6±1.5 kcal/mol) is several kcal/mole more positive than the estimated experimental value. However, modification of the Lennard-Jones parameters for CH_4 and NH_4^+ used in the coupled potential calculations should bring the calculated results into better agreement with the estimated experimental value.

To test the accuracy of the fractional electron method, as well as our ability to account for changes in the number of orbitals in a system three perturbations were carried out. These were the interconversion of ethane to methane, methanol to methane and methane to fluoromethane. The results of these simulations are given in Table 2. How to compare our calculated results to that determined using other approaches is complicated. We could compare to free energies calculated using the appropriate partition functions(50) : however, since we are restraining the molecule in the FEP simulation, the vibrational, rotational and translational contributions are all affected. Given this uncertainty we have presented free energy differences using the partition function approach(50) with all terms considered and with only the vibrational contribution considered. In Table 2 we also present the calculated gas-phase heat of formation difference between the two molecules.

Table 2: Free Energies[a]

	$\Delta G_{gas}^{x \to y}$ (FEP)	$\Delta G_{gas}^{x \to y}$ (partition)[b]	$\Delta G_{gas}^{x \to y}$ (partition)[c]	$\Delta\Delta H_f^{x \to y}$ (gas)
$C_2H_6 \to CH_4$	5.3±01	8.2	5.8	5.1
$CH_3OH \to CH_4$	35.0±0.1	42.6	39.4	39.0
$CH_4 \to CH_3F$	-44.5±0.1	-43.4	-40.8	-40.8

a) All Energies are gin in kcal/mol.
b) Calculated from the vibrational frequencies and the appropriate partition functions (see ref 50).
c) Calculated from the vibration frequencies and the appropriate partition functions (see ref 50) but rotational and translational contributions have been removed.

The free energy difference calculated from the gradual perturbation of ethane to methane in the gas phase was 5.3 kcal/mol. This is in reasonable agreement with the heat of formation difference as well as the partition function approach where the rotational and translational terms have been neglected. The agreement with the full partition function approach is not as good. In the case of the conversion of methane to fluoromethane, the FEP and full partition functions methods were in closer agreement with the calculated free energy changes of -44.5 and -43.4 kcal/mol respectively. However, the FEP value was ~4 kcal/mol higher than the value given by the heat of formation and partial partition function differences. The conversion of ethane to methanol was the last of the test perturbations examined. In this system our FEP value was 35 kcal/mol, while the other calculated vales were 4-7 kcal/mol higher. Overall, the agreement between the various types of calculated values was reasonable and it is clear that the FEP approach is giving results that are comparable to energy differences computed in other ways. Thus, if an accurate QM model is employed, the fractional electron approach should be a reliable method for estimating free energy differences between two molecules. Disagreement between the FEP and partition function energies can be associated with several factors (time-scales, use of MD constraints, etc.) but further study is needed to pinpoint problem areas.

Conclusions

We have demonstrated methods for allowing the direct interconversion of QM systems in FEP calculations. With further refinements of the simulation protocol we believe that these methods will develop into a powerful tool for the accurate calculation of differences in free energies. Only in the past few years has the calculation of extensive QM-MD trajectories and QM-FEP simulations become computationally feasible. It is for this reason that the field remains primarily uninvestigated and in need of significant development and testing. The next decade should bring rapid progress in QM-MD studies on large systems, and through the development of QM-FEP methodologies we will also be able to calculate free energies from condensed phase QM and QM/MM simulations.

Acknowledgments

We would like to thank the NIH (GM-29072) for their generous support of this research. The Pittsburgh Supercomputer Center and The Cornell Theory Center through a MetaCenter grant are acknowledged for generous allocations of computer time.

Literature Cited

1. M. Mezei and D. L. Beveridge *Ann. NY. Acad. Sci.* **1986**, *482*, 1-23.
2. W. L. Jorgensen *Acc. Chem. Res.* **1989**, *22*, 184-189.
3. T. P. Lybrand, I. Ghosh and J. A. McCammon *J. Am. Chem. Soc.* **1985**, *107*, 7793-7794.
4. R. Elber *J. Chem. Phys.* **1990**, *93*, 4312-4321.
5. J. van Eerden, W. J. Briels, S. Harkema and D. Feil *Chem. Phys. Lett.* **1989**, *164*, 370-376.
6. T. J. Marrone and K. M. Merz Jr. *J. Am. Chem. Soc.* **1992**, *114*, 7542.
7. K. M. Merz Jr. and P. A. Kollman *J. Am. Chem. Soc.* **1989**, *111*, 5649-5658.
8. J. Åqvist, M. Carmen and J. Samuelsson *Prot. Eng.* **1994**, *7*, 385-391.
9. P. Cieplak and P. Kollman *J. Chem. Phys.* **1990**, *92*, 6761.
10. J. Åqvist *J. Phys. Chem.* **1990**, *94*, 8021.
11. P. Bash, M. Fields and M. Karplus *J. Am. Chem. Soc.* **1987**, *109*, 8092-8094.
12. L. X. Dang and P. A. Kollman *J. Am . Chem. Soc.* **1990**, *112*, 5716-5720.
13. P. D. J. Grootenhuis and P. A. Kollman *J. Am. Chem. Soc.* **1989**, *111*, 4046-4051.

14. D. Hartsough and K. M. J. Merz *J. Phys. Chem.* **1995**, *99*, 384-390.
15. R. Car and M. Parrinello *Phys. Rev. Lett.* **1985**, *55*, 2471.
16. M. J. Field *J. Phys. Chem.* **1991**, *95*, 5104-5108.
17. B. Hartke and E. A. Carter *J. Chem. Phys.* **1992**, *97*, 6569-6578.
18. X. G. Zhao, C. S. Cramer, B. Wwiner and M. Frenklach *J. Phys. Chem.* **1993**, *97*, 1639-1648.
19. J. Gao *J. Phys. Chem.* **1992**, *96*, 6432-6439.
20. M. J. Field, P. A. Bash and M. Karplus *J. Comput. Chem.* **1990**, *11*, 700-733.
21. R. V. Stanton, D. S. Hartsough and J. K. M. Merz *J. Comput. Chem.* **1995**, *16*, 113-128.
22. U. C. Singh and P. A. Kollman *J. Comput. Chem.* **1986**, *7*, 718-730.
23. D. A. Pearlman *J. Phys. Chem.* **1994**, *98*, 1487-1493.
24. T. Wesolowski and A. Warshel *J. Phys. Chem* **1994**, *98*, 5183-5187.
25. J. J. P. Stewart *J. Comput. Chem.* **1989**, *10*, 209-220.
26. J. J. P. Stewart *J. Comput. Chem.* **1989**, *10*, 221-264.
27. M. J. S. Dewar, J. A. Hashmall and C. G. Venier *J. Am. Chem. Soc.* **1968**, *90*, 1953-1957.
28. J. J. P. Stewart *J. Comput. Chem.* **1991**, *12*, 320-341.
29. M. J. S. Dewar, E. G. Zoebisch, E. F. Healy and J. J. P. Stewart *J. Am. Chem. Soc.* **1985**, *107*, 3902-3909.
30. M. J. S. Dewar and W. Thiel *J. Am. Chem. Soc.* **1977**, *99*, 4899-4907.
31. M. J. S. Dewar and W. Thiel *J. Am. Chem. Soc.* **1977**, *99*, 4907-4917.
32. B. H. Besler, K. M. J. Merz and P. A. Kollman *J. Comput. Chem.* **1990**, *11*, 431-439.
33. K. M. Merz Jr. and B. H. Besler *QCPE Bull.* **1990**, *10*, 15.
34. D. A. Pearlman, D. A. Case, J. C. Caldwell, G. L. Seibel, U. C. Singh, P. Weiner and P. A. Kollman In University of California, San Francisco: 1991; pp .
35. P. A. Kollman *Chem. Rev.* **1993**, *93*, 2395-2417.
36. R. V. Stanton, L. R. Little and K. M. Merz *J. Phys. Chem.* **1995**, *99*, 483-486.
37. H. J. C. Berendsen, J. P. M. Potsma, W. F. van Gunsteren, A. D. DiNola and J. R. Haak *J. Chem. Phys.* **1984**, *81*, 3684-3690.
38. W. F. van Gunsteren and H. J. C. Berendsen *Mol. Phys.* **1977**, *34*, 1311.
39. W. L. Jorgensen, J. Chandrasekhar, J. Madura, R. W. Impey and M. L. Klein *J. Chem. Phys.* **1983**, *79*, 926.
40. M. Born *Z. Phys.* **1920**, *1*, 45.
41. T. P. Straatsma and H. J. C. Berendsen *J. Chem. Phys.* **1988**, *89*, 5876.
42. S. Boudon and G. Wipff *J. Comput. Chem.* **1991**, *12*, 42.
43. R. G. Pearson *J. Am. Chem. Soc.* **1986**, *108*, 6109-6114.
44. H. L. Friedman and C. V. Krishnan In *Water a Comprehensive Treatise*; F. Franks, Ed.; Plenum Press: New York, 1973; Vol. 3; pp 1-118. Thermodynamics of Ion Hydration
45. Y. Marcus *Ion Solvation*; John Wiley: New York, 1985.
46. R. V. Stanton, D. S. Hartsough and K. M. Merz *J. Phys. Chem.* **1993**, *97*, 11868-11870.
47. A. Ben-Naim and Y. Marcus *J. Chem. Phys.* **1984**, *81*, 2016-2027.
48. G. P. Ford and J. D. Scribner *J. Org. Chem.* **1983**, *48*, 2226-2233.
49. D. H. Aue, H. M. Webb and M. T. Bowers *J. Am. Chem. Soc.* **1976**, *98*, 318-329.
50. W. J. Hehre, L. Radom, P. R. van Schleyer and J. A. Pople *Ab Initio Molecular Orbital Theory*; John Wiley: New York, 1986.

Chapter 11

Analytic Second Derivatives of Molecular Energies

Density-Functional Implementation of Perturbations Due to Nuclear Displacements

Heiko Jacobsen, Attila Bérces, David P. Swerhone, and Tom Ziegler[1]

**Department of Chemistry, University of Calgary,
2500 University Drive Northwest, Calgary, Alberta T2N 1N4, Canada**

We report an implementation of analytic second derivatives with respect to nuclear displacements, based on density functional theory within in the Kohn-Sham formalism. The implementation includes the solution of the coupled perturbed Kohn-Sham equations. Our approach is in line with the Amsterdam Density Functional package ADF, and includes the use of numerical integration, density fit as well as the frozen core approximation. The efficiency of the algorithm is tested in comparison with finite difference methods.

1. Introduction

A number of spectroscopic properties can be formulated in terms of second order derivatives of the total electronic energy with respect to two perturbational parameters. When both perturbations are represented by nuclear displacements, the double derivative yields harmonic force fields, which are important in vibrational analyses as well as in molecular dynamics. Compared to finite difference methods, analytical derivatives are superior both with respect to the computational effort as well as the accuracy of the results.

During the last 15 years, the advances in density functional theory (DFT) (1) have made this method very successful in applications to problems in molecular structures and dynamics. Thus, practical DFT implementations of analytic second derivatives have recently been reported by Handy and co-workers in the CADPAC program (2), by Johnson and Frisch in the GAUSSIAN program (3), and by Fitzgerald and co-workers in the DGAUSS program (4).

All these previous implementations have features unique to the way in which the Kohn-Sham (KS) equations (5) are solved in the respective program system. We have reported the implementation of analytic second derivatives into the Amsterdam Density Functional package ADF (6,7). The ADF program system (8,9) has three distinguishing features, namely the use of density fit, the evaluation of most matrix elements by numerical integration as well as the application of the frozen core approximation. A minor deviation is in addition the use of Slater type orbitals (STO) as basis functions, compared to the more commonly used Gaussian type orbitals (GTO).

[1]Corresponding author

0097–6156/96/0629–0154$15.00/0

In this paper, we shall present a short outline of the ADF implementation of the analytical second derivatives. Major emphasis will be given to the special features of the ADF program package. The performance of our algorithm will be judged in a comparison with the finite difference ADF approach.

2. The ADF Implementation of the Kohn-Sham Equations

Since it is not our purpose to review DFT within the KS-formalism, we only present a concise description of this method. For a detailed account, we refer the reader to the literature (1). However, we will focus on the special features of ADF in connection with perturbations due to nuclear displacements. Our notation is such that derivatives are indicated by superscripts in brackets or parentheses.

In ADF, the set of KS-orbitals $\{\psi_i(1); i = 1,n\}$ is determined in such a way that the corresponding KS energy is minimized. This condition is satisfied if the functions are solutions to the following set of spin-unrestricted one-electron KS-equations:

$$h^\gamma(1)\psi_i(1) = \varepsilon_i\psi_i(1) \quad ; \quad \gamma = \alpha, \beta \tag{1}$$

under the condition

$$\int \psi_i^*(1)\psi_j(1)\delta\tau_1 = \delta_{ij} \tag{2}.$$

The one-electron KS operator $h^\gamma(1)$ is given by

$$h^\gamma(1) = T + \sum_A^N V_N^A(1) + V_C(1) + V_{XC}^\gamma \quad ; \quad \gamma = \alpha, \beta \tag{3},$$

where the γ superscript refers to α or β spin, respectively. The terms on the right hand side of equation 3 are the kinetic energy operator, the nucleus-electron attraction operator, the electrostatic potential from the total electronic density, and the exchange correlation potential defined as the functional derivative of the exchange correlation energy with respect to the density.

The KS-orbitals are expanded in terms of STOs as

$$\psi_i = \sum_{\mu=1}^{2M_T} d_{\mu i}\chi_\mu \tag{4}.$$

For the sake of convenience, we write Ω instead of $\Omega(1), \Omega$ being any molecular orbital, basis function or operator, whenever the context remains clear.

2.1 Frozen Core Approximation. Making the assumption that the orbitals describing the inner-shell electrons are unperturbed in going from the free atom to the molecular environment, only the valence electrons have to be treated explicitly in a variational calculation. The KS-orbitals are now expressed by a linear combination of the valence basis set $\{\lambda_\mu\}$

$$\psi_i = \sum_{\mu=1}^{2M} c_{\mu i}\lambda_\mu \tag{5}.$$

However, one has to ensure orthogonality between all molecular orbitals, that is between valence as well as core orbitals. To achieve this, the valence basis set is written as

$$\lambda_\mu = \chi_\mu^{val} + \sum_\tau^{2M_c} b_{\mu\tau}\chi_\tau^{core} \tag{6},$$

where $\{\chi_\mu^{val}\}$ are STOs describing the valence electrons, and $\{\chi_\tau^{core}\}$ is a set of primitive auxiliary core STO functions, one core function for each core orbital kept frozen. The coefficients $b_{\mu\tau}$ are determined under the constrain that all molecular valence orbitals must be orthogonal to all true core orbitals $\{\omega_\tau^{core}; \tau = 1, 2M_c\}$. This requirement is met by making the valence basis set $\{\lambda_\mu; \mu = 1, 2M\}$ orthogonal to the true core:

$$\int \lambda_\mu \omega_\tau^{core} \delta\tau_1 = 0 \tag{7}$$

From equation 7, the coefficients $b_{\mu\tau}$ can be evaluated as

$$\mathbf{B} = -\mathbf{RS}_{core}^{-1} \tag{8a},$$

where the matrix elements are defined as follows:

$$(\mathbf{B})_{\mu\tau} = b_{\mu\tau} \tag{8b}$$

$$(\mathbf{R})_{\mu\tau} = \int \chi_\mu^{val} \omega_\tau^{core} \delta\tau_1 \tag{8c}$$

$$(\mathbf{S}_{core})_{\tau'\tau} = \int \chi_{\tau'}^{core} \omega_\tau^{core} \delta\tau_1 \tag{8d}.$$

It will be necessary to evaluate the derivatives of the core orthogonalization matrix. The first derivative affords

$$\mathbf{B}^{(X_A)} = -\mathbf{R}^{(X_A)}\mathbf{S}_{core}^{-1} - \mathbf{R}(\mathbf{S}_{core}^{-1})^{(X_A)} \tag{9}.$$

The $\mathbf{S}_{core}$ matrix depends on core functions only; therefore overlap elements between functions on different centers become small enough to be neglected. The remaining one center overlap integrals are independent of the molecular structure, and as a consequence the second term of equation 9 can be discarded. Thus,

$$\mathbf{B}^{(X_A)} = -\mathbf{R}^{(X_A)}\mathbf{S}_{core}^{-1} \tag{10}.$$

A similar argument affords for the second derivative

$$\mathbf{B}^{(X_A)(Y_B)} = -\mathbf{R}^{(X_A)(Y_B)}\mathbf{S}_{core}^{-1} \tag{11}.$$

The second derivatives of the $\mathbf{R}$ matrix-elements can readily be obtained from equation 8c.

2.2 Numerical Integration and Density Fit. With the expansion of equation 5 the set of one-electron KS-equations take the form

$$\sum_{\gamma=1}^{2M} [F_{\tau\gamma} - \varepsilon_i(S_{(L)})_{\tau\gamma}]c_{\gamma i} = 0 \quad ; \quad \tau = 1, 2M \tag{12},$$

where the KS-elements, $F_{\tau\gamma}$, and the overlap integral, $(S_{(L)})_{\tau\gamma}$, are defined as

$$F_{\tau\gamma} = \int \lambda_\tau^*(1) h^\gamma(1) \lambda_\gamma(1) \delta\tau_1 \tag{13a}$$

$$(S_{(L)})_{\tau\gamma} = \int \lambda_\tau^*(1) \lambda_\gamma(1) \delta\tau_1 \tag{13b}.$$

The matrix elements $F_{\tau\gamma}$ are calculated in the ADF program by numerical integration as

$$F_{\tau\gamma} = \sum_{k=1}^{N_s} W(\mathbf{r}_k) \lambda_\tau(\mathbf{r}_k) h^\gamma(\mathbf{r}_k) \lambda_\gamma(\mathbf{r}_k) \tag{14},$$

where $W(\mathbf{r}_k)$ is a weight factor associated with each of the N_s integration points, $\mathbf{r}_k$. The evaluation of $F_{\mu\nu}$ requires a precalculation of the electronic potential V_C at each integration point $\mathbf{r}_k$, where V_C is given in terms of the electronic density $\rho(\mathbf{r})$ as

$$V_C(\mathbf{r}_k) = \int \rho(\mathbf{r}_2)/|\mathbf{r}_k - \mathbf{r}_2|\delta\tau_2 \tag{15}.$$

The electrostatic potential $V_C(\mathbf{r}_k)$ is in the ADF program determined by expanding $\rho(\mathbf{r}_k)$ as

$$\rho(\mathbf{r}_k) \approx \tilde{\rho}(\mathbf{r}_k) = \sum_{u=1}^{M_f} a_u f_u(\mathbf{r}_k) \tag{16}.$$

Here, $\{f_u(\mathbf{r}_k)\}$ is a set of M_f auxiliary STO fit functions centered on the different nuclei, and $\{a_u\}$ is a set of coefficients obtained from a least square fit of $\rho(\mathbf{r}_k)$ by $\{f_u(\mathbf{r}_k)\}$. The potential $V_C(\mathbf{r}_k)$ now has the form

$$V_C(\mathbf{r}_k) \approx \tilde{V}_C(\mathbf{r}_k) = \sum_{u=1}^{M_f} a_u \int f_u(\mathbf{r}_1)/|\mathbf{r}_k - \mathbf{r}_1|\delta\tau_1 \tag{17},$$

and can be obtained analytically for each sample point r_k as a relatively small sum of integrals, each involving only a single STO. In contrary, an analytical calculation of $V_c(r_k)$ from equation 15 would have involved a larger number of more cumbersome integrals with two STOs on different centers.

The fit coefficients a_u of equations 16 and 17 can be calculated as

$$a = S_{(f)}^{-1} t + \lambda_{(f)} S_{(f)}^{-1} n \qquad (18),$$

where the terms are defined as:

$$(S_{(f)})_{ij} = \int f_i^* f_j \, \delta\tau \qquad (19a)$$

$$n_i = \int f_i \delta\tau \qquad (19b)$$

$$t_i = \int \rho f_i \delta\tau \qquad (19c)$$

$$\lambda_{(f)} = \frac{N - n^+ S_{(f)}^{-1} t}{n^+ S_{(f)}^{-1} n} \qquad (19d)$$

$$N = \int \rho \delta\tau \qquad (19e).$$

The derivative of V_C with respect to a nuclear displacement writes

$$\tilde{V}_C^{(X_A)} = \sum_{u=1}^{M_f} a_u^{(X_A)} \int f_u(r_1)/|r_k - r_1| \delta\tau_1 + \sum_{u=1}^{M_f} a_u \int f_u^{(X_A)}(r_1)/|r_k - r_1| \delta\tau_1 \qquad (20).$$

Thus, the calculation of $\tilde{V}_C^{(X_A)}$ requires the derivatives of the fit coefficients a_u as well as the derivatives of the potential due to each fit function. The derivatives of a_u can be obtained from equation 18 as

$$a_u^{(X_A)} = (S_{(f)}^{-1})^{(X_A)} t + S_{(f)}^{-1} t^{(X_A)} + \lambda^{(X_A)} S_{(f)}^{-1} n + \lambda (S_{(f)}^{-1})^{(X_A)} n + \lambda S_{(f)}^{-1} n^{(X_A)} \qquad (21)$$

Simple but lengthy manipulations afford the expression

$$a^{(X_A)} = A w + \frac{N^{(X_A)}}{n + S_{(f)}^{-1} n} S_{(f)}^{-1} n \qquad (22)$$

with

$$A = S_{(f)}^{-1} - \frac{S_{(f)}^{-1} n n^+ S_{(f)}^{-1}}{n^+ S_{(f)}^{-1} n} \qquad (23a)$$

$$N^{(X_A)} = \int \rho^{(X_A)}(r) dr \qquad (23b)$$

$$w = t^{(X_A)} - S^{(X_A)} a$$

$$= \int [\rho(r) - a \cdot f(r)] f^{(X_A)}(r) dr + \int [\rho^{(X_A)}(r) - a \cdot f^{(X_A)}(r)] f(r) dr \qquad (23c).$$

The vector w is calculated by a two-center numerical integration scheme using elliptic prolate spheroidal coordinates. The remaining terms in equation 23 are calculated analytically.

3. The Coupled Perturbed Kohn-Sham Equations

Let us consider the first order change in the electron density with respect to the displacement of the nuclear coordinate X_A from a molecular reference conformation characterized by y_0. With the expansion of the KS-orbitals according to equation 5, we get for $\rho^{(X_A)}$

$$\rho^{(X_A)} = \rho^{(X_A)} + \rho^{(B)} + \rho^{(C)}$$

$$= \sum_i^n \left\{ \psi_i^{*(X_A)} \psi_i + \psi_i^* \psi_i^{(X_A)} \right\} + \sum_i^n \left\{ \psi_i^{*(B)} \psi_i + \psi_i^* \psi_i^{(B)} \right\} \qquad (24),$$

$$+ \sum_i^n \left\{ \psi_i^{*(C)} \psi_i + \psi_i^* \psi_i^{(C)} \right\}$$

where all orbital terms are evaluated at the reference geometry y_0. The terms $\psi_i^{(X_A)}$ are readily available from the analytical derivatives of STOs, and the terms $\psi_i^{(B)}$ can be constructed from the derivative of the core orthogonalization matrix $\mathbf{B}$, equation 10. The terms $\psi_i^{(C)}$ require an evaluation of $c_{\mu i}^{(X_A)}$, which involves a solution of the coupled perturbed KS (CPKS) equations. In our treatment outlined here, we follow the original ideas of Gerrat and Mills (10) as well as Pople and co-workers (11).

We assume that the KS-equations, equation 12, have been solved in the unperturbed case characterized by y_0. The corresponding solutions are given by

$$\psi_q(y_o) = \sum_\mu^{2M} c_{\mu q}(y_o) \lambda_\mu(y_o) \quad ; \quad q = 1,2M \qquad (25a).$$

Solutions of the form

$$\psi_q(y) = \sum_\mu^{2M} c_{\mu q}(y) \lambda_\mu(y) \quad ; \quad q = 1,2M \qquad (25b)$$

are now sought in the vicinity of y_0 at y, where $y - y_0$ in our case represents a displacement of the nuclear coordinate X_A. It is expedient to introduce the auxiliary basis

$$\varphi_q(y) = \sum_\mu^{2M} c_{\mu q}(y_0) \lambda_\mu(y) \qquad (25c),$$

and express $\psi_p(y)$ as

$$\psi_p(y) = \sum_q^{2M} u_{qp}(y) \varphi_q(y) \qquad (25d).$$

The problem is now reduced to finding the coefficients u_{pq} which are related to $c_{\mu p}(y)$ by

$$C_{\mu p}(y) = \sum_q^{2M} C_{\mu q}(y_0) u_{qp}(y) \qquad (25e).$$

The relevant elements from the $\mathbf{u}$ matrix are given as (10,11)

$$\mathbf{u}_{qp}^{(1)} = \frac{F_{qp}^{(1)} - (S_{(a)}^{(1)})_{qp} \varepsilon_p^{(0)}}{\varepsilon_p^{(0)} - \varepsilon_q^{(0)}} \qquad (26a),$$

where

$$(S_{(a)}^{(1)})_{qp} = \frac{\delta}{\delta y} \langle \varphi_q | \varphi_p \rangle_{(y=y_o)} \qquad (26b)$$

and

$$F_{qp}^{(1)} = \frac{\delta}{\delta y} \langle \varphi_q | h(y) | \varphi_p \rangle_{(y=y_o)} \qquad (26c).$$

In equation 26, $\varepsilon_p^{(0)}$ is the orbital energy of equation 12 for the unperturbed system. The relation in equation 25e now allows for the calculation of $\rho^{(C)}$:

$$\rho^{(C)} = \rho^{(S_{(a)}^{(1)})} + \rho^{(u_{ai}^{(1)})}$$

$$= \sum_{i}^{occ} \sum_{j}^{occ} \left[u_{ji}^{(1)^*} \varphi_j^*(y_o)\varphi_i(y_o) + u_{ji}^{(1)} \varphi_j(y_o)\varphi_i^*(y_o) \right] \tag{27}.$$

$$+ \sum_{i}^{occ} \sum_{a}^{unocc} \left[u_{ai}^{(1)^*} \varphi_a^*(y_o)\varphi_i(y_o) + u_{ai}^{(1)} \varphi_a(y_o)\varphi_i^*(y_o) \right]$$

It remains to evaluate matrix elements of the type $(F^{(1)})_{ai}$ and $(S_{(a)}^{(1)})_{pi}$. The evaluation of the dependencies through derivatives of the auxiliary basis is somewhat evolved, but straightforward. We will however comment on the derivative of the KS-operator, which is needed in the evaluation of $(F^{(1)})_{ai}$.

$$h^{\gamma(X_A)} = \sum_{A}^{N} V_N^{(X_A)} + V_C^{(X_A)} + \left(V_{XC}^\gamma \right)^{(X_A)} \cdot \rho^{\gamma(X_A)} \quad ; \quad \gamma = \alpha, \beta \tag{28}$$

The derivative of the electron density has been addressed in equation 24. Further, the terms $V_N^{(X_A)}$ and $(V_{XC}^\gamma)^{(X_A)}$ are readily obtained from the analytical form of V_N^A and V_{XC}^γ, respectively. Their evaluation will not further be discussed here. The term $V_C^{(X_A)}$ can be broken down as

$$V_C^{(X_A)} = V_C^{(X_A)} + V_C^{(B)} + V_C^{(c)} \tag{29a}.$$

Using a density fit as described in section **2.2**, we can evaluate $V_C^{(X_A)}$ in each integration point as follows:

$$V_C^{(X_A)}(\mathbf{r}_k) \approx \tilde{V}_C^{(X_A)}(\mathbf{r}_k) = \tilde{V}_C^{(X_A)} + \int \frac{\tilde{\rho}^{(B)}(\mathbf{r}_1)}{|\mathbf{r}_k - \mathbf{r}_1|} d\mathbf{r}_1 + \int \frac{\tilde{\rho}^{(C)}(\mathbf{r}_1)}{|\mathbf{r}_k - \mathbf{r}_1|} d\mathbf{r}_1 \tag{29b}.$$

The last two terms in equation 29b can be obtained from a fit of $\rho^{(B)}$ and $\rho^{(C)}$, respectively. The derivative $\tilde{V}_C^{(X_A)}$ is given in equation 20.

We note that the matrix elements $(F^{(1)})_{ai}$ depend through $h^{\gamma(X_A)}$ on $\mathbf{u}^{(1)}$. Thus, $\mathbf{u}^{(1)}$ has to be determined in a self consistent procedure. The vector $\mathbf{u}^{(1)}$ can be partitioned as

$$\mathbf{u}^{(1)} = \mathbf{u}^0 + \mathbf{w}(\mathbf{u}) \tag{30}$$

In equation 30, the first term is independent of the solution $\mathbf{u}^{(1)}$, whereas the second term depends on $\mathbf{u}^{(1)}$. The solution to equation 30 is expressed as a linear combination of trial vectors

$$\mathbf{u} = \alpha^0 \mathbf{u}^0 + \alpha^1 \mathbf{u}^1 + \alpha^2 \mathbf{u}^2 + \ldots + \alpha^k \mathbf{u}^k \tag{31},$$

where the orthogonal vectors are defined as

$$\mathbf{u}^{n+1} = \mathbf{w}^n - \sum_{l=0}^{n} \frac{\langle \mathbf{u}^l | \mathbf{w}^n \rangle}{\langle \mathbf{u}^l | \mathbf{u}^l \rangle} \mathbf{u}^l \tag{32}.$$

The set of linear combination coefficients $\{\alpha^k\}$ can be determined by solving a set of linear equations obtained by multiplying equation 30 from the left by $\mathbf{u}^l$, and substituting 31 into 30. Typically four to six terms are sufficient to reach convergence.

4. Formulation of the Second Derivatives

Let us express the total electron density $\rho(1)$ as the superposition of spherical atomic densities $\rho^A(1)$ as well as the deformation density $\Delta\rho(1)$, representing the change in density on formation of the molecular system:

$$\rho(1) = \sum_{G}^{N} \rho^G(1) + \Delta\rho(1) \tag{33}.$$

Accordingly, we can write the total KS energy as (12)

$$E_{KS} = \sum_{G}^{N} E_{KS}^{G}(\rho^{G}) + \Delta T + \Delta V + \Delta E_{XC} + V_{Pair} \tag{34},$$

where the terms are defined as

$$\Delta T = \int_{1=1'} \Delta\rho(1,1')T\delta\tau_{1} \tag{35a}$$

$$\Delta V = \int \Delta\rho \left\{ \sum_{G}^{N} V_{N}^{G} + V_{C}^{G} + \frac{1}{2}\Delta V_{C} \right\} \delta\tau_{1} \tag{35b}$$

$$\Delta E_{XC} = E_{XC}(\Delta\rho^{\alpha}, \Delta\rho^{\beta}) \tag{35c}.$$

In equation 31, we have introduced the operator ΔV_C:

$$\Delta V_{C} = \int \Delta\rho(2)/|\mathbf{r}_{1} - \mathbf{r}_{2}|\delta\tau_{2} \tag{36}.$$

Further, V_{Pair} is the electrostatic interaction energy between the atom pair $\{AB\}$ in the combined molecule. Taking the first derivative with respect to a nuclear displacement X_A, we get

$$E_{KS}^{\{X_A\}} = \Delta T^{\{X_A\}} + \Delta V^{\{X_A\}} + \Delta E_{XC}^{\{X_A\}} + V_{Pair}^{\{X_A\}} \tag{37}.$$

The deformation density formalism ensures that, in equation 37, one center terms, which do not contribute to $E_{KS}^{\{X_A\}}$, are explicitly eliminated. This aspect is of major importance since numerical integration rather than analytical integration is used (13,14).

With the expansion of the KS-orbitals according to equation 5, we have for $E_{KS}^{\{X_A\}}$

$$\frac{dE_{KS}}{dX_{A}} = \sum_{\mu} \frac{\partial E_{KS}}{\partial\chi_{\mu}} \cdot \frac{\partial\chi_{\mu}}{\partial X_{A}} + \sum_{\mu}^{2M_{i}} \sum_{\tau}^{2M_{i}} \frac{\partial E_{KS}}{\partial b_{\mu\tau}} \cdot \frac{\partial b_{\mu\tau}}{\partial X_{A}} + \sum_{i} \sum_{\mu} \frac{\partial E_{KS}}{\partial c_{i\mu}} \cdot \frac{\partial c_{i\mu}}{\partial X_{A}} \tag{38a},$$

which we write as

$$E_{KS}^{\{X_A\}} = E_{KS}^{(X_A)} + E_{KS}^{B,X_A} + E_{KS}^{[X_A]} \tag{38b}.$$

The terms on the right hand side of equation 38b describe the explicit dependency of E_{KS} on X_A through the STO basis, the dependency through the derivative of the core orthogonalization matrix as well as the implicit dependency through the change in the expansion coefficients. It is possible (15,11) to express the implicit dependency as

$$E_{KS}^{[X_A]} = -2\sum_{i} \sum_{\mu} \left\langle \psi_{i}^{(X_A)} \middle| \varepsilon_{i}\psi_{i} \right\rangle = R^{(X_A)} \tag{39}.$$

Thus,

$$E_{KS}^{\{X_A\}} = \Delta T^{(X_A)} + \Delta V^{(X_A)} + \Delta E_{XC}^{(X_A)} + V_{Pair}^{(X_A)} + R^{(X_A)} + E_{KS}^{B,X_A} \tag{40}.$$

When taking a second derivative with respect to the nuclear coordinate Y_B, implicit derivatives can no longer be avoided:

$$E_{KS}^{\{X_A\}\{Y_B\}} = \left[E_{KS}^{(X_A)} + E_{KS}^{B,X_A} \right]^{(Y_B)} + \left[E_{KS}^{(X_A)} + E_{KS}^{B,X_A} \right]^{[Y_B]} \tag{41}$$

We shall discuss separately the two terms of equation 41.

4.1 Explicit contributions to the second derivatives. The term $\left[E_{KS}^{(X_A)} + E_{KS}^{B,X_A} \right]^{(Y_B)}$ contains of contributions due to dependencies through the STO basis as well as through the derivative of the KS-operator. The former can be evaluated by straight forward differentiation of the basis functions, whereas the latter can be evaluated as outlined in section 3. However, our implementation makes extensive use of the translational invariance (TI) condition:

$$\sum_{G}^{N} \frac{\partial}{\partial X_{G}} \int f(X_{A}, X_{B}, \cdots, X_{N})d\tau \stackrel{!}{=} 0 \tag{42}$$

The reformulation of certain integrals using TI serves two purposes (13,14). First, terms that although zero by symmetry give rise to spurious non-zero contribution when evaluated by numerical integration, can explicitly be eliminated. Second, the derivative of certain operators can be avoided. As an example, we discuss the evaluation of the term ΔT. The first derivative affords

$$\Delta T^{(X_A)} = 2\sum_i^n \left\langle \psi_i^{(X_A)} \middle| T \middle| \psi_i \right\rangle \tag{43a}.$$

We now have chosen our orbitals to be real, and we will continue to do so throughout the remainder of this paper. Using the TI condition, we can express $\Delta T^{(X_A)}$ as

$$\Delta T^{(X_A)} = \sum_i^n \left\{ \left\langle \psi_i^{(X_A)} \middle| T \middle| \sum_G \psi_i^G \right\rangle - \left\langle \sum_G \psi_i^{(X_G)} \middle| T \middle| \psi_i^A \right\rangle \right\} \tag{43b},$$

where we have explicitly eliminated all one center integrals in the difference to the right. Taking the second derivative, we get

$$\Delta T^{(X_A)(Y_B)} = \sum_i^n \left\{ \left\langle \psi_i^{(X_A)(Y_A)} \middle| T \middle| \sum_G \psi_i^G \right\rangle_{A=B} + \left\langle \psi_i^{(X_A)} \middle| T \middle| \psi_i^{(Y_B)} \right\rangle \right. \tag{44a}.$$
$$\left. - \left\langle \sum_G \psi_i^{(X_G)} \middle| T \middle| \psi_i^{(Y_B)} \right\rangle_{A=B} - \left\langle \psi_i^{(X_B)(Y_B)} \middle| T \middle| \psi_i^A \right\rangle \right\}$$

In equation 44a, the second and third term would require the evaluation of the Laplacian of derivatives of molecular orbitals, $\nabla^2(\partial \psi_i / \partial X)$. To avoid this, we rewrite these terms using TI, and obtain

$$\Delta T^{(X_A)(Y_B)} = \sum_i^n \left\{ \left\langle \psi_i^{(X_A)(Y_A)} \middle| T \middle| \sum_G \psi_i^G \right\rangle_{A=B} - \left\langle \psi_i^{(X_A)(Y_A)} \middle| T \middle| \psi_i^B \right\rangle \right. \tag{44b}.$$
$$\left. - \left\langle \psi_i^{(X_B)(Y_B)} \middle| T \middle| \psi_i^A \right\rangle + \left\langle \sum_G \psi_i^{(X_G)(Y_G)} \middle| T \middle| \psi_i^B \right\rangle_{A=B} \right\}$$

The subscript $A=B$ in equation 44 indicates that the appropriate terms only contribute to one center second derivatives.

The terms ΔV, ΔE_{XC}, and $R^{(x)}$, are treated in a similar fashion. The term V_{Pair} depends on the coordinates for a given atom pair $\{AB\}$ only through the interatomic distance R_{AB} Therefore, the electrostatic interaction energy can be precalculated for each pair of elements on a grid as a function of R_{AB}. Values for V_{Pair} and its derivatives are retrieved by an interpolation procedure.

4.2 Implicit contributions to the second derivatives. The term $\left[E_{KS}^{(X_A)} + E_{KS}^{B,X_A} \right]^{[Y_B]}$ essentially evolves the evaluation of the implicit derivative of the electron density. Writing the electron density in P-matrix formalism

$$\rho = \sum_i^n \psi_i \psi_i = \sum_i^n \left\{ \sum_\mu^{2M} c_{\mu i} \lambda_\mu \sum_v^{2M} c_{v i} \lambda_v \right\} = \sum_\mu^{2M} \sum_v^{2M} P_{\mu v} \cdot \lambda_\mu \lambda_v \tag{45}$$

$$P_{\mu v} = \sum_i^n c_{i\mu} c_{iv} \tag{46},$$

the implicit derivative of the electron density takes on the form

$$\rho^{[Y_B]} = \sum_\mu^{2M} \sum_v^{2M} P_{\mu v}^{[Y_B]} \cdot \lambda_\mu \lambda_v \tag{47}.$$

The implicit derivative of the P-matrix is a result of the solution of the CPKS-equations, and can be calculated as

$$P_{\mu\nu}^{[Y_B]} = \sum_i^{occ} \sum_j^{occ} \left\{ c_{\mu j} u_{ji}^{[Y_B]} c_{\nu i} + c_{\mu i} u_{ij}^{[Y_B]} c_{\nu j} \right\}$$

$$+ \sum_i^{occ} \sum_a^{vir} \left\{ c_{\mu a} u_{ai}^{[Y_B]} c_{\nu i} + c_{\mu i} u_{ia}^{[Y_B]} c_{\nu a} \right\}$$

(48).

The mixed second derivative is easily obtained as

$$\rho^{(X_A)[Y_B]} = \sum_\mu^{2M} \sum_\nu^{2M} P_{\mu\nu}^{[Y_B]} \left\{ \lambda_\mu \lambda_\nu^{(X_A)} + \lambda_\mu^{(X_A)} \lambda_\nu \right\}$$

(49)

We further require the derivatives of the orbital energies ε_i. These are also available from the solution of the CPKS equations, and can be calculated as

$$\varepsilon_p^{Y_B} = F_{pp}^{Y_B} - S_{pp}^{Y_B} \varepsilon_p$$

(50)

5. Performance of the Implementation

Analytical second derivatives are of practical use only if they can compete with a corresponding numerical approach. We will therefore compare the efficiency of our present implementation with that of the finite difference technique. It is important to keep in mind that the execution time greatly depends on the accuracy inflected on the integration scheme. As discussed above, the ADF program uses fully numerical integration for most integrals, and the execution time is linearly proportional to the number of integration points. Here, we employed an integration scheme which ensures six digits of numerical accuracy for scalar matrix elements. In Table I, we present timings for the evaluation of the explicit contributions due to sections 4.1. For all the systems under investigation, the analytical evaluation roughly affords a speed-up by a factor of ten. We also observe that the execution time for this part scales with N^2, N being the number of nuclei in the system under investigation. For this part, the size of the basis set is of minor importance.

Table I: Execution times[a] for the analytical and finite difference evaluation of the explicit contributions to the second derivatives

System[b]	Analytic	Finite Difference[c]	Ratio
NH_3 - dzp[d]	139	1776	12.8
SF_6 - dzp	1728	15834	9.2
C_4H_8O - dz[e]	4963	60606	12.2
$Ni(CO)_4$ tzp[f]/dzp	7113	83268	11.7

[a] In seconds. Timings were obtained on an IBM RS6000 model 375 workstation. [b] All systems were calculated in C_1 symmetry. [c] Based on 2-point numerical differentiation. [d] dzp: double ζ basis with one p,d or f polarization function. [e] dz: double ζ basis. [f] tzp: tripple ζ basis with one p,d or f polarization function.

It is of further interest, how the evaluation of the implicit and explicit contributions compares with time used to solve the CPKS-equations. Appropriate timings are reported in Table II. When a larger basis set is employed, as in the cases of NH_3 and SF_6, the solution of CPKS-equation is the most expensive part in the evaluation of the second derivatives. Thus, the time advantage decreases to a factor of 8.0 and 4.5, respectively. On the other hand, with a small basis set, an excellent speed-up can be achieved for the CPKS-part. For tetrahydrofurane, (C_4H_8O), we found that the analytical evaluation of the derivative of the electronic density should be about 28 times faster than the finite difference approach.

Table II: Analytical vs. numerical Kohn- Sham second derivatives. Execution times [a]

System[b]	CPKS	Implicit	Explicit	Total	Finite Diff.	Ratio
NH_3 - dzp	425	260	139	824	6624	8.0
SF_6 - dzp	9092	1328	1728	12148	54642	4.5
C_4H_8O - dz	3011	4606	4963	12580	193752	15.4

[a] In seconds. [b] For details, compare Table I.

Considering the complete analytic second derivative, one finds that in this case the number of nuclei N dominates, and the evaluation of the implicit and explicit contributions takes more time than the CPKS solution. One reason for this fact is the extensive use of TI in our derivation, which approximately doubles the computational effort. In total, the overall achievement in performance still amounts to a factor of 15.

When making a fair judgment of the timings presented, one should keep in mind that symmetry has not been used in the present calculations. The symmetry advantage for the finite difference calculation is determined by the ratio of all displacements and the symmetry unique displacements. For the analytical evaluation of the second derivatives, the cost of executing some subroutines cannot be reduced when only symmetry unique displacements are considered. It thus can be expected that the speed-up will somewhat decrease when symmetry unique displacements are considered. Nevertheless, the results presented here more than justify our implementation of the analytic second derivatives.

Acknowledgment

This work has been supported by the National Sciences and Engineering Research Council of Canada (NSERC), as well as by the donors of the Petroleum Research Fund, administered by the American Chemical Society (ACS-PRF No 20723-AC3). H. J. acknowledges a scholarship from the chemistry department at the U of C.

References

1. Ziegler, T. *Chem. Rev.* **1991,** *91,* 651.
2. Handy, N. C.; Tozer, D. J.; Murray,C. W.; Laming, G. J.; Amos, R. D. *Isr. J. Chem.* **1993,** *33,* 331.
3. Johnson, B.G., Fisch, M. J. *J. Chem. Phys.* **1994,** *100,* 7429.
4. Komornicki, A.;Fitzgerald, G. J.Chem.Phys. **1993,**98, 1398
5. Kohn, W.; Sham, L.J. *Phys. Rev.* **1965,** *A140,* 1133.
6. Bérces, A.; Dickson, R. M.; Fan, L.; Jacobsen, H.; Swerhone, D. P.; Ziegler, T. *Comp. Phys. Com.* submitted.
7. Jacobsen, H.; Bérces, A.; Swerhone, D. P.; Ziegler, T. *Comp. Phys. Com.* submitted.
8. Baerends, E. J.; Ellis, D. E.; Ros, P. *Chem. Phuys.* **1973,** *2,* 41.
9. teVelde, G.; Baerends, E. J. *J. Comp. Phys.* **1992,** *99,* 84.
10. Gerratt, J.; Mills, I. M. *J. Chem. Phys.* **1966,** *44,* 2480.
11. Pople, J. A.; Krishnan, R.; Schlegel, H. B.; Binkley, J. S. *Int. J. Quantum Chem.* **1979,** *13,* 225.
12. Ziegler, T.; Rauk, A.; *Theor. Chim. Acta* **1977,** *46,* 1.
13. Versluis, L. *The Determination of Molecular Structures by the HFS -Method;* Ph.D. Thesis, The University of Calgary, Calgary, Alberta, 1989.
14. Versluis, L.; Ziegler, T.; *J. Chem. Phys.* **1988,** *88,* 322.
15. Moccia, R. *Theor. Chim. Acta* **1967,** *8,* 8.

DENSITY-FUNCTIONAL METHODS IN STATISTICAL MECHANICS

Chapter 12

Decay of Correlations in Bulk Fluids and at Interfaces: A Density-Functional Perspective

R. Evans and R. J. F. Leote de Carvalho

HH Wills Physics Laboratory, University of Bristol, Bristol BS8 1TL, United Kingdom

Density functional methods can be used to show that for classical fluids with short-ranged interatomic potentials the length scales which describe the asymptotic decay of the one-body density profile at a fluid interface are the same as those which determine the decay of the two-body correlation function $g(r)$ of the bulk fluid. This general result has striking implications for a variety of interfacial properties - the possible occurrence of oscillations in the liquid-vapour density profile, the occurrence of wetting and layering phase transitions at substrate-fluid interfaces and the nature of the decay of the solvation force for a confined fluid. For fluids whose potentials have power-law decay the density profiles and $g(r)$ exhibit power-law decay at longest range. However, for liquid densities the **intermediate** range damped oscillatory decay is governed by the same leading-order pole structure that describes short-ranged forces. Recent work on the decay of correlations in binary liquid mixtures and in ionic fluids is also described.

Consider a simple (atomic) fluid adsorbed at a solid substrate or wall. The average (one-body) density profile $\rho(\mathbf{r})$ will usually exhibit pronounced structure reflecting the ordering or layering that arises from packing effects, ie. from short-range correlations between the atoms. The precise form of the density profile must depend on the details of the wall-fluid potential $V(\mathbf{r})$, the fluid-fluid interatomic potential and the thermodynamic state point. If the bulk fluid is a high density liquid a large number of oscillations develop in the profile, whereas if the bulk is a dilute gas, and there is incomplete wetting of the wall-gas interface by liquid, only a few oscillations, close to the wall, should occur. One might expect that the asymptotic decay of the profile into the bulk fluid far from the wall should be less dependent on the precise form of $V(\mathbf{r})$; rather it should be determined primarily by the properties of the bulk fluid. Recent work has sought to elucidate different classes of asymptotic behaviour and the physical factors which determine these. A key result, obtained from density functional techniques, is that for short-ranged interatomic potentials the length scales which determine the longest range decay of a planar wall-fluid interfacial profile are identical to those which characterise the longest range decay of the radial distribution function $g(r)$ of the bulk liquid. Thus, determining

0097–6156/96/0629–0166$15.00/0

the different classes of asymptotic decay for $g(r)$ is sufficient to determine the decay of planar interfacial profiles. This observation has important consequences for a variety of fluid interfacial phenomena. It leads to predictions for the existence (or non-existence) of oscillations in the density profiles at liquid-vapour and liquid-liquid interfaces and in the solvation force for fluids confined between two parallel walls. It also has implications for wetting and layering transitions at wall-fluid interfaces. In this lecture we review several aspects of the general theoretical framework for describing the decay of inhomogeneous fluid structure in terms of the decay of $g(r)$. The presentation is necessarily brief; further details of the theory and results of calculations can be found in (*1 - 4*).

Asymptotics of the Bulk Pair Correlation Function g(r) for a Pure Fluid.

The asymptotic decay of $g(r)$ is most easily determined from the (bulk) Ornstein-Zernike (OZ) equation, which relates the total pair correlation function $h(r) \equiv g(r) - 1$ to the direct correlation function $c(r)$. In Fourier space this is:

$$\hat{h}(q) = \frac{\hat{c}(q)}{1 - \rho\hat{c}(q)} = \frac{1}{\rho}\left(\frac{1}{1 - \rho\hat{c}(q)} - 1\right) \tag{1}$$

where ρ is the number density and $^\wedge$ denotes the Fourier transform. It follows that

$$rh(r) = \frac{1}{2\pi^2}\int_0^\infty dq\, q \sin qr \frac{\hat{c}(q)}{1 - \rho\hat{c}(q)} \tag{2}$$

We must distinguish between fluids for which $c(r)$ is short ranged (finite ranged or exponentially decaying) and those for which $c(r)$ decays as a power law. Recall that for fluids away from their critical points, diagrammatic analysis shows that $c(r) \rightarrow -\beta\phi(r)$ as $r \rightarrow \infty$, where $\beta = (k_B T)^{-1}$ and $\phi(r)$ is the interatomic pairwise potential.

Short-Ranged Potentials. If $c(r)$ decays faster than a power law it follows (*1 - 4*) from equations 1 and 2 that the asymptotics of $rh(r)$ are determined completely by the poles of $\hat{h}(q)$ at complex $q = \alpha \equiv \alpha_1 + i\alpha_0$ satisfying

$$1 - \rho\hat{c}(\alpha) = 0 \tag{3}$$

There can be no poles lying on the real axis apart from the liquid-vapour spinodals $(\alpha = 0)$ and an infinite ranged oscillatory solution (*4*) often found at very high density $(\alpha_0 = 0, \alpha_1 \neq 0)$. A pole can lie on the imaginary axis where it gives rise to pure exponential decay of $rh(r)$, or poles can lie off the imaginary axis where they yield exponentially damped oscillatory decay. In the latter case the poles occur as conjugate pairs: $\alpha = \pm\alpha_1 + i\alpha_0$. Once the poles have been determined (from knowledge of $\hat{c}(q)$ at a given ρ, T) contour integration can be used (*2, 3*) to obtain

$$rh(r) = \frac{1}{2\pi} \sum_n e^{iq_n r} R_n \tag{4}$$

where q_n is the n th pole and R_n is the residue of $q\hat{c}(q) / (1 - \rho\hat{c}(q))$ at q_n. For the model potentials and the closure approximations that have been studied there is an infinite number of poles but the longest range part of $h(r)$ is determined by the pole or poles with the smallest value of α_0, the inverse decay length. Two scenarios are found: (a) a pole lying on the imaginary axis $q_n = i\alpha_0$ has the lowest value and the ultimate decay is purely exponential

$$rh(r) \sim A e^{-\alpha_0 r}, r \rightarrow \infty \tag{5}$$

or (b) a conjugate pair of poles has a smaller imaginary part $\tilde{\alpha}_0$ than that on the imaginary axis and the ultimate decay is

$$rh(r) \sim \tilde{A} e^{-\tilde{\alpha}_0 r} \cos(\alpha_1 r - \theta), \quad r \rightarrow \infty \tag{6}$$

where θ is a phase angle. The amplitudes A and $\tilde{A}$ and the phase are given in terms of the residues. The line in the ρ, T plane where $\tilde{\alpha}_0 = \alpha_0$ marks the crossover, at longest range, from damped oscillatory to pure exponential decay of $rh(r)$. This is referred to as the Fisher-Widom (FW) line after the authors who first surmised such crossover should occur (5). Crossover arises as a result of competition between repulsive and attractive components of the interatomic potential. At high T and ρ repulsion dominates (the negative portion of a typical $c(r)$) whereas at low T and ρ attraction dominates (the positive portion) (1).

The FW line has been calculated (1) for a $(\sigma, 3\sigma/2)$ square-well fluid of well-depth ε, using a random phase approximation for the direct correlation function:

$$c(r) = c_{hs}(r) - \beta\phi_{att}(r) \tag{7}$$

with $\phi_{att}(r) = -\varepsilon$ for $r < 3\sigma/2$ but zero otherwise. The hard-sphere direct correlation function $c_{hs}(r)$ was obtained from the weighted density approximation of Tarazona. Very similar results are obtained using the Percus-Yevick approximation for $c_{hs}(r)$ (1). The FW line intersects the liquid branch of the liquid-vapour coexistence curve at $T = 0.9T_c$, where T_c is the bulk critical temperature, and it lies above the liquid spinodal (1). As the random phase approximation is a very crude theory of short-range correlations and the square-well model is an extreme idealisation of a fluid pair-potential, subsequent calculations were carried out for the more realistic Lennard-Jones 12-6 potential, truncated and shifted at $R_c = 2.5\sigma$. The calculations (3) were based on the HMSA closure of Zerah and Hansen (6), which interpolates between the hypernetted chain (HNC) and soft-mean-spherical approximation (SMSA). This integral equation approach imposes self-consistency between the virial and compressibility routes to the equation of state and comparisons with simulation data suggest that it is one of the most reliable and accurate modern liquid state theories (see (6) and references in

(*3*)). Figure 1 shows the results for the FW line, plotted along with the coexistence curve obtained from simulation (*7*). The FW line has the same location with respect to the coexistence curve as that found earlier (*1*) for the square-well. It intersects the liquid side at $T = 0.9 T_c$ and at a similar value of ρ / ρ_c, where ρ_c is the critical density. The shapes of the two lines are also quite similar, although the square-well FW line crosses the critical density at a lower temperature ($T/T_c \sim 1.8$) than that inferred from extrapolation of the line in Figure 1. On the basis of these two sets of results it was conjectured that the FW line for any reasonable finite-ranged pair potential model fluid would have much the same form when plotted in terms of reduced variables T/T_c and ρ / ρ_c, ie. it would obey (at least approximately) a law of corresponding states. Further calculations, for different types of potentials, are required to test this conjecture.

The pole structures obtained in the two sets of calculations differ in detail, reflecting the different types of approximation and the fact that the Lennard-Jones potential has a soft core. However, in both cases and for all thermodynamic state points that were considered, the dominant poles are the one on the imaginary axis and a conjugate pair with $|\alpha_1| \sim 2\pi/\sigma$. These have values of α_0 that are much smaller than those of the other poles. The latter are all much further from the real axis than the dominant poles. This observation implies that for large r, an accurate approximation to $rh(r)$ should be obtained by retaining only the contributions of the dominant poles in equation 4. All poles that have been calculated are simple poles so that the residues and, hence, the amplitudes and phase are easily obtained (*2, 3*).

Comparison of numerical results for $h(r)$ from the full solution of the HMSA integral equation with those obtained by retaining only contributions from the pure imaginary and the first conjugate pair of complex poles show the above surmise to be correct. The quality of the fit to the full solution is remarkable (*3*). For densities on either side of the FW line the results from the 'two-pole' approximation are almost indistinguishable from the 'exact' results for $r \geq 3\sigma$. Moreover, at intermediate range, $1.5\sigma \leq r \leq 3\sigma$, the oscillations in $h(r)$ are well-reproduced with errors of only a few percent. A similar level of success is achieved for the square well-fluid in the random phase approximation (*2*). The 'two-pole' approximation

$$rh(r) \sim A e^{-\alpha_0 r} + \tilde{A} e^{-\tilde{\alpha}_0 r} \cos\left(\alpha_1 r - \theta\right) \tag{8}$$

appears to provide an excellent description of the intermediate as well as the long range decay of $h(r)$ in the case of fluids with short-ranged potentials.

Long Ranged Potentials. In real fluids dispersion (London) forces are always present and these have a profound effect on the asymptotics of correlation functions. For simplicity we restrict consideration to the case of a (model) fluid where only the induced-dipole-induced-dipole contribution is retained so that the pairwise potential decays as $\phi(r) \sim -a_6/r^6$, with a_6 positive. Retardation is ignored also. Incorporating extra power-law contributions and crossover effects associated with retardation is possible (*3*) but complicates the analysis.

The presence of the $-a_6/r^6$ tail of the potential gives rise to a positive tail $\beta a_6/r^6$ in $c(r)$. This, in turn, yields a q^3 term in the Fourier transform $\hat{c}(q)$ (*8, 9*), so that for positive real q

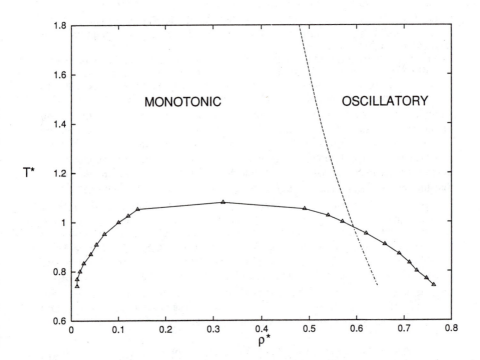

Figure 1. The FW line for a Lennard-Jones fluid (with $R_c = 2.5\sigma$) calculated from the HMSA. On the dashed portion the pressure is positive, whereas on the dash-dotted portion it is negative. The solid curve joining triangles denotes simulation results for the liquid-vapour coexistence curve. $T^* \equiv k_B T / \varepsilon$ and $\rho^* \equiv \rho\sigma^3$ are the reduced temperature and density.

$$\hat{c}(q) = \hat{c}^{sr}(q) + \frac{\beta a_6 \pi^2 q^3}{12} \tag{9}$$

where $\hat{c}^{sr}(q)$ is the Fourier transform of a short-ranged function possessing an expansion in powers of q^2. Equation 9 follows explicitly in the random phase approximation (equation 7); more sophisticated closures could yield further odd powers of q, eg. a term in q^9 (3). The q^3 term is the key signature of dispersion forces. If the potential is truncated, as in computer simulation, there is no q^3 term and the pole analysis described above is appropriate. However, when the q^3 term is present that analysis must be modified and a different contour must be chosen for performing the integration in the complex plane (3). Poles still occur, given by equation 3, but now there are no poles on the imaginary axis. For a specific choice of contour an approximation for the decay of $h(r)$ can be obtained (3):

$$rh(r) \sim S^2(0)\beta \frac{a_6}{r^5} + A e^{-\alpha_0 r} \cos(\alpha_1 r - \theta) \tag{10}$$

where $S(0) = (1 - \rho \hat{c}(0))^{-1}$ is the $q = 0$ limit of the liquid structure factor. Note that $S(0) = \rho \beta^{-1} \kappa_T$, where κ_T is the isothermal compressibility. The ultimate decay of $h(r)$ is always determined by the first term in equation 10 and this reflects directly the q^3 term in $\hat{c}(q)$. For a general power-law potential the ultimate decay is $h(r) \rightarrow -\beta S^2(0)\phi(r)$, as $r \rightarrow \infty$ (8). The second term in equation 10 is the contribution from the (complex) pole with the smallest value of α_0. This pole is expected to have $\alpha_1 \sim 2\pi / \sigma$. Other poles will make further damped oscillatory contributions but, provided they are well-separated, the leading-order pole should give the dominant oscillatory contribution.

Those readers with interest in the history of liquid state science might note that it is thirty years since Enderby et.al. (8) properly identified the ultimate power-law decay of $h(r)$, following an earlier observation by Widom (10).Verlet, in his famous 1968 paper (11) on the molecular dynamics of the Lennard-Jones fluid, was probably the first to enquire just how damped oscillations, which must occur at short and intermediate range, can be separated from the ultimate power-law decay of $h(r)$. We believe equation 10 is the appropriate prescription. How accurate is this approximation? This question was examined at some length in reference (3). Explicit calculations, within the random phase approximation, for a model potential with hard-sphere repulsion and an attractive $-a_6 / r^6$ tail show that equation 10 provides a very accurate fit to the 'exact' $h(r)$, obtained from numerical Fourier transform of $\hat{h}(q)$ given by equation 1, for $r \geq 2\sigma$. At high densities α_0 is not particularly large and $S(0)$ is small, with the result that the second term in equation 10 dominates until $r \sim 25\sigma$, after which the power-law decay takes over. At low densities α_0 is much larger and $S(0)$ is increased, so that the oscillations in $h(r)$ are eroded much faster than at high densities (3).

Comparing equations 8 and 10 it is clear that the most significant difference between short and long-ranged potentials is the replacement of the pure exponentially decaying term by a power-law contribution. As a consequence no sharp FW line can be defined in the long-ranged case, since there can be no crossover from pure exponential to damped oscillatory decay at any range; recall that there is no pole on the imaginary axis. However, this does not mean that **intermediate-range** structure will not reflect the thermodynamic state. As indicated above, at high densities the second term of equation 10 provides a very accurate description of intermediate range oscillatory structure of $h(r)$ and the power-law contribution does not manifest itself until very large separations. At lower densities the damped oscillatory term decays much faster and the power-law contribution has a larger amplitude (controlled by $S(0)$) thereby reducing the number of discernible oscillations. While there is no sharp FW line, there should be a crossover region in the ρ, T plane which marks the erosion of oscillations at intermediate range. Support for this view can be gleaned from two sources: (a) results for $h(r)$ extracted from neutron-diffraction data taken for Ne and Xe along their saturated liquid curves (12) shows that intermediate range damped oscillatory decay persists up to $T \sim 0.95 T_c$, whereas for higher temperatures the oscillatory decay seems to disappear, (b) early calculations of $h(r)$, based on the optimised cluster theory, for the full Lennard-Jones potential indicated crossover from damped oscillatory to monotonic decay at intermediate range as the density was reduced (13). For a given ρ, T one can estimate the separation r at which the oscillations will become indiscernible by equating the magnitudes of the two terms in equation 10.

In computer simulation the pairwise potential is necessarily truncated. For the Lennard-Jones fluid the cut-off separation is often $R_c = 2.5\sigma$ or 3.0σ. Such a short-ranged model will only have the same intermediate range oscillatory structure as the full Lennard-Jones potential if the dominant complex poles (near $\alpha_1 = 2\pi/\sigma$) of both models lie very close together and the corresponding residues are very close. Results, based on the random phase approximation, suggest that this is indeed the case (3).

Asymptotics of $h_{ij}(r)$ for a Binary Fluid Mixture.

The analysis described in the previous section can be extended to mixtures. The generalisation of the OZ equation can be expressed as

$$\hat{h}_{ij}(q) = N_{ij}(q) / D(q) \tag{11}$$

where i, j runs over the species labels. Although the numerator $N_{ij}(q)$ is different for different correlation functions, the denominator $D(q)$ is common. It follows that all the $\hat{h}_{ij}(q)$ exhibit the same pole structure, determined by the zeros of $D(q)$. Thus, for short-ranged potentials,

$$rh_{ij}(r) = \frac{1}{2\pi} \sum_n e^{iq_n r} R_n^{ij} \tag{12}$$

where the (common) n^{th} pole is given by $D(q_n) = 0$. Only the residues R_n^{ij} and hence, the amplitudes, differ for different combinations of species. Since the longest range part of $h_{ij}(r)$ is determined by the q_n with the smallest imaginary part, all $h_{ij}(r)$ will ultimately decay with the **same** exponential decay length and oscillatory wavelength. This fact appears to have been appreciated first by Martynov (*14*) but was first explained in (*2*).

The pole structure in a mixture should be similar to that in a pure fluid so that, for short-ranged potentials, there should be a unique FW surface in thermodynamic phase space separating regions of longest range pure exponential decay from those of exponentially damped oscillatory decay, applicable for **all** the $h_{ij}(r)$. If the poles are simple, the formulae for the residues R_n^{ij} allow one to derive simple relations linking the amplitudes associated with the decay. For the particular case of binary mixtures, where i, j run over the species a and b, one finds for pure exponential decay

$$rh_{ij}(r) \rightarrow A_{ij}e^{-\alpha_0 r} \tag{13}$$

with

$$A_{aa}A_{bb} = A_{ab}^2. \tag{14}$$

Whereas for damped oscillatory decay, where a conjugate pair contribute,

$$rh_{ij}(r) \rightarrow \tilde{A}_{ij}e^{-\tilde{\alpha}_0 r}\cos\left(\alpha_1 r - \theta_{ij}\right) \tag{15}$$

with

$$\tilde{A}_{aa}\tilde{A}_{bb} = \tilde{A}_{ab}^2 \tag{16}$$

and

$$\theta_{aa} + \theta_{bb} = 2\theta_{ab} \tag{17}$$

Explicit formulae for the amplitudes A_{ij} and phases θ_{ij} can be derived (*2*).

The results expressed in equations 13 to 17 are very general. They should apply for the longest range decay of correlations in any binary fluid mixture where the interatomic forces are short-ranged. At first sight, the existence of a common asymptotic form is counterintuitive - especially when one contemplates mixtures with widely differing atomic sizes. It is not obvious that such mixtures should have $h_{ij}(r)$ with a common wavelength and decay length. The accuracy of equation 15 has been examined for hard-sphere mixtures in the Percus-Yevick approximation. For a variety of extreme concentrations and ratios of hard sphere diameters, equation 15, corresponding to a single conjugate pair of complex poles, yields $h_{ij}(r)$ that are very close to the results obtained from numerical Fourier transform of the OZ equations - at least for separations beyond the second maximum. Even the positions of the first and second maxima are given reliably (*2*). In other words the decay and the amplitude and phase relations are obeyed for separations down to second nearest neighbours. Note that the amplitude relations in equations 14 and 16, which have the form of a geometric mixing rule, hold irrespective of the mixing rule for the strength of the ab attractive potential and that for the effective range of the ab repulsion. The details of the chemistry do not matter - provided the relevant poles are simple.

Do these results have relevance for real (atomic) mixtures where dispersion forces are present? If there are well-separated pole structures for binary mixtures (which we expect) the contribution from the pole with the smallest value of α_0 will dominate the intermediate range decay of $h_{ij}(r)$ and the remarkable relations expressed in equations 13 to 17 should hold over a similar (intermediate) range to that found for the pure fluid. Thus, it is feasible that these results could provide some new insight into the interpretation of structural data (from neutron and X-ray diffraction) on binary mixtures.

Ionic Liquids. The analysis described above is not immediately applicable to an ionic liquid. In such systems the existence of the long-ranged Coulomb forces between the ions means that the direct correlation functions $c_{ij}(r)$ decay very slowly, as r^{-1}, for large r. Thus, a priori, it is not obvious that the pairwise correlation functions $h_{ij}(r)$ in a binary ionic liquid, such as a molten salt or a (primitive) electrolyte, should exhibit the same type of decay, with the same amplitude and phase relations, as those (equations 13 to 17) which characterise a binary neutral mixture. That this **is** the case, is a consequence of electrostatic screening. The OZ equations for an ionic mixture are the same as those for any mixture but now the Fourier transforms of the $c_{ij}(r)$ must be separated into Coulombic and non-Coulombic terms:

$$\hat{c}_{ij}(q) = \hat{c}_{ij}^{sr}(q) - \frac{\beta 4\pi Z_i Z_j e^2}{\varepsilon q^2} \qquad (18)$$

where eZ_+ is the charge on the positive ion, eZ_- is that on the negative ion and $\varepsilon \equiv 4\pi\varepsilon_0\varepsilon_r$, with ε_r the relative permittivity. Such a division follows naturally in a density functional approach where the intrinsic free energy functional is written as the total electrostatic energy plus a remainder. $\hat{c}_{ij}^{sr}(q)$, which is analytic in q^2, denotes the short-ranged, non-Coulombic contribution; dispersion forces are not considered here. It is straightforward to show that the OZ equations can be written in the form of equation 11, with suitable definitions of $N_{ij}(q)$ and $D(q)$ (*15*).

The pole structure of an ionic liquid is different from that of its neutral counterpart. This is most clearly illustrated for the special case of a symmetric binary fluid where the ++ interionic potential is equal to the -- potential, so that $h_{++}(r) = h_{--}(r)$. Then there are only two independent pair correlation functions: $h_S \equiv (h_{++} + h_{+-})/2$ and $h_D \equiv (h_{++} - h_{+-})/2$. h_S measures correlations in the total number density and h_D measures correlations in the charge density. The poles of $\hat{h}_S(q)$ (density) and of $\hat{h}_D(q)$ (charge) are given by independent equations and are determined by different physical considerations (*15*). Density poles exhibit similar structure to that found in a single-component neutral fluid and cross-over from pure exponential to damped oscillatory decay of $rh_S(r)$, as the density of ions is increased at fixed T, occurs by the same mechanism as in the neutral fluid giving rise to a FW line. The charge poles behave differently. These depend primarily on the inverse Debye screening length $\kappa_D = \left(4\pi\rho\beta(Ze)^2/\varepsilon\right)^{\frac{1}{2}}$. Crossover from pure exponential to damped oscillatory decay of $rh_D(r)$ occurs at a particular value of κ_D via a coalescence of a pair of pure imaginary poles followed by fission into a conjugate complex pair

(15). Such a mechanism was first described by Kirkwood *(16)* in a discussion of charge oscillations in strong electrolytes. The locus of points in the ρ, T plane where crossover occurs to charge oscillations at longest range is termed the Kirkwood (K) line. Figure 2 shows the results *(15)* of calculations of the FW and the K lines for the restricted primitive model (RPM), ie. charged hard-spheres of equal diameter R, based on the generalised mean spherical approximation (GMSA) which is the simplest thermodynamically self-consistent theory for an ionic liquid. Within the GMSA, crossover to charge oscillations is determined by the condition $\kappa_D R = 1.228$ so the K line is a straight line in the ρ, T plane - see figure 2. It intersects the vapour branch of the coexistence curve close to the critical temperature. The FW line intersects the liquid branch but at a lower value of T / T_c than that in figure 1 for the (truncated) Lennard-Jones fluid. By combining results for the density and charge poles one can ascertain which of these determines the ultimate decay of $h_{++}(r)$ and $h_{+-}(r)$. A variety of crossover lines emerges (see figure 7 of *(15)*). These separate regions of the phase diagram where the decay of ion-ion correlation functions are dominated by monotonic charge, monotonic density, oscillatory charge or oscillatory density poles. The results suggest that 1:1 primitive model electrolytes should exhibit a different sequence of crossover, as the density of ions is increased at fixed T, from 2:2 electrolytes. Since leading order asymptotics provide an accurate description of both $h_D(r)$ and $h_S(r)$ for $r \geq 1.5R$, the results also yield some fresh insight into the nature of density and charge oscillations for molten salts.

Asymptotics of Wall-Fluid Interfacial Profiles.

We now discuss the repercussions of the results obtained for the behaviour of bulk correlation functions for the density profiles of fluids at interfaces. That there should be direct repercussions follows from the fact that the radial distribution of the bulk fluid $g(r) \equiv \rho(r)/\rho$, where $\rho(r)$ is the (inhomogeneous) density profile obtained by fixing an atom at the origin, thereby creating a spherically symmetric 'external' potential $V(r)$ for the other atoms. (In the special case of pairwise potentials $V(r) \equiv \phi(r)$). In other words $g(r)$ can be regarded as a one-body density profile associated with a particular type of inhomogeneity. One might suppose that other types of external potential might give rise to the same type of **asymptotic** decay of the density profile - provided the external potential is sufficiently short-ranged. As an example consider a pure liquid in a bulk state which is on the oscillatory side of the FW line. Then $h(r)$ decays as in equation 6. If the wall-fluid potential $V(z)$, where z is the distance normal to the wall, is of finite range the density profile for this interface should decay as

$$\rho(z) - \rho \sim A_{wf} e^{-\alpha_0 z} \cos\left(\alpha_1 z - \theta_{wf}\right), \quad z \to \infty \tag{19}$$

where ρ is the bulk density. α_0 and α_1 are the same inverse length scales which determine the decay of $rh(r)$. The amplitude A_{wf} and phase θ_{wf} will be different from the corresponding bulk quantities in equation 6; they will depend on the form of $V(z)$.

Equation 19 was derived starting from the exact integral equation for the density profile in planar geometry *(4)*. A somewhat more revealing derivation can be

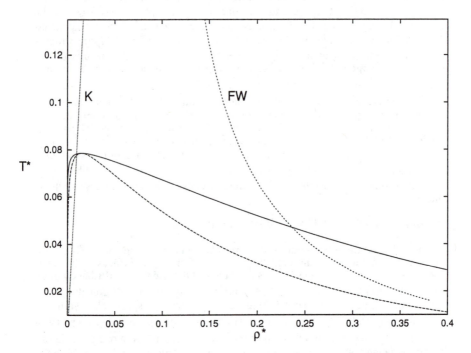

Figure 2. Cross-over lines for the RPM calculated using the GMSA. The short-dashed curve is the FW line where cross-over from monotonic to damped oscillatory decay occurs for the total number density correlation function $h_S(r)$. Onset of charge oscillations, in $h_D(r)$, occurs on the dotted line K. The solid line is the liquid-vapour coexistence curve and the long-dashed line marks the accompanying spinodals. $T^* \equiv k_B T \varepsilon R /(eZ)^2$ and $\rho^* \equiv \rho R^3$ are the reduced temperature and total density.

obtained from density functional theory. The density profile of an inhomogeneous fluid subject to an external potential $V(\mathbf{r})$ satisfies the exact Euler-Lagrange equation (see eg. (17) or (18))

$$\mu = V(\mathbf{r}_1) + \mu_{id}(\rho(\mathbf{r}_1)) - \beta^{-1} c^{(1)}([\rho];\mathbf{r}_1) \qquad (20)$$

where μ is the chemical potential, $\mu_{id}(\rho) = \beta^{-1} \ln(\Lambda^3 \rho)$ is the chemical potential of the ideal classical gas, and $c^{(1)}$, the one-body direct correlation function, is the first derivative of the excess (over ideal) Helmholtz free energy functional $F_{ex}[\rho]$:

$$c^{(1)}([\rho];\mathbf{r}) = -\beta \frac{\delta F_{ex}[\rho]}{\delta \rho(\mathbf{r})} \qquad (21)$$

note that $-\beta^{-1} c^{(1)}$ is the classical analogue of the effective one-body potential that enters the Kohn-Sham theory of the inhomogeneous electron gas. Since we are concerned with the limit where $\rho(\mathbf{r}) \to \rho$, we Taylor expand about the bulk in powers of $\delta \rho(\mathbf{r}) \equiv \rho(\mathbf{r}) - \rho$. Outside the range of the external potential

$$\delta \rho(\mathbf{r}_1) = \rho \int d\mathbf{r}_2 \; \delta \rho(\mathbf{r}_2) c(r_{12}) + 0(\delta \rho)^2 \qquad (22)$$

where

$$c(r_{12}) \equiv \left. \frac{\delta c^{(1)}([\rho];\mathbf{r}_1)}{\delta \rho(\mathbf{r}_2)} \right|_\rho \qquad (23)$$

is the two-body direct correlation function of the homogeneous bulk fluid, $r_{12} = |\mathbf{r}_2 - \mathbf{r}_1|$ and use has been made of the relation $\beta^{-1} c^{(1)}(\rho) = \mu_{id}(\rho) - \mu$. The complex Fourier transform of equation (22) is

$$0 = (1 - \rho \hat{c}(q)) \hat{\delta \rho}(q) + 0(\delta \rho)^2 \qquad (24)$$

It follows that the ultimate asymptotic decay is determined by zeros of $(1 - \rho \hat{c}(q))$ with $q = \alpha$, i.e. the same condition as equation 3, which was derived from the bulk OZ equation. By specialising to planar geometry one can show that equation 19 satisfies equation 22 in lowest order. Similarly for a spherical external potential one can show that this equation has a solution $\delta \rho(r) \sim A_s e^{-\alpha_0 r} \cos(\alpha_1 r - \theta_s)/r$. An equivalent argument goes through for thermodynamic states where the bulk lies on the monotonic side of the FW line; the decay of $\delta \rho(z)$ or $r \delta \rho(r)$ is now pure exponential. Note that equation 24 corresponds to a linear response treatment so on its own it is insufficient to determine amplitudes and phases. These can only be extracted from a full non-linear treatment of the density profile, i.e. by solving equation 20 having obtained $c^{(1)}$ from a specific density functional approximation for $F_{ex}[\rho]$. Moreover the above results are not appropriate for power-law potentials or for exponential wall-fluid potentials whose decay length is longer than that of $c(r)$.

The results generalise to mixtures. Once again the pole structure is determined by the zeros of the same common denominator that enters equation 11 and the asymptotic decay of the density profiles of individual species $\rho_i(z)$, outside the range of the wall-fluid potentials $V_i(z)$, is characterised by the same exponential decay length and oscillatory wavelength. Relations between the amplitudes of $\delta\rho(z_i)$ follow from the residues (2). An alternative procedure, which leads to the same results, makes use of the wall-particle OZ equations obtained from the bulk mixture OZ equations in the limits where the density of one component $\rightarrow 0$ and its diameter $\rightarrow\infty$. Details are given in (2) and (19). The amplitudes are now given formally in terms of the wall-particle direct correlation functions but these can only be obtained by making specific closure approximations.

Density Functional Treatment of the Planar Liquid-Vapour Interface.

In this section we consider the liquid vapour interface of a simple fluid, described by a short-ranged interatomic potential, in the presence of a weak stabilising (gravitational) field. Suppose that the temperature T lies below that for which the FW line intersects the bulk liquid coexistence curve, e.g. below $0.9T_c$ in figure 1. If we regard vapour as exerting a particular type of 'external' field on an atom deep in the liquid, we might expect the density profile $\rho(z)$ at the interface to exhibit damped oscillatory decay into the bulk saturated liquid of density $\rho_l(T)$. The general arguments outlined in earlier sections imply that the exponential decay length and the wavelength of the oscillations should be determined by equation 3, i.e. by $c(r)$ in the saturated liquid. Since vapour always lies on the monotonic side of the FW line, the profile should decay exponentially into vapour for all T, with the decay length set by $c(r)$ of the saturated vapour of density $\rho_v(T)$. On the other hand, for $T_c > T > 0.9T_c$ the saturated liquid lies in the monotonic region and the decay of the profile into the liquid should be pure exponential. Of course, the general arguments make no statement about the amplitude of any oscillatory contribution to $\rho(z)$. We might expect this to be smaller than for $h(r)$ or for wall-fluid interfaces, since a low-density vapour must exert an effective field which is much weaker than that due to a fixed atom or a rigid wall. Moreover, the liquid-vapour interface cannot be regarded as being strictly equivalent to a wall-liquid interface. Capillary-wave fluctuations play an important role in the former but not at the latter - at least in the absence of complete wetting by vapour - and that is why it is necessary to apply a stabilising field. Nevertheless, if one adopts a mean-field description of the 'intrinsic' or 'bare' interface and then attempts to include the effects of fluctuations subsequently, it is possible to investigate oscillatory structure. Density functional theories are the obvious means of calculating the equilibrium density profile. If oscillations are present they should be exposed by those approximations which provide an accurate description of packing effects. This requires a non-local treatment of repulsive forces between atoms. A local treatment will never yield oscillations. Note that all density functional approximations are mean-field-like in that they omit effects of fluctuations on the profile (18).

In order to test the above predictions density functional calculations were performed for the square-well fluid (1) using the weighted density approximation (WDA) to the hard-sphere (hs) free-energy functional developed by Tarazona (20) and treating attractive interactions between atoms in a mean-field approximation so that the intrinsic free energy functional is

$$F[\rho] = F_{hs}[\rho] + \frac{1}{2} \iint d\mathbf{r}_1 d\mathbf{r}_2 \rho(\mathbf{r}_1) \rho(\mathbf{r}_2) \phi_{att}(r_{12}) \tag{25}$$

The two-body direct correlation function obtained by differentiating this functional twice and taking the limit of a bulk homogeneous fluid is that given in equation 7. Special numerical procedures *(4)* were used to obtain very accurate solutions of the Euler-Lagrange equation which results from minimising $F[\rho] - \mu \int d\mathbf{r}\rho(\mathbf{r})$ for this interfacial problem *(1)*. For $T = 0.738T_c$, a temperature for which the saturated liquid is well inside the oscillatory region, $\rho(z)$ was found to exhibit oscillations on the liquid side but these are only visible at about 100x the normal scale for plotting. High magnification data was fitted to the form $\rho(z) - \rho_l \sim A_0 e^{-\alpha_0 z} \cos(\alpha_1 z - \theta)$. The resulting values of α_0 and α_1 are (within errors of the calculations) equal to those obtained from solutions of equation 3 with $c(r)$ calculated consistently for the bulk saturated liquid. For a high temperature state $T = 0.93T_c$, on the monotonic side of the FW line, oscillations could not be discerned even at the highest magnification and the decay of $\rho(z)$ into bulk liquid was pure exponential *(1)*.

Although these results confirm the basic premise regarding the existence of oscillations in the liquid-vapour profile, they will not inspire even the most ardent computer simulator or enthusiastic experimentalist to rush to seek the tiny wiggles. Figure 3 displays some more encouraging results. These refer to $T = 0.64T_c$, a temperature somewhat above the triple point and one that is typical of those investigated in simulations. In this case the maximum amplitude of the oscillations, obtained from the density functional calculations, is about 2% of the bulk liquid density. Were such oscillations to be present at a real fluid interface they should be detectable. However, as noted earlier, the calculations omit the effects of capillary-wave fluctuations which act to smear out the oscillations that arise in the mean-field description. The other curve in figure 3 shows the weighted density profile $\bar{\rho}(z)$ obtained from the solution of the WDA equations. This quantity also exhibits oscillations, but with much reduced amplitude. It corresponds to averaging $\rho(z)$, the dashed curve, over an atomic diameter σ and this mimics the local effects of interface wandering. $\bar{\rho}(z)$ is likely to provide some estimate of what might be observed in a computer simulation of the liquid-vapour density profile at low temperatures *(1)*.

Marrying the effect of fluctuations to an 'intrinsic' (mean-field) profile remains a difficult problem in the physics of interfaces *(18)*. The simplest approach corresponds to a Gaussian smearing over the interfacial roughness $\xi_\perp$. For an oscillatory profile tail with the form of equation 19 the wavelength $2\pi/\alpha_1$ and decay length α_0^{-1} are unaltered but the amplitude is reduced by a factor $\exp\left[-(\alpha_1\xi_\perp)^2/2\right]$ *(2, 21)*. A planar liquid-vapour interface which is at low temperature, and which is of macroscopic extent stabilised by earth's gravity, is usually estimated to have $\xi_\perp \sim$ one or two σ *(17)*. Since $\alpha_1 \approx 2\pi/\sigma$ this factor is $\exp(-2\pi^2)$ or $\exp(-8\pi^2)$, which is not very encouraging! Computer simulations

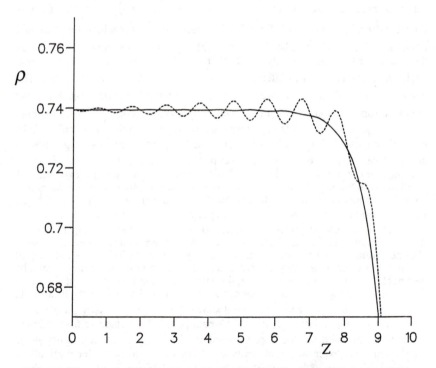

Figure 3. A section of the liquid-vapour density profile obtained from the WDA treatment of the square-well at $T/T_c = 0.64$. The dashed curve shows the equilibrium profile, displaying oscillatory decay into liquid. The solid curve shows the corresponding weighted density profile $\bar{\rho}(z)$. z is given in units of σ and ρ in units of σ^{-3}.

offer a somewhat more attractive route since one can vary $\xi_\perp$ by varying the interfacial area, L_x^2, available in a simulation. The roughness is given by (17)

$$\xi_\perp^2 = (2\pi\beta\gamma)^{-1} \int_{k_{min}}^{k_{max}} dk \frac{k}{k^2} \tag{26}$$

where γ is the surface tension and the upper and lower cut-off are wavenumbers $k_{max} = 2\pi/\sigma$ and $k_{min} = 2\pi/L_x$. This means that the amplitude of the oscillations should be reduced by a factor of $(L_x/\sigma)^{-\pi/(\beta\gamma\sigma^2)}$. The larger the surface tension, the weaker is the dependence on the area.

To the best of our knowledge, systematic simulations aimed at testing this prediction have not been carried out for a pure fluid. (There is a long history of claims of highly pronounced oscillatory structure at a liquid-vapour interface. Most of these were based upon either dubious theories or incomplete simulations - see the comments in reference (1)). However, a very recent molecular dynamics simulation of the planar interface between two **liquid** phases has focused on this issue (22). The model is an equi-molar binary mixture in which the aa and bb interatomic potentials are identical, both are a truncated Lennard-Jones 12 - 6 potential, but phase separation is ensured by making the ab potential purely repulsive. Pronounced oscillations are found in the equilibrium density profiles. These are insensitive to the length of the simulation box L_z (normal to the interface) but their amplitude does depend on the area L_x^2. Increasing the latter reduces the amplitude in a manner that is consistent with the power-law dependence predicted above (22). These results would appear to offer strong support for the physical picture that is presented here.

How would the picture be altered by the presence of dispersion forces, ie. in a real fluid? Experience gleaned from investigating the pole structure associated with the oscillatory decay of $h(r)$ suggests that oscillations, similar to those found for a finite-ranged model potential, would still develop in the density profile on the liquid side of the interface but only at intermediate range. These would be damped by capillary-wave fluctuations in the same fashion as described above. At longest range, in the ultimate tails of the profile, the dispersion forces will give rise to power-law decay of $\rho(z)$. If the pairwise interatomic potential has an attractive $-a_6/r^6$ tail then

$$\rho_l - \rho(z) \sim \pi\beta a_6 (\rho_l - \rho_v)(\rho S(0))_l / 6z^3 \quad, z \to \infty \tag{27}$$

This asymptotic result, like the corresponding result for $h(r)$, dates from the 1960's - see (23) and references therein. It is not known what effects fluctuations have on this result.

Implications for Wetting and Layering Transitions.

Theories of complete and critical wetting at wall-fluid interfaces in dimension $d = 3$ depend on a correct identification of the asymptotic decay of the mean-field liquid-vapour density profile (24). Much of the subtlety and richness of these transitions stems from competition between capillary-wave broadening and underlying mean-field structure. Almost all theories start from the premise that the liquid-vapour density profile is monotonic. Often this is calculated from a square-gradient or Landau type approximation to the free-energy functional. In which case the profile decays into liquid, say, as $\exp(-\kappa z)$, where κ^{-1} is the OZ (second moment) correlation length of the bulk liquid. It is clear from the previous discussion that κ^{-1} is not the appropriate length scale. Even on the monotonic side of the FW line the inverse decay length α_0 does not reduce to κ until very close to the critical point (25, 26). On the other side of the FW line the oscillations play a significant role. The binding potential $W(l)$, which describes the surface excess free energy of a wetting film of thickness l, is damped oscillatory rather than monotonically decaying, as is usually assumed. WDA density functional calculations for the square-well fluid, based on the functional of equation 25, at a square-well wall show that as a consequence of the oscillations in $W(l)$, complete wetting at bulk coexistence is replaced by pseudo wetting (26). Instead of an infinitely thick film of liquid wetting the wall-gas interface, a thick liquid film with $l \sim 12\sigma$ develops below $T = 0.9 T_c$. Only for higher temperatures, where oscillations disappear, does complete wetting occur. If a local density functional approximation were applied, so that oscillations could never develop, complete wetting would occur throughout the temperature range. Of course, this remarkable result is a prediction of a mean-field theory. Including the effects of the capillary-wave fluctuations for the wetting problem is non-trivial but it is likely that these can force the liquid to wet completely, even if $W(l)$ is oscillatory (26, 27). Nevertheless, it is clear that the temperature T_{FW} at which the FW line interesects the coexistence curve does have significance for wetting transitions in systems with short-ranged forces. (The temperature dependence of $W(l)$ and its dependence on the pole structure is discussed further in (26).) Note that the binding potential for real fluids with dispersion forces must contain a term bl^{-2}, with $b > 0$. Ultimately this term will always dominate the exponentially damped oscillatory contribution and lead to complete wetting. T_{FW} also sets an upper bound for the occurrence of a sequence of discrete layering transitions between adsorbed liquid layers (28, 29). Such transitions can only occur if $W(l)$ exhibits oscillations.

The Decay of the Solvation Force for Confined Fluids.

Our work also has implications for the asymptotics of the solvation force $f_s(L)$ for a fluid confined between two identical planar walls, separated by a distance L. The solvation force is the difference between the pressure of the confined fluid and that of a bulk reservoir at the same temperature and chemical potential. It provides an important measure of the effects of confinement on the free-energy of the fluid and has been investigated in many simulations and in calculations based on density functional theories (18,30). It is well-known that f_s exhibits pronounced oscillatory behaviour (with L) for high density liquids confined at small separations. These oscillations are a direct manifestation of packing constraints on the distribution of atoms. In the limit of large separation, $L \to \infty$, one might expect the decay of $f_s(L)$ to reflect the decay of the density profile $\rho(z)$ of the confined fluid. For the particular case of a fluid confined by perfectly hard walls there is an exact result $f_s(L) = \beta^{-1}(\rho_L(0) - \rho_\infty(0))$, where $\rho_L(0)$ is the density at contact. It is possible to argue (1) that for $L \to \infty$, $\rho_L(0)$ must decay in the same fashion as the profile at a single wall decays to its bulk value, which, we have established, is the same decay as that of $rh(r)$. Attard et.al. (19) have analysed the solvation force within the context of wall-particle OZ equations. They show that the one-dimensional Fourier transform of the integral of $f_s(L)$ has exactly the same pole structure as that of $\hat{h}(q)$.

Thus, for finite ranged wall-fluid potentials and fluid-fluid interatomic potentials and $L \to \infty, f_s(L) \sim e^{-\alpha_0 L}$ for states on the monotonic side of the FW line, whereas $f_s(L) \sim e^{-\bar{\alpha}_0 L} \cos(\alpha_1 L - \theta)$ on the oscillatory side - provided the states are sufficiently far from the critical point and there are no complications of phase transitions *(1, 2)*. If dispersion forces are present, as they must be in real systems, $f_s(L)$ will decay ultimately as L^{-3}. However, for sufficiently high densities, there will be exponentially damped oscillations at intermediate separations L whose decay length and oscillatory wavelength are determined by the leading order pole of $\hat{h}(q)$. In the case of confined fluid mixtures the decay of $f_s(L)$ is determined by the condition $D(\alpha) = 0$, ie. the same condition that determines the decay of all the $h_{ij}(r)$ - see equation 11.

It is striking that a thermodynamic quantity (an excess pressure) should exhibit precisely the same decay laws (and crossover) as the interatomic correlation function $h(r)$ and this attests to the universal character of the asymptotic decay of structural correlations - at least for systems with short-ranged forces. On the other hand, one should recall that the solvation force is merely a special case of a fluid mediated particle-particle interaction in which the size of the two particles is made macroscopic.

Conclusions.

The density functional theory of classical inhomogeneous fluids can be viewed as a descendent of that developed by Van der Waals in his 1893 treatment of the liquid-vapour interface *(31)*. Modern weighted density approximations *(18)* go well beyond Van der Waals' ideas in that they include the effects of the short and intermediate-ranged correlations arising from packing of the atoms. This article has summarised some of the striking phenomena that occur when such correlation effects are included. For bulk fluids the main result is the existence of crossover lines - or disorder lines in magnetic terminology. Although no thermodynamic singularity occurs on crossing such a line (no phase transition) the correlation functions do change their character significantly. The repercussions for phase transitions at interfaces are not fully understood and certainly warrant further attention. The structure of the liquid-vapour interface is still of considerable interest. Very recent x-ray reflectivity measurements for liquid mercury indicate pronounced surface layering and point to a density profile with very large amplitude oscillations (much larger than those predicted for model atomic fluids) on the liquid side of the interface *(32)*. It remains to be seen whether the asymptotic approach developed here can be extended to the case of liquid metals and, indeed, to other types of fluids.

Acknowledgements.   Much of the work reported here was carried out in collaboration with J. R. Henderson and D. C. Hoyle. We have also benefitted from many discussions with J. Stecki. RJFLdC is grateful to JNICT/Ciência (Portugal) for financial support.

Literature Cited.

1.   Evans, R., Henderson, J. R., Hoyle, D. C., Parry, A. O. and Sabeur, Z. A. *Mol.Phys.* **1993**, *80*, 755
2.   Evans, R., Leote de Carvalho, R. J. F., Henderson, J. R. and Hoyle, D. C. *J.Chem.Phys.* **1994**, *100*, 591

3.   Leote de Carvalho, R. J. F., Evans, R., Hoyle, D. C. and Henderson, J. R. *J.Phys.Cond.Matt* . **1994**, *6*, 9275
4.   Henderson, J. R. and Sabeur, Z. A. *J.Chem.Phys.* **1992**, *97*, 6750; *Mol.Phys* **1994**, *82*, 765
5.   Fisher, M. E. and Widom, B. *J.Chem.Phys.* **1969**, *50*, 3756
6.   Zerah, G. and Hansen, J-P. *J.Chem.Phys.* **1986**, *84*, 2336
7.   Smit, B. *J.Chem.Phys.* **1992**, *96*, 8639
8.   Enderby, J. E., Gaskell, T. and March, N. H. *Proc.Phys.Soc.* **1965**, *85*, 217
9.   Evans, R. and Sluckin, T. J. *J.Phys.C: Sol.St.Phys* . **1981**, *14*, 2569
10. Widom, B. *J.Chem.Phys.* **1964**, *41*, 74
11. Verlet, L. *Phys.Rev.* **1968**, *165*, 201
12. Bellissent-Funel, M. C., Buontempo, U., Filabozzi, A., Petrillo, C. and Ricci, F. P. *Phys.Rev.B.* **1992**, *45*, 4605
13. Sung, S. H. and Chandler, D. *Phys.Rev.A.* **1974**, *9*, 1688
14. Martynov, G. A. *Fundamental Theory of Liquids: Method of Distribution Functions:* Hilger: Bristol, 1992
15. Leote de Carvalho, R. J. F. and Evans, R. *Mol.Phys.* **1994**, *83*, 619
16. Kirkwood, J. G. *Chem.Rev.* **1936**, *19*, 275
17. Evans, R. *Adv.Phys.* **1979**, *28*, 143
18. Evans, R. *In Fundamentals of Inhomogeneous Fluids;* Henderson, D., Ed; Dekker: New York, 1992; pp 85
19. Attard, P., Ursenbach, C. P. and Patey, G. N. *Phys.Rev.A.* **1992**, *45*, 7621. See also Attard, P., Bérard, D. R., Ursenbach, C. P. and Patey, G. N. ibid, **1991**, *44*, 8224
20. Tarazona, P. *Phys.Rev.A.*, **1985**, *31*, 2672
21. Mikheev, L. V. and Chernov, A. A. *Sov.Phys.JETP*, **1987**, *65*, 971
22. Toxvaerd, S. and Stecki, J. *J.Chem.Phys.* **1995**, *102*, 7163
23. Barker, J. A. and Henderson, J. R. *J.Chem.Phys.* **1982**, *76*, 6303
24. Henderson, J. R. *In Fundamentals of Inhomogeneous Fluids;* Henderson, D., Ed; Dekker: New York, 1992; pp 23
25. Evans, R., Hoyle, D. C. and Parry, A. O. *Phys.Rev.A.*, **1992**, *45*, 3823
26. Henderson, J. R. *Phys. Rev. E.* **1994**, *50*, 4836
27. Chernov, A. A. and Mikheev, L. V. *Physica A*, **1989**, *157*, 1042
28. Ball, P. C. and Evans, R. *J.Chem.Phys.* **1988**, *89*, 4412
29. Fan, Y. and Monson, P. A. *J.Chem.Phys.* **1993**, *99*, 6897
30. Evans, R. *In Liquids at Interfaces;* Charvolin, J., Joanny, J. F. and Zinn-Justin, J., Eds; Les Houches Summer School Session XLVIII; Elsevier: New York, 1990; pp3.
31. Van der Waals, J. D. *Verhik.Akad.Wet. (Sect.1).* **1893**, *1*, No.8. English translation by J. S. Rowlinson, *J.Stat.Phys.* **1979**, *20*, 197
32. Magnussen, O. M., Ocko, B. M., Regan, M. J., Penanen, K., Pershan, P. S. and Deutsch, M. *Phys.Rev.Lett.*, **1995**, *74*, 4444

Chapter 13

Expanded Density Functionals

J. K. Percus

Courant Institute of Mathematical Sciences and Physics Department, New York University, 251 Mercer Street, New York, NY 10012

The free energy density functional format is increasingly used to represent the structure of non-uniform fluids in thermal equilibrium. This representation can often be simplified and extended by introducing in-principle extraneous densities with respect to which the free energy is stationary and, if possible, a minimum. The virtual necessity of such an expanded framework is illustrated for a periodic one-dimensional classical lattice gas, and its practical effectiveness attested to by the exact solution of a one-dimensional classical fluid with nearest neighbor interactions. A number of current approximation methods, classical and quantum, are either in this form or are shown to fall into it under slight reformulation. Identification of large scale excitations makes it possible to include their images in this framework, leaving only local fluctuations to be accounted for, and the possibility is raised of a meaningful density-effective potential joint representation.

Realistic particle systems exhibit phenomenology on multiple scales of space and time. An ideal descriptive vehicle would be one in which one could simultaneously sample at all of these levels. This would of course be an overcomplete description, referring to more degrees of freedom than the system actually possesses. Nonetheless, it is feasible—and I believe ultimately necessary. I would like to discuss this issue in the relatively narrow regime of thermal equilibrium in the grand ensemble, primarily in the format of simple classical fluids, but not necessarily restricted to these.

Although we are thus creating a picture seen at various levels of resolution, the elements of the picture have yet to be spelled out. Traditionally, at least in the past few decades, the pair distribution—certainly reasonable for pair-interacting fluids—was the object of choice, leading to highly effective integral

0097–6156/96/0629–0185$15.00/0

equation approximations (1). For uniform fluids, pair distributions are functions only of the scalar pair separation, rendering quite feasible both analytical and numerical techniques, but in the presence of arbitrary non-uniformity, this comfortable situation does not persist: two full positional arguments are required. On the other hand, the spatial density pattern itself now becomes significant, and can be probed to create all higher distributions (2). Thus, approximate thermodynamics as a functional of the density alone has become a strong competitor (3) to the older integral equation techniques. As we have indicated, this is more of a mathematical device and might rewardingly be broadened to include other locally defined indicators of the physical structure. Wertheim's associated fluid formulation (4) is a well-known example. In fact, the expanded context thus suggested expands the class of "solvable" systems, as well as the class of empirical approximations, and I would like to examine both of these areas, in an introductory fashion.

Prototype

Probably the simplest non-trivial context in which the utility of an expanded description becomes obvious is that of a one-dimensional lattice gas with nearest neighbor exclusion. Here the sites are integers $\{x\}$ and the site occupations $\{\rho_x = 0,1\}$ with the restriction that $\rho_{x-1}\rho_x = 0$ for each x. Non-uniformity is imposed by the external Boltzmann factor $z_x = e^{\beta\mu_x}$ where $\mu_x = \mu - u_x$ is the local chemical potential due to external potential u_x; The *profile relation* between the $\{n_x = \langle\rho_x\rangle\}$ and the μ_x is our way of encompassing the full thermodynamic information. In order to find it, we note that the partition function can be written as

$$\Xi = \sum_{\{\rho_x\}} \prod_x z_x^{\rho_x} \prod_x (1 - \rho_{x-1}\rho_x)$$
$$= Tr \prod_\lambda (z_x e) \tag{2.1a}$$

where

$$z_x = \begin{pmatrix} 1 & 0 \\ 0 & z_x \end{pmatrix}, \quad e = \begin{pmatrix} 1 & 1 \\ 1 & 0 \end{pmatrix}, \tag{2.1b}$$

and then introduce the relative partition function with two adjacent occupations fixed, and their interaction dropped, indicated pictorially as

$$\xi_{\rho\rho'}^{x+\frac{1}{2}} = \cdots \bullet - - - - \bullet - - - - \circ \quad \circ - - - - \bullet - - - - \bullet \cdots / \Xi. \tag{2.2}$$

While site occupations ρ, ρ' at x and $x + 1$ are fixed, the other occupations are summed over. It is clear that

$$n_x = \xi_{10}^{x+\frac{1}{2}}, \qquad n_{x+1} = \xi_{01}^{x+\frac{1}{2}}$$
$$1 = \xi_{00}^{x+\frac{1}{2}} + \xi_{01}^{x+\frac{1}{2}} + \xi_{10}^{x+\frac{1}{2}}. \tag{2.3}$$

But it is also clear on adopting the matrix notation of equation 2.1b that

$$(z_x e)^{-1} \xi^{x+\frac{1}{2}} = \xi^{x-\frac{1}{2}} (e z_x)^{-1}, \tag{2.4}$$

so that

$$\text{Det } \xi^{x+\frac{1}{2}} = \text{Det } \xi^{x-\frac{1}{2}} = \ldots = K \tag{2.5}$$

for some system-wide constant K.

Solving equations 2.3 and 2.5, we readily find

$$\xi^{x+\frac{1}{2}} = \begin{pmatrix} 1 - n_x - n_{x+1} & n_{x+1} \\ n_x & (n_x n_{x+1} + K)/1 - n_x - n_{x+1} \end{pmatrix}, \tag{2.6}$$

and substituting into e.g. the (10) element of equation 2.4 then tells us that

$$\beta \mu_x = \ell n \big(K + n_x (1 - n_x) \big) - \ell n (1 - n_x - n_{x+1}) - \ell n (1 - n_x - n_{x-1}). \tag{2.7}$$

There are now two possibilities we should consider.

a) A volume-constrained open chain, so that $n_x \to 0$ as $x \to \pm\infty$, but $n_x e^{\beta \mu_x} \to 1$. According to equation 2.7, then, $K = 0$, or (5)

$$\beta \mu_x = \ell n\, n_x + \ell n (1 - n_x) - \ell n (1 - n_x - n_{x+1}) - \ell n (1 - n_{x-1} - n_x). \tag{2.8}$$

There is a free energy generating functional $\overline{F}$ in the sense that

$$\mu_x = \partial \overline{F}[n]/\partial n_x , \tag{2.9}$$

and here

$$\beta \overline{F}[n] = \sum_x [n_x \ell n\, n_x - (1 - n_x)\, \ell n (1 - n_x) \tag{2.10}$$
$$+ (1 - n_{x-1} - n_x)\, \ell n (1 - n_{x-1} - n_x)]$$

is *local* in that it and $\beta \mu_x$ only couple sites separated by at most the range of the interaction.

b) A closed chain or ring. Now the only thing one can say about K is that since $\rho_x + \rho_{x+1} \leq 1$, then $n_x + n_{x+1} \leq 1$, so that $K \geq -n_x n_{x+1}$ implies that $K \geq -\frac{1}{4}$. Setting $K = -\frac{1}{4} + C^2$ where $C \geq 0$, we can then write

$$\mu_x = \partial \overline{\overline{F}}[n, C]/\partial n_x|_C \tag{2.11}$$

where

$$\beta \overline{\overline{F}}[n, C] = \sum_x \left[\left(C - \frac{1}{2} + n_x \right) \ell n \left(C - \frac{1}{2} + n_x \right) \right.$$
$$+ \left(C + \frac{1}{2} - n_x \right) \ell n \left(C + \frac{1}{2} - n_x \right) \tag{2.12}$$
$$\left. - (1 - n_{x-1} - n_x) \ell n \left(1 - n_{x-1} - n_x \right) \right].$$

In other words, $\overline{\overline{F}}$ remains local, but in the system-wide environment measured by the collective amplitude C—which however is not yet known. Matters are seen in a different light if we now quote a readily proved theorem (6), valid for a whole set of collective variables:

Theorem. *If for the set $\{C_\alpha\}$, the profile equation $\mu_x = \partial\overline{\overline{F}}[n,C]/\partial n_x|_C$ is valid, then a pure collective contribution $\Delta[C]$ can be found such that on defining*

$$\overline{F}[n,C] = \overline{\overline{F}}[n,C] + \Delta[C],$$

one has

$$\mu_x = \partial\overline{F}[n,C]/\partial n_x|_C$$
$$0 = \partial\overline{F}[n,C]/\partial C|_n$$

and

$$\overline{F}[n] = \overline{F}[n,C[n]]. \tag{2.13}$$

Equivalently, on the expanded space $[n,C]$, one has the stationary principle at fixed $\{\mu_x\}$,

$$\delta\left[\overline{F}[n,C] - \sum_x \mu_x n_x\right] = 0, \tag{2.14}$$

the vanishing C-variations being sufficient in principle to determine the expression $C_\alpha[n]$. For example, in the above nearest neighbor exclusion case, it is not hard to show (e.g. by restricting to the uniform case) that (7)

$$\beta\Delta[C] = \left(C + \frac{1}{2}\right)\ell n\left(C + \frac{1}{2}\right) - \left(C - \frac{1}{2}\right)\ell n\left(C - \frac{1}{2}\right). \tag{2.15}$$

Whether or not $\overline{F}[n,C]$ is convex, leading to a *minimum* principle, is not obvious.

Nearest Neighbor Interaction in One Dimension

Let us take a step upward to a continuum fluid in one dimensional space. From a lattice viewpoint, any interaction of range more than one lattice spacing creates an interaction loop, and so one might expect auxiliary fields to appear here as well. Simplest is an interaction $\varphi(x,y)$ between next neighbors alone, physically meaningful for a sufficiently large hard core, or if the interaction is due to excitation of the medium between two adjacent particles. Now we can order the particles, $x_1 \leq x_2, \cdots \leq x_N$; introducing the external and ordered internal Boltzmann factors

$$z(x) = e^{\beta\mu(x)}, \ e(x,y) = e^{-\beta\varphi(x,y)}\ \theta(y - x) \tag{3.1}$$

where θ is the unit step function, we find at once, in standard Dirac quantum mechanical notation, that the grand partition function is given by

$$\Xi[\mu] = 1 + \langle 1 \mid z(I - ez)^{-1} \mid 1\rangle \tag{3.2}$$

where

$$1(x) = 1$$

and z represents the diagonal matrix with diagonal elements $z(x)$.

Since the density $n(x) = \delta\ell n\Xi/\delta\ell nz(x)$, we then have

$$n(x) = \Xi^-(x)\, z(x)\, \Xi^+(x)/\Xi \;, \tag{3.3}$$

with truncated partition functions, row and column vectors respectively, defined by

$$\Xi^- = \langle 1|(I - ze)^{-1}, \Xi^+ = (I - ez)^{-1}|1\rangle, \text{ or }$$

$$\Xi^- - \Xi^- ze = \langle 1|, \quad \Xi^+ - ez\,\Xi^+ = |1\rangle. \tag{3.4}$$

Equivalently, applying the two-sided inverse e^{-1}, for which one can show that $\langle 1|e^{-1} = 0 = e^{-1}|1\rangle$ (but associativity is not automatic), we have $\Xi^-(e^{-1} - z) = 0$, $(e^{-1} - z)\Xi^+ = 0$, so that (8) on using equation 3.3 to eliminate $z(x)$,

$$\begin{aligned}(\Xi^- e^{-1})(x) - n(x)\, \Xi/\Xi^+(x) = 0 \\ (e^{-1}\Xi^+)(x) - n(x)\, \Xi/\Xi^-(x) = 0.\end{aligned} \tag{3.5}$$

But information has been lost in obtaining equation 3.5, and so boundary conditions have to be resupplied,

$$\Xi^+(\infty) = \Xi^-(-\infty) = 1, \quad \Xi^+(-\infty) = \Xi^-(\infty) = \Xi. \tag{3.6}$$

The key observation is now that equations 3.3, 3.5, and 3.6—which fully determine M, Ξ^-, Ξ^+, and Ξ—all arise from the free energy

$$\beta\overline{F}\,[n, \Xi^-, \Xi^+, \Xi] = \int n(x)\,\left[\ell n\,n(x) - \ell n\,\Xi^-(x) - \ell n\,\Xi^+(x) + \ell n\,\Xi\right]\,dx$$

$$-\frac{1}{2}\left(\ell n\,\Xi^-(\infty) - \ell n\,\Xi^-(-\infty)\right) + \frac{1}{2}\left(\ell n\,\Xi^+(\infty) - \ell n\,\Xi^+(-\infty)\right)$$

$$+\frac{1}{2}\frac{1}{\Xi}\Xi^-(e^{-1}\,\Xi^+) + \frac{1}{2}\frac{1}{\Xi}(\Xi^- e^{-1})\Xi^+$$

$$\tag{3.7}$$

on a highly expanded space. That is, the relations $\beta\mu(x) = \delta\beta\overline{F}/\delta n(x)$, $\delta\beta\overline{F}/\delta\Xi^+(x) = 0 = \delta\beta\overline{F}/\delta\Xi^-(x)$ reproduce equations 3.3 and 3.5, and with a little more care, stationarity with respect to $\Xi^\pm(\pm\infty)$ and Ξ reproduce equation 3.6.

The awkward boundary terms in equation 3.7 are materially simplified by the consistent replacement

$$\Xi^-(x) = e^{-\frac{1}{2}\beta v(x)}\, e^{-\beta\int_{-\infty}^{x}\omega(t)dt}$$

$$\Xi^+(x) = e^{-\frac{1}{2}\beta v(x)}\, e^{-\beta\int_{x}^{\infty}\omega(t)dt} \tag{3.8}$$

$$\Xi = e^{-\beta\int_{-\infty}^{\infty}\omega(t)dt}$$

where $v(x) \to 0$ as $x \to \pm\infty$. Substituting into equation 3.7 then yields the completely equivalent

$$\beta \overline{F}[n, \omega, v] = \int n(x) \, (\ln n(x) - 1 + \beta v(x)) \, dx + \beta \int \omega(x) \, dx$$
$$+ \iint e^{-\frac{\beta}{2}v(x)} \, e^{-\beta \int_y^x \omega(t)\,dt} \, e^{-1}(x, y) \, e^{-\frac{\beta}{2}v(y)} \, dx \, dy, \qquad (3.9)$$

which is much more transparent: $\mu(x) = \delta \overline{F}/\delta n(x)$ identifies $v(x)$ as the effective potential due to the interaction; the complicated last term reduces to $N = \int n(x) \, dx$ when evaluated, thereby canceling the $\int -n(x) \, dx$ in the first term and identifying $\omega(x)$ as a specific grand potential. Furthermore, the form in equation 3.9 is readily—but not uniquely—extrapolated to higher dimensionality.

Empirical expanded functionals

The introduction of auxiliary fluids with respect to which the free energy is stationary is hardly novel. Perhaps simplest is the "rank 1" model (exact for one-dimensional hard cores) discussed some years ago (*9*)

$$\overline{F}[n, \nu] = -\int P(\nu(r)) \, dr + \iint n(r') \, U(\nu(r), r' - r) \, dr \, dr', \qquad (4.1)$$

for which

$$\mu(r; n) = \delta \overline{F}/\delta n(r) = \int U(\nu(r'), r - r') \, dr'$$
$$0 = \delta \overline{F}/\delta \nu(r) = -P'(\nu(r)) + \int n(r') \, U'(\nu(r), r' - r) \, dr', \qquad (4.2)$$

The general strategy employed with such empirical forms is to obtain P and U from assumed knowledge of the uniform system properties, in which state one imposes the condition $\nu = n$. Thus, if $U_k(\nu)$ indicates the Fourier transform of $U(\nu, r)$ with respect to its spatial argument, we first require

$$U_0(n) = \mu(n), \quad P'(n) = nU_0'(n), \qquad (4.3)$$

identifying $P(n)$ as the bulk pressure. Furthermore, differentiation of equation 4.2 readily establishes that the complete direct correlation function is given by

$$C(r, r'; n) = \frac{\delta \beta \mu(r)}{\delta n(r')} = \int \frac{\beta U'(\nu(R), r - R) \, U'(\nu(R), m' - R)}{P''(\nu(R)) - \int n(R') U''(\nu(R), R - R') \, dR'} \, dR$$
$$(4.4)$$

so that, specializing to uniformity and Fourier transforming,

$$\left(U_k'(n)\right)^2 = \frac{P'(n)}{n} \frac{C_k'(n)}{\beta}, \tag{4.5}$$

thereby fixing $U_k(n)$.

The above model is very van der Waals-like in that $C(r, r')$ does not maintain the required singular amplitude $\delta(r - r')/n(r) + \cdots$ when the state is nonuniform. This particular defect is cured by positing the form, equation 4.1, only for the excess free energy

$$\beta \overline{F}^{ex}[n] = \beta \overline{F}[n] - \int n(r) \left(\ell n \, n(r) - 1\right) \, dr. \tag{4.6}$$

Indeed, a number of free energy models have this structure. One of the most effective is the recipe initiated by Meister and Kroll (*10*), in which the mean-field van der Waals

$$\overline{F}^{ex}[n] = \int n(r) f_0\left(n(r)\right) \, dr - \frac{1}{2} \iint n(r) n(r') w(r - r') \, dr \, dr' \tag{4.7}$$

is extended to

$$\overline{F}^{ex}[n, \nu] = \int n(r) f_0(\nu(r)) \, dr - \frac{1}{2} \iint n(r) n(r') \, w(\nu(r), r - r') \, dr \, dr' \tag{4.8}$$

where f_0 and w are determined as in the previous model.

The commonly used weighted density ansatz can be regarded as having as its progenitor the one-dimensional hard core fluid mentioned above, in the form (*11*)

$$\overline{F}[n] = \int \tilde{\nu}(x) \, f\left(\nu(x)\right) \, dx$$

where

$$\nu(x) = \frac{1}{a} \int_{x-a/2}^{x+a/2} n(y) \, dy, \quad \tilde{\nu}(x) = \frac{\partial}{\partial a} a \nu(x). \tag{4.9}$$

Its customary appearance is (*12*)

$$\overline{F}^{ex}[n] = \int n(r) f\left(\nu(r)\right) \, dr \tag{4.10}$$

where

$$\nu(r) = \int n(r') \, w\left(\nu(r), r - r'\right) \, dr'$$

$$\int w(\nu, r) \, dr = 1. \tag{4.10'}$$

Although equation 4.10 is not stationary with respect to ν, it becomes so if equation 4.10' is replaced by its Newton-Raphson (or other) stationary solution, i.e., we need only replace equation 4.10 by

$$\overline{F}^{ex}[n, \nu] = \int n(r) f \left[\frac{\int n(r')\, w^\circ\big(\nu(r),\, r - r'\big)\, dr'}{1 - \int n(r')\, w'\big(\nu(r),\, r - r'\big)\, dr'} \right] dr \qquad (4.11)$$

where

$$w^\circ(\nu) \equiv w(\nu) - \nu w'(\nu)$$

to guarantee equation 4.10' by the condition $\delta \overline{F}/\delta \nu(r) = 0$. Of course, the functions f and w are again to be determined by comparison with uniform fluid singlet and pair densities.

It should also be mentioned in this context that approximation methods outside the domain of classical fluids will often exhibit structural similarities to the above empirical forms. For example, Herring's treatment (13) of the free Fermion kinetic energy density functional $T[n]$ in one-dimensional space can be written in a similar form. Here one takes the Weizsacker energy

$$T_W[n] = \frac{1}{8} \int \big(n'(x)\big)^2 \, dx/n(x) \qquad (4.12)$$

as the low density "ideal gas" contribution, and writes

$$T^{ex}[n] = T[n] - T_W[n]$$
$$= \frac{\pi^2}{6} \int \nu(x) \Big/ \left(\int_{x - \frac{1}{2\nu(x)}}^{x + \frac{1}{2\nu(x)}} dx'/n(x') \right) dx \qquad (4.13)$$

where

$$\int_{x - \frac{1}{2\nu(x)}}^{x + \frac{1}{2\nu(x)}} n(x')\, dx' = 1 \qquad (4.14)$$

as an expression that reduces to the Thomas-Fermi density functional $T_{TF} = (\pi^2/6) \int n(x)^3 \, dx$ for slowly varying density. In the form, equations 4.13 and 4.14, extension to three dimensions is immediate—although not unique—and of course the Newton-Raphson ploy again allows the construction of $T^{ex}[n, \nu]$ if $1/\nu(x)$ in equation 4.13 is replaced by

$$1/\nu(x) \to 1/\nu(x)$$
$$+ 2 \left(n\!\left(x + \frac{1}{2} \frac{1}{\nu(x)} \right) + n\!\left(x - \frac{1}{2} \frac{1}{\nu(x)} \right) \right)^{-1} \left(1 - \int_{x - \frac{1}{2\nu(x)}}^{x + \frac{1}{2\nu(x)}} n(x')\, dx' \right)$$
$$(4.15)$$

Separation of scales

Aside from being a bit ad hoc, the empirical forms noted typically have the property that $\overline{F}[n, v]$ is stationary with respect to v, not a minimum[3], suggesting an unpleasant saddle point character. This need not be the case. A major purpose of the auxiliary field expansion of density space is to allow for the creation of microscopic or macroscopic variables which when fixed leave only local fluctuations of the type that fluid state theory is accustomed to handle. When this motivation is given concrete form, the free energy remains convex, at least before any further approximations. Let us first examine the abstract situation. For internal energy $h[\rho]$, a functional of the microscopic configuration density $\rho(r) = \sum \delta(r - r_j)$, and local chemical potential $\mu(r)$, we write the grand partition function in brief symbolic form as

$$\Xi[\mu] = \int e^{-\beta h[\rho]} \, e^{\beta \rho \cdot \mu} \, D\rho. \tag{5.1}$$

Suppose then that constraining a set $\{v[\rho]\}$ anchors relevant large scale fluctuations, leaving a system only locally correlated, and enforce the constraints to an adjustable extent to values $\{v\}$ by introducing the weight functional $W[[v[\rho]-v]$, normalized so that $\int W[v]Dv = 1$. We can then rewrite

$$\Xi[\mu] = \iint e^{-\beta \mu[\rho]} \, e^{\beta \rho \cdot \mu} \, W[v[\rho] - v] \, D\rho \, Dv \tag{5.2}$$

$$= \int \Xi[\mu|v] \, Dv$$

where

$$\Xi[\mu|v] = \int W[v(\rho) - v] \, e^{\beta \rho \cdot \mu} \, e^{-\beta \mu[\rho]} \, D\rho. \tag{5.3}$$

$\Xi[\mu|v]$ is the constrained partition function that we assume has large scale fluctuations quenched, and $H[\mu|v] = -\frac{1}{\beta}\ell n\, \Xi[\mu|v]$ the corresponding effective energy in v-space. The approximate evaluation of $H[\mu|v]$ is then a standard way—a la Landau-Lifshitz—of treating systems whose description calls for more than one space-time scale. However, we will go one step further. Introduce $\{\xi\}$ conjugate to $\{v\}$, and generalize equation 5.2 to

$$\Xi[\mu, \zeta] = \int \Xi[\mu|v] \, e^{-\beta \zeta \cdot v} \, Dv, \tag{5.4}$$

to be evaluated at $\zeta = 0$. If indeed v plays the role of order parameter, we can anticipate that the integral will be restricted to a small region of v-space, depending of course on ζ. Since $\Xi[\mu, \zeta]$ is the Laplace transform (2-sided) of a positive kernel,

$$\Omega[\mu, \zeta] = -\frac{1}{\beta} \, \ell n\, \Xi[\mu, \zeta] \tag{5.5}$$

will be concave in the pair of fields. Defining

$$n = -\delta\Omega/\delta\mu, \quad \nu = \delta\Omega/\delta\zeta, \tag{5.6}$$

we can hence Legendre transform to a convex free energy

$$\overline{F}[n, \nu] = n \cdot \mu[n, \nu] - \nu \cdot \zeta[n, \nu] + \Omega\left[\mu[n, \nu], \; \zeta[n, \nu]\right] \tag{5.7}$$

and this in fact will have the $\zeta = 0$ specialization given by the now familiar

$$\mu = \delta\overline{F}/\delta n, \quad 0 = \delta\overline{F}/\delta\nu. \tag{5.8}$$

It is important to observe that the final expression equation 5.4 can be written as

$$\Xi[\mu, \zeta] = \int e^{-\beta h[\rho]} \, e^{\beta\rho\cdot\mu} \, e^{-\beta\zeta\cdot\nu[\rho]} \, D\rho, \tag{5.9}$$

independent of the specific form of W, which has been replaced by a Legendre transform for the purpose of fixing the $\{v[\rho]\}$.

For example, in the Ramakrishnan-Yussouff theory (14) of crystallization, signaled by the macroscopic appearance of a set of wave vectors $K = \{k\}$, the order parameters are $(1/N)\sum_j e^{ik\cdot r_j}$ where N estimates the mean particle number, and one starts with

$$W\left[\frac{1}{N}\sum_j e^{ik\cdot r_j} - \nu_k\right] = \exp{-\beta\frac{N}{2}\sum_{k\in K}\lambda_k \left|\frac{1}{N}\sum_j e^{ik\cdot r_j} - \nu_k\right|^2}, \tag{5.10}$$

suitably normalized. Integration over ν in equation 5.4 results in insertion of the factor
$\exp(-\beta/N \sum \zeta_k e^{ik\cdot r_j})$ into equation 5.2, precisely as in the usual analysis.

Density-potential hybrid

The simultaneous advantage and disadvantage of the formulation, equations 5.9, 5.7, is that it makes use of explicit expressions for the order parameters which one believes are excited under the circumstances of interest. If they were determined only implicitly, as in the Gervais-Sakita prescription (15) for the instantaneous interface $z = \eta(x, y)$ of a nominally planar two-phase separation:

$$\int n_I'\left(z' - \eta(x, y)\right) w(x - x', \, y - y')\big(\rho(r') - \eta_I(0)\big) \, dr' = 0, \tag{6.1}$$

where n_I is a reference interfacial profile and w a smoothing function, matters wold be much more complicated. It would clearly be more satisfactory if the separation of scales entered as an inescapable aspect of the structure.

One way of invoking this behavior is by means of the Kac-Siegert (*16*) or Hubbard-Stratonovich procedure applied to interactions composed of short range "reference" φ_0 and long-range attraction $-\varphi$. The former are responsible for local particle-like fluctuations, while the latter play a direct role in the wave-like collective motions associated for example with interfaces. The separation is made quite literally:

$$\Xi[\mu] = \int e^{-\beta h_0[\rho]}\, e^{\beta \mu \cdot \rho}\, e^{\frac{1}{2}\beta \rho \cdot \varphi \rho}\, D\rho, \qquad (6.2)$$

where interaction self-energy has been included for convenience, and then represented as a functional Laplace transform

$$\Xi[\mu] = \int e^{-\beta h_0[\rho]}\, e^{\beta \mu \cdot \rho}\, e^{-\beta \rho \cdot v}\, e^{-\frac{1}{2}\beta v \cdot \varphi^{-1} v}\, D\rho\, Dv. \qquad (6.3)$$

For present purposes, we then generalize to

$$\Xi[\mu, \zeta] = \int e^{-\beta h_0[\rho]}\, e^{\beta \mu \cdot \rho} e^{-\beta(\zeta + \rho)\cdot v}\, e^{-\frac{1}{2}\beta v \cdot \varphi^{-1} v}\, D\rho\, Dv, \qquad (6.4)$$

$$\Omega[\mu, \zeta] = -\frac{1}{\beta}\, \ell n\, \Xi[\mu, \zeta],$$

to be evaluated at $\zeta = 0$.

Let us now introduce the conjugate $n = \langle \rho \rangle = -\delta\Omega/\delta\mu$, $\nu = \langle v \rangle = \delta\Omega/\delta\zeta$, and the corresponding convex $\overline{F} = \Omega + n \cdot \mu - \nu \cdot \zeta$. If we wanted to model $\overline{F}[n, \nu]$ by an empirical form, we would fit functional parameters, as in Sec. 4, by insisting upon reproducing uniform system correlations. The expanded space correlations are readily found if we observe on making the transformation $v \to v - \varphi$ in equation 6.4 that $\Xi[\mu, \zeta] = e^{\frac{1}{2}\beta\zeta \cdot \varphi\zeta}\, \Xi[\mu + \varphi\zeta]$, or

$$\Omega[\mu, \zeta] = \Omega[\mu + \varphi\zeta] - \frac{1}{2}\zeta \cdot \varphi\zeta. \qquad (6.5)$$

Hence, continuing our obvious condensed notation,

$$\begin{aligned} n[\mu, \zeta] &= -\delta\Omega/\delta\mu = n[\mu + \varphi\zeta] \\ \nu[\mu, \zeta] &= \delta\Omega/\delta\zeta = -\varphi\, n[\mu + \varphi\zeta] + \zeta, \end{aligned} \qquad (6.6)$$

Identifying ν as the effective potential due to $n + \zeta$, while the compound structure factor matrix is given by

$$\begin{aligned} S_{nn}[\mu, \zeta] &= \delta n[\mu, \zeta]/\delta\beta\mu = S[\mu + \varphi\zeta] \\ S_{n\nu}[\mu, \zeta] &= -\delta n[\mu, \zeta]/\delta\beta\zeta = -S[\mu + \varphi\zeta]\varphi \\ S_{\nu\nu}[\mu, \zeta] &= -\delta\nu[\mu, \zeta]/\delta\beta\zeta = \varphi S[\mu + \varphi\zeta]\varphi + \varphi/\beta. \end{aligned} \qquad (6.7)$$

Consequently, we have for the direct correlation matrix,

$$\underset{\approx}{C}[n,\nu] = \underset{\approx}{S}[\mu,\zeta]^{-1} = \begin{pmatrix} S^{-1}[\mu + \varphi\zeta] + \beta\varphi & \beta I \\ \beta I & \beta\varphi^{-1} \end{pmatrix}. \tag{6.8}$$

But

$$\underset{\approx}{C}[n,\nu] = \begin{pmatrix} \delta^2\beta\overline{F}[n,n]/\delta n\delta n & -\delta^2\beta\overline{F}[n,\nu]/\delta n\delta\nu \\ -\delta^2\beta\overline{F}[n,\nu]/\delta\nu\delta n & \delta^2\beta\overline{F}[n,\nu]/\delta\nu\delta\nu \end{pmatrix} \tag{6.9}$$

and (6.6) implies

$$\begin{aligned} -\delta\overline{F}[n,\nu]/\delta\nu &= \zeta[n,\nu] = -\varphi^{-1}\nu - n \\ \delta\overline{F}[n,\nu]/\delta n &= \mu[n,\nu] = \mu[n] + \nu + \varphi n. \end{aligned} \tag{6.10}$$

From either (6.9) or (6.10), we find the unsurprising

$$\overline{F}[n,\nu] = \overline{F}[n] + \frac{1}{2} n \cdot \varphi n + \nu \cdot n + \frac{1}{2} \nu \cdot \varphi^{-1}\nu : \tag{6.11}$$

minimizing over the effective potential ν simply says that $\nu = -\varphi n$ and $\overline{F}[n,\nu] = \overline{F}[n]$. However, equation 6.11 is itself a special case of the empirical model

$$\overline{F}[n,\nu] = F_0[n] + \nu \cdot n + F_1[\nu], \tag{6.12}$$

representing two interacting fluids. And according to our standard model procedure, we only have to make sure that equation 6.8 is satisfied at uniformity:

$$\begin{aligned} \frac{\delta^2\beta F_0[n]}{\delta n\delta n} &= C[n] + \beta\varphi \\ \frac{\delta^2\beta F_1[\nu]}{\delta\nu\delta\nu} &= \beta\varphi^{-1} \end{aligned} \tag{6.13}$$

for uniform n and ν. The possibilities inherent in equations 6.12 and 6.13 remain to be investigated.

Acknowledgment
This study was supported in part by a grant from the National Science Foundation.

Literature Cited
1. Hansen, J. P. and McDonald, I. R. *Theory of Simple Liquids,* (Academic Press, London, 1986).
2. Lebowitz, J. L.; Percus, J. K. *J. Math. Phys.* **1963**, *vol. 4*, pp. 116–123.
3. Evans, R. In *Inhomogeneous Fluids;* Henderson, D., Ed.: (Dekker, New York, 1992).
4. Wertheim, M. S. *J. Stat. Phys.* **1984**, *vol. 35*, pp. 19–47.
5. Percus, J. K. *J. Stat. Phys.* **1977**, *vol. 16* pp. 299–309.
6. Percus J. K.; Zhang, M. Q. *Phys. Rev.* **1988**, *vol. B38*, pp. 11737–11740.

7. Percus, J. K. *Inverse Prob.* **1990**, *vol. 6*, pp. 789–796.

8. Percus, J. K. *J. Stat. Phys.* **1982**, *vol 28*, pp. 67–81.

9. Percus, J. K. *J. Stat. Phys.* **1988**, *vol. 52*, pp. 1157–1178.

10. Meister, T. F.; Kroll, D. M. *Phys. Rev.* **1987**, *vol. A36*, pp. 4356–4363.

11. Percus, J. K. *J. Chem. Phys.* **1981**, *vol. 75*, pp. 1316–1319.

12. Curtin W. A.; Ashcroft, N. W. *Phys. Rev.* **1985**, *vol. A32*, pp. 2909–2919.

13. Herring, C. *Phys. Rev.* **1986**, *vol. A34*, pp. 2614–2622.

14. Ramakrishan, T. V.; Yussouff, M. *Phys. Rev.* **1979**, *vol. B19*, pp. 2775–2786.

15. Gervais, J. L.; Sakita, B. *Phys. Rev.* **1975**, *vol. D11*, pp. 2943–2950.

16. Kac, M. In *Applied Probability*; MacColl, L. A., Ed.; (McGraw-Hill, New York, 1957).

    Siegert, A. J. F. In *Statistical Mechanics at the Turn of the Decade*, Cohen, E. G. D., Ed., (Dekker, New York 1970).

Chapter 14

Geometrically Based Density-Functional Theory for Confined Fluids of Asymmetric ("Complex") Molecules

Yaakov Rosenfeld

Nuclear Research Center Negev, P.O. Box 9001, Beer-Sheva 84190, Israel

By capturing the correct geometrical features, the fundamental-measure free energy density functional leads to accurate description of the structure of the general inhomogeneous simple ("atomic") fluid. The initial hard-sphere functional utilizes weighted-densities, which are system-averages of the true density profiles weighted by the individual particles' geometries. It then yields explicit expressions for the "universal bridge functional" which is applicable for arbitrary pair interactions. The key for the derivation of the hard-sphere functional is the convolution decomposition of the excluded volume for a pair of spheres in terms of characteristic functions for the geometry of the two individual spheres. The recently found relation of that decomposition with the Gauss-Bonnet theorem for the geometry of convex bodies enabled to extend the hard-sphere functional to hard-body liquid crystals. Like the hard-sphere functional for simple fluids, the fundamental-measure functional for hard-bodies is an initial step towards a comprehensive free energy density functional for complex fluids of asymmetric molecules, which keeps the geometric features to the forefront.

Density functional methods have received increasing attention in recent years, and achieved a fair amount of success and sophistication in applications to inhomogeneous classical fluids *(1)*. They played a key role in providing the now emerging comprehensive picture of the complex thermodynamic behavior of fluids in confined geometries *(2-4)*. As a quite general approach to the equilibrium properties of non uniform fluids *(5-7)*, the density functional method has proven *(1-7)* to be one of the more successful and widely applicable approaches to a variety of interfacial phenomena like adsorption, wetting, and freezing. The idea is to express the free energy as a functional of the average one-body densities $\{\rho_i(\vec{r})\}$, of the various species $\{i\}$ of particles, from which all the relevant thermodynamic functions can be calculated. This enables to investigate confined fluids with all sorts of inhomogeneities. The central quantity in the density functional theory for nonuniform fluids is the excess free energy (over the "ideal - gas" contributions)

0097–6156/96/0629–0198$15.00/0
© 1996 American Chemical Society

, $F_{ex}[\{\rho_i(\overrightarrow{r})\}]$. This (generally unknown!) quantity originates in interparticle interactions, and many equillibrium properties of the fluid (e.g. tensions at interfaces, solvation forces for confined fluids, phase transitions for different inhomogeneities) can be derived from it. In contrast to the many developments in density functional theory of simple ("atomic") fluids *(1)*, the corresponding theory for molecular fluids is at a more rudimentary stage *(8-11)*, as expected in view of the increase in complexity *(12)*.

The more sophisticated versions of the density functional theory, namely those based on *non-local* excess free energy functionals, employ (coarse grained) weighted densities (obtained by weighted average of the true density profiles) which are constructed to *fit* available structural and thermodynamic properties of the homogeneous (bulk) fluid *(1),(13-19)*. Some of these require to solve complicated non linear equations which relate the weight functions to the bulk direct correlation functions. By construction, these theories encounter escalating computational (if not conceptual) difficulties upon moving from one component systems to mixtures, and to non-spherical molecules. Indeed, ingenious ad-hoc modifications *(20-22)* were required of even the simplest hard-sphere functionals in order to make them applicable for hard-body liquid crystals with, never the less, little predictive power.

This review describes the fundamental-(geometric)-measure functional which is being developed in recent years *(23-30)*. This geometrically-based approach to inhomogeneous fluids is formulated apriori for mixtures of non-spherical molecules, and derives the uniform fluid properties as a special case, rather than employ them as input. It is comprehensive, yet its application requires the minimal computational complexity.

Basic idea: Interpolation using natural "basis functions"

The idea for this kind of theory came from earlier work on integral equation theories for liquid structure. Analysis *(40-42)* of all major present day approximate theories of the structure and thermodynamics of simple liquids *(43-45),(46-47)* revealed that, in effect, they interpolate between the standard "ideal gas", low density, and a high density ,"ideal liquid", limits *(48-53)*. The role of the "ideal liquid" is played by the asymptotic high-density limit of the hypernetted-chain integral equation (denoted the Onsager limit) which has been later proposed as a reference ideal state (replacing the ideal-gas reference state) for developing a systematic theory of the liquid structure. This Onsager "ideal liquid" limit for all systems with repulsive interactions, maps universally *(50)* onto the corresponding limit for hard-spheres, for which it is obtained from the solution of the Percus-Yevick integral equation *(54-56)*. The Onsager limit corresponds to an exact lower bound for the potential energy of the system, and is characterized by single particle geometries *(51-53)*. For charged hard particles it can be achieved by immersing the entire system in an infinite conductor, which isolates the individual particles. The expansion around this state is fastly convergent at liquid densities, involving mathematical constructs (i.e. liquid-like basis functions) that enable analytic connection to functions described by low order diagrams *(48-53)*. These results suggest that an explicit description in terms of "natural" basis functions should be the starting point for developing the excess free energy functional of the inhomogeneous fluid.

Following this approach, a new kind of general functional was derived with a unique result for the inhomogeneous hard sphere fluid mixture *(23-29)*, which keeps the geometric features to the forefront. The basic idea is to interpolate between the "ideal liquid", high density, limit where the pair direct correlation function is dominated by convolutions of single particle geometries, i.e. overlap volume and overlap surface area, and the limit of low density where it is given by the pair exclusion volume *(53)*. The key for the realization of this idea is the convolution decomposition of the excluded volume for a pair of convex hard-bodies in terms of characteristic functions for the geometry of the two individual bodies. On the basis of a unique convolution decomposition for spheres it was possible to derive *(23)* the fundamental-measure free energy functional for hard-sphere mixtures, in which the weight functions represent the geometry of the individual particles. The fundamental-measure bridge functional, which is derived from this free-energy functional, can then be utilized for *arbitrary* pair interactions *(27)*. After many tests of the theory *(23-29),(31-39)* it is fair to say that by capturing the correct geometrical features, the fundamental-measure free energy density functional leads to a highly accurate description of the structure of the general inhomogeneous simple ("atomic") fluid.

The relation of the convolution decomposition for spheres with the Gauss-Bonnet theorem for general convex bodies, was found very recently *(30)*. It enables to extend the fundamental-measure functional to ("complex") fluids of asymmetric hard-bodies, from which we can begin to develop a geometrically-based free energy functional for molecular fluids. The extension to molecular fluids thus follows naturally as the next step in a continuing systematic development (described below in more detail) within fundamental liquid theory, aiming to provide physical insights, simplicity, and predictive power which are not always provided *(8)* by other approaches.

Fundamental-measure free-energy functional for hard-particles: Unique result for hard-spheres

Consider a general fluid of hard convex bodies with one-particle densities $\{\rho_i(\vec{r})\}$. For notational simplicity adopt the *descrete representation for polydispersity* where an object i is considered distinct from j if they differ in any of their physically relevant characteristics, like *size, shape* or *orientation* in space. Let $\vec{R}_i(\theta, \varphi)$ be the radius-vector from the "center" of particle i to its surface, $R_i = |\vec{R}_i(\theta, \varphi)|$ (= constant for spheres), and let $\vec{r_i}$ denote the radius-vector to the "center". The interaction potential $\phi_{ij}(\vec{r_{ij}})$ between two hard bodies i and j is infinite if they overlap and zero otherwise, and the Mayer f-function $f_{ij}(\vec{r_{ij}}) = exp(-\phi_{ij}(\vec{r_{ij}})/k_B T)$ characterizes the pair excluded volume ,

$$f_{ij}(\vec{r_{ij}}) = \quad 0 \quad \text{for } i \cap j = \emptyset$$
$$f_{ij}(\vec{r_{ij}}) = \quad -1 \quad \text{for } i \cap j \neq \emptyset \tag{1}$$

Here $\vec{r_{ij}} = \vec{r_j} - \vec{r_i}$, $i \cap j$ is the intersection of the bodies, and $\emptyset$ denotes the empty set. For spheres, $f_{ij}(\vec{r_{ij}}) = -\Theta(|\vec{r_{ij}}| - (R_i + R_j))$, where $\Theta(x)$ is the unit step function, $\Theta(x > 0) = 0$, $\Theta(x \leq 0) = 1$.

In order to *interpolate (53)* between the low density (near *ideal gas*) limit described by the *pair excluded volume* (2-particle diagram) , $\left(\frac{F_{ex}}{k_B T}\right)_{\rho \to 0} \to$

$\frac{1}{2}\sum_{i,j} \bullet_i - - - \bullet_j = \frac{1}{2}\sum_{i,j} \int d\vec{r} d\vec{r'} \rho_i(\vec{r})\rho_j(\vec{r'}) f_{ij}(\vec{r} - \vec{r'})$, and the *ideal liquid asymptotic limit (40-42),(51-52)* characterized by *1- particle geometries*, we postulated *(23)* the following general excess (over ideal gas) free energy functional

$$\frac{F_{ex}[\{\rho_i(\overrightarrow{r})\}]}{k_B T} = \int d\overrightarrow{x} \, \Phi[\{n_\alpha(\overrightarrow{x})\}] \tag{2}$$

where it is assumed that the *excess free energy density* Φ is a function of only the system averaged fundamental geometric measures of the particles,

$$n_\alpha(\overrightarrow{x}) = \sum_i \int \rho_i(\overrightarrow{x'}) w_i^{(\alpha)}(\overrightarrow{x'} - \overrightarrow{x}) d\overrightarrow{x'} \tag{3}$$

The *weighted densities* $n_\alpha(\overrightarrow{x})$ are dimensional quantities with dimensions $[n_\alpha] = (\text{volume})^{(\alpha-3)/3}$ where $0 \le \alpha \le 3$, and provide a functional basis set for expanding the function Φ which has dimension $(\text{volume})^{-1}$. The *weight functions* $w_i^{(\alpha)}$ are characteristic functions for the geometry of the particles, and are determined *(23)* by expanding the Mayer 2-particle function in terms of characteristic functions for the individual particles.

A *unique* solution was found for the special case of *spheres* with a *convolution decomposition* involving a *minimal number* of different weight functions *(23),(27)*:

$$
\begin{aligned}
-f_{ij}(\overrightarrow{r_{ij}}) = & \; w_i^{(0)} \otimes w_j^{(3)} + w_j^{(0)} \otimes w_i^{(3)} + w_i^{(1)} \otimes w_j^{(2)} + w_j^{(1)} \otimes w_i^{(2)} \\
& - \overrightarrow{w}_i^{(V1)} \otimes \overrightarrow{w}_j^{(V2)} - \overrightarrow{w}_j^{(V1)} \otimes \overrightarrow{w}_i^{(V2)}
\end{aligned}
\tag{4}
$$

where the convolution product

$$w_i^{(\alpha)} \otimes w_j^{(\gamma)} = \int w_i^{(\alpha)}(\overrightarrow{x} - \overrightarrow{r_i}) \cdot w_j^{(\gamma)}(\overrightarrow{x} - \overrightarrow{r_j}) d\overrightarrow{x} \tag{5}$$

also implies the scalar product between vectors. This minimal weight-function space contains only three functions, two scalar functions representing the characteristic functions for the volume and the surface of a particle and a surface vector function,

$$
\begin{aligned}
& w_i^{(3)}(r) = \Theta(r - R_i) \; ; \; w_i^{(2)}(\overrightarrow{r}) = | \overrightarrow{\nabla} w_i^{(3)}(r) | = \delta(r - R_i) \\
& \overrightarrow{w}_i^{(V2)}(\overrightarrow{r}) = \overrightarrow{\nabla} w_i^{(3)}(r) = \frac{\overrightarrow{r}}{r} \delta(r - R_i)
\end{aligned}
\tag{6}
$$

The other weight functions appearing in equation 4 are proportional to these three, and are given by

$$w_i^{(0)}(\overrightarrow{r}) = \frac{w_i^{(2)}(\overrightarrow{r})}{4\pi R_i^2} \; ; \; w_i^{(1)}(\overrightarrow{r}) = \frac{w_i^{(2)}(\overrightarrow{r})}{4\pi R_i} \; ; \; \overrightarrow{w}_i^{(V1)}(\overrightarrow{r}) = \frac{\overrightarrow{w}_i^{(V2)}(\overrightarrow{r})}{4\pi R_i} \tag{7}$$

There are only 5 positive-power combinations of the weighted densities which are scalars of dimension $(\text{volume})^{-1}$. For the isotropic uniform fluid these weighted densities should correspond to the fundamental-measure scaled particle variables *(53)*, and the excess chemical potential should feature the scaled-particle analytic interpolation between the exact limits of *small* and *large* particle size. This corresponds to the "scaled particle" differential equation *(23),(53)*, $\frac{\partial \Phi}{\partial n_3} = \sum_\alpha n_\alpha \frac{\partial \Phi}{\partial n_\alpha}$, for the free energy density in terms of the weighted densities. This equation yields the forms of the five dimensionless (n_3-dependent) coefficients, and the integration constants are determined from known low density properties for the uniform fluid. The following excess free energy density is thus derived *(23),(27)*:

$$\Phi = -n_0 \ln(1 - n_3) + \frac{n_1 n_2}{1 - n_3} + \frac{n_2^3}{24\pi(1 - n_3)^2} - \left(\frac{\vec{n}_{V1} \cdot \vec{n}_{V2}}{1 - n_3} + \frac{n_2(\vec{n}_{V2} \cdot \vec{n}_{V2})}{8\pi(1 - n_3)^2} \right)$$

(8)

The application of this procedure for one dimensional "spheres" leads to the exact result for hard rods as obtained earlier by Percus *(13)*. In two dimensions it leads to accurate analytic structure factors for hard disks *(26)*. This new free energy model, based on the fundamental geometric measures of the particles, provides the first unified derivation of the scaled-particle and Percus-Yevick theories for hard spheres, as a special case. The scaled-particle *(57-58)* and Percus-Yevick *(54-56)* theories provide the most comprehensive available analytic description of the bulk hard sphere thermodynamics and structure, and serve as the standard input for other weighted density models. This new model yields analytic expressions for pair and higher order direct correlation functions of a uniform (homogeneous) fluid, in good agreement with all available simulation results for the thermodynamics, structure, and adsorption of hard sphere mixtures. Starting from the original fundamental-measure functional *(23)*, a "simplified" representation (without the vector-type weights) was derived *(31),(34)* for the hard-spheres. It was then proved *(37)* to be equivalent to the original functional. This "simplified", fully scalar representation provides a useful computational aide for the special case of hard-spheres, which comes, however, at the cost of abandoning the geometric meaning of some of the weight functions; it works for spheres but it does not seem to be extendable to general hard-body fluids.

Fundamental-measure universal bridge-functional: Optimized free energy functional for general simple ("atomic") fluids

The starting point for the application of the density functional method for both uniform and non-uniform fluids are the density-profile equations *(1-2),(5)*. The density profiles $\{\rho_m(\vec{r})\}$ for the fluid subject to external potentials $\{u_m(\vec{r})\}$ which couple to the particles of type $\{m$; $m = 1, 2, ..., M\}$ are obtained by solving the Euler-Lagrange equations

$$\frac{\delta\Omega[\{\rho_m(\vec{r})\}]}{\delta\rho_i(\vec{r})} = 0 \ , \ i = 1, 2, ..., M$$

(9)

which correspond to the minimization of the grand potential $\Omega[\{\rho_m(\vec{r})\}] = F_{id}[\{\rho_m(\vec{r})\}] + F_{ex}[\{\rho_m(\vec{r})\}] + \sum_i \int d\vec{r} \rho_i(\vec{r})[u_i(\vec{r}) - \mu_i]$, where μ_i are the chemical potentials. The ideal-gas free energy is given by the exact relation $F_{id}[\{\rho_m(\vec{r})\}] = k_B T \sum_i \int d\vec{r} \rho_i(\vec{r})\{\ln[\rho_i(\vec{r})\lambda_i^3] - 1\}$, where $\lambda_i = (\frac{h^2}{2\pi m_i k_B T})^{1/2}$ are the de Broglie wave-lengths. A hierarchy of direct correlation functions $c^{(n,FD)}$ is given by *functional derivatives* (FD) of the excess free energy functional. For a fluid in contact with a reservoir bulk fluid, of average densities $\{\rho_{m,0}\}$, the density profile equations 9 can be written in the following form *(27-28)*

$$\ln g_i(\overrightarrow{r}) = -\frac{u_i(\overrightarrow{r})}{k_B T} - B_i[\{\rho_{m,0}\}; \{\rho_m(\overrightarrow{r})\}; \overrightarrow{r}]$$
$$+ \sum_j \rho_{j,0} \int d\overrightarrow{r'} \, c_{ij}^{(2,FD)}[\{\rho_{m,0}\}; (\mid \overrightarrow{r} - \overrightarrow{r'} \mid)](g_j(\overrightarrow{r'}) - 1) \tag{10}$$

where $g_i(\overrightarrow{r}) = \frac{\rho_i(\overrightarrow{r})}{\rho_{i,0}}$, and $c_{ij}^{(2,FD)}[\{\rho_{m,0}\}; (\mid \overrightarrow{r} - \overrightarrow{r'} \mid)]$ are the bulk direct correlation functions as obtained from the second functional derivatives of the excess free energy functional, $k_B T c_{ij}^{(2,FD)}(\overrightarrow{r}_1, \overrightarrow{r}_2) = -\frac{\delta F_{ex}[\{\rho_m(\overrightarrow{r})\}]}{\delta \rho_i(\overrightarrow{r}_1)\delta \rho_j(\overrightarrow{r}_2)}$, in the uniform fluid limit. The bridge functional $B_i[\{\rho_{m,0}\}; \{\rho_m(\overrightarrow{r})\}; \overrightarrow{r}]$ is related to the sum of *all terms beyond second order* in the corresponding functional Taylor expansion of the *excess free energy*.

The fundamental measure functional provides explicit simple expressions for the bridge functional, involving only integrations of known functions. This enables to apply the hard-sphere functional for fluids with arbitrary interactions, going beyond the (standard) van der Waals approximation *(59-61)* for the attractive interactions. A new free energy density functional *(27-28)* for general inhomogeneous simple ("atomic") fluid mixtures is based on that for the hard spheres, and is operationally equivalent to an ansatz of *universality* of the bridge functional: The bridge functional, namely the sum of all terms beyond second order in the functional Taylor expansion, is approximated by that for the hard-spheres. The *second-order* functional can be *optimized* by imposing the *test-particle self-consistency*, namely by considering the density profile equations (i.e. the Euler-Lagrange equations for minimizing the Grand Potential), with the same universal functional, but in the special case when the external potential is generated by a test-particle at the origin of coordinates. These equations are solved coupled with the Ornstein -Zernike relations, to generate the bulk pair correlation functions, which are the required input for the second-order functional. The choice of the hard-sphere diameters (for fluids with arbitrarily soft potentials) is such that an error term in the free energy is minimized. There is no attempt to impose any specific structural-thermodynamic consistency relations, *everything is predetermined* by the quality of the approximation for the initial free energy functional for the inhomogeneous hard sphere fluid mixture, from which the "universal" bridge functional is obtained. The application of this new general method for non-uniform fluids to the special case of the bulk fluid, corresponds to the well established, successful thermodynamically-consistent modified-hypernetted-chain theory *(43-45)* (but with the bridge functions now generated by an explicit, "universal", hard-sphere bridge functional). Indeed, using this new method accurate results are obtained for the bulk pair correlation functions for a large variety of potentials, for both one component systems and mixtures *(27),(38-39)*. The systems considered include *(38)* non-additive hard sphere mixtures, Lennard-Jones mixtures, and strongly coupled binary plasma mixtures.

Unless the initial hard-sphere functional is already the exact one, the optimized hard-sphere functional is more accurate than the initial one. In particular, the fundamental-measure initial hard-sphere functional predicts the Percus-Yevick closure result that hard-sphere fluid binary mixtures never phase separate, but when optimized it does predict phase separation for large size ratios *(28)*. The hard-sphere "universal" bridge-functional, and the corresponding optimized free energy functional have been tested (directly and also implicitly) very successfully, for a variety of hard and soft pair interactions and external potentials, by comparison with computer simulations of density profiles for a large variety of situations where size or packing effects play an important role *(23-29),(31-39)* and by comparison with experiments on colloids and emulsions *(62)* which address the challenging question of phase separation in asymmetric binary hard-sphere mixtures *(28)*. The cases considered include among others: (1) wetting transitions and capillary condensation of argon atoms in carbon slit-like pores, (2) adsorption of binary methane-argon mixtures on a graphite surface, (3) argon-krypton mixtures in carbon slit-like pores, (4) selective adsorption from binary mixtures, and (5) strong electrolytes near a charged electrode. One of the most striking tests for soft potentials is provided by the plasma (point charges!) near a wall. It appears from all these investigations that, by capturing the correct geometrical features, the fundamental-measure hard-sphere functional leads to accurate description of the structure of the inhomogeneous simple fluid.

Extensive calculations have revealed *(39)* that while the structural results of this theory are highly accurate, its results for the phase boundaries (e.g. for the wetting transition) of Lennard-Jones type potentials are less accurate than those obtained from the simple van-der-Waals mean field theory (that employs the same accurate hard-sphere functional). The technical reason for this is the following *(39)*: relying on expansion from reference bulk density, the functional of the test-particle consistent theory does not satisfy the Gibbs adsorption equation relating between the excess adsorption and the derivative of the excess grand free energy with respect to the chemical potential, while the simple mean-field theory does. A remedy can be provided by improving the "reference" state, e.g. by providing the test-particle consistent functional for the repulsive part of the Lennard-Jones type potential, and by adding the attractive part in the simple mean form. Recall that the test-particle consistent theory does not impose any particular thermodynamic consistency relation, and a slight inconsistency (which is inevitable for an approximation) is apparently augmented when it comes to locating the phase boundaries for inhomogeneous fluids. It should be noted, however, that the statement of universality of the bridge functional, with the approximate functional given by the fundamental-measure theory, is the overall most accurate available general approximation for the structure of all types of fluids, and thus remains viable.

Free-energy functional for molecular fluids :
Application of the Gauss-Bonnet theorem for convex bodies

Consider a body i, denote its surface by ∂i, and let $\hat{\mathbf{n}}_i$ be the outward unit normal to that surface *(63)*. From the *principal curvatures* $\kappa_a^{(i)}$, $\kappa_b^{(i)}$ of the surface of the body i, obtain the mean, $H_i = \frac{1}{2}\left(\kappa_a^{(i)} + \kappa_b^{(i)}\right)$, and the *Gaussian*, $K_i = \kappa_a^{(i)}\kappa_b^{(i)}$, curvatures. For spheres the curvatures $\kappa_a^{(i)} = \kappa_b^{(i)} = H_i = \frac{1}{R_i}$, $K_i = \frac{1}{R_i^2}$ are constant on the surface. Let S be a simply connected portion of a surface whose boundary is the *closed* curve C with arc length s. Let κ_g be the geodesic curvature of C at a given point on the surface and let K be the Gaussian curvature of S, then *(63-64)* (*Gauss-Bonnet theorem*)

$$2\pi = \int\int_S K dA + \int_C \kappa_g ds \tag{11}$$

where dA is the element of area, and ds the element of arc. The integral Gaussian curvature for any convex body i is equal to 4π, $G(i) = \int\int_{\partial i} K_i dA_i = 4\pi$. The intersection of two convex bodies is a single convex body, so that $f_{ij}(\overrightarrow{r_i} - \overrightarrow{r_j}) = -G(i\cap j)/4\pi$. The intersection $\partial i \cap j$ is the surface of i which is inside j, and the intersection $\partial i \cap \partial j$ are closed curves which are shared by the surfaces of i and of j. In the simple case when the intersection of i and j produces only one intersection curve connecting the two pieces of $i \cap j$, we apply the Gauss-Bonnet theorem (equation 11) to each piece separately, we then combine the results to get

$$G(i \cap j) = 4\pi = \int\!\!\int_{\partial i \cap j} K_i dA_i + \int\!\!\int_{\partial j \cap i} K_j dA_j + \int_{\partial i \cap \partial j} \left(\kappa_g^{(i)} + \kappa_g^{(j)} \right) ds \quad (12)$$

The convolution decomposition for spheres (equation 4) turns out *(30)* to be just a special case of the Gauss-Bonnet theorem (equation 12) for convex bodies, which can be written as

$$-4\pi f_{ij}(\overrightarrow{r_i} - \overrightarrow{r_j}) = G(i \cap j) = G(i \cap j)_{CA} + \int_{\partial i \cap \partial j} \frac{1 - \hat{\mathbf{n}}_i \cdot \hat{\mathbf{n}}_j}{|\hat{\mathbf{n}}_i \times \hat{\mathbf{n}}_j|} (\Lambda_i + \Lambda_j) ds \quad (13)$$

Here

$$G(i \cap j)_{CA} = \int\!\!\int_{\partial i \cap j} K_i dA_i + \int\!\!\int_{\partial j \cap i} K_j dA_j$$

$$+ \int_{\partial i \cap \partial j} H_i \frac{ds}{|\hat{\mathbf{n}}_i \times \hat{\mathbf{n}}_j|} + \int_{\partial i \cap \partial j} H_j \frac{ds}{|\hat{\mathbf{n}}_i \times \hat{\mathbf{n}}_j|} - \int_{\partial i \cap \partial j} H_i \frac{\hat{\mathbf{n}}_i \cdot \hat{\mathbf{n}}_j ds}{|\hat{\mathbf{n}}_i \times \hat{\mathbf{n}}_j|} - \int_{\partial i \cap \partial j} H_j \frac{\hat{\mathbf{n}}_i \cdot \hat{\mathbf{n}}_j ds}{|\hat{\mathbf{n}}_i \times \hat{\mathbf{n}}_j|}$$

$$(14)$$

denotes the convolution approximation (CA). For spheres, equation 14 corresponds, term by term, with equation 4 multiplied by 4π. If $\kappa_a^{(i)}$ is the *normal* curvature of the surface at a point on the curve $\partial i \cap \partial j$ in the direction of its tangent, and $\kappa_b^{(i)}$ is that in direction at right angle to it, then the *curvature asymmetry is given by*, $\Lambda_i = \frac{1}{2}(\kappa_a^{(i)} - \kappa_b^{(i)})$. The curvature asymmetry term in equation 13 cannot be expressed as a convolution. The normal curvature of a sphere is the same in all directions so that the curvature asymmetry vanishes, and the *convolution approximation,*

$$G(i \cap j) = G(i \cap j)_{CA} \quad (15)$$

, becomes exact. The curvature asymmetry term becomes more significant with increasing deviations from sphericity.

As first step we adopt *(30)* the convolution approximation equation 15 for *general* hard bodies. It is equivalent to the convolution decomposition equation 4 provided that the following set of six distinct weight functions for the general hard body i is employed:

$$w_i^{(3)}(\overrightarrow{r}) = \Theta(\left|\overrightarrow{r} - \overrightarrow{R}_i(\theta, \varphi)\right|) \; ; \; w_i^{(2)}(\overrightarrow{r}) = |\overrightarrow{\nabla} w_i^{(3)}(\overrightarrow{r})| = \delta(\overrightarrow{r} - \overrightarrow{R}_i(\theta, \varphi))$$

$$\overrightarrow{w}_i^{(V2)}(\overrightarrow{r}) = \overrightarrow{\nabla} w_i^{(3)}(\overrightarrow{r}) = \hat{\mathbf{n}}_i \delta(\overrightarrow{r} - \overrightarrow{R}_i(\theta, \varphi))$$

$$w_i^{(0)}(\overrightarrow{r}) = \frac{K_i}{4\pi}w_i^{(2)}(\overrightarrow{r}) \; ; \; w_i^{(1)}(\overrightarrow{r}) = \frac{H_i}{4\pi}w_i^{(2)}(\overrightarrow{r}) \; ; \; \overrightarrow{w}_i^{(V1)}(\overrightarrow{r}) = \frac{H_i}{4\pi}\overrightarrow{w}_i^{(V2)}(\overrightarrow{r})$$

(16)

For spheres this set reduces to the set given by equations 6-7. For non-convex bodies we should (see equations 4.7,4.8 in *(53)*) replace these $w_i^{(1)}(\overrightarrow{r})$ and $\overrightarrow{w}_i^{(V1)}(\overrightarrow{r})$ by $\frac{H_i}{4\pi}\delta(\overrightarrow{r} - \overrightarrow{R}_i^{(CE)}(\theta,\varphi))$ and $\frac{H_i}{4\pi}\hat{\mathbf{n}}_i\delta(\overrightarrow{r} - \overrightarrow{R}_i^{(CE)}(\theta,\varphi))$, respectively, where $\overrightarrow{R}_i^{(CE)}(\theta,\varphi)$ is the radius vector to the surface of the convex envelope (CE) of the particle i. The mean and Gaussian curvatures, H_i , K_i , are not constant on the surface of the general hard body, and six different weighted densities have to be calculated separately (compared with the three for spheres). Equations 2,3,8, and 16 thus define the fundamental measure (FM) excess free energy functional for the general hard-body fluid, f_{FM}^{ex} , which is an initial geometrically-based free energy functional for hard-body fluids.

Expectations and preliminary results

General. The optimized free energy, based on the universality of the bridge functional, is applicable to *molecular fluids*. It requires only an initial accurate free energy functional for hard-body fluids, from which the bridge functional for *non-spherical hard particles* will be derived. The hard-body functionals, starting with the one derived above using the hard-sphere paradigm, are expected to provide initial accurate functionals for general molecular ("complex") fluids. It is expected that the relative insensitivity of the bridge functional to the density profiles, as exhibited for atomic fluids, will hold also for molecular fluids. In particular, it should not be very sensistive to the distribution of orientations, so that bridge functionals derived from free energy functionals for isotropic or nematic hard body fluids, may have wider validity in the context of the optimized theory. In parallel with hard-body fluids we may consider also charged (with Coulomb or Yukawa charges) hard-body fluids which provide a starting point for the statistical thermodynamic analysis of a large variety of interesting physical systems *(65-68),(52)* including ,e.g. , water, electrolytes, molten salts, liquid metals, dense plasmas, colloidal dispersions, microemulsions, and micelles. Within the optimized theory, the charge contributions to the direct correlation functions, in particular for the Onsager limit *(27),(51-52)*, can also be discussed *(27)* in terms of the fundamental-measures. Recall that the strong coupling limit of the direct correlation functions of charged fluids is given by the electrostatic interaction between the charges when they are smeared on the surface or in the volume of the particles, i.e. a convolution form that involves the geometric weighted densities. It should be emphasized that the geometric approach keeps the comuputational complexity at its lowest possible level, namely at the level of the description of the pair interactions and the geometry of the container. There are various methods *(69)* for solving the density profile equations. When the geometry of the bodies is specified by a limited number of mesh-points on its surface, it seems that the finite-element method might be particularly useful.

Isotropic hard-body fluid. When applied to the homogeneous (bulk) hard-body fluid f_{FM}^{ex} is independent of the distribution of orientations and is equal to the form obtained from scaled particle theory *(53)*. This indicates that although derived for *arbitrary* inhomogeneous hard-body fluids, f_{FM}^{ex} is better suited for *isotropic* fluids. Remarkably, eventhough the convolution approximation equation 15 is generally approximate, f_{FM}^{ex} yields the *exact* second virial coefficient, B_{ij}, for the homogeneous isotropic hard convex body bulk fluid. Indeed, using equation 16 integrate equation 4 to obtain the exact result *(70),(53)* $B_{ij} = \frac{1}{2}\left[V(i) + S(i)\overline{R}(j) + \overline{R}(i)S(j) + V(j) \right]$ where $V(i)$, $S(i)$, and $\overline{R}(i) = \frac{1}{4\pi}\int_{\partial i} H_i dA_i$ are, respectively, the volume, surface area, and *mean-radius* of the body i. The functional f_{FM}^{ex} predicts a third virial coefficient for the one-component isotropic hard spherocylinder bulk fluid which is only 10% smaller than the exact coefficient, even for a large length over width ratio, $L/D = 6$. As can be gleaned from related work *(53),(71),(9)*, f_{FM}^{ex} yields an accurate equation of state for isotropic bulk fluids. Using the second functional derivatives of f_{FM}^{ex} the direct correlation functions are expressed in terms of convolutions of the weight functions, and for the special case of spheres they are identical *(23),(53)* to the exact solution of the Percus-Yevick equation. Thus, when truncated at second order, f_{FM}^{ex} reduces to a version of a functional which was recently employed successfuly *(11)* for the inhomogeneous fluid of hard linear molecules (spherocylinders, ellipsoids). As can be gleaned from these results and related work *(53),(71),(72),(9)*, as long as the aspect ratio of the molecules (e.g. length/width) is not too large (e.g. smaller than about five), f_{FM}^{ex} for *isotropic* hard-particle fluids will be of accuracy comparable to that it demonstrated for the spheres. The accuracy of f_{FM}^{ex} can be estimated apriori by the extent to which the convolution approximation holds. Note that the convolution approximation is approximate also for spheres in even dimensions (e.g. D=2) *(24-26)*, yet the fundamental measure functional for hard-disks proved accurate *(26)*. Existing programs for solving the the hypernetted-chain equations using expansions in appropriate angle-dependent basis set *(73)* require only a subroutine which evaluates the bridge functional.

Aligned hard-body fluid. The system of parallel hard ellipsoids (PHE's) is related to the hard-spheres by an anisotropic mapping *(74-75)*. By applying that mapping on the weight functions 6-7, we obtain the PHE-weight-functions. By applying that mapping further on equation 4 we find that the convolution decomposition holds exactly also for PHE's. The PHE-weight-functions thus define the fundamental measure (FM) excess free energy functional for the PHE's fluid, denoted $f_{FM,PHE}^{ex}$. It yields the exact analytic solution of the Percus-Yevick equations for the DCF's of the PHE fluid, and thus *predicts correctly* the absence of a smectic phase for the PHE fluid. The PHE's provide a very useful reference system for fluids of aligned particles *(8),(11),(20-21)* , and $f_{FM,PHE}^{ex}$ provides a paradigm as well as the starting point for the fundamental-measure functional for fully aligned particles (by, e.g. perturbation theory for nematic and smectic phases).The insensitivity of the bridge functions to the shape of the molecules enables to utilize the (easier to calculate) PHE weight functions (parametized) in the bridge functional expression for other shapes of molecules. General theoretical techniques for liquides of orientationally ordered particles were developed *(76)* which are particularly useful for solving the (hypernetted-chain type) integral equations for the density profiles, which arise from the present theory. The existing programs *(76)* need only be supplemented with a routine that evaluates the bridge functional. Binary mixtures of aligned particles *(77)* do not require any modification of the present formalism.

Hard-body fluid of arbitrary orientations. It is clear that without incorporating somehow the curvature asymmetry contributions, the uniform fluid limit of f_{FM}^{ex} does not depend on the distribution of orientations. The comparison of the PHE-weight-functions with the set 16 is expected to provide hints for how to improve the weight functions for *the general distribution of orientations*. Pending these calculations, one possibility is to treat the H_i and K_i as "free" orientation-dependent parameters on the surface, which can be determined by imposing (approximately) the equality 15 for the bodies in question (i.e. by imposing the exact second virial coefficient for arbitrary distribution of orientations for the *homogeneous* fluid). Another possibility which might be explored is to consider orientaion-dependent weighted densities (specifically the dimensionless n_3-dependent coefficients) in equation 3, and correspondingly to allow Φ to be orientation-dependent in equation 2. It is not yet clear under what constraints it will be possible to determine the functional form of Φ. The general functionals as obtained from the different methods should have the property that in the appropriate limits they reduce to the functionals f_{FM}^{ex} , $f_{FM,PHE}^{ex}$ as obtained above for the isotropic fluid, and for the parallel hard ellipsoids, respectively.

Acknowledgments

This review has taken its final form during the author's stay in the Physics Department of Bristol University as Benjamin Meaker Visiting Professor. The author thanks the members of the Physics Department, in particular Professor Bob Evans, for their warm hospitality, and The Benjamin Meaker Foundation for support. This research was supported by The Basic Research Foundation administered by The Israel Academy of Sciences and Humanities.

Literature Cited

1. *Fundamentals of Inhomogeneous Fluids*; Henderson, D., Ed.; Dekker , 1992; and in particular the review by Evans, R.
2. Rowlinson, J.S. ; Widom, B. *Molecular Theory of Capilarity*; Clarendon Press, Oxford ,1982.
3. Evans, R. *J.Phys.Condens.Matter* **1990**, 2, 8989 ; Oxtoby, D.W. In *Liquids, Freezing and the Glass Transition*; Editors, Hansen, J.P.; Levesque, D.;Zinn-Justin, J.; Les Houches Session 51; Elsevier, New York ,1990.
4. Lowen, H. *Phys.Rep.* **1994**, 5, 249.
5. Evans, R. *Adv.Phys.* **1979**, 28, 143.
6. Percus, J.K. In *The Liquid State of Matter: Fluids Simple and Complex*; Editors, Montroll, E.W. ; Lebowitz, J.L.; North- Holland ,1982.
7. Baus, M. *J.Phys.Cond.Matter* **1990**, 2, 2111.
8. Vroege, G.J. ; Lekkerkerker, H.N.W. *Rep.Prog.Phys.* **1992**, 55, 1241.
9. Cuesta, J.A.; Tejero, C.F.; Baus, M. *Phys.Rev.A* **1992**, 45 7395.
10. Osipov, M.A. ; Hess, S. J.Chem.Phys. **1993**, 99, 4181. Samborski, A.; Evans, G.T. *J.Chem.Phys.* **1994**, 101, 6005.
11. Rickayzen, G. *Mol.Phys.* **1992**, 75, 333.; Kalpaxis, P.; Rickayzen, G. *Mol.Phys.* **1993**, 80, 391. Calleja, M.; G. Rickayzen, G. *Phys.Rev.Letters* **1995**, 74, 4452.
12. Gray, C.G.; Gubbins, K.E. *Theory of Molecular Fluids*; Clarendon Press, Oxford, 1984.
13. Percus, J.K., *J.Stat.Phys.* **1976**, 15, 505. Percus, J.K. *J.Chem.Phys.* **1981**, 75, 1316. Robledo, A.; Varea, C. *J.Stat. Phys.* **1981**, 26, 513. Fischer, J. ; Heinbuch, U. *J.Chem.Phys.* **1988**, 88, 1909 .

14. (a) Nordholm, S.; Johnson, M.; Freasier, B.C. *Aust.J.Chem.* **1980**, 33, 2139.
(b) S. Nordholm, S.; Penfold, R. *J.Chem.Phys.* **1992**, 96, 3022.
15. Tarazona, P. *Mol.Phys.* **1984**, 52, 81. Tarazona, P.; Evans, R. Mol.Phys.
1984, 52, 847. Tarazona, P. *Phys.Rev.A* **1985**, 31, 2672.
16. T.F. Meister, T.F.; Kroll, D.M. *Phys.Rev. A* **1985**, 31, 4055. Groot, R.D.;
van der Eerden, J.P. *Phys.Rev.A* **1987**, 36, 4356. Sokolowski, S.; Fischer, J.
Mol.Phys. **1989**, 68, 647.
17. W.A. Curtin, W.A.; Ashcroft, N.W. *Phys.Rev. A* **1985**, 32, 2909. Denton,
A.R.; Ashcroft, N.W. *Phys.Rev.A* **1989**, 39, 426, 4701. Leidl, R.; Wagner, H.
J.Chem.Phys. **1993**, 98, 4142. C.N.Likos, C.N.; Ashcroft, N.W. *J.Chem.Phys.*
1993, 99 9090.
18. Vanderlich, T.K.; Scriven, L.E.; Davis, H.T. *J.Chem.Phys.* **1989**, 90, 2422.
Kroll, D.M.; Laird, B.B. *Phys.Rev.A* **1990**, 42 4806.
19. Lutsko, J.F.; Baus, M. *Phys.Rev.Letters* **1990**, 64 761. Zeng, X.C.; Oxtoby,
D.W. *Phys.Rev.A* **1990**, 41, 7094.
20. Somoza, A.M.; Tarazona, P. *Phys.Rev.Letters* **1988**, 61, 2566. Somoza,
A.M.; Tarazona, P. *J.Chem.Phys.* **1989**, 91, 517. Somoza, A.M.; Tarazona,
P. *Phys.Rev.A* **1990**, 41, 965.
21. Mederos, L.; Sullivan, D.E. *Phys.Rev.A* **1989**, 39, 854. Somoza, A.M.;
Mederos, L.; Sullivan, D.E. *Phys.Rev. Letters* **1994**, 72, 3674.
22. Poniewierski, A.; Holyst, R. *Phys.Rev. Letters* **1988**, 61, 2461. Holyst, R.;
Poniewierski, A. *Phys.Rev.A* **1989**, 39, 2742.
23. Rosenfeld, Y. *Phys.Rev. Letters* **1989**, 63, 980.
24. Rosenfeld, Y.; Levesque, D; Weis, J.J. *J.Chem.Phys.* **1990**, 92, 6818.
25. Rosenfeld, Y. *J.Chem.Phys.* **1990**, 93, 4305.
26. Rosenfeld, Y. *Phys.Rev. A* **1990**, 42, 5978 .
27. Rosenfeld, Y. *J.Chem.Phys.* **1993**, 98, 8126 .
28. Rosenfeld, Y. *Phys.Rev. Letters* **1994**, 72, 3831. Rosenfeld, Y. *J. Phys.
Chem.* **1995**, 99, 2857.
29. Rosenfeld, Y. In *Condensed Matter Theories* Vol. 8; Editors Blum, L.; Malik,
F.B.; Plenum Press, NY 1993; pp 411-425. Rosenfeld, Y. In *Strongly Coupled
Plasma Physics*; Editors, VanHorn, H.; Ichimaru, S.; University of Rochester
Press, 1993; pp 313-322.
30. Rosenfeld, Y. *Phys.Rev.E* **1994**, 50, R3318. Rosenfeld, Y. *Molec.Phys*, Issue
in honor of D. Henderson, in print.
31. Kierlik, E.; Rosinberg, M.L. *Phys.Rev.A* **1990**, 42, 3382.
32. Kierlik, E.; Rosinberg, M.L. *Phys.Rev.A* **1991**, 44, 5025.
33. Kozak, E.; Sokolowski, S. *J.Chem.Soc.Faraday Trans.* **1991**, 87, 3415.
34. Kierlik, E.; Rosinberg, M.L.; Finn, J.E.; Monson, P.A. *Molec.Phys.* **1992**,
75, 1435.
35. Kierlik, E.; Rosinberg, M.L.; Fan, Y.; Monson, P.A. *J.Chem.Phys.* **1994**, 101,
10947 .
36. Kierlik, E.; Fan, Y.; Monson, P.A.; Rosinberg, M.L. *J.Chem.Phys.* **1995**,
102, 3712.
37. Phan, S.; Kierlik, E.; Rosinberg, M.L.; Bildstein, B.; Kahl, G. *Phys.Rev.E*
1993, 48, 618 .
38. Bildstein, B.; Kahl, G.; Rosenfeld, Y. unpublished.
39. Sweatman, M., July **1995**, Ph.D. Thesis, Bristol University ; Sweatman M.;
Evans, R. unpublished.
40. Rosenfeld, Y. *Phys. Rev. A* **1982**, 25, 1206.
41. Rosenfeld, Y. *Phys. Rev. A* **1985**, 32, 1834.
42. Rosenfeld, Y. *Phys. Rev. A* **1986**, 33, 2025 .
43. Rosenfeld, Y.; Ashcroft, N.W. *Phys. Rev. A* **1979**, 20, 1208 . Rosenfeld, Y.;
Ashcroft, N.W. *Phys. Letters* **1979**, 73A, 31. Rosenfeld, Y. *J. de Phys. (Paris)*
1980, Colloque C2 (suplement to vol. 41) 77.

44. Rosenfeld, Y; Blum, L. *J. Chem.Phys.* **1986**, 85, 2197 . M. Llano-Restrepo, M.; Chapman, W.G. *J.Chem.Phys.* **1994**, 100, 5139.
45. Rosenfeld, Y. *J. Stat. Phys.* **1986**, 42, 437. Gonzalez, L.E.; Mayer, A.; Iniguez, M.P.; Gonzalez, D.J.; Silbert, M. *Phys.Rev.E* **1993**, 47, 4120.
46. Barker, J.A.; Henderson, D. *Rev.Mod.Phys.* **1976**, 48, 587.
47. Hansen, J.P.; McDonald, *I.R. Theory of Simple Liquids*; 2'nd edition; Academic Press, London, **1986**.
48. Rosenfeld, Y. *Phys. Rev. A* **1987**, 35, 938. Rosenfeld, Y. *Phys.Rev.A* **1988**, 37, 3403. Rosenfeld, Y.; Levesque, D.; Weis, J.J. *Phys.Rev.A* **1989**, 39, 3079.
49. Rosenfeld, Y. In *High-Pressure Equations of State : Theory and Applications*; Editors, Eliezer, S.; Rici, R.; North-Holland, Amsterdam, 1991; pp 285-326.
50. Rosenfeld, Y. In *The Equation of State in Astrophysics*; Editors, Chabrier, G.; Schatzman, E.; Cambridge University Press, 1994, pp 78-105.
51. Rosenfeld, Y.; Blum, L. *J. Phys. Chem.* **1985**, 89, 5149. Rosenfeld, Y.; Blum, L. *J. Chem.Phys.* **1986**, 85, 1556. Rosenfeld, Y.; Gelbart, W.M. *J. Chem. Phys.* **1984**, 81, 4574. Blum, L.; Rosenfeld, Y. *J.Stat.Phys.* **1991**, 63, 1177.
52. Rosenfeld, Y. *Phys.Rev.E* **1993**, 47, 2676.
53. Rosenfeld, Y. *J.Chem.Phys.* **1988**, 89, 4272.
54. Percus, J.K.; Yevick, G.J. *Phys.Rev.* **1958**, 110, 1.
55. Wertheim, M.S. *Phys.Rev. Letters* **1963**, 10, 321. Thiele, E. *J.Chem.Phys.* **1963**, 39, 474.
56. Lebowitz, J.L. *Phys.Rev.A* **1964**, 133, 895. Lebowitz, J.L.; Rowlinson, J.S. *J.Chem.Phys.* **1964**, 41, 133.
57. Reiss, H.; Frisch, H.; Lebowitz, J.L. *J.Chem.Phys.* **1959**, 31, 369. Reiss, H. *Adv.Chem.Phys.* **1965**, IX, 1.
58. Reiss, H. *J.Phys.Chem.* **1992**, 96, 4736.
59. Curtin, W.A.; Ashcroft, N.W. *Phys.Rev.Letters* **1986**, 56, 2775. Mederos, L.; Navascues, G.; Tarazona, P. *Phys.Rev.E* **1994**, 49, 2161.
60. Mier-y-Teran, L.; Suh, S.H.; White, H.S.; Davis, H.T. *J.Chem.Phys.* **1990**, 92, 5087.
61. Tang, Z.; Scriven, L.E.; Davis, H.T *J.Chem.Phys.* **1991**, 95, 2659. Sokolowski, S.; Fischer, J. *J.Chem.Phys.* **1992**, 96, 5441.
62. Steiner, U.; Meller, A.; Stavans, J. *Phys.Rev. Letters* **1995**, 74, 4750. See updated list of references therein.
63. Weatherburn, C.E. *Differential Geometry of Three Dimensions*; Cambridge University Press, Vol.I , 1961. Abelson, H.; diSessa, A. *Turtle Geometry*, MIT Press, Cambridge, Massachusetts, 1980; Santalo, L.A. *Integral geometry and geometric probability*, Addison-Wesley, 1976.
64. Examples of other recent applications of the Gauss-Bonnet theorem are : Robledo,A; Varea, C. In *Condensed Matter Theories* Vol. 8; Editors Blum, L.; F.B. Malik, F.B.; Plenum Press, NY, 1993, pp 595-602. Wertheim, M.S. *ibid* , pp 435-442. Interesting applications of fundamental geometric measures ("Minkowski functionals") are: Mecke, K.R.; Buchert, T.; Wagner, H. *Astronomy & Astrophysics* **1994**, 288, 697(1994). Likos, C.N.; Mecke, K.R.; Wagner, H. *submitted to J.Chem.Phys.*, **1995**. Wertheim, M.S. *Molec.Phys.* **1994**, 83 519.
65. Rowlinson, J.S. *Physica A* **1989**, 156, 15.
66. Alexander, S.; Chaikin, P.M.; Grant, P.; Morales, G.J.; Pincus, P.; Hone, D. *J.Chem.Phys.* **1984**, 80, 5776.
67. Robbins, M.O.; Kremer, K.; Grest, G.S. *J.Chem.Phys.* **1988**, 88, 3286. Meijer, E.J.; Frenkel, D. *J.Chem.Phys.* **1991**, 94, 2269. M.J.Stevens, M.J.; Robbins, M.O. *J.Chem.Phys.* **1993**, 98, 2319.
68. Rosenfeld, Y. *Phys.Rev.E* **1994**, 49, 4425.
69. The finite-element method for solving the density profile equations is imple-

mented by J.R. Henderson, J.R.; Sabeur, Z.A. *J.Chem.Phys.* **1992**, 97, 6750; *Mol.Phys.* **1994**, 82, 765. Various other methods for solving the density profile equations are discussed by Nordholm and Penfold, ref.(14b) above.

70. Ishihara, A. *J.Chem.Phys.* **1950**, 18, 1446.

71. P.G. Bolhuis, P.G.; Lekkerkerker, H.N.W. *Physica A* **1993**, 196, 375. Buitenhuis, J.; Dhont, J.K.G.; Lekkerkerker, H.N.W. *Macromolecules*, **1994**, 27, 7267 .

72. A. Chamoux, A.; and A. Perera, A. *submitted to J.Chem.Phys.*, **1995**.

73. Fries, P.H.; Patey, G.N. *J.Chem.Phys.* **1985**, 82, 429. Perera, A.; Kusalik, P.G.; Patey, G.N. *J.Chem.Phys.* **1987**, 87, 1295.

74. Lebowitz, J.L.; Perram, J.W. *Mol.Phys.* **1983**, 50, 1207.

75. Harrowell, P.; Oxtoby, D. *Mol.Phys.* **1985**, 54, 1325 .

76. Caillol, J.M.; Weis, J.J.; Patey, G.N. *Phys.Rev.A* **1988**, 38, 4772. Caillol, J.M.; Weis, J.J. *J.Chem.Phys.* **1989**, 90, 7403. Caillol, J.M.; Weis, J.J. *J.Chem.Phys.* **1990**, 92, 3197. Azzouz, H.; Caillol, J.M.; Levesque, D.; Weis, J.J. J.Chem.Phys. **1992**, 96, 4551.

77. Sear, R.P.; Jackson, G. *J.Chem.Phys.* **1995** , 102, 2622.

Chapter 15

Density-Functional Theory for Nonuniform Polyatomic Fluids

E. Kierlik, S. Phan, and M. L. Rosinberg

Laboratoire de Physique Théorique des Liquides, Unité de Recherche
Associée 765, Université Pierre et Marie Curie, 4 Place Jussieu,
75252 Paris Cedex 05, France

The structure of molecular and polymer fluids near surfaces and
in thin films is a topic of great fundamental and practical interest
which is still not well understood. We present a density functional
theory which is a generalization to inhomogeneous polyatomic fluids of Wertheim's thermodynamic perturbation theory for associating fluids. In the local density approximation, this theory takes
a very simple form which can be used to study the structure and
the thermodynamics of long chains at the free surface. As an application, we compute the variations of the surface tension with
temperature and chain length and we investigate the surface segregation effects due to side branching, segment size, or isotopic
substitution.

In the case of classical simple fluids, it is known that density functional theory in the van der Waals approximation where attractive forces are treated
in a mean-field fashion while repulsive forces are represented by an equivalent
hard sphere interaction, captures the essential physics of interfacial phenomena, such as adsorption or wetting at solid/fluid or liquid/vapor interfaces (*1*).
The simplest Local Density Approximation (LDA) where the free energy of the
inhomogeneous hard sphere fluid is just the spatial integral of the local free
energy density of the bulk fluid already gives the gross features of the structure
and phase equilibria. Somewhat better results are obtained when using one of
the so-called "weighted-density approximations" which have flourished in the
past few years (*1*). In some instances, one can get quantitative predictions for
experimental quantities like the surface tension, although the results seem to
be rather sensitive to the choice of the interatomic potential or to the mixing
rules in the case of mixtures.

The question that we want to investigate in this paper is the following: can we build the same type of theory and reach the same level of accuracy in the case of inhomogeneous liquid hydrocarbons or long chain polymeric fluids ? Let us first briefly summarize the present theoretical situation. Leaving apart the scaling approaches which only focus on the universal large scale behavior (*2*), we have on one hand *self-consistent field* (SCF) type of calculations based on coarse-grained lattice models, and on the other hand *Landau-Ginzburg* treatments which use a Cahn-Hilliard-de Gennes free energy functional gradient expansion in analogy with similar theories of simple liquids. The SCF methods (*3*) which are extensions of the Flory mean-field picture to inhomogeneous situations, may be quite successfull in describing qualitatively the structure and the thermodynamics of the interfaces. But the use of a lattice model is a very crude representation of a real fluid, especially when packing or free-volume effects are important, as they are near a solid surface or at the free interface, respectively. In the Landau-Ginzburg approaches (*4,5*), one usually assumes that the surface terms contain unknown phenomenological parameters which are supposed to describe both entropic and enthalpic contributions. Moreover, the homogeneous part of the free energy is usually described by the simplest Flory-Huggins expression which cannot distinguish between different molecular architectures or orientational effects.

Therefore, there is a need for a theory which could be used to interpret the experimental results in terms of fundamental quantities such as *interacting* parameters, rather than phenomenological parameters. By "interacting" parameters, we mean parameters which have a physical meaning at the level of the structural units and are reasonably independent of temperature and concentration. What are the ingredients that should contain this theory ?
a) It should be a continuum (i.e. off-lattice) theory in order to describe packing and compressibility effects (what polymerists call equation-of-state effects),
b) it should be able to describe the influence of chain length and chemical architecture on the various interfacial properties,
c) it should describe the conformational changes of the molecules near a surface,
d) and of course, it should contain energetic effects, at least in a mean-field fashion, as done for simple fluids.

Density functional theory seems to be a good candidate for that and several theories have been proposed in the last few years (*6-10*). The one that we shall discuss in the following is called Perturbation Density Functional Theory (PDFT) (*11-13*). Indeed, we shall first explain how we can treat perturbatively the influence of connectivity in the Helmholtz free energy by generalizing Wertheim's theory of chemical association (*14*) to inhogeneous fluids. It has been shown elsewhere (*13,15,16*) that for purely repulsive or athermal chains, this yields good predictions for the structure near a solid. Here we shall add attractive interactions in a van der Waals fashion and use the local density approximation to study liquid/vapor interfaces. We shall see that the theory is able to give the correct trends for the surface tension as a function of temperature and chain length. Finally, we shall study the influence of segment size and side branching and consider the surface behavior of isotopic blends.

Perturbation density functional theory

If one wants to extend DFT to polyatomic molecules, the first thing to be noted is that there are different ways of expressing the minimization principle because one can define several local densities. For instance, for a molecule composed of M monomers or units or interaction sites, one can first define the single molecule density $\rho_M(r_1, r_2, .., r_M)$ which is the M-point joint probability distribution function for finding atom 1 at r_1, atom 2 at r_2, etc... But one can also define the individual site densities $\rho_i(r) = \int dr_1 dr_2...dr_M \delta(r - r_i)\rho_M(r_1, r_2, .., r_M), i = 1, ..., M$, the average site or monomer density $\rho(r) = \sum_{i=1}^{M} \rho_i(r)$, the joint probability distribution function for 2 sites, 3 sites, etc...It is clear that these contracted distribution functions contain less information about the conformational structure than the single molecule density ρ_M.

It turns out that one can express the free energy either as a functional of the site densities $\rho_i(r)$ or as a functional of the single molecule density $\rho_M(r_1, r_2, .., r_M)$. The problem with the first version which has been introduced by Chandler, Mc Coy, and Singer some years ago (6) is that even the computation of the ideal part of the free energy is a nontrivial problem. This difficulty does not occur with a free energy expressed as a functional of the single molecule density. The general formulation has been stated by Pratt and Chandler in 1976 (17,18) in the framework of an interaction site cluster expansion. They showed that one can write formally the intrinsic Helmholtz free energy of a homonuclear chain fluid as

$$\beta F = \int d1_M \rho_M(1_M)[\ln \rho_M(1_M) - 1 + \beta w_M(1_M)] + \beta F^{ex}[\rho_M], \qquad (1)$$

where 1_M denotes the positions $r_1, r_2, .., r_M$ of the monomers collectively and $w_M(1_M)$ is the intramolecular energy for an isolated molecule. The first term is just the ideal part of the free energy and $F^{ex}[\rho_M]$ is the excess part, expressed in terms of an infinite set of cluster diagrams built with the contracted intramolecular distribution functions and the Mayer function associated with the interaction potential between sites in different molecules. Of course, this is a formal expression and one needs some rule to classify the diagrams in a systematic and sensible way. Such an approximation scheme is provided to us by Wertheim's work on associating fluids. What we shall use is essentially a generalization of Wertheim's thermodynamic perturbation theory (TPT) (14) to non-uniform fluids, in the limit of complete association where all molecules are fully formed.

What is the perturbation scheme ? Consider for instance a linear chain of M monomers. The main idea consists in building progressively the molecule from a reference hypothetical fluid in which all monomers are totally dissociated at the same temperature and monomer density as the real system. For instance, in the case of pearl-necklace hard-sphere chain, the reference fluid is just the hard-sphere fluid at the same packing fraction. Now the perturbation procedure consists in computing the work which is required to build up chain segments of increasing length in the reference system. More precisely, one can order the infinite sum of graphs F^{ex} as follows

$$F^{ex}[\rho_M] = F^{ex}_{ref}[\rho] + F^{ex}_1[\rho_M] + F^{ex}_2[\rho_M] + ..., \tag{2}$$

where the first term is the excess Helmholtz free energy of the reference monomeric fluid, F^{ex}_1 represents the work to build a single chain in the reference fluid, $F^{ex}_1 + F^{ex}_2$ the work to build two chains together, etc... Keeping only the first two terms corresponds to what is called the "single chain approximation" (SCA): this means that the atoms on the chain see the rest of the fluid as an atomic system. Usually, one stops at this level. It can be shown that F^{ex}_1 has the following expression (*13*),

$$\beta F^{ex}_1[\rho_M] = - \int d1_M \rho_M(1_M) \ln(y_{ref}(1_M; [\rho_M])), \tag{3}$$

where y_{ref} is the M-point cavity function in the reference system. Then it is clear that all phenomena involving interactions between two or more chains cannot be described. This is the case for instance of the isotropic to nematic transition in the case of semi-flexible molecules. If one wants to study this problem, one must work at least at the level of the two-chain approximation. However, even staying at the level of the SCA does not provide a tractable theory because the intramolecular potential w_M still contains all the M-body excluded volume effects. One then breaks the chain into smaller segments and considers the work needed to build these segments independently. At the lowest order, the chain is just a succession of $M - 1$ dimers and this is Wertheim's TPT1 theory. Then, the density functional has the following expression,

$$\beta F_{TPT1}[\rho_M] = \int d1_M \rho_M(1_M)[\ln(\rho_M(1_M)) - 1 + \beta w^*_M(1_M)] + \beta F^{ex}_{ref}[\rho]$$

$$- \int d1_M \rho_M(1_M) \ln[y_{ref}(1,2) y_{ref}(2,3)...y_{ref}(M-1,M)], \tag{4}$$

where $y_{ref}(i,j)$ is the reference pair cavity function and w^*_M contains the bonding potential, but not anymore the intramolecular van der Waals interactions which are responsible for the intra-chain excluded volume effects. Of course, both F^{ex}_{ref} and y_{ref} must be treated as functionals of the average monomer density $\rho(\mathbf{r})$ and this may be a complicated problem *per se*. On the other hand, considerable simplification occurs within the local density approximation which consists in evaluating the excess part of the Helmholtz free energy (i.e. the last two terms in Eq. (4)) by integrating over the whole volume the excess free energy density of the uniform fluid with density $\rho(\mathbf{r})$.

What is the physics behind all this and when should this perturbative approach work ? First, it is known from Flory that in concentrated polymeric fluids, long range intramolecular excluded volume effects are screened out by intermolecular interactions so that one can take into account only local inter-actions (*19*). This is the justification, for instance, of the Rotational Isomeric State model. More generally, this gives some rationale to the TPT scheme at high densities. In counterpart, it is clear that this type of theory is unable to reproduce the correct scaling behavior in the high molecular weight limit for diluted or semi-diluted solutions. Secondly, we note that this type of approach is quite common in polymer physics and is related to the self-consistent field theory introduced by Edwards (*20*) which is an extension of the Flory-Huggins mean-field picture to non-uniform situations. This can be better understood by looking at the Euler-Lagrange equations obtained by minimization of the TPT1 grand potential in the LDA (for a repulsive homonuclear chain of M segments with bonding length d),

$$\rho_M(1_M) = \exp \beta[\mu_M - w^*(1_M) - \sum_{i=1}^{M} \psi(\mathbf{r}_i)]$$

$$\psi(\mathbf{r}) = v(\mathbf{r}) + \frac{\partial f_{ref}^{ex}(\rho(\mathbf{r}))}{\partial \rho}$$
$$- \frac{M-1}{M} k_B T [\ln y_{ref}(d, \rho(\mathbf{r})) + \rho(\mathbf{r}) \frac{\partial \ln y_{ref}(d, \rho(\mathbf{r}))}{\partial \rho}], \qquad (5)$$

where μ_M is the polymer chemical potential and $y_{ref}(d, \rho(\mathbf{r}))$ is the pair cavity function at distance d in a homogeneous system with local density $\rho(\mathbf{r})$. Eq. (5) shows that this approximation corresponds to an ideal polymer melt submitted to an external potential ψ which is the sum of the real external potential $v(\mathbf{r})$ plus an effective potential field which arises self-consistently from the *average* monomer density profile $\rho(\mathbf{r})$. We know that this type of approximation gives its best results when monomer density fluctuations can be ignored, which is the case in concentrated solutions and melts. To be complete, we must note that the full version of the theory (i.e. with no LDA) is more elaborated than the usual SCF theories because the basic variables in the self-consistent equations are the two-site densities (*13*).

Finally, in the long chain limit, the SCF equations which essentially de-scribe a biased random walk can be transformed as usual into a diffusion equation for the propagator (*20*)

$$G_M(\mathbf{r}, \mathbf{r}') = \int d1_M \rho(1_M) \delta(\mathbf{r} - \mathbf{r}_1) \delta(\mathbf{r}' - \mathbf{r}_M) \qquad (6)$$

which is the unnormalized probability for finding monomer 1 at $\mathbf{r}$ and monomer M at $\mathbf{r}'$. However, we would like to point out that with today's powerful computers, it is easier to calculate directly the path integral for the ideal chain than to solve the diffusion equation. This is also true when dealing with branched molecules.

Results

It is shown elsewhere that the full version of the TPT1 theory with a weighted density approximation for the hard-sphere part of the free energy provides a faithfull representation of the site density profile of semi-flexible chains of tangent hard spheres in the vicinity of a hard wall (*15,16*). The agreement with the simulation improves as the stiffness of the molecule increases. In the following, we shall only focus on the liquid/vapor interface where we can safely use the LDA. Therefore, we add an attractive Lennard-Jones interaction between non bonded sites in order to get phase separation. This attraction is treated in a mean-field fashion and we use the Barker-Henderson division of the potential (*21*) . For the sake of simplicity, the equivalent hard-sphere diameter and the bonding length are both taken equal to σ. So we still consider tangent units and molecules are supposed to be totally flexible.

Now, we can solve the equation of state for the liquid/vapor equilibrium and compute the surface tension to plot universal curves in reduced units as a function of temperature and chain length. In qualitative agreement with experiments, we see in Fig. 1 that for a homologous series of liquids, the surface tension γ increases and the surface entropy $-d\gamma/dT$ decreases with molecular weight. These results are similar to those predicted by the widely used Poser-Sanchez lattice fluid theory (*22*). When the chain length goes to infinity the surface tension approaches a limiting value and Fig. 2 shows a clear $1/M$ dependence, which is also in agreement with experiment for high molecular weight polymers (*23*).

The advantage of our DFT over previous theories is that we can study quite easily the effect of the molecular architecture or segment size on surface phenomena. Let us first consider the effect of side branching. In Fig. 3, we show the surface segregation in an equimolar blend composed of linear chains of 49 units and branched chains which have the same total number of segments but with a side group hanging off every three backbone units. The segment size and the attractive interactions are the same. This can be related to recent experiments which have been performed for copolymer blends of polyethylene (PE) and polyethylenepropylene (PEP) (*24*), although we have not tried to make any adjustment of the parameters. It has been observed experimentally that the more branched component segregates to the surface. This is indeed what we find: in Fig. 3.a we show the segment density profiles at the liquid/vapor interface and in Fig. 3.b the relative volume fraction of the branched component (the Gibbs dividing surface is approximately located at $z = 0$). There has been some controversy in the literature about the physical origin of this surface segregation. It has been suggested that it was a purely entropic effect due to the difference in the statistical lengths of the components (*25*). Our study of this system as a function of temperature seems to show that actually there is a subtle interplay between entropic and enthalpic contributions, a conclusion which has also been reached by Yethiraj in a simulation of the same system at the solid/fluid interface (*26*).

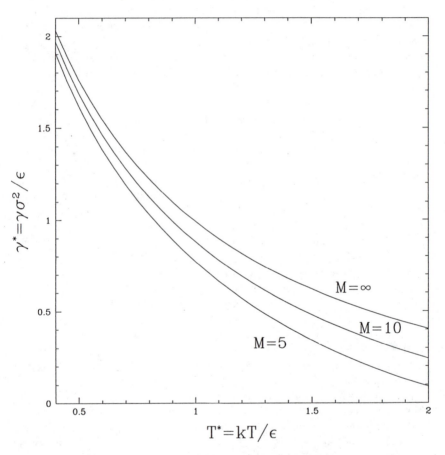

Figure 1. Polymer surface tension of the liquid/vapor interface as a function of temperature and chain length in reduced units.

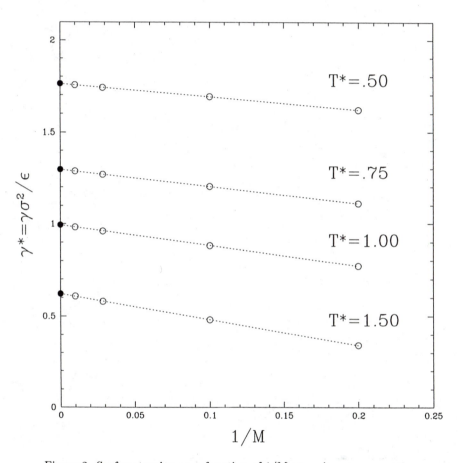

Figure 2. Surface tension as a function of 1/M at various temperatures.

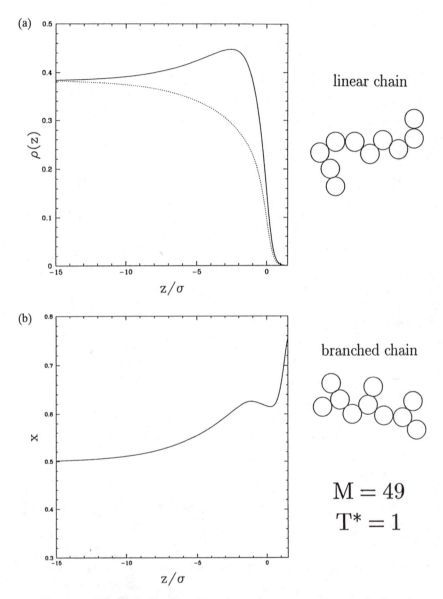

Figure 3. Effect of side branching on surface segregation at the liquid-vapor interface ($M = 49$, $T^* = 1$).
3a) Segment density profiles of linear (dotted line) and branched chains (solid line).
3b) Relative volume fraction of the branched component.

The effect of segment size disparity on surface segregation phenomena has not yet been studied in the litterature because it is difficult to include this feature in a lattice model or in a phenomenological square gradient theory. In Fig. 4, we show what happens for an equimolar mixture when the two chains have the same number of units ($M = 100$) and the same van der Waals interactions, but the sizes of their monomers differ by 2%. We find rather surprisingly that there is a strong surface enhancement for the chain with the largest size. The effect becomes less pronounced as the temperature increases and again there is probably an interplay between entropic and energetic effects.

So far, we have not tried to make quantitative comparison between our predictions and experimental data. For that purpose, we must choose some numerical values for the interaction parameters. In principle, three parameters can be adjusted, the chain length M, the segment size σ and the energy parameter of the LJ interaction ϵ. Since the freely jointed sphere model is a rather coarse-grained representation of a real molecule, it is not clear that a sphere corresponds to a single repeat unit or to several monomers. For clarity, we shall decide that M indeed corresponds to the real number of repeat units. It remains the problem of determining σ and ϵ. Since a calculation from first principles seems totally hopeless, we have to treat these quantities as adjustable parameters, hoping that the theory has sufficient realism at the microscopic level to warrant meaningful comparison with the experiment. We know that even for simple fluids it is very difficult to reproduce both bulk and surface properties with the same values of σ and ϵ (*27*). Therefore, we shall make the choice of adjusting these parameters to a surface property, namely the surface tension. In Fig. 5, for instance, are shown the results for the variation of the surface tension of PDMS (polydimethylsiloxane) as a function of temperature for two different molecular weights corresponding respectively to chains of 10 and 432 units. It can be seen that reasonable agreement between theoretical predictions and experiments (*28*) can be reached with the values $\sigma = 4.85\text{Å}$ and $\epsilon/k = 324K$. Now, we can use these values to study the surface tension of a blend composed of these two chains and vary the composition of the shortest one. Fig. 6 shows a good agreement with the experiments (*29*), which proves that the theory is able to describe accurately polydispersity effects.

As a final example, we present some results for isotopic polystyrene blends. These blends have attracted considerable interest in the last few years because the replacement of the hydrogen atom by deuterium causes enough change in the zero-point energy of the macromolecule that a mixture of sufficiently high molecular weight can phase separate (*30*). Even in the one-phase region, there is also significant segregation at the free surface that can be studied by neutron reflectivity or secondary ion mass spectrometry (SIMS). When the chains, protonated and deuterated, have the same length, it is found that the deuterated component always partitions preferentially to the surface because of energetic considerations (ϵ_{DD} is slightly lower than ϵ_{HH}) (*31*). But when the protonated chain is much shorter than the deuterated one, it goes to the surface

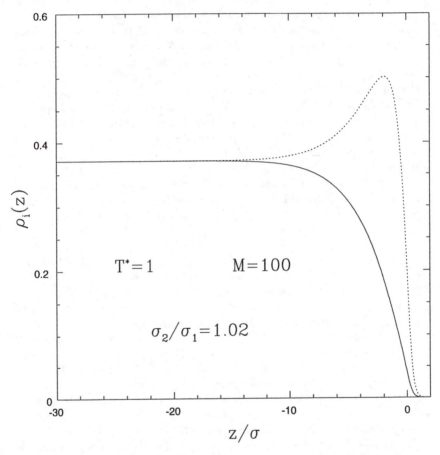

Figure 4. Influence of segment size disparity on surface segregation at the liquid/vapor interface. The two chains have the same number of units ($M = 100$). The dotted line represents the segment density profile of the chain with the largest segment size.

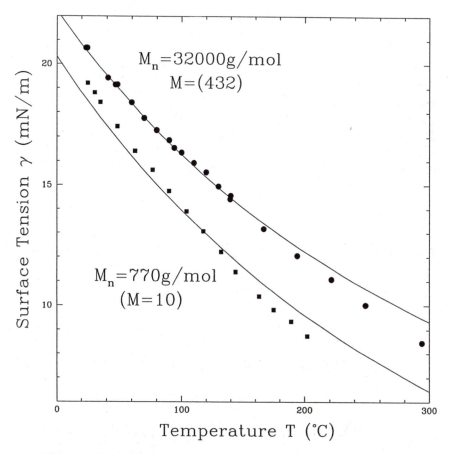

Figure 5. Surface tension as a function of temperature for two samples of PDMS with Number Average Molecular Weights $M_n = 32000$ g/mol and $M_n = 770$ g/mol. The solid lines are the predictions of PDFT. The symbols are experimental results of Dee and Sauer (*28*).

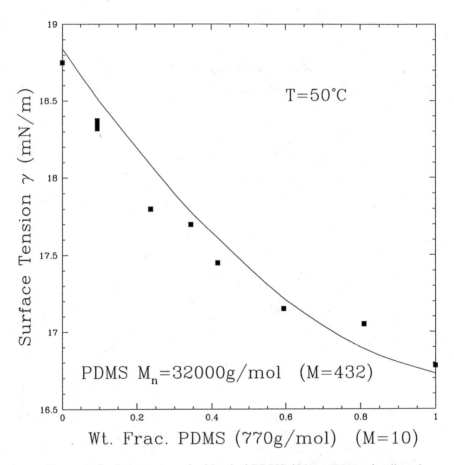

Figure 6. Surface tension of a blend of PDMS (M_n = 32000 g/mol) and PDMS (M_n = 770 g/mol) as a function of the weight fraction of the low molecular weight conponent. The solid lines are the predictions of PDFT. The symbols are experimental results of Dee and Sauer (*29*).

because of entropic effects (*32*). This has been termed as "the reversal of the isotopic effect". This effect can be predicted rather accurately by our theory. In our calculation (for a bulk composition $x(PSD) = 0.48$ and at $T = 433K$), the length of the deuterated chains is kept constant and equal to 1000 units and the length of the protonated chains is varied from 20 to 500 units. The ϵ_{HH} and σ parameters for the protonated polystyrene have been adjusted on pure surface tension data (*28*) and ϵ_{DD} for the deuterated component has been adjusted to reproduce the experimental results for isotopic blends with chains of the same length (*32*). We clearly see in Fig. 7 the reversal of the isotopic effect as the length of the protonated chain decreases. We have plotted in Fig. 8 the relative surface enhancement of the deuterated chain as a function of the

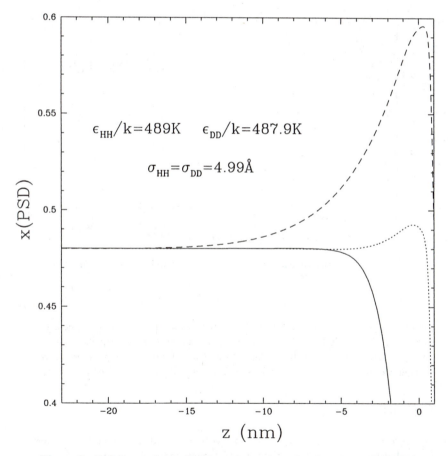

Figure 7. Relative volume fraction of deuterated polystyrene (M(PSD) =1000) as a function of the polymerization degree of the protonated component. Solid line : M(PSH) =20; dotted line : M(PSH)=200; dashed line: M(PSH)=500.

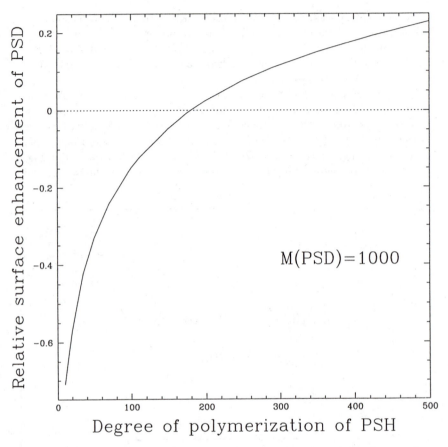

Figure 8. Reversal of the isotopic effect as a function of the degree of polymerization of PSH.

degree of polymerization of the protonated component. The cross-over occurs near $M(PSH) \simeq 200$ which is in much better agreement with the experimental value $\simeq 500$ than lattice fluid calculations (33). One may expect to improve further the predictions by introducing some rigidity in the molecular model.

Conclusion

We have presented a density functional theory for inhomogeneous polymeric fluids, based on perturbation theory, which can predict surface properties with a level of accuracy comparable to what is reached in simple fluids. The use of the LDA for the liquid/vapor interface allows to perform calculations for chains with several hundreds of units. This theory is best suited for studying surface segregation effects in blends. Extension to copolymer systems is straightforward.

Acknowledgments. We are grateful to G. T. Dee and B. B. Sauer for having sent us their experimental data for PDMS.

Literature Cited.

1. Evans R. In *Inhomogeneous Fluids*; Henderson D.; Publisher: Dekker, New York, 1992.
2. de Gennes P. G. *Scaling Concepts in Polymer Physics*; Cornell University Press, Ithaca, 1979.
3. e.g. Scheutjens J.M. H. M.; Fleer G. J. *J. Phys. Chem.* **1979**, 83, 1619.
4. Nakanishi H.; Pincus P. *J. Chem. Phys.* **1983**, 79, 997.
5. Schmidt I.; Binder K. *J. Physique* **1985**, 46, 1631.
6. Chandler D.; McCoy J. D.; Singer S. J. *J. Chem. Phys.* **1986**, 85, 5971.
7. McMullen W. E.; Freed K. F. *J. Chem. Phys.* **1990**, 92, 1413.
8. Tang H.; Freed K. F. *J. Chem. Phys.* **1991**, 94, 1572.
9. Woodward C. E. *J. Chem. Phys.* **1992**, 97, 4525.
10. Yethiraj A.; Woodward C.E. *J. Chem. Phys.* **1995**, 102, 5499.
11. Kierlik E.; Rosinberg M.L. *J. Chem. Phys.* **1992**, 97, 9222.
12. Kierlik E.; Rosinberg M.L. *J. Chem. Phys.* **1993**, 99, 3950.
13. Kierlik E.; Rosinberg M.L. *J. Chem. Phys.* **1994**, 100, 1716.
14. Wertheim M.S. *J. Chem. Phys.* **1987**, 87, 7323.
15. Phan S.; Kierlik E.; Rosinberg M.L.; Yethiraj A.; Dickman R. *J. Chem. Phys.* **1995**, 102, 2141.
16. Yethiraj A. this book.
17. Chandler D.; Pratt L. R. *J. Chem. Phys.* **1976**, 65, 2925.
18. Pratt L. R.; Chandler D. *J. Chem. Phys.* **1977**, 66, 147.
19. Flory P. *Principles of Polymer Chemistry*; Cornell University Press, Ithaca, New York, 1971.
20. Edwards S. F. *Proc. Phys. Soc. (London)* **1965**, 85, 613.
21. Hansen J. P.; McDonald I. R. *Theory of Simple Liquids*; Academic, New York, 1976.
22. Poser C. I.; Sanchez I. C. *J. Colloid. Interface. Sci.* **1979**, 69, 539.
23. Sauer B. B.; Dee G. T. *J. Colloid. Interface. Sci.* **1994**, 162, 25.
24. Sikka M.; Singh N.; Karim A.; Bates F. S.; Satija S. K.; Majkrzak C. F. *Phys. Rev. Lett.* **1993**, 70, 307.
25. Fredrickson G. H.; Donley J. P. *J. Chem. Phys.* **1992**, 97, 8941.
26. Yethiraj A. *Phys. Rev. Lett.* **1995**, 74, 2018.
27. Winkelmann J.; Brodrecht U.; Kreft I. *Ber. Bunsenges. Phys. Chem.* **1994**, 98, 912.
28. Dee G. T.; Sauer B. B. *J. Colloid. Interface. Sci.* **1992**, 152, 85.
29. Dee G. T.; Sauer B. B. *Macomolecules* **1993**, 26, 2771.
30. Bates F. S. et Wignall G. D. *Phys. Rev. Lett.* **1986**, 57, 1429.

31. Jones R. A. L.; Kramer E. J.; Rafailovich M. H.; Sokolov J.; Schwarz S.
    A. *Phys. Rev. Lett.* **1989**, 62, 280.
32. Hariharan A. ; Kumar S. K.; Russell T. P. *J. Chem. Phys.* **1993**, 98,
    4163.
33. Hariharan A.; Kumar S. K.; Russell T. P. *J. Chem. Phys.* **1993**, 99,
    4041.

Chapter 16

A Density-Functional Approach to Investigation of Solid–Fluid Interfacial Properties

D. W. M. Marr[1] and A. P. Gast[2]

[1]Chemical Engineering and Petroleum Refining Department,
Colorado School of Mines, Golden, CO 80401–1887
[2]Department of Chemical Engineering, Stanford University,
Stanford, CA 94305–5025

We present a formulation of density-functional theory ideally suited for investigation of solid-fluid interfacial properties. We use this approach to investigate the role of interactions in determining both the energy and structure of the interface by examining a number of systems including hard-sphere, adhesive-sphere, and Lennard-Jones fluids. In addition, we study the orientational dependence of interfacial properties in the adhesive-sphere system.

Despite its ubiquity in nature, remarkably little is known of the solid-fluid interfacial structural and energetic properties. The reason for this lies in the difficulty in experimentally assessing the interface; most materials of technological importance (e.g. metals) are not transparent making observation extremely difficult [1]. Efforts to experimentally study the solid-fluid interface often center around removal of the solid phase from the equilibrium fluid but lead to a modified interface, frustrating efforts to examine the equilibrium structure. One can also employ a rapid temperature quench, some kind of sectioning, and then microscopy but this method is applicable only to multi-component systems. Methods of experimentally determining solid-fluid interfacial tensions can involve the measurement of grain boundary intersection angles or the study of grooves in the crystal surface but are normally applicable only to a few systems.

Results from computer simulation are also limited. The complexity of the solid-fluid interface makes computational approaches extremely costly and has limited study to only a few systems, for which a nice review is available [2]. Because of the limited amount of computational or experimental study, little is known of the influence of the interaction potential on the structure and energy of the interface. In general, we expect that the interface to become extremely sharp at low temperatures due to its low entropy. As temperature increases, however, the interface will widen and the transition from solid to fluid will occur over a larger distance. The interface width will be limited, though, by the high

0097–6156/96/0629–0229$15.00/0

energetic cost of producing regions having a density different from that of either coexisting equilibrium phase. At a given temperature for a specific interaction these factors combine to give a minimum in the free energy of the interface at a particular interfacial width. A number of questions immediately arise: How will the interactions between particles influence the broadening of the interface? Will attractions have a strong influence or do repulsions alone determine interfacial structure? How do these influences change with interface orientation? Answering these questions requires a technique that can readily determine interfacial thermodynamic properties as a function of interaction potential in a computationally tractable manner.

Theoretical Approach

The thermodynamic properties of a variety of model fluid systems are well established. In general, liquid-state theory allows one, with knowledge of the interaction potential, to determine correlations between particles within the homogeneous fluid phase. Combined with a closure appropriate for the interaction potential of interest, one can obtain thermodynamic properties such as the pressure and chemical potential of the homogeneous fluid.

Determining the thermodynamic properties of the solid phase is much more difficult. One cannot apply the equations of liquid-state theory directly and the use of computer "experiments" such as Monte Carlo or molecular-dynamics simulations is both difficult and computationally demanding. Due to the success of liquid-state theory, there has been much effort in describing the solid state with liquid properties, leading to the development of density-functional theory. Recently, density-functional theory has provided a means to describe the structure and energetics of the solid phase. In addition to facility with the homogeneous solid phase via the principles of liquid-state theory, density-functional theory has allowed description of interfaces and other inhomogeneous systems.

While there are a number of implementations of density-functional theory for studying phase transitions, all of them seek to describe the structure and properties of the solid phase from information about the fluid. Several excellent reviews of the different density-functional approaches have recently appeared [3–8]. Basically, we can place the density-functional theories into two categories: i) The description of the solid phase through a truncated functional Taylor expansion of the n-particle direct correlation function [9,10] and ii) the description of the solid phase through appropriate choice of an effective liquid approximating its thermodynamics [11–13]. The latter approach, while somewhat *ad hoc*, is quite successful in the description of hard sphere solids [11, 12, 14–20]. Once the effective liquid density is chosen, the description of the solid becomes a matter of applying information available for the liquid state. There are a range of approaches for choice of the effective liquid from the early effective liquid approximation (ELA) of Baus and Colot [12], where an *ad hoc* but physically appealing comparison of structure was invoked, to the computationally demanding weighted density approximation (WDA) of Curtin and Ashcroft [13]. More recent criteria for choosing an effective liquid density bring the Baus and Colot effective liquid approximation (GELA and SCELA [11]) into accord with the weighted-density approaches [8]. We focus on the approach of Curtin and Ashcroft [13] who define a weighting function which links the solid and liquid states. This approach, known as the weighted density approximation (WDA), involves the determination of a spatially *variant* weighted density where fluid properties approximate those of the solid. This method has been effective in predicting solid properties and phase coexistence but its application to more complex problems has been hindered by the computational requirements in the determination of weighted densities. To overcome these difficulties, Denton and Ashcroft have developed the modified weighted density approximation (MWDA) [15]. In contrast to the WDA, this approach requires only calculation of a spatially *invariant* weighted density and significantly lowers computation time.

In order to study the structural details and energetics of the solid-liquid interface, one must determine the appropriate weighted density to model each density through the interface. This can be done with the WDA [16, 17]; however, the calculation requires tremendous computational effort making it impractical for complex situations. We are interested in systems including finite interparticle interactions where the densities of coexisting phases will depend on temperature. In this situation, the interfacial structure and energy must be determined for a variety of temperatures, significantly increasing the required amount of computation and motivating the development of a tractable approach to describe the interface. Encouraged by the success of the MWDA in decreasing the computational requirements of the WDA, we developed a planar weighted density approximation (PWDA) to describe the interface [21].

One begins by separating the total Helmholtz free energy of the solid phase into two components

$$F[\rho] = F_{\mathrm{id}}[\rho] + F_{\mathrm{ex}}[\rho] \tag{1}$$

representing the ideal and excess contributions to the total free energy. The ideal term can be calculated for any given density distribution $\rho(\mathbf{r})$ from

$$F_{\mathrm{id}}[\rho] = \beta^{-1} \int d\mathbf{r}\rho(\mathbf{r})[\ln\left(\rho(\mathbf{r})\Lambda^3\right) - 1] \tag{2}$$

where $\beta = 1/kT$ and Λ is the de Broglie wavelength, and the total excess free energy can be expressed as the sum of local contributions

$$F_{\mathrm{ex}}[\rho] = \int d\mathbf{r}\rho(\mathbf{r})\psi(\mathbf{r}; [\rho]) \tag{3}$$

where ψ is the local excess free energy per particle.

Weighted-Density Approximation. Curtin and Ashcroft [13] approximated the local excess solid free energy per particle with the excess free energy per particle of a homogeneous fluid, indicated by a subscript 0, evaluated at some effective liquid density $\bar{\rho}$,

$$F_{\mathrm{ex}}^{\mathrm{WDA}}[\rho] \equiv \int d\mathbf{r}\rho(\mathbf{r})\psi_0(\bar{\rho}(\mathbf{r})) \tag{4}$$

where the effective liquid density is a weighted average of the local solid densities in the vicinity of $\mathbf{r}$. In the WDA, the spatially *variant* effective liquid density is defined by

$$\overline{\rho}(\mathbf{r}) \equiv \int d\mathbf{r}'\rho(\mathbf{r}')w(\mathbf{r} - \mathbf{r}'; \overline{\rho}(\mathbf{r})) \tag{5}$$

where the weighting function w is introduced with the normalization requirement

$$\int d\mathbf{r}w(\mathbf{r} - \mathbf{r}'; \rho) = 1 \tag{6}$$

and is determined by requiring the exact 2-particle direct correlation function c to be recovered in the homogeneous limit,

$$-\beta^{-1}c_0(\mathbf{r} - \mathbf{r}'; \rho_0) = \lim_{\rho \to \rho_0} \frac{\delta^2 F_{\text{ex}}}{\delta\rho(\mathbf{r})\delta\rho(\mathbf{r}')}. \tag{7}$$

Solving for w is most readily done in Fourier space, leading to the following differential equation

$$-\beta^{-1}c_0(k; \rho_0) = 2\frac{\partial\psi_0}{\partial\rho}w(k; \rho_0) + \rho_0\frac{\partial}{\partial\rho}\left[\frac{\partial\psi_0}{\partial\rho}w^2(k; \rho_0)\right]. \tag{8}$$

Modified Weighted-Density Approximation. The MWDA [15] differs from this approach in that the excess Helmholtz free energy is calculated from a spatially invariant weighted density,

$$F_{\text{ex}}^{\text{MWDA}}[\rho] \equiv N\psi_0(\hat{\rho}) \tag{9}$$

where N is the number of particles and the weighted density determined from

$$\hat{\rho} \equiv \frac{1}{N}\int d\mathbf{r}\rho(\mathbf{r})\int d\mathbf{r}'\rho(\mathbf{r}')w(\mathbf{r} - \mathbf{r}'; \hat{\rho}). \tag{10}$$

Using the normalization condition (equation 6) and imposing the exact fluid free energy in the homogeneous limit (equation 7) one obtains

$$-\beta^{-1}c_0(k; \rho_0) = 2\frac{\partial\psi_0}{\partial\rho}w(k; \rho_0) + \delta_{k,0}\rho_0\frac{\partial^2\psi_0}{\partial\rho^2}. \tag{11}$$

This equation for $w(k, \rho)$ is easier to solve than the WDA; it is proportional to the direct correlation function (for non-zero k) and does not involve the solution of a non-linear differential equation. The MWDA requires a great deal less computation and is therefore the preferred approach for the study of the transition from liquid to solid [4, 7, 8, 19, 22–24].

Planar Weighted-Density Approximation. In systems such as the solid-fluid interface where the bulk density varies with position, one cannot apply the MWDA because of the need to retain a spatially-varying weighted density.

Curtin [17] has applied the WDA to the interfacial problem with good results but after significant computational effort. One can, however, lower these requirements while still retaining the physical approach to the problem by incorporating the MWDA into the interface [21]. This is done by realizing that the bulk density parallel to a planar interface remains constant. In the spirit of Denton and Ashcroft's reduction of the computational requirements for WDA models of bulk systems, one may approach the interface with a planar-averaged spatially-variant weighted density that will significantly decrease calculation costs. Other authors have succeeded in modelling systems such as a hard-sphere fluid next to a hard wall [25–28] with a one-dimensional weighted density; however, efforts to describe the freezing transition with such a weighted density have not led to a stable solid.

One begins by expressing the excess free energy in terms of the planar-averaged density $\hat{\rho}(z)$ and a planar-averaged free energy

$$F_{\text{ex}}[\rho] = \int d\mathbf{r} \hat{\rho}(z) \psi(z) \tag{12}$$

$$\hat{\rho}(z) = \frac{1}{A} \int dx\, dy\, \rho(\mathbf{r}). \tag{13}$$

This 'local' free energy is now approximated with that of a homogeneous fluid evaluated at some planar weighted density $\overline{\rho}(z)$

$$F_{\text{ex}}^{\text{PWDA}}[\rho] \equiv \int d\mathbf{r} \hat{\rho}(z) \psi_0(\overline{\rho}(z)), \tag{14}$$

determined self-consistently from

$$\overline{\rho}(z) \equiv \frac{\int dx\, dy\, \rho(\mathbf{r}) \int d\mathbf{r}' \rho(\mathbf{r}') w(\mathbf{r} - \mathbf{r}'; \overline{\rho}(z))}{\int dx\, dy\, \rho(\mathbf{r})} \tag{15}$$

where once again the weighting function is determined from the normalization condition (equation 6) and the requirement on the limiting behavior (equation 7). In Fourier space one obtains

$$-\beta^{-1} c_0(k; \rho_0) = 2 \frac{\partial \psi_0}{\partial \rho} w(k; \rho_0) + \delta_{\mathbf{k}_{\parallel}, 0} \rho_0 \frac{\partial}{\partial \rho} \left[\frac{\partial \psi_0}{\partial \rho} w^2(k; \rho_0) \right], \tag{16}$$

reducing to the WDA weighting function in the $\mathbf{k}_{\parallel} = 0$ case and the MWDA weighting function when $\mathbf{k}_{\parallel}$ is non-zero. This approach incorporates many of the computational savings inherent in the MWDA and yet is applicable to the determination of solid-fluid interfacial properties.

Bulk Properties. In order to solve these equations for the solid free energy one must first model the solid structure. As proposed by Tarazona [20], the solid phase density distribution can be represented as the sum of normalized Gaussians

$$\rho_s(\mathbf{r}) = \left(\frac{\alpha}{\pi} \right)^{3/2} \sum_{\mathbf{R}} e^{-\alpha(\mathbf{r} - \mathbf{R})^2} \tag{17}$$

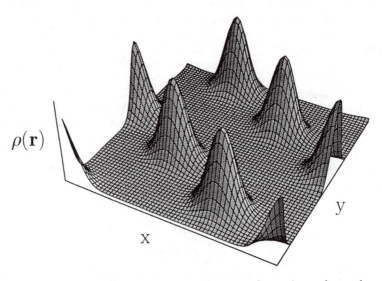

Figure 1. Solid phase represented as the sum of gaussians; the peaks correspond to particles located about their lattice positions.

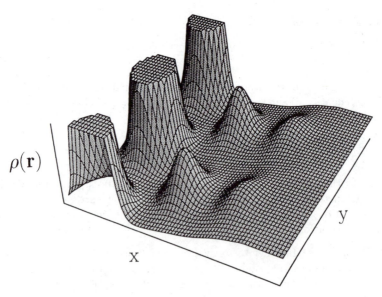

Figure 2. Example solid-fluid interface (The peaks have been cut short to better illustrate the transition).

where **R** are the Bravais lattice vectors, or in Fourier space as

$$\rho_s(\mathbf{r}) = \rho_s + \sum_{G \neq 0} \rho_G e^{i\mathbf{G} \cdot \mathbf{r}} \tag{18}$$

where **G** are the reciprocal lattice vectors and $\rho_G \equiv \rho_s e^{-G^2/4\alpha}$. The parameter α describes the structure of the solid; the higher the value of α the more localized the structure and a value of zero corresponds to the homogeneous fluid. Figure 1 illustrates this solid-phase parameterization.

To determine the solid-phase thermodynamic properties one minimizes the total free energy for a given solid density with respect to α. A global minimum occurring at a non-zero α indicates a stable solid phase, determining both the stable α and excess free energy corresponding to a given solid density. The total free energy is found by adding the ideal and excess contributions. Phase coexistence occurs when the chemical potential μ and pressure P of the solid and fluid phases are identical.

$$\mu = \frac{\partial(\rho \frac{F[\rho]}{N})}{\partial \rho} \tag{19}$$

$$P = \rho(\mu - F[\rho]/N) = \rho^2 \frac{\partial(\frac{F[\rho]}{N})}{\partial \rho} \tag{20}$$

Interfacial Properties. Curtin [16, 17] has developed a convenient two parameter model in his application of the WDA to the solid-fluid interface. He represents the solid as the sum of Fourier components as before but now allows these components to decay as one makes the transition from solid to liquid along the z direction across the interface:

$$\rho(\mathbf{r}) = \rho_l + (\rho_s - \rho_l)f_0(z) + \sum_G \rho_G f_G(z) e^{i\mathbf{G} \cdot \mathbf{r}} \tag{21}$$

where

$$f_G(z) = \begin{cases} 1 & |z| < z_0 \\ \frac{1}{2}(1 + \cos(\pi \frac{z-z_0}{\Delta z_G})) & z_0 < |z| < z_G \\ 0 & z_G < |z| \end{cases} \tag{22}$$

z_0 is the position of the solid-fluid interface boundary, Δz the interface width, $f_0(z) = f_{G_1}(z)$, $\Delta z_G = (G_1/G)^\nu \Delta z = z_G - z_0$, and ν the decay rate of the higher order Fourier components. Figure 2 shows an example of this interfacial profile parameterization where the peaks have been truncated for clarity. One can now express the weighted density in the PWDA in terms of the reciprocal lattice vectors and the Fourier transforms $w(k; \rho)$ and $f_G(k)$ as

$$\overline{\rho}(z) = \rho_l + (\rho_s - \rho_l)\frac{1}{2\pi} \int dk e^{ikz} w(k; \overline{\rho}(z))f_0(-k) \tag{23}$$

$$+ \frac{(\rho_l + (\rho_s - \rho_l)f_0(z))}{2\pi\hat{\rho}(z)} \int dk w(k; \overline{\rho}(z)) e^{ikz} \sum_G \delta_{G_{||},0} \rho_G f_G(k + G_z)$$

$$+ \frac{1}{2\pi\hat{\rho}(z)} \int dk \sum_{G_i} \rho_{G_i} f_{G_i}(z) e^{i(G_{iz} - k)z}$$

$$\times \sum_{G_j} \delta_{G_{i||},-G_{j||}} \rho_{G_j} w((G_{j||}^2 + k^2)^{1/2}; \overline{\rho}(z)) f_{G_j}(k + G_{jz})$$

where $\mathbf{G}_{||} = (G_x, G_y)$. Though this equation appears quite complex, the delta functions in the second and third terms evaluate to zero for many of the reciprocal lattice vectors used in the summation.

In order to determine interfacial properties one must first calculate bulk properties including the pressure, chemical potential, solid localization parameter and the densities of the coexisting bulk phases following the procedure outlined in the previous section. While μ and P remain constant throughout the interface, the coexisting densities define the boundary conditions on the interfacial profile. Minimizing the excess grand potential $\Delta\Omega$

$$\Delta\Omega[\rho(\mathbf{r})] = F[\rho(\mathbf{r})] - \mu \int d\mathbf{r}\rho(\mathbf{r}) + PV \tag{24}$$

determines both interfacial structure and energy. Applying equations 1 and 2 and the PWDA one obtains

$$\begin{aligned} \Delta\Omega[\rho(\mathbf{r})] &= F_{ex}^{PWDA}[\rho(\mathbf{r})] + PV \\ &+ \int d\mathbf{r}\rho(\mathbf{r})[\beta^{-1}\{\ln(\rho(\mathbf{r})\Lambda^3) - 1\} - \mu]. \end{aligned} \tag{25}$$

Finally, after applying equation 14 and defining

$$\hat{f}_{id}(z) = \frac{1}{A} \int dxdy\rho(\mathbf{r})\ln\rho(\mathbf{r}) \tag{26}$$

$$\hat{\mu} = \beta\mu - (\ln\Lambda^3 - 1) \tag{27}$$

one obtains the interfacial tension as

$$\gamma = \frac{\Delta\Omega}{A}\bigg|_{min} = \frac{1}{\beta} \int dz \{\hat{\rho}(z)[\beta\psi_0(\overline{\rho}(z)) - \hat{\mu}] + \hat{f}_{id}(z) + \beta P\}\bigg|_{min}. \tag{28}$$

Interfacial Properties of Model Systems

The equilibrium phase behavior of a system is dictated by the interaction potential between individual particles. Phase transitions induced by attractions depend on the depth of the attractive minimum as well as the range of the interaction. Generally speaking, systems having a deep attractive well separate into a dense solid phase and a dilute vapor. It has been postulated that a long-range attraction is required to produce a triple point and vapor-liquid coexistence. It

Table I: Hard-sphere phase coexistence predicted by the various weighted density approaches compared to simulation results

	$\rho_f \sigma^3$	$\rho_s \sigma^3$
WDA [13]	0.881	1.02
MWDA [15,18]	0.912	1.044
PWDA	0.882	1.026
MC [33]	0.943	1.041

is unclear, however, how the individual factors of range and depth contribute to the phase behavior since most interactions include both and their influence is highly coupled. This is a question of current and continuing interest [29–31]. One can extend this and ask how these factors influence not only the phase behavior but also the interfacial properties between phases. In an effort to answer this question, we look at the solid-fluid interfaces of a number of model systems, beginning with one of the simplest models available and then gradually adding complexity to the interactions.

Hard Spheres. The hard-sphere system is a purely repulsive model often employed in statistical mechanics and described by the following interaction potential

$$u(r)/kT = \begin{cases} \infty & 0 < r < \sigma \\ 0 & \sigma < r \end{cases} \tag{29}$$

with σ the hard-sphere diameter. Because of its simplicity there exist analytic solutions for the fluid-state properties $\psi_0(\rho)$ and $c_0(k;\rho)$ in the Percus-Yevick approximation [32], making application of the PWDA relatively straightforward.

From simulation [33,34] the properties of the coexisting solid and fluid phases are well known. It has been found that, at equilibrium, the hard-sphere solid phase preferentially organizes into a close-packed structure. We study the interface along the densest face (111) of the close packed face centered cubic (fcc) lattice, where the distance between planes is $\delta_{111} = a/\sqrt{3}$, a being the fcc lattice constant $= (4/\rho_s)^{1/3}$. We see excellent agreement between the PWDA and both the WDA and the MWDA in a calculation of the hard-sphere order-disorder transition (see Table I). All three of the theories agree reasonably well with Monte Carlo simulations justifying their use in describing the hard-sphere system.

There exist few studies of the hard-sphere interface; however, we compare results from the planar-averaged approach to those obtained using the fully three-dimensional WDA. The two theories agree extremely well with $\gamma^{\mathrm{WDA}} \sigma^2/kT = 0.63 \pm 0.02$, $\Delta z^{\mathrm{WDA}}/\delta_{111} = 3 - 4$ and $\gamma^{\mathrm{PWDA}} \sigma^2/kT = 0.60 \pm 0.02$, $\Delta z^{\mathrm{PWDA}}/\delta_{111} = 3 - 4$. Also, the absolute magnitudes of the interfacial tensions determined for this system agree fairly well with an experimental estimate for the hard-sphere interfacial tension of $0.55 kT/\sigma^2$ [35].

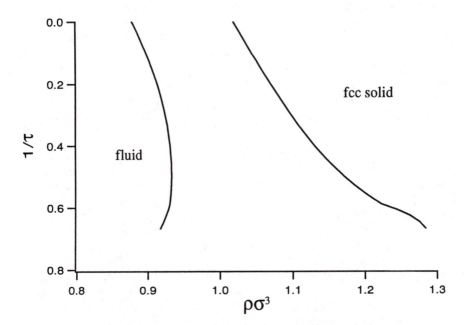

Figure 3. Adhesive-sphere phase diagram.

Adhesive Spheres. First introduced by Baxter [36], the adhesive hard-sphere system has zero range but a variable attractive strength represented by

$$u(r)/kT = \lim_{\sigma' \to \sigma} \begin{cases} \infty & 0 < r < \sigma' \\ \ln\left[\frac{12\tau(\sigma-\sigma')}{\sigma}\right] & \sigma' < r < \sigma \\ 0 & \sigma < r \end{cases} \tag{30}$$

where the parameter τ is a dimensionless measure of the well depth. A value of $\tau = 0$ corresponds to infinite attraction or zero temperature while a value of $\tau \to \infty$ corresponds to infinite temperature and hard-sphere behavior. The interaction parameter τ has been related to well depth in the square-well system [37] and temperature in experimental systems [38] by equating second virial coefficients.

The adhesive hard sphere is an excellent system to study for two reasons. First, it provides a system for study of the influence of attractive strength on phase behavior by decoupling the influence of interaction range. The second, and far more practical reason, is that there exist analytic solutions in the Percus Yevick (PY) approximation for the direct correlation function and the excess free energy per particle; having analytic solutions simplifies calculation significantly. Both Seaton and Glandt [39] and Kranendonk and Frenkel [40] have performed Monte Carlo simulations on the adhesive-sphere system and found good agreement with the PY solution, at least to moderate densities. Its rich phase behavior [36, 41–46] has stimulated interest in adhesive spheres as a model system to describe colloidal interactions including interactions between micelles [47,48] and those involving colloidal silica with surface grafted octadecyl chains in marginal solvents [38, 49–55].

Bulk Properties. Figure 3 shows the results of our calculations for the adhesive sphere phase diagram. We obtain identical coexisting solid and liquid phase densities, within computational error, for the fcc and hexagonal close packed (hcp) lattices and find the body centered cubic (bcc) phase to be metastable. We were unable to continue the phase diagram below a value of $\tau = 1.3$ for the fcc/hcp lattices and $\tau = 1.1$ for the bcc lattice; below this point, we could not find a self-consistent solution for the weighted density. As the interactions are increased, solid phase coexistence occurs at higher densities and correspondingly larger values of α, resulting in smaller values of the weighted density. As τ is lowered below 1.3 for the fcc or below 1.1 for the bcc lattice, no weighted density can be found.

Despite its relative simplicity, we find a broadening in the density difference between the coexisting fluid and solid phases with decreasing temperature τ characteristic of more complex interaction potentials. Unfortunately we were unable to calculate the solid-fluid phase behavior down to τ low enough test the existence of a triple point and hence the relevance of the vapor-liquid phase envelope.

Interfacial Properties. As with the hard-sphere study, we study the interface along the densest face (111) of the fcc lattice. As in the phase diagram, the

Table II: Results for the fcc solid-fluid adhesive-sphere interface

τ	$\Delta z/\delta_{111}$	$\gamma\sigma^2/kT$	$\gamma/\Delta H\rho_s^{2/3}$
∞	3-4	0.71 ± 0.02	0.48
10.0	3-4	0.74 ± 0.02	0.49
5.0	3-4	0.75 ± 0.02	0.47
3.0	3	0.80 ± 0.02	0.46
2.0	3	0.91 ± 0.02	0.44
1.7	3	1.12 ± 0.01	0.48
1.5	2	1.25 ± 0.01	0.49

adhesive sphere system mimics the interfacial behavior of more complex systems. As we lower the interaction parameter τ, the interface narrows and the interfacial tension increases. One rather striking result appears when one normalizes the calculated interfacial tensions as suggested by the empirical expression of Turnbull [56] who found that $\gamma/\Delta H\rho_s^{2/3} \sim 0.45$ for metallic elements. We find that the calculated interfacial tensions follow $\gamma/\Delta H\rho_s^{2/3} \sim 0.47$ as summarized in Table II. This rule persists as τ is decreased and the density difference between coexisting phases increases.

The Lennard-Jones Fluid. To extend the treatment of the interface to include systems having both attractive interactions and range the Lennard-Jones potential

$$u_{lj}(r) = \frac{4}{T^*}\left[\left(\frac{\sigma}{r}\right)^{12} - \left(\frac{\sigma}{r}\right)^6\right] \tag{31}$$

is useful where $T^* = kT/\epsilon$, ϵ is the attractive well depth and σ the distance r where the potential equals zero.

Theoretical Approach. It is difficult, however, to apply the previous approach directly to studies of the Lennard-Jones potential due to the lack of analytic expressions for the fluid phase direct correlation function $c(r)$ and excess free energy per particle $\psi(\rho)$. We therefore follow Barker-Henderson perturbation theory [57] separating the potential into structure determining repulsions and perturbative attractions that modify system energetics. We model this system as hard spheres of an effective diameter d $(= \int_0^\sigma dr[1 - e^{-u(r)}])$ and include attractions as a mean-field perturbation whose magnitude is determined from the effective hard-sphere system structure and the interaction potential. To first order in the attractive perturbation, the resultant free energy is

$$F(\rho) = F_{hs}(\rho, d) + 2\pi N\rho\beta^{-1}\int_\sigma^\infty dr g_{hs}(r, \rho, d)u_{lj}(r)r^2 \tag{32}$$

where g_{hs} is the effective hard-sphere radial distribution function. One can express this in the language of density-functional theory by dividing the excess free energy from equation 1 into a hard sphere and an attractive term

$$F[\rho] = F_{\rm id}[\rho] + F_{\rm ex}^{\rm hs}[\rho] + F_{\rm ex}^{\rm att}[\rho]. \tag{33}$$

The ideal and hard sphere excess terms can be calculated as shown previously and the attractive functional approximated as a density dependent function with equation 32 giving

$$F_{\rm ex}^{\rm att}[\rho] \sim 2\pi A \beta^{-1} \int dz \hat{\rho}(z) \rho(z) \int_{\sigma}^{\infty} dr' g_0(r'; \rho(z); d) u_{\rm lj}(r') r'^2. \tag{34}$$

Interfacial Properties. Using this approach we have computed the phase diagram, determining bulk properties, μ, P, and the solid localization parameter, α, as a function of temperature. We then use the PWDA with equations 33 and 34 to determine the interfacial weighted densities and their associated free energies. We summarize the resulting interfacial tensions in Table III.

Table III: Calculated interfacial tensions for the Lennard-Jones solid-liquid (S-L) and solid-vapor (S-V) systems

T^*	transition	interface width $\Delta z/\delta_{111}$	fluid density $\rho_f \sigma^3$	solid density $\rho_s \sigma^3$	interfacial tension $\gamma \sigma^2/kT$	$\gamma/\Delta H \rho_s^{2/3}$
1.15	S-L	3-4	0.992	1.104	0.87 ± .02	0.66
0.617	S-L	3-4	0.962	1.063	0.82 ± .02	0.69
0.44	S-L	3-4	0.950	1.046	0.83 ± .02	0.76
0.44	S-V	2	$1.7 \cdot 10^{-5}$	1.046	2.38 ± .01	0.14
0.40	S-V	2	$3.6 \cdot 10^{-6}$	1.060	2.72 ± .01	0.11
0.36	S-V	2	$5.1 \cdot 10^{-7}$	1.073	3.15 ± .01	0.10

It is interesting to note the failure of the Turnbull empiricism for this interaction potential. In fact, as the triple point is approached from above, agreement becomes progressively worse, indicating the influence of strong long-range attractions.

Interfacial Orientation

As discussed previously, the density-functional approach can be used to examine the various crystal structures and their solid-fluid interfaces. One issue not yet addressed is how and whether the various possible interfacial structures, for a given lattice, will have different surface energies. How these structural differences will impact the interfacial energetics is a question now accessible using density-functional theory.

There has been relatively little investigation of the orientational dependence of interfacial properties in solid-fluid model systems because of the tremendous computational requirements of traditional approaches. Because of these limitations only the hard-sphere system has been investigated using density-functional techniques. One may expect *a priori*, little orientational dependence of the interfacial tension in this system because of its entropy dominated, high temperature nature [58]. It is clear from previous studies [17, 59–61] however, that there is little consensus on the variation of the interfacial tension with orientation as well as its absolute magnitude, properties that both strongly influence the equilibrium crystal structure. For comparison to these theoretical predictions the only simulation study available is the molecular dynamics investigation of Broughton and Gilmer [62] who studied the interfacial free energy in the Lennard-Jones system and found that the crystal-melt interface at the triple point is nearly isotropic.

From the work of Wulff we know that the relative values of the interfacial tension determine the crystal shape, including facet size and stability. Ideally, to determine the hard-sphere equilibrium crystal shape one would calculate $\gamma(\theta)$ at all interaction strengths τ. This would allow determination of both the state of faceting and the roughening transition via a Wulff-type construction. Unfortunately, and despite the reduced dimensionality inherent in our approximation, our approach to determining interfacial properties still requires a large computational effort making such a calculation impractical. One can, however, look at higher index interfaces and ask the question of their stability as a function of τ. For example, from a Wulff-type construction we know that the 211 interface will be stable (that is, not facet into a combination of 100 and 111 faces) if $\gamma_{211} < 0.388\gamma_{100} + 0.672\gamma_{111}$. Similarly for the 311 interface the stability condition is $\gamma_{311} < 0.603\gamma_{100} + 0.522\gamma_{111}$.

Adhesive Spheres. As discussed previously, the adhesive-sphere system is a convenient model because the attractive strength can be varied from the purely repulsive hard-sphere limit to a potential which includes a deep attractive well. The phase behavior in the adhesive-sphere system is a strong function of the strength of interaction (see Figure 3), resulting in a large increase of the interfacial tension in the fcc(111) direction with increasing attraction strength. The question however remains: How will the various crystalline orientations influence this behavior?

We begin by first examining the hard-sphere limit and then gradually increase the attractive strength, determining the structure and energy of the resulting equilibrium interfaces. As seen previously for the fcc (111) interface, increasing the strength of attraction in the system causes the fcc 110 and 100 interfaces to increase their interfacial free energy and decrease their interfacial thickness. Structurally, in fact, these interface becomes sharper ($\Delta z/a = 1.95$ and $\nu = 0.33$ at $\tau = \infty$, decreasing to $\Delta z/a = 1.30$ and $\nu = 0.13$ at $\tau = 1.5$).

There appears to be little dependence on crystalline orientation in this system; the surface free energies are nearly identical as τ is decreased from the

Table IV: Interfacial tensions γ for the low and high-index crystal face orientations at various adhesive-sphere strengths τ

τ	$\gamma_{111}^{PWDA}\sigma^2/kT$	$\gamma_{110}^{PWDA}\sigma^2/kT$	$\gamma_{100}^{PWDA}\sigma^2/kT$	$\gamma_{211}^{PWDA}\sigma^2/kT$	$\gamma_{311}^{PWDA}\sigma^2/kT$
∞	0.70 ±.01	0.70 ± .01	0.70 ± .01	0.70 ± .01	0.70 ± .01
3	0.79 ± .01	0.80 ± .01	0.79 ± .01	0.80 ± .01	0.80 ± .01
2	0.91 ± .01	0.92 ± .01	0.92 ± .01	0.91 ± .01	0.91 ± .01
1.7	1.01 ± .01	1.02 ± .01	1.03 ± .01	0.99 ± .01	1.01 ± .01
1.5	1.05 ± .01	1.07 ± .01	1.09 ± .01	1.03 ± .01	1.04 ± .01

hard-sphere limit but begin to show a small amount of anisotropy under conditions of the strongest attractions studied. The origin of this anisotropy remains unclear as it does not appear to directly correlate with the surface density ϱ, where $\varrho_{111} = 4/\sqrt{3}a^2$, $\varrho_{100} = 2/a^2$, and $\varrho_{110} = \sqrt{2}/a^2$. One thing to note is that the lowest tension corresponds to that interface with the highest surface density suggesting the importance of interplanar interactions in determining interfacial tension.

We list in Table IV the values we calculate for both the low and high index interfaces where, once again, there is little anisotropy. According to the stability condition developed in the previous section, both the higher order 211 and 311 interfaces are stable under the stengths of attraction studied here, indicating that the adhesive sphere crystal structure is nonfaceted. This apparent lack of a transition from a spherical to a faceted equilibrium crystal shape as the attractive interactions increases may suggest that the interaction potential must have range in order to have a nonroughened, faceted equilibrium crystal structure. One must be careful however not to generalize since we are unable to study attractions stronger than those found at τ of 1.5. Returning to the work of Broughton & Gilmer on the Lennard-Jones system, they obtain nearly isotropic values for the interfacial free energy at the triple point ($T^* = 0.617$). Equating virial coefficients [63] allows us to approximate an equivalent τ via $\tau_{equiv} = 1 + 2/T^*$, giving $\tau_{equiv} \sim 0.2$, a value significantly lower than that investigated here.

Summary

Density-functional theory and the PWDA allow one to examine both the interaction strength and orientational dependence of solid-fluid interfaces. We have investigated the influence of interactions on interfacial properties, including hard spheres, adhesive spheres, and the Lennard-Jones system. We have also used the adhesive-sphere system to investigate both low and high index surfaces and found a small amount of anisotropy in the interfacial tension at the highest attraction strengths. We see no direct evidence of faceting in the adhesive-sphere system for the conditions investigated here.

Literature Cited

[1] Woodruff, D.P. *The Solid-Fluid Interface*; Cambridge University Press, 1973.

[2] Laird, B.B.; Haymet, A.D.J. *Chem. Rev.* **1992**, *92*, 1819.

[3] Oxtoby, D.W. *Liquids, Freezing, and the Glass Transition*, Les Houches; Elseveir: New York, **1991**; Vol. 51.

[4] Baus, M. *J. Phys.: Condensed Matter* **1990**, *2*, 2111.

[5] Singh, Y. *Physics Reports* **1991**, *207*, 352.

[6] Evans, R. *Liquids at Interfaces*; Les Houches; Elseveir: New York, **1989**; Vol. 48.

[7] Lutsko, J.F.; Baus, M. *Phys. Rev. A* **1990**, *41*, 6647.

[8] Lutsko, J.F. *Phys. Rev. A* **1991**, *43*, 4124.

[9] Haymet, A.D.J.; Oxtoby, D.W. *J. Chem. Phys.* **1981**, *74*, 2559.

[10] Ramakrishnan, T.V.; Yussouff, M. *Phys. Rev. B* **1979**, *19*, 2775.

[11] Lutsko, J.F.; Baus, M. *Phys. Rev. Lett.* **1990**, *64*, 761.

[12] Baus, M.; Colot, J.L. *Mol. Phys.* **1985**, *55*, 653.

[13] Curtin, W.A.; Ashcroft, N.W. *Phys. Rev. A* **1985**, *32*, 2909.

[14] Curtin, W.A.; Ashcroft, N.W. *Phys. Rev. Lett.* **1986**, *56*, 2775.

[15] Denton, A.R.; Ashcroft, N.W. *Phys. Rev. A* **1989**, *39*, 4701.

[16] Curtin, W.A. *Phys. Rev. Lett.* **1987**, *59*, 1228.

[17] Curtin, W.A. *Phys. Rev. B* **1989**, *39*, 6775.

[18] Denton, A.R.; Ashcroft, N.W. *Phys. Rev. A* **1990**, *42*, 7312.

[19] Laird, B.B.; Kroll, D.M. *Phys. Rev. A* **1990**, *42*, 4810.

[20] Tarazona, P. *Mol. Phys.* **1984**, *52*, 81.

[21] Marr, D.W.; Gast, A.P. *Phys. Rev. E* **1993**, *47*, 1212.

[22] de Kuijper, A.; Vos, W.L.; Barrat, J.L.; Hansen, J.P.; Schouten, J.A. *J. Chem. Phys.* **1990**, *93*, 5187.

[23] Rosenfeld, Y. *Phys. Rev. A* **1991**, *44*, 8141.

[24] Kyrlidis, A.; Brown, R.A. *Phys. Rev. A* **1991**, *44*, 5424.

[25] Kierlik, E.; Rosinberg, M.L. *Phys. Rev. A* **1990**, *42*, 3382.

[26] Kierlik, E.; Rosinberg, M.L. *Phys. Rev. A* **1991**, *44*, 5025.

[27] Denton, A.R.; Ashcroft, N.W. *Phys. Rev. A* **1991**, *44*, 8242.

[28] Kroll, D.M.; Laird, B.B. *Phys. Rev. A* **1990**, *42*, 4806.

[29] Tejero, C.F.; Daanoun, A.; Lekkerkerker, H.N.W.; Baus, M. *Phys. Rev. Lett.* **1994**, *73*, 725.

[30] Hagen, M.H.J.; Meijer, E.J.; Mooij, G.C.A.M.; Frenkel, D.; Lekerkerker, H.N.W. *Nature* **1993**, *365*, 425.

[31] Coussaert, T.; Baus, M. *to appear, Phys. Rev. E.*

[32] Hansen, J.P.; McDonald, I.R. *Theory of Simple Liquids*; Academic Press, 1986.

[33] Hoover, W.G.; Ree, F.H. *J. Chem. Phys.* **1968**, *49*, 3609.

[34] Alder, B.J.; Hoover, W.G.; Young, D.A. *J. Chem. Phys.* **1968**, *49*, 3688.

[35] Marr, D.W.M. *J. Chem. Phys.* **1995**, *102*, 8283.

[36] Baxter, R.J. *J. Chem. Phys.* **1968**, *49*, 2770.

[37] Menon, S.V.G.; Manohar, C.; Rao, K.S. *J. Chem. Phys.* **1991**, *95*, 9186.

[38] Rouw, P.W.; Vrij, A.; de Kruif, C.G. *Colloids and Surfaces* **1988**, *31*, 299.

[39] Seaton, N.A.; Glandt, E.D. *J. Chem. Phys.* **1987**, *86*, 4668.

[40] Kranendonk, W.G.T.; Frenkel, D. *Mol. Phys.* **1988**, *64*, 403.

[41] Smithline, S.J.; Haymet, A.D.J. *J. Chem. Phys.* **1985**, *83*, 4103.

[42] Zeng, X.C.; Oxtoby, D.W. *J. Chem. Phys.* **1990**, *93*, 2692.

[43] Marr, D.W.; Gast, A.P. *J. Chem. Phys.* **1993**, *99*, 2024.

[44] Tejero, C.F.; Baus, M. *Phys. Rev. E* **1993**, *48*, 3793.

[45] Chiew, Y.C.; Glandt, E.D. *J. Phys. A: Math. Gen.* **1983**, *16*, 2599.

[46] Seaton, N.A.; Glandt, E.D. *J. Chem. Phys.* **1987**, *86*, 4668.

[47] Regnaut, C.; Ravey, J.C. *J. Chem. Phys.* **1989**, *91*, 1211.

[48] deKruif, C.G. *Langmuir* **1992**, *8*, 2931.

[49] Penders, M.H.G.M.; Vrij, A. *J. Chem. Phys.* **1990**, *93*, 3704.

[50] Rouw, P.W.; de Kruif, C.G. *J. Chem. Phys.* **1988**, *88*, 7799.

[51] Rouw, P.W.; de Kruif, C.G. *Phys. Rev. A* **1989**, *39*, 5399.

[52] Rouw, P.W.; Woutersen, A.T.J.M.; Ackerson, B.J.; deKruif, C.G. *Physica A* **1989**, *156*, 876.

[53] Gopala Rao, R.V.; Debnath, D. *Ind. J. Phys. A* **1991**, *3*, 204.

[54] Grant, M.C.; Russel, W.B. *Phys. Rev. E* **1993**, *47*, 2606.

[55] Vrij, A.; Penders, M.H.G.M.; Rouw, P.W.; de Kruif, C.G.; Dhont, J.K.G.; Smits, C.; Lekkerkerker, H.N.W. *Faraday Discuss. Chem. Soc.* **1990**, *90*, 1.

[56] Turnbull, D. *J. Appl. Phys.* **1950**, *21*, 1022.

[57] Barker, J.A.; Henderson, D. *J. Chem. Phys.* **1967**, *47*, 4714.

[58] Conrad, E.H. *Prog. Surf. Sci.* **1992**, *39*, 65.

[59] Kyrlidis, A.; Ph.D. Thesis, MIT **1993**.

[60] Ohnesorge, R.; Löwen, H.; Wagner, H. *Phys. Rev. E* **1994**, *50*, 4801.

[61] McMullen, W.E.; Oxtoby, D.W. *J. Chem. Phys.* **1988**, *88*, 1967.

[62] Broughton, J.Q.; Gilmer, G.H. *J. Chem. Phys.* **1986**, *84*, 5759.

[63] Barboy, B. *J. Chem. Phys.* **1974**, *61*, 3194.

Chapter 17

Inhomogeneous Rotational Isomeric State Polyethylene and Alkane Systems

John D. McCoy and Shyamal K. Nath

Department of Materials and Metallurgical Engineering, New Mexico Institute of Mining and Technology, Socorro, NM 87801

The density functional modeling of Rotational Isomeric State (RIS) chains is reviewed. Two cases are considered. First, the freezing of polyethylene is investigated, and the melt and solid densities at the transition are predicted. New results are reported which incorporate the attractive as well as the repulsive contribution of the site-site potential. Good agreement is found with experimental measurements. Second, the structure of a tridecane melt near a hard wall is considered. Both the site density profile and the distortion of the backbone structure are predicted. Good agreement is found with the results of simulation and equation-of-state predictions.

Recently there has been considerable interest in applying density functional (DF) methodology to inhomogeneous polymeric systems [1-18]. Here we focus on the Chandler-McCoy-Singer (CMS) formulation of molecular DF theory [19-21] where bonding constraints are explicitly retained in the "ideal" system. In addition, we restrict ourselves to the case where the homogeneous liquid state input is included through site-site correlation functions as opposed to being introduced through the equation-of-state. Related work on inhomogeneous polymeric systems is reviewed by McMullen [10], Rosenberg [14], and Yethiraj [18].

Ubiquitous to all density functional theories is the expression of a free energy, usually the grand potential, $\Omega = -PV$, as a functional of the inhomogeneous density distribution, $\rho(r)$, as well as of more traditional variables such as the temperature, T; the volume, V; the chemical potential, μ; and the external field, $U(r)$. Since the chemical potential and the external field conveniently couple as $\psi(r) = \mu - U(r)$, the grand potential functional can be denoted as $\Omega[T, V, \psi(r); \rho(r)]$. Of course, because $\rho(r)$ itself is a functional of T, V, and $\psi(r)$, only three of the variables in the brackets are independent and including $\rho(r)$ in the expression for Ω

0097–6156/96/0629–0246$15.00/0

appears to be no more than a computational convenience. On the other hand, a new functional, $W[T, V, \psi(r); \rho(r)]$ can be considered which is identical to $\Omega[T, V, \psi(r); \rho(r)]$ except that all four variables are independent. Physically, one can think of $W[T, V, \psi(r); \rho(r)]$ as the grand potential where constraints on the density have been imposed at each point in space. By virtue of the second law, the minimization of $W[T, V, \psi(r); \rho(r)]$ with respect to $\rho(r)$ (or, equivalently, the removal of the internal constraints) results in the grand potential. The $\rho(r)$ which minimizes the functional is the equilibrium density profile and is often the quantity of principle interest.

In and of itself this "free energy" route to finding the properties of an inhomogeneous system is exact; however, approximations are necessary in order to generate an explicit form of $W[T, V, \psi(r); \rho(r)]$ which can then be minimized. The approximations considered here result from the Taylor expansion (in $\rho(r)$) of the excess Helmholtz free energy about the homogeneous liquid (which has the same T,V, and μ as the inhomogeneous system, but with $U(r) = 0$). By convention, the "excess free energy" is defined as the free energy in excess of that of an exactly solvable, or "ideal", system. For atomic systems, the ideal system used is that of non-interacting atoms which leads to the ideal-mixture-like contribution to the free energy of $\int d\underline{r}\, \rho(\underline{r}) \ln[\rho(\underline{r})]$.

For polymeric systems, the selection of an ideal system is less straightforward than in the atomic case. Although physically unappealing, it is mathematically tempting to use an ideal system of non-interacting *sites*. This is an exceptionally poor choice since the entropy of mixing implicit in the resulting $\Sigma \int d\underline{r}\, \rho_i(\underline{r}) \ln[\rho_i(\underline{r})]$ contribution to the free energy is far too large to be representative of the polymeric system, leaving a large correction for the excess free energy. Indeed, from a Flory-Huggins viewpoint, this contribution is about N times too large where N is the number of sites in a chain. Simplifying approximations based on this idea have been developed [22], and appear to work well.

Vastly better are ideal systems of non-interacting *chains*. The only subtlety concerns the role of long ranged interactions between sites on the same chain. Because we have been studying polymer melts, we have, so far, assumed that the long ranged interactions are screened out by virtue of the Flory ideality hypothesis, and, consequently, we have neglected such interactions in our ideal systems. In those cases where we have been able to compare to single chain structures generated by full computer simulations, we find that the predictions of density functional theory based on such an ideal system are either accurate to within the error-bars of the simulation or are more strongly influenced by errors introduced by other simplifying assumptions. If one were interested in polymers in solutions where the long ranged excluded volume forces are not screened, better results would be achieved by retaining these interactions in the ideal system.

The structural consequences of such a DF theory can be viewed as a balance between single-chain and many-body contributions to the free energy. The single-chain contribution contains both a center-of-mass, Flory-Huggins-like, ideal-mixing component, and an entropy of chain conformation component. For atomic systems, the $\rho\ln\rho$ and the many-body terms balance to give reasonably good results in some cases (hard sphere and Lennard-Jones systems) and poor results in others (repulsive $1/r^n$ potentials and the liquid-gas transition).

For polymeric systems, the balance shifts to be between the chain conformation and the many-body terms. Consequently, conclusions concerning the accuracy or inaccuracy of a polymeric DF theory based upon atomic system results can be tentative at best. Polymers differ qualitatively from atoms, and, while we do not expect the simple form we adopt for the many-body contribution to the DF theory to be accurate for all polymeric applications, for the cases we have investigated, the theory has produced excellent results. On the other hand, our recent investigations [23] of systems with non-negligible attractive interactions indicate that a delicate balance exists between attractions and repulsions, and, consequently, care must be taken in the application of DF theory to such systems.

System Model.

In the series of studies reviewed here [1-4], the intrachain interactions were described by the Rotational Isomeric State (RIS) model. In the freezing studies [1,2], polyethylene in the long chain limit was modeled, and, in the wall / polymer studies [3,4], tridecane was modeled between smooth, hard walls which were adequately separated so that bulk behavior occurred in the center of the slit. In both cases, the RIS parameters were as in Flory [24]. The gauche-trans energy was 500cal/mole, the "pentane-effect" was enforced with an additional 2000cal/mole for adjacent gauche bonds of opposite handedness. The C-C-C bond angle is 112°; the gauche states are located at 120° from the trans state; the carbon-carbon bond length is 1.54Å. The temperature was 300K for the wall studies and, for the freezing studies, a range of temperatures centered about 430 K was investigated.

Since very few theoretical or simulation studies have been performed on RIS chains, conducting benchmark studies for the DF theory of such chains is difficult. We have compared the predictions of our studies to experimental results [25,26] for the freezing of polyethylene, and, for the wall studies, to the results of both Monte Carlo simulation [27] and Generalized Flory Dimer (GFD) theory [28-30].

The primary reason for these comparisons was to test the DF theory itself. Since the theoretical predictions depend upon both the DF theory and the liquid state information which is required by DF theory as input, efforts were made to ensure the accuracy of the homogeneous liquid state information.

Polymer reference interaction site model (PRISM) [32-36] liquid state theory was used to generate this input. PRISM is accurate in predicting chain structure from site-site potentials; however, comparisons with DF theory when there is an uncertainty in the interaction potentials, as is always the case with experiential data, require the agreement between PRISM predictions and the experimental values of selected liquid state properties. Only properties associated with the homogeneous phase are used in the selection of an interaction potential: the properties of the inhomogeneous system are predicted by the DF theory.

The interchain interactions are chosen to keep the number of parameters to a minimum. In the bulk of the work, the sites were modeled as united atom, hard spheres where the interaction potential for separations greater than the site diameter is zero and, for smaller separations, is infinite. The single parameter of the model - the site diameter - was varied in order to describe the different systems studied. Attractions, when used, were included in the DF theory by perturbation techniques.

For the comparison with freezing, the hard site diameter is selected to agree with experimental measurements [31] on the polymer melt. By choosing the hard site diameter to be 3.90Å, there is excellent agreement between PRISM and X-ray scattering results for polyethylene at the experimental temperature of 430 K [31]. In addition, this value of 3.90Å can also be motivated through excluded volume arguments. Interestingly, a diameter of about 3.75Å is required to describe the structure factor of alkane melts at room temperature.

Recently [37], we have revisited the freezing problem. Our new evaluation is based upon a Lennard-Jones site-site interaction potential

$$u(r) = 4\varepsilon\left[\left(\frac{\sigma}{r}\right)^{12} - \left(\frac{\sigma}{r}\right)^{6}\right] \qquad (2.1)$$

where the ε = 45.4 K and σ= 4.423 Å. This potential was found by requiring that the compressibility of polyethylene be described by PRISM theory over a range of temperatures [38]. At 430K, this potential gives a hard site diameter of 3.99Å which, while larger than the value found from a comparison with X-ray results, is not unreasonably so.

For the comparison of the predictions of the DF theory of melts near a wall with those of simulation and GFD theory [28-30], a number of different hard site diameters were used. The potential used in the simulation [39] implies that a diameter of about 3.39 Å should be used. Other potentials (such as the Ryckaert and Bellemans [40] potential) suggest that a diameter of about 3.73 Å would be more representative of true tridecane and, as mentioned above, a similar value of the site diameter permits PRISM theory to predict the X-ray structure factor in alkanes. Finally, a diameter of zero is considered for comparative purposes. The number density of the bulk melt (which is in equilibrium with the inhomogeneous melt near the wall) was held fixed at the experimental (and simulation) value corresponding to 0.750 kg/m^3.

Theory.

We approximate the excess Helmholtz free energy by a Taylor expansion in the density distribution $\rho(r)$ about the homogeneous state [19-21]. A Legendre transform results in the grand potential functional

$$\Delta W = \int d\underline{r}\big[\psi_L - \psi(\underline{r})\big]\rho(\underline{r}) + \int d\underline{r}\big[\psi^o(\underline{r}) - \psi_L^o\big]\rho(\underline{r})$$
$$-\frac{1}{N}\int d\underline{r}\,\Delta\rho(\underline{r}) - \frac{1}{2}\int\int d\underline{r}'\,d\underline{r}\,c\big(|\underline{r}'-\underline{r}|\big)\Delta\rho(\underline{r}')\Delta\rho(\underline{r}) \tag{3.1}$$

where $\Delta W = W - W_L$; $\Delta\rho(r) = \rho(r) - \rho_L$; "L" refers to the homogeneous liquid state; "o" refers to the ideal system which has the same density profile as the fully interacting system; and $c(r)$ is the direct correlation function in the homogeneous liquid. All energies are in units of kT where k is the Boltzmann constant and T is the temperature. The ideal field $\psi^o(r)$ is related to the density by

$$\rho(\underline{r}) = \sum_{i=1}^{N}\int\ldots\int d\underline{r}_1\ldots d\underline{r}_{i-1}d\underline{r}_{i+1}\ldots d\underline{r}_N\exp\Big[\sum_j\psi^o(\underline{r}_j)\Big]S(\underline{r}_1,\ldots,\underline{r}_N) \tag{3.2}$$

where $S(\underline{r}_1,\ldots,\underline{r}_N)$ is the N body correlation function for an ideal chain and N is the number of sites in the chain.

Unlike the atomic case, the ideal field cannot, in general, be removed from the expression for the grand potential functional: the minimization procedure must take into account that the ideal field is a functional of $\rho(r)$. Formally, this can be done with the method of undetermined multipliers where $\rho(r)$ and $\psi^o(r)$ are treated as independent functions and equation (3.2), as the constraint. The resulting formalism is cumbersome; however, matters can be improved somewhat by combining equations (3.1) and (3.2) as

$$\Delta W = \int d\underline{r}\big[\psi_L - \psi(\underline{r})\big]\rho(\underline{r}) + \int d\underline{r}\big[\psi^o(\underline{r}) - \psi_L^o\big]\rho(\underline{r})$$
$$-\int\ldots\int d\underline{r}_1\ldots d\underline{r}_N\exp\Big[\sum_j\psi^o(\underline{r}_j)\Big]S(\underline{r}_1,\ldots,\underline{r}_N) + \int d\underline{r}\,\rho_L \tag{3.3}$$
$$-\frac{1}{2}\int\int d\underline{r}'\,d\underline{r}\,c\big(|\underline{r}'-\underline{r}|\big)\Delta\rho(\underline{r}')\Delta\rho(\underline{r}).$$

The undetermined multiplier can now be taken to be zero since the constraint is enforced through setting the derivative with respect to $\psi^o(r)$ to zero. Unfortunately, the resulting conditions

$$\left(\frac{\delta\Delta W}{\delta\rho(\underline{r})}\right)_{\psi^o(r)} = 0$$
$$\left(\frac{\delta\Delta W}{\delta\psi^o(\underline{r})}\right)_{\rho(r)} = 0 \tag{3.4}$$

represent a saddle point in $\rho(r)$ - $\psi^o(r)$ space which is more difficult to treat numerically than a minimum would be. The introduction of such auxiliary thermodynamic variables is discussed in more detail by Percus [41].

The density and the ideal field can be solved for on the differential level, and their values used to evaluate the free energy difference in equation (3.1). The maximization with respect to $\psi^0(r)$ results in the constraint equation (3.2). The minimization with respect to $\rho(r)$ yields

$$\psi^\circ(\underline{r}) = \psi(\underline{r}) - \psi_L + \psi_L^\circ + \int d\underline{r}\, c\big(|\underline{r} - \underline{r}'|\big)\Delta\rho(\underline{r}') \qquad (3.5)$$

or

$$U^\circ(\underline{r}) = U(\underline{r}) - \int d\underline{r}\, c\big(|\underline{r} - \underline{r}'|\big)\Delta\rho(\underline{r}') + \text{constant} \qquad (3.6)$$

where the chemical potential has been removed from the "fields". By straightforward analogy with liquid state theory, this is a Hypernetted Chain (HNC) type of relationship, and, still by analogy, a Percus-Yevick (PY) type field can be proposed as well [42]. Given $\rho(r)$, the external ideal field is easily found through equation (3.6); however, finding $\rho(r)$ given the field (for a self-consistent solution) through equation (3.2) is not straightforward. Because of this, the partial minimization of the free energy functional is appealing in cases such as freezing where great simplification results.

By rewriting equation (3.2) as

$$\rho(\underline{r}) = (\text{constant})\sum_{i=1}^{N} \int \dots \int d\underline{r}_1 \dots d\underline{r}_{i-1} d\underline{r}_{i+1} \dots d\underline{r}_N \exp\left[-\sum_j U^\circ(\underline{r}_j)\right] S(\underline{r}_1, \dots, \underline{r}_N) \quad (3.7)$$

it is clear that, as one might expect, the density is related to the external field by a Boltzmann weighted average of a single chain over all space where $S(\underline{r}_1, \dots, \underline{r}_N)$ enforces the bonding constraints. The constant can be determined by a single condition on the inhomogeneous density such as requiring the average density to have a particular value.

The implementation of DF theory was different in the two classes of problems we have considered. In the work on the freezing of (RIS) chains [1,2], the ideal fields were parameterized so that a (constrained) minimization with respect to density was simple. The field for the ideal liquid, ψ^0_L, can be expressed (from equation (3.2)) in terms of the partition function, Z_L, of a single, unconstrained, ideal chain;

$$\psi_L^\circ = \frac{1}{N}\ln\left[\frac{\rho_L}{N}\right] - \frac{1}{N}\ln\left[\frac{Z_L}{V}\right] \qquad (3.8)$$

where V is the volume. Since, for long chains, $\ln[Z_L/V]$ is proportional to the chain length, ψ^0_L will contribute in that limit. For the crystal, only a restricted class of ideal fields were considered. The field $\psi^0(r)$ is required to be of a finite value, ψ^0, within small spherical volumes centered about the set of points {R}; otherwise, it is taken to be infinite. The points are selected so that the chains must be in an all-trans configuration and, for long, RIS chains, the volumes can be, in effect, shrunk to points. The resulting expression for ψ^0 is, to good approximation,

$$\psi^\circ = \frac{1}{N}\ln\left[\frac{\rho(\underline{r})}{N}\right] \tag{3.9}$$

where the single chain entropy has been set to zero. The simplicity of this relationship is a peculiarity of the RIS model. Rigid rotations of the entire chain, which are unimportant for long chains, have been neglected.

The above relationships are more easily understood if the polymers are viewed as molecules rather than collections of sites. The net field felt by a chain is $N\psi^0$ (or $N\psi^0{}_L$) and the molecular densities are $\rho_m(r) = \rho(r)/N$ and $\rho_{m,L} = \rho_L/N$. Consequently, the difference in "molecular" fields would be

$$N\psi^\circ - N\psi_L^\circ = \ln\left[\frac{\rho_m(\underline{r})}{\rho_{m,L}}\right] + \ln\left[\frac{Z_L}{V}\right]. \tag{3.10}$$

This is reasonable: in addition to the center of mass contribution familiar from atomic DF theories, there is also a contribution from the loss of "single chain entropy" in going from disordered liquid chains to the low entropy, all-trans chains of the crystal. As the chains become large, the center of mass entropy, $\ln[\rho_m(r)/\rho_{m,L}]$, becomes of secondary importance to the single chain entropy, $\ln[Z_L/V]$ because of the latter's proportionality to the chain length. The (constrained) minimization procedure is, simply, to vary the locations of the non-infinite values of the $\psi^0(r)$ or, equivalently, to vary the lattice parameters.

Finding the density distribution of RIS chains near a wall required the full solution of equation (3.7). This was done as follows. A simple step function density profile was guessed. The field was calculated through equation (3.6) from this $\rho(r)$. A new density profile was then found through a Monte Carlo simulation of a single chain in this external field. The resulting density, $\rho(r)$, was multiplied by a factor which forced the density far from the wall to be the bulk density. The new and old density profiles were "mixed", a new field calculated, and the procedure repeated until convergence was achieved.

Freezing.

As discussed above, for the freezing of polyethylene [1,2], the RIS model permits simplification of equation (3.1). In the crystal, the chains are fully extended and, since the torsional vibrations about the trans state are non-existent in the RIS model, it can be assumed that the sites in an infinitely long chain in the crystalline state are localized. In the long chain limit, equation (3.1) becomes

$$\Delta W = \int d\underline{r} \ln \lambda \, \rho(\underline{r}) - \frac{1}{2} \int \int d\underline{r}' \, d\underline{r} \, c\left(\left|\underline{r}' - \underline{r}\right|\right) \Delta \rho(\underline{r}') \Delta \rho(\underline{r}) \tag{4.1}$$

where λ is related to the single chain partition function in the liquid, Z_L, by

$$Z_L = \lambda^{N-1} V \tag{4.2}$$

and is well known for the RIS chain [24]. In Fourier space, equation (4.1) becomes

$$\Delta W^* = \frac{\Delta W}{\rho_L V}$$

$$= \frac{\rho_s}{\rho_L} \left\{ \ln \lambda - \hat{c}(0) \frac{(\rho_s - \rho_L)^2}{\rho_s} - \frac{1}{2} \left[\sum_R c(|\mathbf{R}|) - \rho_s \hat{c}(0) \right] \right\} \tag{4.3}$$

where ρ_s is the bulk density in the solid, $\hat{c}(0)$ is the Fourier transform of the direct correlation function evaluated at $k=0$, and the sum is over all lattice spacings of separations less than one hard site diameter. The crystal density, $\rho(r)$, has been treated as a collection of delta functions. This treatment of the density (and, consequently, equation (4.3)) is only appropriate for the RIS model in the long chain limit where there is no vibration about the trans state and where the center-of-mass entropy is overwhelmed by the single chain entropy.

It is worth emphasizing that the reduction in entropy associated with freezing in polyethylene is one of reduced backbone rather than spatial disorder. Because of this, the freezing transition is sensitive to the type of backbone used to describe the polymer. Gaussian chains, for instance, have, on a per site basis, a large amount of entropy in the melt. This entropy is so large that the packing effects which are adequate to stabilize the extended chain structure in RIS chains cannot, in the Gaussian case, compensate for the entropy loss upon freezing.

In earlier work [1,2] on the freezing of polyethylene, the interaction between sites was taken to be hard in nature. Consequently, the density of the crystal was the close packed density. This is higher than seen experimentally. On the other hand, the liquid coexistence density was found to be close to the experimental value of 0.78 g/cc. By using an empirical equation-of-state along with the density functional results, the melting temperature was found to be 427 K which is in good agreement with the accepted range of 415 - 420 K.

The main physical feature which is neglected in the hard site model of polymer freezing is, one expects, the attractive well. In order to include this feature in the formalism, an attractive potential based upon a PRISM description of the melt should be used. Recently [38], we have used PRISM theory to calculate the isothermal compressibility and adjusted the site-site interaction potential so that this quantity is well described over a range of temperatures.

The attractive component of the potential is introduced through perturbation theory. We divide the grand potential functional $\Delta W[\rho(r)]$ into two parts: the first is the grand potential functional due to the hard sphere reference potential, ΔW_d, and the second is the contribution from the long range attractive perturbation $\phi(r)$. Taking only the first order term in the perturbation theory and approximating the radial distribution function of the hard sphere system outside one hard sphere diameter as unity, the functional $\Delta W[\rho(r)]$ is,

$$\Delta W[\rho(\underline{r})] = \Delta W_d[\rho(\underline{r})] + \frac{1}{2} \int \int d\underline{r} d\underline{r}' \phi(|\underline{r} - \underline{r}'|) \Delta\rho(\underline{r}) \Delta\rho(\underline{r}'). \tag{4.4}$$

The density functional of this form was first introduced by Sullivan [43,44] to study a variety of interfacial phenomena, and later used by Curtin and Ashcroft [45] to predict the freezing of Lennard-Jones atoms from a hard sphere reference system. In this formulation of the perturbation theory, the additional contribution to the hard site grand potential (equation (4.3)) is,

$$\Delta W_{pert}^* = \frac{1}{2\rho_L V} \int \int d\underline{r} d\underline{r}' \phi(|\underline{r} - \underline{r}'|) \Delta\rho(\underline{r}) \Delta\rho(\underline{r}')$$

$$= \frac{\rho_s}{\rho_L} \left\{ \hat{\phi}(0) \frac{(\rho_s - \rho_L)^2}{2\rho_s} + \frac{1}{2} \left[\sum_R \phi(|R|) - \rho_s \hat{\phi}(0) \right] \right\} \tag{4.5}$$

where $\hat{\phi}(0)$ is the Fourier transform of $\phi(r)$ evaluated at $k = 0$.

As seen in Table I, both the solid and liquid coexistence densities are seen to be in good agreement with experimental results.

Table I. Crystallization of Polyethylene at Atmospheric Pressure

	Liquid density (g/cm^3)	Solid density (g/cm^3)	lattice parameters a (Å)	b (Å)
Experiment	0.7834[a]	0.9673[a]	7.706[b]	4.936[b]
DF theory				
Hard Sites	0.778	1.13	7.61	4.25
L-J Sites	0.80	0.9504	7.79	4.92

a.   reference 54.
b.   reference 55.

Near a Wall.

The structure of polymeric systems near a hard wall is determined largely by a balance between two competing entropic contributions. A chain which is near the wall has fewer available conformations than a chain far from the wall: a depletion of the chains in the wall region is favored. On the other hand, as in the case of atomic hard sphere liquids, high packing fractions result in an enhancement of the density near the wall since effective occupation of space is the primary consideration. In figure 1, this interplay of conformational and packing entropies can be seen [3,4]. At low packing fractions (d=0), there is a large density depletion near the wall. As the packing fraction is increased, the emerging dominance of the packing entropy is evidenced by the increase of the density near the wall. Interestingly, a short ranged depletion region remains even at high packing fraction.

Because the chains were explicitly simulated [3,4], more detailed information concerning the chain structure, the distribution of site type, etc. can be obtained. Chains near the wall are seen to stretch a small amount. For example, the average end-to-end separation increases in the wall region by about 6% for a chain with a site diameter of 3.39 Å. The DF prediction for this quantity is in good agreement with simulation results. In addition, this suggests that the use of an ideal system with random walk scaling is reasonable - at least for alkanes.

Also in agreement with simulation results is the distribution of site types. At low densities, the chain ends strongly segregate towards the wall since they are less entropicly repelled by the wall than a center site would be. However, as the packing fraction is increased, this effect is diminished.

The short ranged depletion region is an intriguing feature of the density profile predicted by DF theory. In order to verify the accuracy of DF theory in this region, the value of the contact density can be independently verified. The contact site density is well known [46-48] to be proportional to the pressure. Consequently, the equation-of-state can be evaluated from DF theory and compared to GFD theory. This comparison is shown in figure 2. We would expect this level of agreement to hold for other alkanes and, perhaps, for other models where the ratio of bond length to site diameter is about 0.5. In general, one expects the PY relation for the ideal field to be more accurate for structural calculations. It is difficult to predict how generally applicable this methodology of calculating the pressure is; its robustness as attractions are incorporated [23] is of particular importance.

Discussion.

The application of the CMS formulation of molecular DF theory to hard-site polymeric systems appears to be more accurate than the atomic theory of which it is an extension. In part this is, no doubt, because the open crystal structures which are problematic for the atomic theory are not of importance in polymeric systems. However, of more relevance is the change in the role of the ideal system. This variation is well illustrated by the DF theory of freezing of RIS chains. As seen in equation (3.10), for

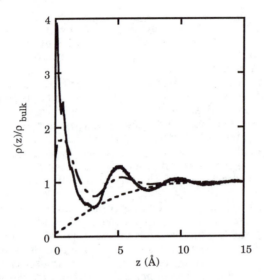

Figure 1. Density profiles of a 13-mer RIS chain near a hard surface with d=0 (dotted curve), d=3.39Å (dashed curve), and d=3.73Å (solid curve).

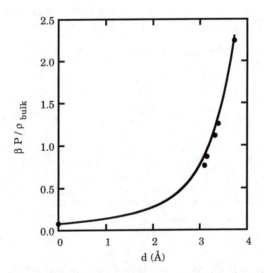

Figure 2. Pressure as a function of hard site diameter with temperature equal to 300 K and the bulk site density equal to 0.03192 (1/Å^3). The solid curve was obtained from the GFD equation of state, while the points are the result of density functional theory.

atomic and small molecular systems, the center-of-mass, entropy-of-mixing term dominates the ideal system contribution to the overall free energy. For polymeric systems, on the other hand, the entropy-of-mixing term is overwhelmed by the conformational entropy, and, just as in the Flory-Huggins theory, the underlying physics of a polymeric system is expected to differ from that of an atomic system. Finally, the magnitude of the direct correlation function for a polymeric system is an order of magnitude smaller than that of the related hard sphere system [1,2]. This implies that the functional Taylor series would be more rapidly convergent for polymeric systems. In other words, as the chains become more polymeric, more of the problem is contained in the single chain calculation, and the DF theory becomes more accurate. Indeed, in the long chain limit where there is but a single chain, the DF theory becomes exact in a trivial manner.

The DF approach for polymeric systems is closely related to the Kohn-Sham DF theory [49] for inhomogeneous electronic systems. In both cases the ideal field is found from the density profile by an approximate relationship (in our case, equation (3.6)), and the density distribution for the ideal system is evaluated exactly in the presence of this ideal field. In the Kohn-Sham case, the Schrodinger equation is solved for the non-interacting system while, in the polymer case, the classical density (equation (3.7)) is found through a Monte Carlo simulation.

The structure of the DF approach adopted here is easily confused with the self-consistent-field (SCF) theory of Helfand and Tagami [50] since the DF theory is generically of the self-consistent-field type. Moreover, DF theory in the "string" limit collapses to SCF theory and can be couched in the usual propagator language of SCF theory [51]. On the other hand, the term "SCF theory" carries with it connotations which do not apply to DF theory. First, the ideal field in SCF theory is local in the density. That is, the field at a particular point depends only upon the density, the gradient of the density, etc. at that particular spatial location. In DF theory, the field at a point depends upon, in general, all spatial positions and strongly depends upon densities within a site diameter. Second, rather than evaluating the density through equation (3.7) or the equivalent, SCF theory computes the density distribution from the ideal field in the continuum chain limit. In other words, the non-interacting chains are treated as quantum particles, and, in effect, the Schrodinger equation is solved for the density. Third, except in a very few applications, SCF theory requires that the *total* density at each point in space be a constant. Consequently, SCF theory, as applied, is a theory of composition fluctuations rather than one of density variations.

A modified form of SCF theory has recently been explored [52] which retained the quantum calculation of the density, but the DF relation for the ideal field was used. It captured most of the qualitative features of the full DF solution for Gaussian chains near a hard wall. Such good agreement between the two methods would not be expected for chain models with a number of length scales.

While simpler models such as freely jointed and Gaussian chains do not contain the short wavelength detail that the RIS model does, they are of computational convenience. The calculation of the density distribution near a hard, smooth wall is no exception to this. Such simple Markoff chains do not require a simulation in order to calculate the density distribution from the ideal field, instead, a propagator formalism can be implemented [51]. It seems reasonable to expect that the more complicated models lacking long-ranged interactions (i.e., RIS chains) will be able to be treated with a generalized propagator approach in the spirit of Flory's matrix treatment for the average end-to-end separation of a RIS chain.

The site-site pair correlation function, g(r), can also be evaluated from DF theory. The external field is taken to be generated by a molecule, and the inhomogeneous density is the bulk density times g(r). The pair correlation function has been calculated and compared to simulation results for hard site diatomics [42]. From DF theory with a simple HNC field (equation (3.6)), the contact g(r) was found to be overestimated for tangent sites, much as one would expect from the behavior of the HNC liquid state theory of hard spheres. In analogy with liquid state theory, a PY form of the ideal field functional results in very good values for the contact g's. Interestingly, if the sites are highly overlapped (i.e., a bond length of half the site diameter which is about what it would be in RIS polyethylene), then both HNC and PY fields produce contact g's which are in good agreement with simulation results.

The only system in contact with a wall that has been studied in any detail with the DF methods discussed here is RIS chains near a smooth, hard wall. Both the contact density and the density profile are in good agreement with simulation and equation-of-state information. Recent results for a tangent site model indicate [53] that the wall contact density is overestimated which is in keeping with the results for g(r) at contact for a HNC field.

The hard wall contact density can be used to predict the equation-of-state. As shown in section 5, the equation-of-state for a 13-site alkane melt generated through DF theory is in surprisingly good agreement with GFD theory. It is too early to tell how well this route to the equation-of-state will work over a range of system types. If it is safe to draw general conclusions from the g(r) calculations, the PY field would need to be used with tangent site models.

There are a large number of future applications for DF theory to polymer problems. A number of these are relatively straightforward extensions: softening the wall interaction; calculating the surface tension; investigating the behavior of blends near a wall; and finding the density distribution in simple confining geometries. More computationally intensive applications such as studying surface roughness and tethered chains will require the breaking of the x-y symmetry of the ideal field.

Literature Cited

1. McCoy, J. D.; Honnell, K. G.; Schweizer, K. S.; Curro, J. G., *Chem. Phys. Lett.* **1991**, *179*, 374.
2. McCoy, J.D.; Honnell, K.G.; Schweizer, K.S.; Curro, J.G., *J. Chem. Phys.*, **1991**, *95*, 9348.
3. Sen, S.; Cohen, J.; McCoy, J.D.; Curro, J.G. *J. Chem. Phys.* **1994**, *101*, 5971.

4. Sen, S.; McCoy, J.D.; Nath, S.K.; Donley, J.P.; Curro, J.G. *J. Chem. Phys.* **1995**, *102*, 3431.
5. McMullen, W.E.;Freed, *J. Chem. Phys.* **1990**,*92*, 1413.
6. Tang, H.; Freed, K.F., *J. Chem. Phys.* **1991**, *94*, 1592.
7. Wong, K. Y.; Trache, M.; McMullen, W.E., *J. Chem. Phys.* **1994**,*101*, 5372.
8. Tang, H.; Freed, K.F., *J. Chem. Phys.* **1991**, *94*, 6307.
9. Brazhnik, P. K.; Freed, K.F.; Tang, H., *J. Chem. Phys.* **1994**, *101*, 9143.
10. McMullen, W.E., This symposium.
11. Kierlik, E.; Rosinberg, M.L., *J. Chem. Phys.* **1992**,*97*, 9222.
12. Kierlik, E.; Rosinberg, M.L., *J. Chem. Phys.* **1993**,*99*, 3950.
13. Kierlik, E.; Rosinberg, M.L., *J. Chem. Phys.* **1994**,*100*, 1716.
14. Rosinberg, M.L., This symposium.
15. Woodward, C.E., *J.Chem.Phys.* **1991**, *94*, 3183.
16. Woodward, C.E., *J. Chem. Phys.* **1992**,*97*, 4525.
17. Woodward, C.E.; Yethiraj, A.; *J. Chem.Phys.* **1994**,*100*, 3181.
18. Yethiraj, A., This symposium.
19. Chandler,D.; McCoy, J.D.; Singer,S.J., *J. Chem. Phys.* **1986**,*85*, 5971.
20. Chandler,D.; McCoy, J.D.; Singer,S.J., *J. Chem. Phys.* **1986**,*85*, 5977.
21. McCoy, J.D.; Singer, S.J.; Chandler, D., *J. Chem. Phys.* **1987**, *87*, 4953.
22. Ding, K.; Chandler, D.; Smithline, S.J.; Haymet, A.D.J., *Phys. Rev. Lett.* **1987**, *59*, 1698.
23. Nath, S.; Rottach, D.; McCoy, J.D., (In preparation).
24. Flory,P. J. *Statistical Mechanics of Chain Molecules* (Wiley, NY, 1969).
25. Wunderlich, B.; Czornj, G., *Macromolecules* **1977**,*10*, 906.
26. Swan, P.R., *J. Polym. Sci.* **1962**,*56*, 403.
27. Vacatello, M.; Yoon, D.Y.; Laskowski, B.C., *J. Chem. Phys.* **1990**, *93*, 779.
28. Dickman, R.; Hall, C. K., *J. Chem. Phys.* **1986**,*85*, 4108.
29. Honnell, K. G.; Hall, C. K., *J. Chem. Phys.*, **1989**,*90*, 1841.
30. Yethiraj, A.; Curro, J. G.; Schweizer, K. S.; McCoy, J. D. *J. Chem. Phys.* **1993**, *98*, 1635.
31. Honnell, K. G.; McCoy, J. D.; Curro, J. G.; Schweizer, K. S.; Narten, A.H.; Habenschuss, A., *J. Chem Phys.* **1991**, *94*, 4659.
32. Chandler, D.; Andersen, H.C., *J. Chem. Phys.* **1972**,*57*, 1930.
33. Chandler, D. in *Studies in Statistical Mechanics VIII*", ed. Montroll, E.; Lebowitz, J. (North Holland-Amsterdam, 1982), p. 284.
34. Schweizer, K. S.; Curro, J. G. *Phys. Rev. Lett.* **1987**,*58*, 246.
35. Honnell, K. G.; Curro, J. G.; Schweizer, K. S., *Macromolecules* **1990**, *23*, 3496.
36. Schweizer, K. S.; Curro, J. G., *Adv. Polym. Sci.* (in press) and references cited therein.
37. Nath, S.K.; McCoy, J.D.; Curro, J.G., (In preparation)
38. Nath, S.K.; McCoy, J.D., (In preparation)
39. Yoon, D.Y.; Suter, U.W.; Sundararajan, P.R.; Flory, P.J., *Macromolecules* **1975**,*8*, 784.
40. Ryckaert, J.-P.; Bellmans, A., *Faraday Discuss. Chem. Soc.* **1978**,*66*, 95.

41. Percus, J.K., This symposium.
42. Donley, J. P.; Curro, J.G.; McCoy, J.D., *J. Chem. Phys.* **1994,***101*, 3205.
43. Sullivan, D. E. *Phys. Rev. B*, **1979**, 20, 3991.
44. Sullivan, D. E.*J. Chem. Phys.*, **1981**, 74, 2604.
45. Curtin, W.A.; Ashcroft, N.W., *Phys. Rev. Lett.* **1986,***56*, 2775.
46. Lebowitz,J.L. *Phys. Fluids* **1960,***3*, 64.
47. Percus, J.K., *J. Stat. Phys.* **1976,***15*, 423.
48. Dickman,R.; Hall, C.K. *J. Chem. Phys.* **1988,***89*, 3168.
49. Parr,R. G.; Yang, W., *Density-Functional Theory of Atoms and Molecules* (Oxford, NY, 1989).
50. Helfand, E.; Tagami, Y., J. Chem. Phys. **1978,***56*, 3592.
51. Donley, J.P.; Rajasekaran, J.J.; McCoy, J.D.; Curro, J.G., *J. Chem. Phys.* (In press).
52. Nath, S.K.; McCoy, J.D.; Donley, J.P.; Curro, J.G., *J. Chem. Phys.* **1995,** *103*, 1635.
53. Yethiraj, A., Private communication.
54. Wunderlich, B.; Czornj, G., *Macromolecules*, **1977,** *10*, 906.
55. Swan, P.R., *J. Polym. Sci.* **1962,***56*, 403.

Chapter 18

Density Functionals for Polymers at Surfaces

William E. McMullen

Department of Chemistry, Texas A & M University,
College Station, TX 77843–3255

We derive an expression for the external field necessary to produce an arbitrary monomeric density near a planar surface. This
result becomes exact in the limit of weak external fields. We illustrate the utility of the formalism by applying it to a polymer
blend interacting with a surface. Due to monomer-surface correlations, the monomer densities decay to the bulk composition
more slowly than in previous phenomenological theories of polymer adsorption. For our choice of monomer-surface Hamiltonian,
we observe only first-order wetting.

Density functional theories of dense systems often separate important thermodynamic potentials into ideal and nonideal contributions. For classical, atomic
fluids, the division is obvious since the partition functions of noninteracting,
monatomic species avail themselves to exact analyses. For example, researchers
customarily define the ideal free energy functional $F_{id}[\rho]$ and the interaction free
energy Φ so that, in terms of the total free energy $F[\rho]$

$$F[\rho] = F_{id}[\rho] - \Phi[\rho]. \tag{1}$$

The analytical expression for $F_{id}[\rho]$ (1)

$$F_{id}[\rho] = \int d1\rho(1) \left[\ln\rho(1)\Lambda^3 - 1\right] \tag{2}$$

and the well-known identity

$$\delta F_{id}[\rho]/\delta\rho(1) = -\beta v(1) \tag{3}$$

lead to

$$\rho(1) = \exp[c(1) - \beta v(1)]/\Lambda^3 \tag{4}$$

0097–6156/96/0629–0261$15.00/0

with Λ an atomic length scale, and

$$c(1; [\rho]) = \delta\Phi[\rho]/\delta\rho(1).\tag{5}$$

The combination $c(1) - \beta v(1)$ plays the role of an effective external field acting on an ideal gas of atoms.

Whereas equation 4 hardly solves the atomic density functional problem, it does suggest some reasonable schemes (*2*) for approximating the interaction term $\Phi[\rho]$. Some of these yield surprisingly good, mean-field descriptions of dense systems. Part of the successes of these theories trace back to equations 2 and 4 which automatically incorporate the ideal, translational free energy of atoms into the underlying formalism. Whatever mistakes the theories make in approximating $\Phi[\rho]$, at least they describe ideal gases correcly. Extensions of the formalism to complicated polyatomic species like polymer fluids are not readily accomplished. Even for an idealized model polymer in which monomers do not interact (e.g., freely jointed chains, Gaussian random walks, continuum chains, etc.), the monomeric density does not, in general, reduce to equation 4 with $c(1) = 0$. Imagine, for instance, a linear polymer subjected to an external field that acts only on monomers at one end of the chain. The covalent bonds defining the polymer transfer the response of the monomers at that end to all other monomers. This induces a density variation even in regions of space where the external field does not act. In the language of atomic density functionals, we say that from a monomeric point of view, the external field induces a nonlocal density response. Only in the limit that the external field varies imperceptibly over the volume occupied by a chain can we propose a simple form for the monomer density. In this case, for a chain of N monomers that each interact with an external field v (*3*),

$$\rho(1) \propto \exp[-N\beta v(1)]\tag{6}$$

where we determine the proportionality constant from the chemical potential or the average density.

Generally speaking, the field-density relation for an ideal polymer fluid is far more complicated than equation 6, and before attempting to construct a density functional for interacting polymers, we must develop methods for describing the single-polymer limit. At this stage, density functional theories exist for describing weakly perturbed, bulk polymer chains. We wish to extend those theories to polymer fluids which, even in the absence of applied fields, are highly inhomogeneous. In the next part of this chapter, we study how individual chains respond to an external field while in the presence of a surface. The surface makes the problem difficult since it breaks the translational invariance of the reference, field-free system. In the absence of a surface, our formalism reduces to the problem of single polymers in bulk and reproduces the existing approaches to translationally invariant polymer fluids (*3,4*). On the other hand, our formalism is readily adaptable to other inhomogeneous polymeric fluids (e.g., polymers near a corrugated boundary or near the surface of a small, spherical, colloidal particle) whenever we possess sufficient statistical information about the field-free problem. We note that many of the most important scientific and engineering applications of polymers involve surfaces (*5*), so a practical motivation transcending theoretical or mathematical curiosity exists for focussing on the planar polymer-surface geometry. The third section of this chapter describes such an application to binary polymer blends. We show that correlations between monomers and a surface result in a surface composition profile that decays to the bulk more slowly than previously predicted. This feature of our results brings density functional theory into better agreement with existing experimental studies (*6*). The last part of this chapter summarizes our methodology and results and outlines some of the limitations of the theory.

Derivation of Ideal Density Functionals

Consider a fluid composed of noninteracting polymers. We do not discount the possibility of nonbonded, *intra*molecular, monomer-monomer interactions although the most straightforward applications of our theory will involve simple models lacking even those interactions. Besides the usual kinetic- and potential-energy contributions, the Hamiltonian includes an external field that acts on the monomers. Use μ to denote the chemical potential of a chain. In terms of the single-chain density operator

$$\hat{\rho}_1(\mathbf{R}) = \sum_{i=1}^{N} \delta(\mathbf{R} - \mathbf{r_i}), \tag{7}$$

the grand partition function reduces to (*7*)

$$\ln \Xi = Z \tag{8}$$

where

$$Z = \frac{e^{\beta\mu}}{\Lambda^{3N}} \int d\mathbf{r}^N \exp\left[-\beta H_1 + \int d\mathbf{R} w(\mathbf{R})\hat{\rho}_1(\mathbf{R})\right], \tag{9}$$

H_1 is the Hamitonian of a single chain, Λ—with units of length—results from integrations over the monomer momenta, and $w = -\beta v$.

H_1 contains *all* interactions of the chain with the surface and any intramolecular interactions. We seek a relation expressing external field $w(\mathbf{r})$ in terms of the average density given by

$$\rho(1) = \frac{\delta \ln \Xi}{\delta w(1)} = \frac{\delta Z}{\delta w(1)}, \tag{10}$$

this result following from equations 7-9. Anticipating that equation 10 will reduce to equation 6 when the field varies slowly, we express the density as

$$\rho(1) = \frac{e^{\beta\mu}}{\Lambda^{3N}} \int d\mathbf{r}^N e^{-\beta H_1} \exp\left[\int d\mathbf{r}w(\mathbf{r})\hat{\rho}_1(\mathbf{r})\right]\hat{\rho}_1(1) \qquad (11)$$

$$= \frac{e^{Nw(1)}e^{\beta\mu}}{\Lambda^{3N}} \int d\mathbf{r}^N e^{-\beta H_1}\hat{\rho}_1(1)\exp\left\{\int d\mathbf{r}\hat{\rho}_1(\mathbf{r})\left[w(\mathbf{r}) - w(1)\right]\right\}.$$

When w does not vary over the volume of the polymer, the second line indeed simplifies to equation 6. More generally, we ask: How does the presence of a slowly varying field alter the form of equation 6? Ultimately, the answer to this question will enable us to approximately invert the field-density relation in order that we can address the problem of determining the external field that leads to a particular density profile.

We expand the second line of equation 11 as

$$\rho(1) = \frac{e^{\beta\mu}e^{Nw(1)}}{\Lambda^{3N}} \int d\mathbf{r}^N e^{-\beta H_1}\hat{\rho}_1(1)\sum_{n=0}^{\infty}\frac{1}{n!}\int d\mathbf{R}_1 \rightarrow d\mathbf{R}_n \qquad (12)$$

$$\times \prod_{i=1}^{n}[w(\mathbf{R}_i) - w(1)]\hat{\rho}_1(\mathbf{R}_i)$$

$$= e^{Nw(1)}\left\{\rho_0(1) + \sum_{n=1}^{\infty}\int d2 \rightarrow d(n+1)\frac{g_{n+1}(1,\ldots,n+1)}{n!}\right.$$

$$\left. \times \prod_{n=1}^{n}[w(i+1) - w(1)]\right\}$$

in which $\rho_0(1)$ is the density in the absence of the field $w(1)$, and the $g_n(1,\ldots,n)$'s are monomer-monomer distribution functions. The passage from the first to second equalities of equation 12 defines the distribution functions. Our goal in this analysis is to invert equation 12 and determine $w(1)$ as a functional of $\rho(1)$. For a translationally invariant, reference state (the reference state corresponds to the case $w(1) = 0$), one can accomplish this when w varies slowly by using iteration and a gradient expansion. This method (4) assumes that $\nabla\rho(1)$ and ∇w vary over similar length scales. However, near a surface, $\rho_0(1)$—the field-free monomer density—changes sensibly over length scales comparable to a radius of gyration. Upon turning on the field $w(1)$, the reference and perturbed densities will likely vary over similar length scales even if $w(1)$ varies continuously and much more slowly than $\rho_0(1)$. It follows that simple iteration will not facilitate the inversion of equation 12.

Consider, however, the quantity $\ln\rho(1)/\rho_0(1)$ which vanishes in the absence of an imposed field, is constant for $w(1) \rightarrow$ constant, and changes slowly in space when $w(1)$ varies slowly. Now cast equation 12 in the form

$$\ln \frac{\rho(1)}{\rho_0(1)} = Y(1) + \ln \left\{ 1 + \sum_{n=1}^{\infty} \int d2 \to d(n+1) \frac{h_{n+1}(1,\ldots,n+1)}{n!} \right. \tag{13}$$

$$\left. \times \prod_{i=1}^{n} [Y(i+1) - Y(1)] \right\}$$

where $Y = Nw$ and

$$h_n(1,\ldots,n) = \frac{g_n(1,\ldots,n)}{N^{n-1}\rho_0(1)}. \tag{14}$$

To simplify the notation a little further, we define

$$P(1) = \ln \frac{\rho(1)}{\rho_0(1)}. \tag{15}$$

At this stage, we specialize our analysis to a planar surface and a field that varies only in the direction perpendicular to the surface. The system is translationally invariant in the directions parallel to the surface. To effect the field-density inversion, we postulate

$$Y(z) = P(z) + f_1(z) \frac{dP}{dz} + f_2(z) \frac{d^2P}{dz^2} + f_3(z) \left(\frac{dP}{dz} \right)^2 + \ldots \tag{16}$$

where $f_1(z)$, $f_2(z)$, and $f_3(z)$ etc. vary on a length scale ξ_s comparable to a radius of gyration whereas we imagine that $P(z)$ varies on a length scale $\xi_\rho \gg \xi_s$. These properties follow from the fact that f_1, f_2, f_3, etc. are related to correlations of the monomers with the wall for the zeroth-order problem. These correlations decay over distances comparable to the radius of gyration R_G. On the other hand, we expect that $P(z)$ will vary slowly for weak external fields. Equation 16 contains terms of order ξ_ρ^{-2} and larger.

The results of the present section hinge on the existence of the expansion given in equation 16. The ultimate justification for this assumption can be found in reference (7) which presents more details of the analysis and an independent derivation of the functional dependence of Y on ρ. Substituting equation 16 into 13 and using

$$\frac{d^nY}{dz_1^n} = (\delta_{n1} + \delta_{n2}) \frac{d^nP}{dz_1^n} + \frac{d^nf_1}{dz_1^n} \frac{dP}{dz_1} + n \frac{d^{n-1}f_1}{dz_1^{n-1}} \frac{d^2P}{dz_1^2} + \frac{d^nf_2}{dz_1^n} \frac{d^2P}{dz_1^2} \tag{17}$$

$$+ \frac{d^nf_3}{dz_1^n} \left(\frac{dP}{dz_1} \right)^2 + \mathcal{O}(\xi_\rho^{-3})$$

and

$$Y(z_2) - Y(z_1) = \sum_{n=1}^{\infty} \frac{(z_2 - z_1)^n}{n!} \frac{d^nY}{dz_1^n}, \tag{18}$$

we find, after considerable analysis, that f_1, f_2, and f_3 obey the linear, integral equations

$$\int_0^\infty dz_2 f_1(z_2) g_2(z_1, z_2) = -\int_0^\infty dz_2(z_2 - z_1) g_2(z_1, z_2), \qquad (19a)$$

$$\int_0^\infty dz_2[f_1(z_2)(z_2 - z_1) + f_2(z_2)]g_2(z_1, z_2) = -\frac{1}{2}\int_0^\infty dz_2(z_2 - z_1)^2 \qquad (19b)$$
$$\times g_2(z_1, z_2),$$

and

$$\int_0^\infty dz_2 f_3(z_2) g_2(z_1, z_2) = -\frac{1}{2N}\int_0^\infty dz_2 \int_0^\infty dz_3(f_1(z_2) + z_2 - z_1) \qquad (19c)$$
$$\times (f_1(z_3) + z_3 - z_1)g_3(z_1, z_2, z_3).$$

We outline a numerical procedure for solving equations 19a-c for f_1 through f_3 in (7).

Having determined f_1, f_2, and f_3 numerically or otherwise, we can solve equation 16 to determine $P(z)$ and, through equation 15, the perturbed density given an imposed external field. In many approximate theories, the field arises self-consistently from the interactions of different types of monomers (8). For instance, in the random-mixing approximation (9), the intrinsic free energy of a binary blend near a surface becomes

$$F[\rho_A, \rho_B] = F_{id}^A[\rho_A] + F_{id}^B[\rho_B] + \chi^* \int_0^\infty dz \rho_A(z)\rho_B(z) \qquad (20)$$

upon invoking the conventional incompressibity assumption

$$\rho_A(z) + \rho_B(z) = \rho = \text{constant} \qquad (21)$$

to approximate the long-wavelength effects of short-range, monomer-monomer repulsions. In the absence of an external field, the functional derivative of equation 20 with respect to either $\rho_A(z)$ or $\rho_B(z)$ equals a constant as we have disregarded certain inconsequential linear terms in the expression for F. [The linear terms contribute the constant to the derivative. Had we included the linear terms in F, its derivative would equal zero in accord with equation 3.] Then, for a symmetric polymer blend ($N_A = N_B = N$),

$$\frac{\delta F[\rho_A(z)]}{\delta \rho_A(z)} = \frac{\delta F_A^{id}[\rho_A(z)]}{\delta \rho_A(z)} + \frac{\delta F_B^{id}[\rho - \rho_A(z)]}{\delta \rho_A(z)} + \chi^*(\rho - 2\rho_A(z)) - \lambda \qquad (22)$$

$$= \frac{Y_A(z)}{N} - \frac{Y_B(z)}{N} + \chi^*(\rho - 2\rho_A(z)) - \lambda = 0$$

where equation 16 gives the functionals Y_A and Y_B. In equation 22, we can interpret Y_A as the response of A chains to an effective field. The effective field consists of the random-mixing term, λ, and the other single-chain functional Y_B. Solving the second-order, nonlinear differential equation of equation 22 amounts to a self-consistent determination of the field.

To eliminate λ, we demand that Y_A and Y_B vanish in the bulk. This step and the introduction of the volume fraction $\phi(z) = \rho_A(z)/\rho$ transforms equation 22 into

$$
P_A(z) + f_1^A(z)\frac{dP_A(z)}{dz} + f_2^A(z)\frac{d^2 P_A(z)}{dz^2} + f_3^A(z)\left(\frac{dP_A(z)}{dz}\right)^2 \tag{23}
$$
$$
-\left[P_B(z) + f_1^B(z)\frac{dP_B(z)}{dz} + f_2^B(z)\frac{d^2 P_B(z)}{dz^2} + f_3^B(z)\left(\frac{dP_B(z)}{dz}\right)^2\right]
$$
$$
+2\chi N(\phi_\infty - \phi(z)) = 0
$$

in which $\chi = \rho\chi^*$,

$$
P_A(z) = \ln\frac{\phi(z)}{\phi_0(z)}, \tag{24a}
$$

$$
P_B(z) = \ln\frac{1 - \phi(z)}{1 - \phi_0(z)}, \tag{24b}
$$

ϕ_∞ denotes $\phi(z \to \infty)$, and $\phi_0(z)$ is the density profile of the unperturbed blend (i.e., a blend with $\chi = 0$).

With the solution of the composition profile $\phi(z)$ from equation 23 in hand, we determine the excess surface free energy γ from (8).

$$
\frac{\beta\gamma N}{\rho} = \int_0^\infty dz\left[\phi(z)P_A(z) + (1 - \phi(z))P_B(z) + \left(f_3^A(z) - f_2^A(z)\right)\phi(z)\right. \tag{25}
$$
$$
\left. \times \left(\frac{dP_A}{dz}\right)^2 + \left(f_3^B(z) - f_2^B(z)\right)(1 - \phi(z))\left(\frac{dP_B}{dz}\right)^2 - \chi N(\phi(z) - \phi_\infty)^2\right].
$$

In most interfacial density functional theories, expressions for the densities analogous to equation 23 can be derived from a variational principle on a potential related to γ. However, the coupling of the length scales ξ_s and ξ_ρ in the present analysis and our application of the expansion 16 destroys this feature ($7,8$) of the exact theory. In general, we must first solve equation 23 for $\phi(z)$ and then substitute the result into equation 25 to obtain γ. Numerical analysis of equation 23 using standard relaxation or shooting methods proceeds readily, so in most instances, this feature of our theory does not pose a problem.

An Application to a Binary Melt near a Surface

We consider an application of the formalism described in the previous section to a symmetric, binary blend and an impenetrable surface. Besides the random-mixing and incompressibility approximations described in Sec. II, we approximate chains as continuum, random walks and model the interaction of monomers with the surface as a delta-function pseudopotential. For chains consisting of N statistical segments of average length l, we employ (10)

$$\beta H_1 = \frac{3}{2l^2} \int_0^N d\tau \left| \frac{d\mathbf{r}}{d\tau} \right|^2 + c \int_0^N d\tau \, \delta(z(\tau)) \tag{26}$$

where c parametrizes the surface-monomer interaction. The surface repels monomers when $c > 0$ and tends to adsorb them if $c < 0$. All monomers are confined to the half-space $z > 0$.

For this model, the correlation functions $g_2(z_1, z_2)$ and $g_3(z_1, z_2, z_3)$ follow straightforwardly from the solution to the modified diffusion equation (9,10)

$$\frac{\partial g(z_1, z_2; \tau)}{\partial \tau} = \frac{l^2}{6} \frac{\partial^2 g(z_1, z_2; \tau)}{\partial z_1^2} \; ; \; z_1, \, z_2 > 0 \tag{27}$$

subject to the boundary condition (10)

$$\frac{\partial g(z_1, z_2; \tau)}{\partial z_1} \bigg|_{z_1=0} = \frac{6c}{l^2} g(0, z_2; \tau). \tag{28}$$

The quantity $g(z_1, z_2; \tau)$ is the probability that a chain of length τ has one end at z_1 and the other at z_2. Reference (8) describes the derivation of the correlation functions. The solution of equation 19a reveals that f_1 has a delta-funtion singularity at the origin so that

$$f_1(z) = f_1^0(z) - R_G^2 \delta(z) \tag{29}$$

where we use f_1^0 to denote the nonsingular part of f_1. For finite c, both f_2 and f_3 are well behaved near $z = 0$. Substitution of equation 29 for each blend component into equation 23 leads to the boundary condition

$$\frac{dP_A(z)}{dz} - \frac{dP_B(z)}{dz} = 0 \quad \text{at } z = 0. \tag{30}$$

From the definition of P_A and P_B—equation 15—and a boundary condition (7,8)

$$\frac{1}{\rho_0(z)} \frac{d\rho_0(z)}{dz} \bigg|_{z=0} = \frac{12c}{l^2} \tag{31}$$

on the unperturbed densities, we find

$$\frac{d\phi}{dz} = \frac{12}{l^2}(c_A - c_B)\phi(1 - \phi) \; ; \; z = 0 \tag{32}$$

for the perturbed blend. A second boundary condition

$$\phi(z \to \infty) = \phi_\infty \tag{33}$$

ensures that $\phi(z)$ relaxes to its bulk value far from the surface. These two boundary conditions and f_1^0, f_2, and f_3 as functions of c enable us to solve equation 23.

Numerically we observe that the surface correlation functions f_1, f_2, and f_3 decay to their asymptotic values over a range comparable to R_G—the radius of gyration of a single chain. Figure 1 plots these functions for a representative value of $\bar{c} = N^{1/2}c/l$. All spatial variations in the surface correlation functions occur within $2R_G$ of the surface, and f_2 and f_3 vary less dramatically than f_1. The simplest density functional theories (*11,12*) of polymer adsorption employ asymptotic values for the surface correlation functions and a boundary condition at $z = 0$ derived from an assumed, local surface contribution to the free energy. To assess the effects of the surface correlations, we compare, in Figure 2, the composition profiles obtained from 23 and from the older theory. Generally speaking, the improved theory predicts profiles that decay to their bulk values more slowly than those predicted by the older theory. Although the effect is not pronounced, it accounts, in part, for a systematic deviation of experimentally measured profiles (*6*) from the simpler density functional theory. Our calculations agree, in this sense, with self-consistent-field studies of polymer adsorption (*13*) where surface correlations also dilate the surface profile.

Superficially, equation 23 exhibits some rather profound differences from analogous density functional theories (*11,12*) that ignore surface correlations. This motivates a brief discussion of the predictions of our theory with regards to surface phase transitions. Imagine a binary fluid mixture on its bulk coexistence curve so that the bulk fluid consists of only one of the pair of phases. Suppose that the surface adsorbs most strongly the phase not present in bulk. We refer to this as the A phase. At low temperature, the surface fluid consists of droplets of A adsorbed to the surface surrounded by the other phase—the B phase. Simple scaling arguments (*14*) predict that as one follows the coexistence curve to higher temperatures, the droplets disappear in favor of a macroscopically thick layer of phase A. The transition from droplets to a thick layer occurs at the wetting temperature T_w where the relation (*14*)

$$\gamma_{SB} = \gamma_{SA} + \gamma_{AB} \tag{34}$$

holds between the surface tension γ_{AB} of the AB interface, the free energy γ_{SA} of the surface-A-phase interface, and the free energy γ_{SB} of the surface-B-phase interface.

For our model, the wetting transition is always first order. Figure 3 plots the adsorption layer thickness (*12*)

$$D = \frac{1}{1 - 2\phi_\infty} \int_0^\infty dz(\phi(z) - \phi_\infty) \tag{35}$$

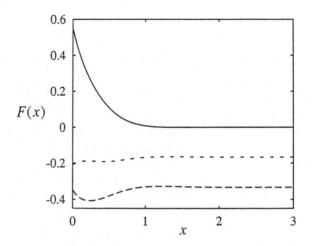

Figure 1. The surface correlation functions f_1, f_2, and f_3 (denoted by $F(x)$) for $\bar{c} = $ -0.01. R_G is an ideal-chain radius of gyration and $x = z/(2R_G)$. Solid curve f_1, long-dashed curve f_2, and short-dashed curve f_3.

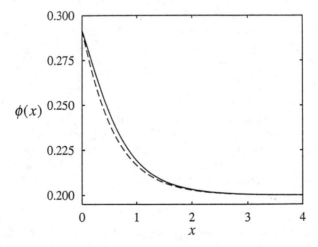

Figure 2. Composition profiles near a planar surface of a blend in the one-phase region. $\chi N = 2.31$ and $x = z/(2R_G)$. Solid curve: Theory of the present article using $\bar{c}_A = -\bar{c}_B = $ -0.025. Dashed curve: Phenomenological theory that ignores surface correlations. $\phi(0)$ of the dashed curve is chosen to equal that of the solid curve.

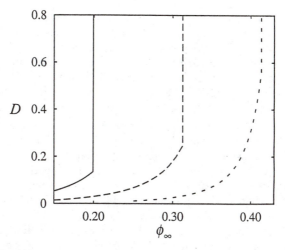

Figure 3. Interfacial thickness D, as defined by Eq. (35), on the B-rich branch of the coexistence curve. $\phi_\infty = 0.5$ at the bulk critical point. For each plot, $\bar{c}_A = -\bar{c}_B$. Solid curve, $\bar{c}_A = -0.03$; long-dash curve $\bar{c}_A = -0.01$; short-dash curve $\bar{c}_A = -0.002$.

versus the composition of the bulk B ϕ_∞ for a series of surface-interaction parameters $\bar{c}_A$ and $\bar{c}_B$ where $\bar{c}_i = c_i N^{1/2}/l$. Recall that $\phi_\infty \to 0.5$ at the bulk critical point. We observe that D always jumps discontinuously to ∞ for any finite value of the surface parameters. Changing the absolute value of the ratio $|\bar{c}_A/\bar{c}_B|$ predictably alters the wetting temperature, but the observation of a first-order transition does not vary except when $\phi_\infty \to 0.5$ at the bulk critical point.

Summary and Conclusions

This article outlines the derivation of an approximate external-field density functional for single polymer chains near a surface. This result provides a connection between an applied or self-consistent external field and the perturbed density. An application to a binary polymer blend interacting with a surface illustrates the utility of the formalism. Our analysis requires the solution of a simple, albeit nonlinear, ordinary differential equation for the composition profile. This compares to alternative self-consistent-field (SCF) theories (*9,13*) that require iterative numerical solutions of a partial differential equation. Rigorous application of our functional demands that the external or self-consistent field varies only slightly over distances comparable to a chain radius of gyration. In the context of our application to a blend, this requires that $\bar{c}_A$ and $\bar{c}_B \ll 1$, a condition satisfied whenever wetting occurs near the critical point or, in one-phase fluids, near weakly interacting surfaces. We regard our theory as a compromise between accuracy and convenience. Certainly, the formalism will lead to significant quantitative errors at surfaces that strongly segregate monomers. In weakly segregated systems, though, it represents an improvement on phenomenological theories of polymer adsorption without the computational overhead of the more accurate SCF methods. We note that the phenonomenolgical theories (11,12), in their present form, do not account for surface correlations accurately even in the weak-segregation limit.

Applications to other polymer-surface systems also appear feasible. Most of the calculational effort goes into determining $g_2(z_1, z_2)$ and $g_3(z_1, z_2, z_3)$. Once these are available, inversion of the integral equations for f_1, f_2, and f_3 proceeds readily. This suggests that more sophisticated, even atomistic, models of chains could be simulated to obtain g_2 and g_3. In that case, our formalism would enable a description of surface-polymer correlations that dispenses entirely with the somewhat overly simplified Edwards' Hamiltonian. Of course, we cannot expect the theory to provide nontrivial short-wavelength information about such systems beyond that contained in the unperturbed density and correlation functions.

Returning to our specific application to a blend, we recall that most phenomenological theories (11,12) predict second- as well as first-order wetting transitions. We suspect that the absence of second-order wetting in our calculations arises, in part, from our single-chain-surface Hamiltonian. Phenomenological theories of surface adsorption generally assume a more highly parametrized surface Hamiltonian that accounts for binary- as well as single-monomer interactions with the surface. This Hamiltonian can exhibit a minimum in the surface composition $\phi(0)$. When this minimum coincides with the concentration of one of the bulk phases, the wetting transition can become second order. The Hamiltonian employed in our theory contains only a purely repulsive or attractive contribution and exhibits no such energy minimum. However, the differential equation 23 for the concentration profile contains surface correlations and thus terms that do not appear in the phenomenological analysis. Inspection of equation 23 does not immediately suggest how these can affect the wetting transition. Our calculations show that surface correlations primarily dilate the interfacial profile, but this dilation cannot, by itself, induce second-order wetting. Future applications of the theory should explore the parameter space of surface Hamiltonians for second-order wetting more carefully. We would also hope to advance our treatment of intermonomer interactions in those calculations since collective effects may significantly impact surface boundary conditions.

Literature Cited

(1) Evans, R. Adv. Phys. (1989), 28, 143.
(2) Evans, R. In Microscopic Theories of Simple Liquids and their Interfaces; Charvolin, J.; Joanny, J.F.; Zinn-Justin, J., Ed.; Elsevier: Les Houches, 1989.
(3) McMullen, W.E. In Physics of Polymer Surfaces and Interfaces; Sanchez, I.C., Ed.; Butterworth: London, 1992.

(*4*) Tang, H.T.; Freed, K.F. *J. Chem. Phys.* (**1991**), *94*, 3183.

(*5*) Wu, S. Polymer *Polymer Interface and Adhesion*; Marcel Dekker: New York, 1982.

(*6*) Jones, R.A.L.; Norton, L.J.; Kramer, E.J.; Composto, R.J.; Stein, R.S.; Russell, T.P.; Mansour, A.; Karim, A.; Felcher, G.P.; Rafailovich, M.H.; Sokolov, J.; Zhao, X.; Schwarz, S.A. *Europhys. Lett.* (**1990**), *12*, 41.

(*7*) McMullen, W.E.; Trache, M. *J. Chem. Phys.* (**1995**), *102*, 1449.

(*8*) Wong, K.Y.; Trache, M.; McMullen, W.E. *J. Chem. Phys.* in press (**1995**).

(*9*) de Gennes, P.G. *Scaling Concepts in Polymer Physics*; Cornell University: Ithaca, NY, 1979.

(*10*) Nemirovsky, A.; Freed, K.F.; *J. Chem. Phys.* (**1985**), *83*, 4166.

(*11*) Nakanishi, H.; Pincus, P. *J. Chem. Phys.* (**1983**), *79*, 997.

(*12*) Schmidt, I.; Binder, K. *J. Phys.* (Paris) (**1985**), *46*, 1631.

(*13*) Carmesin, I.; Noolandi, J. *Macromolecules* (**1989**), *22*, 1689.

(*14*) Cahn, J.W. *J. Chem. Phys.* (**1977**), *66*, 3367.

Chapter 19

Weighted Density Approximation for Polymer Melts

Arun Yethiraj

Theoretical Chemistry Institute and Department of Chemistry, University of Wisconsin, 1101 University Avenue, Madison, WI 53706–1396

This chapter describes a density functional theory for polymer melts which combines a weighted density approximation for the excess free energy functional with an exact treatment of the ideal gas free energy functional. The theory employs a Monte Carlo simulation to calculate the properties of a single chain in an arbitrary external field. The bulk fluid properties required in the theory are obtained from a generalized Flory equation of state. The theory is found to be accurate for the density profiles of semiflexible polymer melts confined between flat plates for several densities and molecular stiffnesses. The theory is also compared to other theories for nonuniform polymer melts.

In many applications, such as the processing of polymer melts, the behaviour of polymer molecules within a few Angstroms of a surface is of critical importance. Consequently there have been a number of recent studies devoted to obtaining an understanding of polymers near surfaces using liquid state methods such as computer simulation (1-3) integral equations (4), and density functional theories (5-11). In this chapter I describe the density functional theory recently proposed by Yethiraj and Woodward (10) and compare the predictions of this theory to Monte Carlo simulations (12, 13) and to other theories.

The quantity of primary interest in nonuniform fluids is the density profile of the fluid. For simple liquids, the two theoretical approaches that have been employed to calculate the density profile are integral equations and density functional theories. Integral equations are based on the growing adsorbent model (14, 15) where the equations are solved for a mixture of the fluid and an adsorbent in the limit as the adsorbent species becomes infinitely dilute and infinitely large. The fluid density profile is simply related to the adsorbent-fluid pair correlation function. Although theories of this nature are in good agreement with simulations of hard spheres at a hard wall, they can be in qualitative error when fluid-fluid and fluid-wall attractive forces are present (16). In recent years attention has largely shifted to density functional theories (17). In density functional theory one starts with a free energy functional which gives the free energy of the fluid for any given density profile. Minimization of this functional with respect to the density profile then gives the equilibrium density profile. This functional is not known exactly except for an ideal gas, and approximations must be invoked for real fluids. A popular and accurate approach is the weighted density approximation (WDA) (18). This chapter is concerned with the extension of the WDA to inhomogeneous *polymer* melts and solutions.

Both integral equations and density functional theories can be extended to nonuniform polymer melts, but there are some complications that arise because of the internal degrees of freedom in polymer molecules. In the case of the growing adsorbent approach it is not possible to account for the change in the conformations of the molecules in the vicinity of the surface. In the case of density functional theory the ideal gas free energy functional, which is trivial for simple liquids, cannot be obtained in a simple form. Therefore, applications of density functional theory to polymers either use very crude approximations to the ideal gas free energy functional or, as is done in this work, treat the ideal gas free energy functional exactly via a computer simulation.

The rest of the chapter is organized as follows. The molecular model is described in section 2, the integral equation and density functional theories and their implementation are described in section 3, the theoretical predictions are compared to computer simulations in section 4, and some conclusions are presented in section 5.

2   Molecular Model

The theories are tested by comparing them to exact many chain Monte Carlo simulations for a simple model polymer melt. The model chosen is one for which many chain simulation data are already available, i.e. a melt of semiflexible hard chains confined between two infinite parallel walls. The chains are composed of a string of tangent hard spheres of diameter σ. In addition to excluded volume, a bending potential is introduced in order to make the chains stiff. This bending potential, E_B, is given by

$$E_B = \epsilon(1 + \cos\theta) \tag{1}$$

where θ is the bond angle between two consecutive bonds, and ϵ is the parameter that controls the stiffness of the molecules. If $\epsilon=0$ the molecules are freely-jointed and if $\epsilon=\infty$ the molecules are rods. The intramolecular potential, V, is given by

$$V(\mathbf{R}) = \sum_{j=2}^{N-1} \epsilon(1 + \cos\theta_j) + \sum_{i=3}^{N}\sum_{j=1}^{i-2} v_{HS}(|\mathbf{r}_i - \mathbf{r}_j|) + \sum_{j=2}^{N} v_b(|\mathbf{r}_j - \mathbf{r}_{j-1}|) \tag{2}$$

where $\mathbf{r}_i$ is the position of the ith bead of the polymer molecule, $\mathbf{R}$ denotes the position of all N monomers on the polymer molecule, $\cos\theta_j = (\mathbf{r}_{j-1} - \mathbf{r}_j)\cdot(\mathbf{r}_{j+1} - \mathbf{r}_j)$, $v_{HS}(r) = \infty$ if $r < \sigma$ and zero otherwise, and v_b constrains adjacent beads to a fixed separation σ. The external field $\Phi(\mathbf{R}) = \sum_{i=1}^{N} \phi(\mathbf{r}_i)$, where

$$
\begin{aligned}
\phi(z) &= u_w(z) + u_w(H - z), & 0 < z < H\sigma \\
&= \infty & \text{otherwise,}
\end{aligned}
\tag{3}
$$

$$
\beta u_w(z) = \epsilon_w \left\{ 1 + c \left[\frac{1}{(1+z/d)^9} - \frac{1}{(1+z/d)^3} \right] \right\} \Theta(p\sigma - z),
\tag{4}
$$

$c = 3^{3/2}/2$, $d = p\sigma/(3^{1/6} - 1)$, p=0.5, and $\Theta(x) = 1$ for $x > 0$. Most of the results presented are for ϵ_w=0. All the results presented are for H=10.

3 Density Functional Theory

In density functional theory, one starts with an expression for the grand free energy, Ω, as a functional of the density profile of the fluid. (At equilibrium $\Omega = -PV$ where P is the pressure and V is the volume.) If $\mathbf{R}$ denotes the positions of all the N monomers on a polymer molecule and $\rho_M(\mathbf{R})$ is the molecular density as a function of these positions, then the functional $\Omega[\rho_M(\mathbf{R})]$ gives the value of Ω for a given density profile $\rho_M(\mathbf{R})$. At equilibrium Ω is stationary with respect to changes in the density distribution, i.e.

$$
\frac{\delta\Omega}{\delta\rho_M(\mathbf{R})} = 0,
\tag{5}
$$

and this condition is used to determine both $\rho_M(\mathbf{R})$ and the equilibrium free energy.

The functional Ω is related to the Helmholtz free energy functional, $F[\rho_M]$, via a Legendre transform:

$$
\Omega[\rho_M(\mathbf{R})] = F[\rho_M(\mathbf{R})] + \int [\Phi(\mathbf{R}) - \mu]\,\rho_M(\mathbf{R})d\mathbf{R},
\tag{6}
$$

where μ is the chemical potential and $\Phi(\mathbf{R})$ is the external field. The functional $F[\rho_M]$ is generally expressed as the sum of an ideal (F^{id}) and excess (F^{ex}) part,

$$
F[\rho_M(\mathbf{R})] = F^{id}[\rho_M(\mathbf{R})] + F^{ex}[\rho_M(\mathbf{R})].
\tag{7}
$$

The ideal functional is known exactly:

$$
F^{id}[\rho_M(\mathbf{R})] = kT \int d\mathbf{R}\rho_M(\mathbf{R})\left[\ln\rho_M(\mathbf{R}) - 1\right] + \int d\mathbf{R}V(\mathbf{R})\rho_M(\mathbf{R}),
\tag{8}
$$

where $V(\mathbf{R})$ describes all the intramolecular interactions (including bonding and long-range excluded volume).

In principle, with a judicious choice of $F^{ex}[\rho_M]$, the free energy functional can be decomposed as

$$F[\rho_M(\mathbf{R})] = F^{id}[\rho_M(\mathbf{R})] + F^{ex}[\rho(\mathbf{r})], \qquad (9)$$

where

$$\rho(\mathbf{r}) = \int d\mathbf{R} \sum_{i=1}^{N} \delta(\mathbf{r} - \mathbf{r}_i)\rho_M(\mathbf{R}) \qquad (10)$$

is the average density of sites, $\mathbf{r}_i$ are the positions of the beads ($\mathbf{R} = \{\mathbf{r}_i\}$), and N is the number of sites (or beads) on each polymer molecule. In this case, F^{id} will have the interpretation of being the exact free energy of the ideal chain constrained to have a site density $\rho(\mathbf{r})$. (It can be shown that Ω can be expressed as a unique functional of $\rho(\mathbf{r})$ which is minimal at equilibrium (6).)

The excess functional is not known exactly, and two broad classes of approximations have been attempted. In one approach the free energy is expanded in a functional series in the density (about some uniform fluid density); the kernels of successive terms in the expansion are direct correlation functions of increasing order. This is the approach followed by Chandler and coworkers (5) and Sen et al. (11) and is described later. The other approach is more phenomenological. Generally, the free energy of the nonuniform fluid is expressed as an integral over all space of the local free energy density of the fluid evaluated at some coarse-grained density. This functional incorporates bulk thermodynamics via an equation of state and fluid correlations via the definition of the coarse-grained density.

3.1 Weighted Density Approximation (WDA)

In this work, an approximate form for $F^{ex}[\rho(\mathbf{r})]$ is obtained via the weighted density approximation (18):

$$F^{ex}[\rho(\mathbf{r})] = \int \rho(\mathbf{r})f(\bar{\rho})d\mathbf{r} \qquad (11)$$

where $f(\rho)$ is the excess (over ideal gas) free energy per site of the bulk hard chain fluid evaluated at a site density ρ. The weighted density is given by

$$\bar{\rho}(\mathbf{r}) = \int \rho(\mathbf{r}')w(|\mathbf{r} - \mathbf{r}'|)d\mathbf{r}' \qquad (12)$$

where $w(r)$ is the weighting function, and is normalized so that $\int w(\mathbf{r})d\mathbf{r} = 1$. In this work the simplest choice for $w(r)$ is employed, i.e.

$$w(r) = \frac{3}{4\pi\sigma^3} \qquad r < \sigma$$
$$= 0 \qquad r > \sigma \qquad (13)$$

where σ is the diameter of a bead on the chain. The range of the direct correlations in the fluid are expected to be of the order of the bead diameter; this approximation sets this range to be identically equal to the bead diameter. The function $f(\rho)$ is obtained from an equation of state as described in reference (10).

A formal minimization of the grand free energy gives,

$$\rho_M(\mathbf{R}) = \exp[-\beta V(\mathbf{R}) + \beta\mu - \beta\Phi(\mathbf{R}) - \beta\Lambda(\mathbf{R})] \tag{14}$$

where $\beta = 1/k_B T$, and

$$\Lambda(\mathbf{R}) = \frac{\delta F^{ex}}{\delta\rho_M(\mathbf{R})} \tag{15}$$

is the field due to the other molecules. With the approximation embodied in equation (9),

$$\Lambda(\mathbf{R}) = \sum_{i=1}^{N} \frac{\delta F^{ex}}{\delta\rho(\mathbf{r}_i)} = \sum_{i=1}^{N} \lambda(\mathbf{r}_i), \tag{16}$$

with

$$\lambda(\mathbf{r}) = f(\bar{\rho}(\mathbf{r})) + \int d\mathbf{r}'\rho(\mathbf{r}')w(|\mathbf{r} - \mathbf{r}'|)f'(\bar{\rho}(\mathbf{r}')), \tag{17}$$

and $f' = df/d\rho$. Equations (14) and (17) form a set of two equations that must be solved simultaneously to obtain the density profile for a fixed value of μ. In simulations of polymers the average density in the system rather than μ is normally treated as the independent variable. This this is because maintaining a constant chemical potential requires the successful insertion of chain molecules, which is difficult. In order to compare to simulations, the average density rather than μ is fixed in these calculations.

3.2   Truncated Functional Expansion (TFE)

Recently Sen et al. (11) have presented a theory that is very similar in spirit to the one described above; the only difference is that the excess free energy functional is given by,

$$F^{ex}[\rho(\mathbf{r})] = \int d\mathbf{r}'c(|\mathbf{r}' - \mathbf{r}|; \rho_b)[\rho(\mathbf{r}) - \rho_b] \tag{18}$$

where $c(\mathbf{r}' - \mathbf{r}; \rho_b)$ is the direct correlation of the bulk fluid evaluated at the bulk density ρ_b. This corresponds to a functional expansion of the free energy about the uniform fluid truncated after the second order term. This theory suffers from some drawbacks which do not plague the weighted density approximation. For example, it is thermodynamically inconsistent and does not satisfy the exact wall sum rule, $\beta P = \rho(z = 0)$, where P is the bulk pressure. Furthermore, for a uniform fluid it does not satisfy the exact relation

$$\int d\mathbf{r}_n c^{(n)}(\rho; \mathbf{r}_1, ..., \mathbf{r}_n) = \frac{\partial c^{(n-1)}(\rho; \mathbf{r}_1, ..., \mathbf{r}_{n-1})}{\partial\rho} \tag{19}$$

between direct correlation functions of successive order. However, for alkanes at surfaces this theory has been found to be quite accurate, and it is of interest to compare it to the weighted density approximation.

The bulk fluid direct correlation function required in the TFE theory is obtained from the PRISM (polymer reference interaction site model) integral equation theory (19). This theory relates the intermolecular correlation functions to the intramolecular correlation functions via a nonlinear integral equation which is given by:

$$\hat{h}(k) = \hat{\omega}(k)\hat{c}(k)\hat{\omega}(k) + \rho_b\hat{\omega}(k)\hat{c}(k)\hat{h}(k) \tag{20}$$

where $h(r)$ is the total correlation function, $\hat{\omega}(k)$ is the single chain structure factor, and the carets denote Fourier transforms. Given an expression for $\hat{\omega}(k)$ and another closure relation between $h(r)$ and $c(r)$ one can solve the above equation for the direct correlation function $c(r)$. In this work the chains are treated as ideal on long length scales and $\hat{\omega}(k)$ is calculated from a single chain model where excluded volume is ignored for beads separated by three bonds or more. The single chain structure factor is calculated exactly for beads separated by one or two bonds (20) and approximately using the discrete Koyama distribution for beads separated by more than two bonds (21). An approximate density overlap correction is also employed to account for the unphysical intramolecular overlaps present in the semiflexible chain model which tends to decrease the effective packing fraction of the system. The Percus-Yevick closure, $h(r)$=-1 for $r < \sigma$ and $c(r) = 0$ for $r > \sigma$, is used to close the PRISM equation.

3.3   Integral Equation Theory

The wall-PRISM theory (4) is an extension of the growing adsorbent model (14, 15) to polymers. In this case the PRISM equations are solved for a mixture of polymers and spheres in the limit as the spheres become infinitely dilute and infinitely large. The resulting Ornstein-Zernike like equation for the wall-fluid correlation functions is

$$\tilde{h}_w(k) = \hat{S}(k)\tilde{c}_w(k) \tag{21}$$

where the tilde denote one dimensional Fourier transforms and $\hat{S}(k)$ is the bulk fluid static structure factor obtained by solving the bulk PRISM equations. The density profile is given by $\rho(z) = \rho_b(1 + h_w(z))$. The above equation is supplemented with the Percus-Yevick like closures: $h_w(z) = -1$ for $z < 0$ and $z > H$ and $c_w(z) = 0$ for $0 < z < H$. The reader is referred to reference (4) for details. An advantage of the integral equation theory is that computational requirements are trivial when compared to the density functional theories. However, the theory cannot describe the change in chain conformations near the surface and, like the TFE theory, is not thermodynamically consistent. Furthermore, like other integral equation theories, it is not expected to be reliable when wall-fluid and fluid-fluid attractive interactions are present.

3.4   Numerical procedure

The equations that describe the WDA density functional theory are solved via a Newton-Raphson procedure described in reference (10). Briefly, the region between the surfaces is divided into a number of bins and the value of the density profile in each bin is treated as an independent variable. If there are M bins, then there are M simultaneous equations given by

$$\rho(z_k) - < \sum_{i=1}^{N} \delta(z - z_i) > = 0 \qquad (22)$$

where $\rho(z_k)$ is the value of the density at point z_k, and the term in brackets is the density profile calculated from the single chain simulation using the fields generated using the density profile $\rho(z_k)$. Given a guess for the density profile, the next guess is obtained following the standard Newton-Raphson recipe.

The implementation of the TFE theory has an additional feature not present in the implementation of the WDA theory: the bulk density is, in general, not known. (Recall that only the average density in the cell is fixed in the computer simulations.) In this work an iterative procedure is used to obtain the bulk density. A value for the bulk density is first guessed, the direct correlation function is then obtained using the PRISM theory (19), and the density profile of the confined fluid is calculated using the Newton Raphson procedure described above. In all the cases investigated the profile is flat in center of the region between the walls. It is therefore reasonable to assume that the fluid is bulk-like in this region and use this density as the next guess for the bulk density. The procedure is continued until convergence (normally 3 or 4 iterations are required). The implementation of this theory is therefore about 3 or 4 times more computationally intensive than the WDA theory. The integral equation theory is solved via a standard Picard iteration procedure (16).

4   Comparison to Monte Carlo Simulations

Figure 1 compares the density profiles predicted by the various theories to Monte Carlo simulations (13) of hard freely-jointed trimers ($\epsilon=0$) confined between hard walls. In these calculations (and simulations) the average volume fraction, η, in the system is fixed. (The position $z=0$ represents the distance of closest approach of the chain sites to the wall.) At the lower density ($\eta=0.1$, see inset) both the density functional theories are in excellent agreement with the simulation data. The wall-PRISM theory overestimates the value of the density at the surface, but for distance greater than about 0.5 σ is also in good agreement with the simulations. At the higher density ($\eta=0.4$) differences between the performance of the various theories are apparent. The most accurate theory in this case is the wall-PRISM theory. The WDA is very accurate at the surface but predicts oscillations in the density profile that are slightly out of phase with the simulations. The TFE is accurate for distances greater than about 0.5 σ but significantly overestimates the value of the density at the surface (by about 30 - 40 %). This suggests that the accurate contact values reported by Sen et al. (11) for alkanes at surfaces are fortuitous (recall that the theory is not thermodynamically consistent) and not expected for all models of the polymer molecules.

The density profiles of semiflexible hard chains at surfaces are governed by a competition between packing and configurational entropic effects (1, 12). For fully flexible chains, the loss in configurational entropy suffered by a single chains promotes a depletion of chain sites near the surface, whereas the packing of the molecules against the surface promotes an enhancement of chain sites near the surface. As the chains become stiffer, the restrictions on configurational entropy increase and this tends to decrease the density close to the wall. On the other hand, the packing effects help align the chain segments parallel to the wall and this tends to increase the density close to the wall. The observed density profiles are a result of a superposition of these effects.

Figure 2 and 3 compare theoretical predictions for the density profiles of hard 20-mers between hard walls to simulation results (12) for values of $\epsilon=0$ and 5, respectively. In all cases the agreement between the WDA theory and simulation is quite good. In fact it is in better agreement with the simulations for 20-mers than it is for trimers. This could be because the single chain entropy plays a more important role in long chains, especially stiff chains, and this aspect is treated exactly in the density functional theories. At low densities the WDA is more accurate than the TFE in all cases. At high densities, for $\epsilon=0$ the two theories are about equally accurate, the WDA is more accurate near contact and the TFE is more accurate slightly further away. For $\epsilon=5$ the WDA appears to be more accurate at all densities. The wall-PRISM theory is in good agreement with the simulations for $\epsilon=0$, but for higher stiffnesses it misses even the shape of the density profile. It appears that the wall-PRISM theory is accurate when the fluid structure is dominated by packing effects; the performance of this theory diminishes when single chain entropy effects become important.

Figure 4 compares the WDA theory to simulations for $\epsilon_w \neq 0$, i.e. a repulsion or attraction between the walls and the beads is present, for $N=20, \epsilon=0$ and $\eta=0.35$. An attractive potential results in a large increase in the density at the surface whereas a repulsive wall-fluid potential results in a significant depletion in the density at the surface, as expected. Again, the theory is in good agreement with the simulations.

5   Summary and Discussion

A density functional theory for nonuniform polymer melts is presented which combines a single molecule simulation with the weighted density approximation. When compared to simulations of semiflexible polymer melts, the theory is found to be in good agreement over a range of chain lengths, stiffnesses, and densities. The weighted density approximation and the truncated functional expansion theories are both quite accurate, although the WDA is a little more accurate in the immediate vicinity of the surface (especially for the stiffer chains) while the TFE is a little more accurate at intermediate distances for the shorter chains. The integral equation theory is very accurate for freely jointed hard chains. At liquid

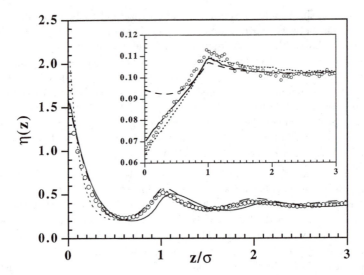

Figure 1. Comparison of theoretical predictions (WDA: solid, TFE: short dashes, wall-PRISM: long-dashes) to Monte Carlo simulations (symbols) for the density profile of 3-mers between hard walls ($\epsilon_w = 0$) for $\epsilon=0$ and $\eta=0.4$ and 0.1 (inset).

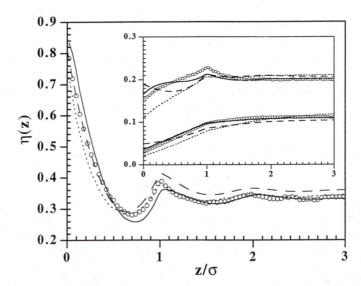

Figure 2. Comparison of theoretical predictions to Monte Carlo simulations for the density profile of 20-mers between hard walls ($\epsilon_w = 0$) for $\epsilon=0$ and $\eta=0.1$ and 0.2 (inset) and 0.35. See figure 1 for description of symbols and lines.

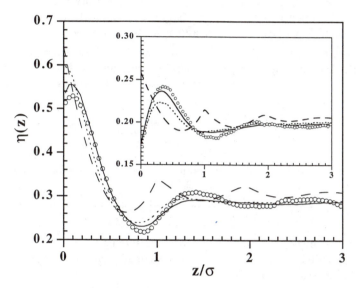

Figure 3. Comparison of theoretical predictions to Monte Carlo simulations for the density profile of 20-mers between hard walls ($\epsilon_w = 0$) for $\epsilon=5$ and $\eta=0.2$ (inset) and 0.3. See figure 1 for description of symbols and lines.

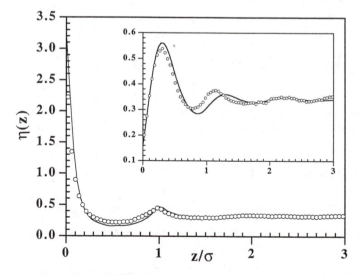

Figure 4. Comparison of WDA predictions (lines) to Monte Carlo simulations (symbols) for the density profile of hard 20-mers for $\epsilon_w=-2$ and $\epsilon_w=2$ (inset) for $\epsilon=0$ and $\eta=0.35$.

like densities it is more accurate than either of the density functional theories. For stiff chains, however, this theory misses the shape of the density profile and predicts a significant layering that is absent in the simulations.

Both the versions of the density functional theory can be improved upon. The WDA is easily improved by incorporating information about the pair correlations of the bulk fluid into the weighting function. To do this one choose $w(r)$ so that the identity

$$-kTc(|\mathbf{r} - \mathbf{r}'|) = N^2 \left. \frac{\delta^2 F^{ex}}{\delta\rho(\mathbf{r})\delta\rho(\mathbf{r}')} \right|_{\rho(\mathbf{r})=\rho(\mathbf{r}')=\rho}, \tag{23}$$

is automatically satisfied. This results in an integral equation for $w(r)$ given by,

$$-\frac{kT}{N^2}c(r) = 2\frac{\partial f}{\partial \rho}w(r) + \rho\frac{\partial^2 f}{\partial \rho^2} \int w(\mathbf{r}')w(|\mathbf{r}-\mathbf{r}'|)d\mathbf{r}' + 2\rho\frac{\partial f}{\partial \rho} \int \frac{\partial w(\mathbf{r}')}{\partial \rho}w(|\mathbf{r}-\mathbf{r}'|)d\mathbf{r}' \tag{24}$$

which can be solved iteratively or directly by assuming the last term is small. A similar approximation results in very accurate predictions for the density profiles of hard spheres at a hard wall (17).

A systematic improvement of the TFE is more difficult. The version of this theory tested corresponds to a hypernetted-chain style self-consistent field, and for the tangent chains of this work better results might be expected with a Percus-Yevick style self-consistent field (9). Like integral equation theories, the accuracy of any particular closure relation is only known *post facto*, and often difficult to justify *a priori*.

6   Acknowledgments

I would like to thank Professor C. E. Woodward for useful discussions. Acknowledgment is made to the donors of the Petroleum Research Fund administered by the American Chemical Society for partial support of this research.

Literature Cited

(1) Dickman, R.; Hall, C. K. *J. Chem. Phys.* **1988**, *89*, 3168.

(2) Kumar, S. K.; Vacatello, M.; Yoon, D. Y. *J. Chem. Phys.* **1988**, *89*, 5209; *Macromolecules* **1990**, *23*, 2189.

(3) Yethiraj, A.; Hall, C. K. *J. Chem. Phys.* **1989**, *91*, 4827; *Macromolecules* **1990**, *23*, 1635; *Macromolecules* **1991**, *24*, 709; *Molec. Phys.* **1991**, *73*, 503.

(4) Yethiraj, A.;Hall, C. K. *J. Chem. Phys.* **1991**, *95*, 3749.

(5) Chandler, D; McCoy, J. D.; Singer, S. J. *J. Chem. Phys.* **1986**, *85*, 5971 and 5977.

(6) Woodward, C. E. *J. Chem. Phys.* **1991**, *94*, 3183; *J. Chem. Phys.* **1992**, *97*, 695; *J. Chem. Phys.* **1992**, *97*, 4525.

(7) Woodward, C. E.; Yethiraj, A. *J. Chem. Phys.* **1994**, *100*, 3181.

(8) Kierlik, E.; Rosinberg, M. L. *J. Chem. Phys.* **1994**, *100*, 1716.

(9) Donley, J. P.; Curro, J. G.; McCoy, J. D *J. Chem. Phys.* **1994**, *101*, 3205.

(10) Yethiraj, A.; Woodward, C. E. *J. Chem. Phys.* **1995**, *102*, 5499.

(11) Sen, S.; Cohen, J. M.; McCoy, J. D.; Curro, J. G. *J. Chem. Phys.* **1994**, *101*, 9010.

(12) Yethiraj, A. *J. Chem. Phys.* **1994**, *101*, 2489.

(13) Phan, S.; Rosinberg, M. L.; Kierlik, E.; Yethiraj, A.; Dickman, R. *J. Chem. Phys.* **1995**, *102*, 2141.

(14) Henderson, D.; Abraham, F. F.; Barker, J. A.; *Mol. Phys.* **1976**, *31*, 1291.

(15) Zhou, Y.; Stell, G. *Mol. Phys.*, **1988**, *66*, 797.

(16) Hansen, J.P.; McDonald, I. R. *Theory of Simple Liquids*, Second Edition, Academic Press, New York, NY, 1986.

(17) Evans, R. In *Fundamentals of Inhomogeneous Fluids*, Editor, Henderson, D., Dekker, 1992.

(18) Nordholm, S.; Johnson, M.; Freasier, B. C. *Aust. J. Chem.* **1980**, *33*, 2139; Tarazona, P. *Mol. Phys.* **1984**, *52*, 81; Tarazona, P.; Evans, R. *Mol. Phys.* **1984**, *52*, 847.

(19) Curro, J. G.; Schweizer, K. S. *J. Chem. Phys.* **1987**, *87*, 1842.

(20) Schweizer, K. S.; Honnell, K. G.; Curro, J. G. *J. Chem. Phys.* **1992**, *96*, 3211.

(21) Honnell, K. G.; Curro, J. G.; Schweizer, K. S. *Macromolecules* **1990**, *23*, 3496.

Chapter 20

Density-Functional Theory of Quantum Freezing and the Helium Isotopes

Steven W. Rick[1], John D. McCoy[2], and A. D. J. Haymet[3]

[1]Structural Biochemistry Program, Frederick Biomedical Supercomputing Center, National Cancer Institute–Frederick Cancer Research and Development Center, Frederick, MD 21702
[2]Department of Materials Engineering, New Mexico Institute of Mining and Technology, Socorro, NM 87801
[3]School of Chemistry, University of Sydney, Sydney, New South Wales 2006, Australia

Over the last fifteen years density functional theory has lead to new understanding of many properties of the freezing transition. The essence of the theory is a functional Taylor series expansion of the free energy of the solid about that of a reference liquid. In the simplest version of the theory, the reference state is taken to be the coexisting liquid. The series is typically truncated after second order in the density difference between the two phases. In the classical atomic— and original—version of the theory, all the information specific to the system is contained in the second order expansion coefficient, which is related to pair correlation functions of the liquid (1, 2). The theory demonstrated that liquids freeze when they are strongly correlated, in agreement with empirical freezing rules observed in computer simulations (3, 4).

An early, successful application of density functional freezing theory was to the Lennard-Jones fluid (5). The phase diagram for many of the rare gas elements, when plotted in reduced variables obtained by scaling with the Lennard-Jones parameters which contain the length and energy scales of the particular element, are remarkably similar. For example, the triple points of neon, argon, krypton, and xenon all lie in the same region of phase space. Furthermore, the phase diagram for the Lennard-Jones system, calculated both from simulation and density functional theory, agrees well with the rare gas data. The exception is helium. In terms of its scaled variables, helium freezes at a much *lower* density than the more classical rare gases (6). Additionally, the less massive and therefore more quantum mechanical isotope, ^{3}He, freezes at a lower density than ^{4}He. Quantum effects are therefore promoting freezing and stablizing the solid phase. On the other hand, quantum effects decrease correlations between particles. This last fact is problematic for density functional freezing theory, which gets all its information about the system from liquid-state correlation functions. The influence of quantum mechanics, if only interparticle correlations are considered, will be to increase the freezing density.

0097–6156/96/0629–0286$15.00/0

Therefore, from these two facts—quantum effects decrease the freezing density and also decrease interparticle correlations—we know *a priori* that classical atomic freezing theory will fail to predict the proper trend among ^{3}He, ^{4}He and the classical limit. Recognizing that the free energy Taylor series is an expansion of the excess free energy, defined as the difference between the system of interest and an ideal system at the same thermodynamic conditions, improvements can be made to the theory by choosing a better ideal system. Classically, it suffices to choose a set of non-interacting classical particles. For the quantum case, then, it is sensible to choose a set of non-interacting quantum particles. This choice should make the series more rapidly convergent, hopefully by the second order term, which is where we typically truncate. By using the Feynman path integral representation of quantum mechanics, with its well-known isomorphism between the quantum particle and a ring polymer, the quantum freezing theory has many features in common with the density functional theory as applied to actual polymers (*7*).

Quantum Density Functional Theory of Freezing

This derivation of the quantum freezing functional follows closely that found in Reference 8, which the interested reader should consult for further details. The first step in the derivation of the free energy is a functional Taylor series expansion of the excess Helmholtz free energy about the liquid state in powers of the singlet density difference,

$$(A - A^0)_S = (A - A^0)_L + \int_V dr\, \Delta\rho(r) \left(\frac{\partial(A - A^0)}{\partial\rho(r)} \right)_{\rho(r)=\rho_L}$$

$$+ \frac{1}{2} \iint_V dr\, dr' \left(\frac{\partial^2(A - A^0)}{\partial\rho(r)\partial\rho(r')} \right)_{\rho(r)=\rho_L} \Delta\rho(r)\, \Delta\rho(r') \,, \qquad (2.1)$$

where the subscripts S and L refer to the solid and liquid states, ρ_L is the bulk liquid density, $\rho(r)$ is the solid singlet density $\Delta\rho(r) = \rho(r) - \rho_L$ and A^0 is the free energy of the ideal system. The natural variables of the Helmholtz ensemble are the temperature, T, the volume, V, and ρ. The ideal system has the same values of T,V, and ρ as the interacting system; this means that for the ideal system of the solid phase, which has a spatially varying density, an external field must be placed on the ideal system to produce the same density.

In order to study phase coexistence, it is more convenient to use the Grand ensemble, which is related to the Helmholtz potential through the Legendre transform,

$$\Omega = A - \int_V dr\, \Psi(r)\rho(r) \,, \qquad (2.2)$$

where Ω is the Grand potential, $\Psi(r) = \mu - U(r)$, μ is the chemical potential, and $U(r)$ is an external field. The variables ρ and Ψ are conjugate and the following relations hold,

$$\frac{\partial A}{\partial\rho(r)} = \Psi(r) \qquad (2.3)$$

and

$$\frac{\partial \Omega}{\partial \Psi(r)} = -\rho(r). \tag{2.4}$$

Using Equations (2.2) and (2.3) and defining

$$c(|\mathbf{r} - \mathbf{r}'|) = -\beta \left(\frac{\partial^2 (A - A^0)}{\partial \rho(\mathbf{r}) \partial \rho(\mathbf{r}')} \right)_{\rho(r)=\rho_L}, \tag{2.5}$$

where $\beta^{-1} = kT$, leads to the following functional for the Grand potential free energy difference,

$$\begin{aligned}
\Delta\Omega^*[\rho(\mathbf{r})] &= \Omega^* - \Omega_L^* \\
&= A^0 - A_L^0 + \int_V d\mathbf{r} \, [\Psi_L - \Psi(\mathbf{r})]\rho(\mathbf{r}) - \int_V d\mathbf{r} \, \Psi_L^0 \Delta\rho(\mathbf{r}) \\
&\quad - \frac{1}{2} kT \iint_V d\mathbf{r} \, d\mathbf{r}' \, c(|\mathbf{r} - \mathbf{r}'|)\Delta\rho(\mathbf{r})\Delta\rho(\mathbf{r}') \, .
\end{aligned} \tag{2.6}$$

Since the natural variables of Ω are T, V, and $\Psi(r)$, the free energy should be a functional of these variables only. However, rather than eliminate $\rho(r)$ in favor of these variables, it is more convenient to consider Ω as a functional of T,V, $\Psi(r)$, and $\rho(r)$, with an additional condition which fixes the value of $\rho(r)$. This condition is that $\rho(r)$ is given by the value which minimizes Ω^* for fixed T,V, and $\Psi(r)$. The asterisk on the $\Delta\Omega^*$ functional denotes that $\rho(r)$ is a free variable, only when the functional is minimized does it equal the the grand potential difference, $\Delta\Omega$.

It is now necessary to choose the ideal system. For classical atomic liquids, the ideal system is commonly chosen to be classical non-interacting particles. For this ideal system, c(r) is the Ornstein-Zernike direct correlation function, $\rho(r) = \exp(\beta\Psi(r))$ and

$$\beta A^0 = \int_V d\mathbf{r} \, \rho(\mathbf{r})[\ln(\rho(\mathbf{r}) - 1] \, . \tag{2.7}$$

For a quantum mechanical ideal system, using the Feynman path integral representation, the singlet density of the ideal non-interacting system is related to $\Psi^0(r)$ through

$$\begin{aligned}
\rho(\mathbf{r}) = \lim_{P \to \infty} \left(\frac{P}{\lambda^2} \right)^{3(P-1)/2} \int \cdots \int_V d\mathbf{r}_2 \ldots d\mathbf{r}_P \, e^{(\beta/P)[\Psi^0(\mathbf{r}) + \cdots + \Psi^0(\mathbf{r}_P)]} \\
\times \, e^{-(\pi P/\lambda^2)[(\mathbf{r} - \mathbf{r}_2)^2 + \cdots + (\mathbf{r}_P - \mathbf{r})^2]} \, ,
\end{aligned} \tag{2.8}$$

where $\lambda = (2\pi mkT/h^2)^{-1/2}$, m is the mass of the particle, h is Planck's constant, and P is the number of discretizations of the path integral (9, 10). From Equations (2.2) and (2.4),

$$\beta A^0 = \int_V d\mathbf{r} \, \rho(\mathbf{r})[\Psi^0(\mathbf{r}) - 1] \, . \tag{2.9}$$

For the liquid phase Ψ is a constant and therefore $\rho_L = \exp(\beta\Psi_L^0)$ just as in he classical case. Substituting Equation (2.9) into Equation (2.6) leads to

$$\beta\Delta\Omega^* = \rho_L V + \beta \int_V dr \left[\Psi_L - \Psi(r) - \ln\rho_L + \Psi^0(r) - 1\right]\rho(r)$$

$$- \frac{1}{2} kT \iint_V dr\, dr'\, c(|r - r'|)\Delta\rho(r)\Delta\rho(r') . \tag{2.10}$$

As already noted, $\beta\Delta\Omega^*$ is only the free energy difference when minimized with respect to $\rho(r)$, but now we have the added function, $\Psi^0(r)$, which cannot be eliminated in favor of $\rho(r)$ because Equation (2.8) cannot be inverted to give $\Psi^0(r)$ as a functional of $\rho(r)$. Therefore, the $\beta\Delta\Omega^*$ from Equation (2.10) is a functional of both $\rho(r)$ and $\Psi^0(r)$. The functional so defined is the actual Grand potential free energy difference when

$$\frac{\partial\beta\Delta\Omega^*}{\partial\rho(r)} = 0 \tag{2.11}$$

and

$$\frac{\partial\beta\Delta\Omega^*}{\partial\Psi^0(r)} = 0. \tag{2.12}$$

The solution of Equation (2.10) together with Equations (2.11) and (2.12) is a difficult computational problem. For this reason it is convenient to parameterize the solid density. One common assumption is to assume that the solid singlet density is of Gaussian form (and hence spherically symmetric) about each lattice site (*11-13*). This assumption has been tested against exact parameterizations for classical systems and is a good approximation for close packed crystals (*5*). The Gaussian approximation for the $\rho(r)$ is equivalent to assuming that $\Psi^0(r)$ to has a following parabolic form in each unit cell

$$\Psi^0(r) = a - br^2 , \tag{2.13}$$

where r is the distance from a lattice site. Putting this parameterization of $\Psi^0(r)$ into Equation (2.8), leads to this parameterization of the singlet density (*9*)

$$\rho(r) = (2\gamma(C - 1)/\pi)^{3/2} e^{-2\gamma(C-1)r^2} , \tag{2.14}$$

where $C = \cosh(\lambda(b/\pi)^{1/2})$ and

$$\gamma = \frac{\sqrt{b\pi}}{\lambda\sinh(\lambda(b/\pi)^{1/2})} . \tag{2.15}$$

The density is therefore a Gaussian distribution around each lattice site, with a determined by normalization of $\rho(r)$ ($a = 3\ln[2(C - 1)/\lambda^2]/2$) and b a variational parameter.

The double integral in Equation (2.10) is best evaluated by using the periodic nature of the density and expanding $\rho(r)$ in a Fourier series

$$\Delta\rho(\mathbf{r}) = \sum_{\mathbf{k}} \rho(\mathbf{k}) e^{i\mathbf{k}\cdot\mathbf{r}} . \tag{2.16}$$

The Fourier coeficients for the density from Equation (2.14) are given by

$$\rho(\mathbf{k}) = \begin{cases} \rho_S - \rho_L & \text{for } \mathbf{k} = 0 \\ \rho_S \, e^{-k^2/(8\gamma(C-1))} & \text{for } \mathbf{k} \neq 0 \end{cases} , \tag{2.17}$$

where ρ_S is the bulk solid density. Then,

$$\frac{1}{2} \iint_V d\mathbf{r} \, d\mathbf{r}' \, c(|\mathbf{r} - \mathbf{r}'|)\Delta\rho(\mathbf{r})\Delta\rho(\mathbf{r}') = \frac{1}{2} V \sum_{\mathbf{k}} [\rho(\mathbf{k})]^2 c(k) , \tag{2.18}$$

where $c(k)$ is the Fourier transform of $c(r)$.

Using Equations (2.13, 2.14 and 2.18) in Equation (2.10) and $\Psi_L = \Psi(r)$ (since both phases are to be at the same chemical potential) leads to

$$\frac{\beta}{V}\Delta\Omega^*[\rho(r)] = \rho_L - \rho_S \left(1 + \ln(\rho_L) - \frac{3}{2}\ln(2(C-1)/\lambda^2) + \frac{3b}{4\gamma(C-1)}\right)$$

$$- \frac{1}{2} \sum_{\mathbf{k}} [\rho(\mathbf{k})]^2 c(k) , \tag{2.19}$$

where we have assumed that $\rho(r)$ centered on one lattice site does not overlap adjacent sites. This equation has the correct classical limit. By minimizing Equation (2.19) with respect to $\rho(r)$ (which through Equation (2.14) means minimizing with respect to ρ_S and b) for a given crystal symmetry, one obtains the grand potential difference between the liquid and solid phases. One then varies the chemical potential of the liquid by varying ρ_L to find the point where $\Delta\Omega = 0$, thereby obtaining the phase coexistence point.

One final point concerns the effect of quantum mechanics on the liquid correlation functions. The correlation function

$$\chi(\mathbf{r}_1, \mathbf{r}_2) = \frac{\partial\rho(\mathbf{r}_1)}{\partial\beta\Psi(\mathbf{r}_2)} \tag{2.20}$$

is the probability [minus a mean field contribution of $\rho(\mathbf{r}_1)\rho(\mathbf{r}_2)$] that a particle exists at a position $\mathbf{r}_1$ and simultaneously a particle, either the same particle or different, exists at $\mathbf{r}_2$. $\chi(\mathbf{r}_1, \mathbf{r}_2)$ is written as a sum of self and distinct particle terms as

$$\chi(\mathbf{r}_1, \mathbf{r}_2) = S(\mathbf{r}_1, \mathbf{r}_2) + \rho(\mathbf{r}_1)\rho(\mathbf{r}_2)h(\mathbf{r}_1, \mathbf{r}_2) , \tag{2.21}$$

where $S(\mathbf{r}_1, \mathbf{r}_2)$ is the self-correlation function and $h(\mathbf{r}_1, \mathbf{r}_2)$ is the total correlation function. In conventional path integral studies, correlation functions are calculated between particles at the same imaginary time. The correlation functions in Equation (2.21), however, include correlations between different imaginary times as well as the same imaginary time. For liquid helium the total correlation function calculated either way are very similar.

For a classical system, if a particle exists at r_1, then that same particle only exists at r_1, so the self correlation function is proportional to a delta function. However, quantum mechanically a particle at r_1 also may have a finite probability of being at r_2 at the same time. For the ideal non-interacting system,

$$S^0(r_1, r_2) = \rho_L \frac{2}{\lambda^2} \frac{1}{|r_1 - r_2|} e^{-4\pi(r_1 - r_2)^2/\lambda^2} . \tag{2.22}$$

From Equations (2.3) and (2.20),

$$\chi^{-1}(r_1, r_2) = \frac{\partial^2 \beta A}{\partial \rho(r_1) \partial \rho(r_2)} \tag{2.23}$$

where the superscript "-1" indicates the functional inverse, defined by

$$\delta(r_1 - r_2) = \int_V dr_3 \, \chi(r_1, r_3) \, \chi^{-1}(r_3, r_2). \tag{2.24}$$

This equation can be combined with Equation (2.21) to form a quantum mechanical Ornstein-Zernike equation. For the liquid, the direct correlation function needed as input into the freezing theory is most easily expressed in Fourier space as

$$c(k) = S^{0\,-1}(k) - S^{-1}(k) + \frac{S^{-1}(k)\,\rho_L^2\,h(k)}{S(k) + \rho_L^2\,h(k)} . \tag{2.25}$$

In the classical limit, this definition of $c(r)$ reduces to the usual direct correlation function and Equation (2.25) becomes exactly the Ornstein-Zernike equation. This result can be thought of as the RISM integral theory applied to a ring polymer liquid where each ring has an infinite number of sites with an infinitesimal bond length (*14, 15*).

The self-correlation function for interacting systems is in general different than $S^0(r)$ since interactions compress the extent of quantum delocalization. However, $S(r)$ is slow to converge with path discretizations, P, and for this reason it is difficult to calculate from simulations. For the results presented in the next section, we approximate $S(r)$ as $S^0(r)$ (*6*). Another approximation, involving giving the ideal system an effective mass so that $S^0(r)$ is equal to an estimated $S(r)$. The use of an effective mass leads only to small differences between the results using $S(r) = S^0(r)$, except at temperatures below 50 Kelvin where the use of an effective mass did not lead to freezing solutions. Other input needed for the freezing theory, namely the liquid correlation function, $h(r)$, was calculated using path integral Monte Carlo (PIMC) simulations, with 500 particles and P=10′ path integral discretizations for temperatures below 21 K and P=3 at higher temperatures. The potential used was the Lennard-Jones potential with ϵ/k_B=10.22 K and σ=2.556Å (*16*).

The Freezing of Helium

Crystal-liquid phase points were calculated for ^{4}He at temperatures ranging from 8 to 204.4 K. In Figure 1, those results are compared to the experimental phase diagram. Experimentally, the stable crystal is face centered cubic (fcc) above 14.9 K and hexagonal close packed (hcp) for lower temperatures (18, 17). At temperatures above 290 K, there is evidence that the fluid again freezes into an hcp crystal (22). The theory predicts fcc to be stable at all temperatures studied, even at 8 K. For quantum and well as classical systems, the resolution of the free energy differences between these two close packed crystals is a difficult problem. Otherwise, the agreement is very good. The predicted solid density is overestimated, this also tends to be the case in the classical theory. The solid density is dependent on the liquid pair correlation function, $g(r)$. Other results, using a $g(r)$ not from the Lennard-Jones potential and PIMC simulations but from classical pair pseudo-potentials and integral equation theory, predict a smaller density change upon freezing (23).

Phase coexistence points were calculated for ^{3}He as well. The isotopic substitutions provide a means to judge the importance of the things left out of the theory, such as particle statistics. For a range of temperatures (20, 102.2 and 204.4 K), the freezing theory predicts that ^{3}He freezes at a lower density than ^{4}He, in agreement with experiment. The predicted crystal phase is again found to be fcc. Experimentally, the stable phase in this temperature range is believed to be fcc (24); the fcc to hcp transition occurs at 17.73 K (25). The theory gives a difference in the liquid density at the freezing point between ^{4}He and ^{3}He of 0.0011 Å^{-3} at 20 K, in very good agreement with the experimental value of 0.0009 Å^{-3} (17). Furthermore, at this temperature density functional theory predicts that the classical Lennard-Jones fluid freezes at a density of 0.0088 Å^{-3} higher than ^{4}He (5). Therefore, quantum effects stablize the solid phase.

One interesting feature of the ^{3}He and ^{4}He phase is the isotopic shift in the pressure at the freezing point, P_F (Figure 2). At low temperatures, both the theoretical (shown by the filled squares) and the experimental (shown by the circles) (17) results indicate that ^{3}He freezes at a higher pressure than ^{4}He, even though it freezes at a lower density. At higher temperatures, the theory predicts the opposite isotopic shift: ^{3}He freezes at a lower pressure than ^{4}He. The high pressure diamond-anvil cell experiments of Loubeyre, et al. , find that (except perhaps at 175 K) ^{3}He freezes at a higher pressure than ^{4}He (24). Also shown in Figure 2 are simulation studies of the isotopic shift (triangles) (26). These results are generated using the experimental values of the freezing pressures and densities of ^{4}He and then using free energy perturbation theory to predict P_F of ^{3}He. This requires performing PIMC simulations at the two phases of both isotopes. The results of Reference 26 used the exp-6 pair potential, similar to the Lennard-Jones potential used in the DFT calculations but with the parameters used it has a well which is slightly deeper and at larger separations (27). The DFT and the simulation results show the same trend for the isotopic shift. The disagreement with the experimental results is due to deficiencies with the pair potentials. The

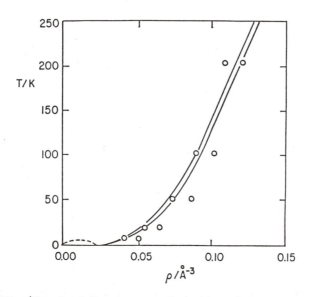

Figure 1: The ⁴He phase diagram: experimental results for the liquid-crystal coexistence curve (solid line) (*17-20*) and the gas-liquid coexistence line (dashed line) (*21*), and the DFT results (circles) (*6*).

pressures needed to make helium form a solid at temperatures above 150 K are in the GPa range. At these pressures, many-body interactions become important. Simulations similar to those of Reference 26 but with three-body potential terms added perturbatively give results which are in close agreement with experiment (the diamond in Figure 2) (*28*).

The crystal density profiles are displayed in Figure 3, which compares $\rho(r)$ for helium with the classical Lennard-Jones result (the solid line) (*5*). Quantum effects cause the density to become much broader, even when scaled by the nearest neighbor distance, d_{nn}. The ⁴He density at 20 K (the dashed line) is much broader than the classical result near the same temperature and the ³He density (the dotted line) is broader still. In addition, the density gets broader as the temperature is lowered (the dot-dashed line). The width of the density is customarily measured by the Sutherland(*29*)/Lindemann(*30*) ratio, which is the average root mean square deviation of a particle in the crystal from its lattice site, measured in units of d_{nn},

$$L = \frac{1}{d_{nn}} \left[\int_{peak} d\mathbf{r} \, r^2 \, \rho(\mathbf{r}) \right]^{1/2} = \frac{1}{d_{nn}} \left[\frac{3}{4} \frac{1}{\gamma(C-1)} \right]^{1/2}. \tag{2.1}$$

For ⁴He and ³He at 20 K, L is 0.116 and 0.124, respectively; in the classical limit at this temperature, L is 0.076 (*5*). For ⁴He at 8 K, L is 0.145. The

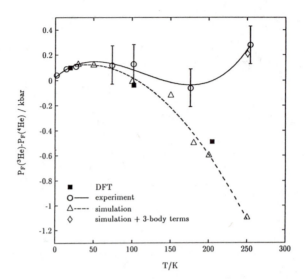

Figure 2: Isotopic shift in the pressure at the freezing point, P_F, as a function of temperature comparing experimental results (circles, the solid line is a polynomial fit) (*17, 24*), simulations using the exp-6 pair potential (triangles and dashed line fit) (*26*), simulations with additional three-body energy terms (diamond) (*28*), and DFT with the Lennard-Jones pair potential (filled squares) (*6*).

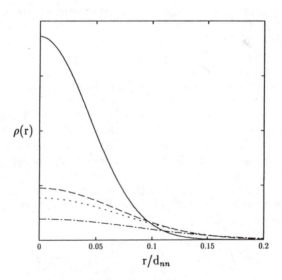

Figure 3: Crystal singlet density profiles for classical Lennard-Jones particles at T=20.44 K (solid line) (*5*), ^{4}He at 20 K (dashed line), ^{3}He at 20 K (dotted line), and ^{4}He at 8 K (dot-dashed line), as a function of distance from a lattice site, scaled by the nearest neighbor distance, d_{nn}.

Sutherland/Lindemann empirical rule states that a crystal will melt when L is greater than 0.1. This holds for a number of simple crystals but is clearly not true for helium, for which L can be as large as 0.38 (*31*).

In conclusion, we have presented a theory of freezing which successfully extends the classical theory to systems where quantum effects are very large (^{3}He and ^{4}He). Quantum mechanics influences the freezing transition by introducing a new length scale, so that the length scale of the system is not determined by that of the intermolecular interactions alone, but also by the de Broglie wavelength λ. The quantum freezing theory as given by Equation (2.19) is successful because the convolution of $S(r)$ and $h(r)$ that is in the input (Equation (2.25)) contains both these length scales. Quantum mechanics increase the length scale and lowers the freezing density in part because quantum delocalization causes the particles to be larger than the classical limit. The isotopic shifts in the freezing points are correctly predicted by the theory (see Figure 2), which implies that the quantum effects left out of the theory, namely exchange, are not important at these temperatures. Exchange could be included in the theory to second order through the correlation function, although exchange only effect $h(r)$ at temperatures below 2 K (*32*). Freezing theories have also been developed for systems for which exchange effects are important, including fluids with internal quantum states at non-zero temperatures (*33*) and a variety of systems at zero temperature: electrons (*34, 35*), quantum hard-spheres (*36*), and superfluid ^{4}He (*37, 38*). For a review see Reference 39.

Acknowledgments: The content of this publication do not necessarily reflect the views or policies of the Department of Health and Human Services, nor does mention of trade names, commercial products, or organizations imply endorsement by the U.S. government. This research was supported in part by NSF Grant CHEM-8913006.

Literature Cited
1. Ramakrishnan, T.V.; Yussouff, M. *Phys. Rev. B*, **1979**, *74*, 2775.
2. Haymet, A.D.J.; Oxtoby, D.W. *J. Chem. Phys.*, **1981**, *74*, 2559.
3. Verlet, L. *J. Chem. Phys*, **1969**, *7*, 591.
4. Hansen, J.P.; Verlet, L. *Phys Rev.*, **1969**, *184*, 1501.
5. Laird, B.B.; McCoy, J.D.; Haymet, A.D.J. *J. Chem. Phys.*, **1987**, *87*, 5449.
6. Rick, S.W.;McCoy, J.D.; Haymet, A.D.J. *J. Chem. Phys.*, **1990**, *92*, 3040.
7. McCoy, J.D.; Honnell, K.G.; Schweizer, K.S.; Curro, J.G. *J. Chem. Phys.*, **1991**, *95*, 9348.
8. McCoy, J.D.; Rick, S.W.; Haymet, A.D.J. *J. Chem. Phys.*, **1990**, *92*, 3034.
9. Feynman, R.P. *Statistical Mechanics: A Set of Lectures*; Benjamin: Reading, MA, 1972.
10. Chandler, D.; Wolynes, P.G. *J. Chem. Phys.*, **1981**, *74*, 4078.
11. Jacobs, R.L. *J. Phys.*, **1983**, *16*, 273.
12. Jones, G.; Mohanty, U. *Mol. Phys.*, **1985**, *54*, 1241.
13. Tarazona, P. *Phys. Rev. A*, **1985**, *31*, 2672; **1985**, *32*, 3148(E).
14. Chandler, D,; Singh, Y.; Richardson, D.M. *J. Chem. Phys.*, **1981**, *81*, 1975.
15. Laria, D.; Wu, D.; Chandler, D. *J. Chem. Phys.*, **1991**, *95*, 4444.

16. Pollock, E.L.; Ceperley, D.M. *Phys. Rev. B*, **1984**, *30*, 2555.
17. Grilly, E.R.; Mills, R.L. *Ann. Phys. (NY)*, **1959**, *8*, 1.
18. Dugdale, J.S.; Simon, F.E. *Proc. R. Soc. (London) Ser. A*, **1953**, *218*, 291.
19. Mills, R.L.; Liebenberg, D.H.; Bronson, J.C. *Phys. Rev. B*, **1980**, *21*, 5137.
20. Loubeyre, P. In *Simple Molecular Systems at High Density*; Polian, A; Loubeyre, P.; Boccara, N., Ed.; Plenum: New York, NY, 1989.
21. Roach, P.R.; Douglas, Jr, D.H. *Phys. Rev. Lett.*, **1967**, *19*, 287.
22. Mao, H.K *Phys. Rev. Lett.*, **1988**, *60*, 2649.
23. McCoy, J.D.; Rick, S.W.; Haymet, A.D.J. *J. Chem. Phys.*, **1989**, *90*, 4622.
24. Loubeyre, P.; Letoullec, R.; Pinceaux, J-P *Phys. Rev. Lett.*, **1992**, *69*, 1216.
25. Franck, J.P. *Phys. Rev. Lett.*, **1961**, *7*, 435.
26. Barrat, J-L; Loubeyre, P.; Klein, M.L. *J. Chem. Phys.*, **1989**, *90*, 5645.
27. Young, D.A.; McMahan, A.K.; Ross, M. *Phys. Rev. B*, **1981**, *24*, 5119.
28. Boninsegni, M.; Pierleoni, C.; Ceperley, D.M. *Phys. Rev. Lett.*, **1994**, *72*, 1854.
29. Sutherland, W. *Phils. Mag.*, **1890**, *30*, 318.
30. Lindemann, F.A. *Phys. Z.*, **1910**, *11*, 609.
31. Daniels, W.B. In *Simple Molecular Systems at High Density*; Polian, A; Loubeyre, P.; Boccara, N., Ed; Plenum: New York, NY, 1989.
32. Ceperley, D.M.; Pollock, E.L. *Phys. Rev. Lett.*, **1986**, *56*, 351.
33. Sengupta, S.; Marx, D.; Nielaba, P. *Europhys. Lett.*, **1992**, *20*, 383.
34. Senatore, G.; Moroni, S. *Phys. Rev. Lett.*, **1990**, *64*, 303.
35. Moroni, S.; Senatore, G. *Phys Rev. B*, **1991**, *44*, 9864.
36. Denton, A.R.; Nielaba, P.; Runge, K.J.; Ashcroft, N.W. *Phys. Rev. Lett.*, **1990**, *64*, 1529.
37. Dalfovo, F.; Dupont-Roc, J.; Pavloff, N.; Stringari, S.; Treiner, J. *Europhys. Lett.*, **1991**, *16*, 205.
38. Moroni, S.; Senatore, G., *Europhys. Lett.*, **1991**, *16*, 373.
39. Moroni, S.; Senatore, G., *Phil. Mag.*, **1994**, *69*, 957.

Chapter 21

Freezing of Colloidal Simple Fluids

C. F. Tejero

Facultad de Ciencias Fisicas, Universidad Complutense de Madrid,
28040 Madrid, Spain

It is well-known that the Lennard-Jones potential provides a fair description of the phase diagram of a simple atomic fluid with a fluid-fluid critical point and a fluid-fluid-solid triple point. In this chapter we analyze the modifications occurring in the phase diagram of a simple fluid when the range of the attractions is reduced. We consider two different theoretical approaches based on the Gibbs-Bogoliubov inequality in order to evaluate the fluid and solid free energies. It is shown that for intermediate attractions the liquid phase disappears and the phase diagram only exhibits a fluid-solid transition. A further reduction of the range of the attractions leads to a phase diagram with a solid-solid critical point and a fluid-solid-solid triple point. These phase diagrams could be of relevance to the phase behavior of colloidal simple fluids.

The modern theory of freezing has been formulated within the density functional theory of nonuniform phases (1-6). Its application to the freezing of hard spheres into perfect crystals gives an accurate description of the stability and the thermodynamics of the hard-sphere (HS) solid (6). The extension of the above formalism to the freezing of particles interacting with more realistic potentials has however encountered difficulties (7-8) which have been overcome by means of different theoretical approaches. We are concerned here with two approximate theories for the description of the phase diagram of systems interacting with a pairwise potential consisting of a repulsive part at short distances and an attractive part at long distances. The prototype of one such interaction is the Lennard-Jones (LJ) potential which is known to provide a good description of the phase diagram of simple atomic substances such as argon (9). The development of new experimental techniques for the preparation of model colloids has provided a growing area of research in which new phase diagrams have been observed. One of the main advantages of colloidal dispersions as compared to atomic systems is that in the former case the range and depth of the interactions can be controlled.

0097–6156/96/0629–0297$15.00/0

In this chapter we consider the modifications occurring in the phase diagram of a simple fluid when the range of the attractive part is strongly reduced. Our theoretical predictions agree with the experimental results found in monodisperse suspensions of spherical colloidal particles (10) and with the simulation results performed on systems with short-ranged attractions (11,12).

The chapter is organized as follows. We first review two nonperturbative density functional theories, the modified weighted density approximation (5) and the generalized effective liquid approximation (6). Their implementation to the freezing of hard spheres, soft potentials and adhesive hard spheres is briefly summarized. We then consider the depletion potential in colloidal dispersions and the resulting phase diagrams obtained by reducing the range of the attractions. We finally develop two simple theoretical approaches for the description of these new phase diagrams.

Density Functional Theory of Freezing

In the density functional theory of freezing (13-15) the local equilibrium density of the solid is determined by minimizing at constant average density the variational Helmholtz free energy of the solid, $F[\rho]$, which is a unique functional of the local density $\rho(\mathbf{r})$. $F[\rho]$ can be split into an ideal part, $F_{id}[\rho]$, and an excess term originating from the particle interactions, $F_{ex}[\rho]$, i.e.:

$$F[\rho] = F_{id}[\rho] + F_{ex}[\rho] \tag{1}$$

where the ideal contribution is expressed as:

$$\beta F_{id}[\rho] = \int d\mathbf{r}\, \rho(\mathbf{r})\, [\ln(\Lambda^3 \rho(\mathbf{r})) - 1] \tag{2}$$

where $\beta = 1/k_B T$ is the inverse temperature and Λ is the thermal de Broglie wavelength, whereas the excess term is given by:

$$\beta F_{ex}[\rho] = -\int d\mathbf{r}\, \rho(\mathbf{r}) \int d\mathbf{r}'\, \rho(\mathbf{r}') \int_0^1 d\lambda\, (1 - \lambda)\, c(\mathbf{r}, \mathbf{r}'; [\lambda\rho]) \tag{3}$$

In equation 3, $c(\mathbf{r}, \mathbf{r}'; [\lambda\rho])$ denotes the direct correlation function of the solid and λ $(0 < \lambda < 1)$ is a parameter defining a linear path of integration in the space of density functions $\rho_\lambda(\mathbf{r}) = \lambda\rho(\mathbf{r})$ connecting the zero reference density to the local density of the solid. The variational principle involves the direct correlation function of the solid which is the only unknown in equations 2-3 and hence some explicit approximation for the excess contribution $F_{ex}[\rho]$ is required.

Although a variety of forms for $F_{ex}[\rho]$ have been proposed in the literature (13-15) we will only consider here the modified weighted density approximation (MWDA) of Denton and Ashcroft (5) and the generalized effective liquid approximation (GELA) of Lutsko and Baus (6). Both approximations are based on the similarity of the thermodynamic properties of the solid and fluid phases to map the excess free energy per particle of the solid $f_{ex}[\rho] \equiv F_{ex}[\rho]/N$, where $N = \int d\mathbf{r}\rho(\mathbf{r})$ is the number of particles, onto that of an effective uniform fluid:

$$\beta f_{\text{ex}}[\rho] = \beta f_{\text{ex}}(\hat{\rho}) = -\hat{\rho} \int d\mathbf{r} \int_0^1 d\lambda \, (1 - \lambda) \, c(|\mathbf{r}|; \lambda\hat{\rho}) \tag{4}$$

with $f_{\text{ex}}(\hat{\rho})$ and $c(|\mathbf{r}|; \lambda\hat{\rho})$ denoting the free energy per particle and the direct correlation function of a uniform fluid, respectively, and $\hat{\rho}$ being the effective liquid density which is used to represent the solid of density $\rho(\mathbf{r})$.

In the MWDA the effective liquid density $\hat{\rho}$ is defined as a doubly averaged solid density with a certain weighting function (see (5) for details) whereas in the GELA $\hat{\rho}$ is determined from the structural mapping (see (6) for details). In both approximations a self-consistent equation for the determination of $\hat{\rho}$ in terms of the local density of the solid $\rho(\mathbf{r})$ and the direct correlation function of the liquid $c(|\mathbf{r}|; \rho)$ is obtained. The resulting equations for the determination of $\hat{\rho}$ can be simplified if the local density of the solid is parametrized as a sum of normalized Gaussians centered about the lattice sites:

$$\rho(\mathbf{r}) = \left(\frac{\alpha}{\pi}\right)^{3/2} \sum_j e^{-\alpha(\mathbf{r}-\mathbf{r}_j)^2} \tag{5}$$

where α denotes the inverse width of the Gaussians and the sum runs over the sites of the Bravais lattice. With this approximation, the effective liquid density is an ordinary function $\hat{\rho}(\rho, \alpha)$ of the average density of the solid, ρ, and of the Gaussian width parameter, α, and hence the variational solid free energy per particle can then be written as:

$$\beta \tilde{f}(\rho, \alpha) = \frac{3}{2} \ln\left(\frac{\alpha}{\pi}\right) + 3 \ln \Lambda - 1 + \beta f_{\text{ex}}(\hat{\rho}(\rho, \alpha)) \tag{6}$$

where the asymptotic large-α form of the ideal free energy has been used. The equilibrium solids are found by minimizing equation 6 with respect to α at constant average density.

Hard spheres. The implementation of the density functional theory of freezing for the HS solid can be easily undertaken since analytic expressions for the structure (the direct correlation function) and the thermodynamics (the excess free energy) of the HS fluid are known from different theories (16). For the structure of the fluid phase the solution of the Percus-Yevick (PY) equation is usually considered, whereas the thermodynamics is obtained by integration of either the Carnahan-Starling (CS) or the PY equation of state. In Table I we compare the coexistence data for the freezing of hard spheres into a perfect fcc solid, using the CS equation of state for the fluid phase, as obtained from the MWDA, the GELA, and by Monte Carlo (MC) simulations (17). ρ_F and ρ_S denote the fluid and solid coexisting densities, P is the pressure at coexistence, σ is the HS diameter, and L is the Lindemann ratio. It is seen that both theories gives good estimates for the coexisting densities, the best prediction for the pressure at coexistence being that of the GELA. In the two approaches the Lindemann ratio is underestimated, expressing that they predict a stronger localization of the hard spheres in the solid phase than that obtained by MC simulations.

Table I. HS fcc solid-fluid coexistence data

	$\rho_F \sigma^3$	$\rho_S \sigma^3$	$\beta P \sigma^3$	L
MC	0.943	1.041	11.7	0.126
MWDA	0.909	1.035	10.1	0.097
GELA	0.945	1.041	11.9	0.100

Soft interactions. In view of the accurate description of the first-order fluid-solid transition for hard spheres, a natural question concerns the extension of the density functional theory of freezing to continuous interactions. Such a program was undertaken by Laird and Kroll (8) by considering the freezing properties of particles interacting with repulsive inverse-power potentials:

$$V(r) = \epsilon \left(\frac{\sigma}{r} \right)^n \tag{7}$$

and with the Yukawa potential:

$$V(r) = \epsilon \frac{e^{-\kappa r}}{r} \tag{8}$$

Equation 7 reproduces the HS potential for $n = \infty$, σ being the HS diameter, and the Coulomb potential for $n = 1$, i.e.: a one-component plasma of ions of charge $e^2 = \epsilon\sigma$. For intermediate n-values, computer simulations have shown that for $n > 6$ there is a fluid-fcc transition whereas for $n \leq 6$ there is a fluid-bcc transition followed at high densities by a bcc-fcc transition (18).

Equation 8 is a simplified model of the interactions in monodisperse charge-stabilized colloids. These interactions are described by the Derjaguin-Landau-Verwey-Overbeek (DLVO) (19,20) potential which is determined by electrical double layer repulsion and van der Waals attraction. When the attractive contribution is neglected and the point-particle limit of the repulsive contribution is taken, the DLVO potential reduces to Equation 8. Computer simulations have shown that particles interacting with the Yukawa potential can also freeze into a fcc or a bcc crystal (21-23). The bcc (fcc) phase is stable at low (high) inverse screening lengths κ and there exists an structural bcc-fcc transition at low temperatures.

The question is therefore how the density functional theories work for systems where the thermodynamically stable solid phase is not necessarily the close packed structure. The results of Laird and Kroll indicate that the freezing of inverse-power and Yukawa interactions into a fcc crystal as described by the MWDA leads to an overestimation of the liquid-solid densities at coexistence and that the error increases with the range of the potential. On the other hand, the MWDA fails in predicting any fluid-bcc transition. The situation worsens for the GELA because no stable or metastable fcc or bcc solids are found for inverse-power potentials, i.e.: there is no minimum in the variational solid free energies as a function of the inverse width of the Gaussians α for any value of the solid density.

The different phase transitions for particles interacting with the pair potentials (7) and (8) can however be described using a HS perturbation theory (24) for the solid phase similar to the one used for real fluids. The theory requires as input the fcc and bcc HS density functional theory results and gives a good agreement with the computer simulations for soft spheres (*24*) and for the Yukawa potential (*25*).

Adhesive hard spheres.   As shown in the previous section, the straightforward application of the MWDA and the GELA to soft potentials has encountered difficulties. A particular pair potential which can be situated in between the HS potential and the real interactions is the adhesive hard sphere (AHS) interaction, first introduced by Baxter (*26*). The AHS pair potential can be understood by considering particles interacting with a square well (SW) potential:

$$V(r) = \begin{cases} \infty & 0 < r < \sigma \\ -\epsilon & \sigma < r < \sigma(1 + \gamma) \\ 0 & \sigma(1 + \gamma) < r \end{cases} \tag{9}$$

consisting of a HS part (σ being the HS diameter) and an attractive ($\epsilon > 0$) well of width $\gamma\sigma$ and depth ϵ. As shown by Baxter, the direct correlation function and the equation of state of the fluid phase can be obtained analytically within the PY approximation if one takes in equation 9 the limit of zero width ($\gamma \to 0$) and infinite depth ($\epsilon \to \infty$) in such a way as to keep finite (but nonzero) the contribution of the well to the second virial coefficient. By denoting this finite value by $1/4\tau$, where τ is known as the adhesiveness parameter, the second virial coefficient of the AHS model can be written as:

$$B_2^{AHS} = B_2^{HS} \left(1 - \frac{1}{4\tau}\right) \tag{10}$$

where $B_2^{HS} = 2\pi\sigma^3/3$ is the second virial coefficient of hard spheres. When $\tau \to \infty$ the AHS model reduces to the HS system, while for finite values of τ the AHS model can be viewed as a HS system plus an infinitely narrow attractive well.

The fcc solid-liquid coexistence in the AHS model within the MWDA was analyzed by Zeng and Oxtoby (*27*) and by Marr and Gast (*28*). By increasing the interactions (lowering τ) the fluid-solid coexistence was found at higher densities and correspondingly large values of α, resulting in smaller values of the effective liquid density. At low temperatures there is no solution for the effective liquid density $\hat{\rho}$ for the fcc lattice at $\tau = 1.3$ (*28*).

Although the GELA gives results similar to the MWDA, we consider here an important feature of the AHS model which has been described in detail elsewhere (*29*). At high temperatures ($\tau > 3$) the solid free energy is a convex function of the density. At $\tau = 3$ a transition occurs in such a way that for $\tau < 3$ the solid free energy becomes a concave function of the density; that is, for $\tau < 3$ the solid becomes mechanically unstable (negative compressibility). This instability of the AHS solid at low temperatures suggests the formation of an incipient van der Waals loop similar to the one found for the fluid phase using a mean-field theory. Within the GELA it is not possible to complete the loop, because the effective liquid density becomes negative at high densities. Note that if we were able to complete the van der Waals loop of the solid free energy, we would find, by performing Maxwell's double tangent construction on such a loop, a solid-solid transition. This solid-solid transition would correspond to the coexistence of two solids with the same symmetry but differing only in their density. The resulting solid-solid transition could therefore end in a critical point because both solids have the same symmetry.

Colloidal Dispersions

The addition of non-adsorbing polymer to a suspension of colloidal particles can cause an effective attraction between the particles by the depletion mechanism. This phenomenon can be interpreted in terms of a volume restriction whereby the exclusion of polymer particles between two neighboring colloidal particles produces a net attraction between them. Let us assume for simplicity that the colloidal particles and the polymer molecules are hard spheres of diameters σ and σ', respectively. On an isolated colloidal particle the polymer suspension exerts a uniform isotropic osmotic pressure Π. But if two colloidal particles approach each other so that the center-to-center separation r is smaller than $\sigma + \sigma'$, polymer molecules will be excluded from a well-defined region between the particles. The resulting effect is an unbalanced osmotic pressure driving the particles together. Integration of this osmotic pressure over the portion of available surface area of the two particles gives rise to the depletion potential, which can be expressed as (30):

$$V(r) = \begin{cases} \infty & 0 < r < \sigma \\ -\Pi\,\Omega(r; \sigma + \sigma') & \sigma < r < \sigma + \sigma' \\ 0 & \sigma + \sigma' < r \end{cases} \quad (11)$$

where $\Omega(r; x)$ is the volume of the overlapping depletion zones, i.e.:

$$\Omega(r; x) = \frac{\pi x^3}{6}\left[1 - \frac{3r}{2x} + \frac{r^3}{2x^3}\right] \quad (12)$$

The depletion potential (equation 11) consists therefore of a HS part plus an attractive part. The range of the attractive part is determined by σ', and provided the ratio $\psi = \sigma'/\sigma << 1$ (i.e. if the radius of gyration of the polymer is smaller than the radius of the colloidal particle) we have a picture similar to the AHS model but, in this case, the interparticle potential is separately tunable in range and depth. The range is determined by the molecular weight of the added polymer while the depth is essentially controlled by the polymer concentration.

Intermediate attractions. It has been shown experimentally (10) that for large size ratios ($\psi \simeq 0.6$) the phase diagram is similar to the LJ phase diagram of simple atomic substances with a liquid-vapor critical point and a solid-liquid-vapor triple point. For intermediate size-ratios ($\psi \simeq 0.33$), the vapor-liquid coexistence region is very small and the critical point almost merges into the triple line. Finally, for size ratios $\psi \leq 0.3$ the liquid phase disappears and the only effect of adding polymer is to expand the solid-fluid coexistence region. These experimental results agree with previous theoretical findings by Gast et al. (*31*) and Lekkerkerker et al. (*32*).

As shown above, a system of hard spheres does not have a vapor-liquid critical point. Therefore, it has been widely assumed that a necessary condition for the critical point is the presence of attractive interactions. In atomic systems, where the range of the attractions is larger or comparable to the range of the repulsions, the prevailing situation is a LJ-like phase diagram. In colloidal systems, where the range of the attractions can be experimentally reduced, we have seen that the presence of attractive interactions is not a sufficient condition for the appearance of a liquid phase.

Computer simulations have been performed to investigate the existence of a liquid phase in the phase diagram. For hard spheres with an attractive Yukawa potential it has been shown (*33*) that when the range of the attractive part is $\simeq \sigma/6$, with σ being the HS diameter, the liquid-vapor coexistence curve disappears (is metastable). For C_{60} molecules, which are represented by spheres interacting via LJ potentials summed over all 60 carbon atoms, it has also been shown by MC simulations that there is no liquid phase(*34*). In this case, the pair potential of C_{60} differs significantly from the LJ form, since the ratio of the width of the attractive part to the diameter of the repulsive core of the potential is much less for C_{60} than for noble gases.

Recently, on the basis of a van der Waals theory (see below) the relation between the nature of the phase diagram and the pair potential has been investigated yielding a necessary and sufficient condition for the occurrence of a liquid phase (*35*).

Short-ranged attractions.   Once the liquid phase disappears, a further reduction of the range of the attractions leads to a new phase diagram obtained by MC simulations (*11,12*) which however has not yet been observed experimentally. This new phase diagram exhibits an isostructural solid-solid transition and is in many ways the specular image of the LJ phase diagram. The model investigated by Bohluis and Frenkel is a SW potential which, although simple, provides an adequate description for uncharged colloidal particles (see equation 11).

An intuitive argument for the appearance of the isostructural solid-solid transition was given by Bolhuis and Frenkel by comparing the situation of an expanded solid close to melting and a dense solid near close packing. Using the ideas of a simple uncorrelated cell model in which each particle moves independently in the cell formed by its neighbors, they considered the two following situations: 1) If the width of the attractive well γ is smaller than the radius of the cell R, a given particle can have at most three neighbors within the range γ, though the average will be far less. 2) If $R < \gamma$ then a particle interacts with all its nearest neighbors simultaneously. For the latter situation the potential energy is much lower than for the former, and at low temperatures this decrease of the energy will overweigh the loss of entropy originated by the decrease of the free volume. Therefore, the free energy will have an inflection point leading to a first-order transition to a dense solid. For large-γ values, the solid-solid transition disappears (is metastable) because it is preempted by the melting transition. Indeed, the resulting diagram corresponding to a SW potential with a very narrow width should be similar to the one found for AHS. In order to compare both systems, Bolhuis and Frenkel defined a parameter τ related to the second virial coefficient of SW particles as:

$$B_2^{SW} = B_2^{HS}\left(1 - \frac{1}{4\tau}\right) \tag{13}$$

By comparing the phase diagram obtained for AHS (*29*) to the SW phase diagram, they found that al low τ the solid-solid transition crosses the melting line of the AHS model. This implies that, in this case, the expanded solid is no longer stable and the fluid coexists with the dense solid. This could explain why the AHS solid becomes mechanically unstable at low τ.

Theoretical Approaches

Two theoretical approaches for the description of the isostructural solid-solid transition are analyzed in this section. In both studies some general aspects for the determination of the phase diagram of a simple fluid have been used. They can be summarized as follows. Suppose that we construct a (mean-field) theoretical model for the determination of the fluid and solid free energies of a system of particles interacting with a central pair-potential $V(r)$, consisting of a repulsive part at short distances plus an attractive part at long distances. At high temperatures, both free energies will be convex functions of the density. At low temperatures, each free energy will develop a van der Waals loop separating the free energy into a low-density branch and a high-density branch. As usual, a Maxwell's double tangent construction on such loops will be performed. For the fluid phase the two branches will correspond to the vapor-liquid transition, whereas for the solid phase they will correspond to the isostructural solid-solid transition. Needless to say, the critical temperatures below which the free energies become concave functions of the density are different for the fluid and the solid phases. Notice, moreover, that it is always possible to perform a double tangent construction between the fluid branch and the solid branch to obtain a fluid-solid transition. In order to obtain the phase diagram of the system, we will finally construct the convex envelope of the total free energy in order to separate the stable from the metastable phase transitions.

As it will be shown below the different phase diagrams (from atomic to colloidal systems) described in the preceding sections can be obtained from these two simple theoretical approaches. Both studies start with the Gibbs-Bogoliubov inequality (*15*):

$$F \leq \tilde{F} = F_0 + \frac{1}{2}\int d\mathbf{r}_1 \int d\mathbf{r}_2\, \rho_0(\mathbf{r}_1, \mathbf{r}_2)[V(r_{12}) - V_0(r_{12})]. \tag{14}$$

which gives an upper bound, $\tilde{F}$, of the Helmholtz free energy of a system, F, in terms of the Helmholtz free energy, F_0, and the pair density, ρ_0, of a reference system. $V(r)$ is the pair potential of the system and $V_0(r)$ that of the reference system.

A simple model.   In the first approach, the reference system for the fluid and solid phases and the pair potential are chosen in such a way that $\tilde{F}$ can be determined analytically *(36,37)*. The best estimate for F can then be obtained by minimizing $\tilde{F}$ with respect to a free parameter of the reference system. To this end, we consider the following double-Yukawa (DY) pair potential $V(r) = \epsilon\phi(r)$, ϵ being a measure of the strength of the interactions, and

$$\phi(r) = c\frac{\sigma}{r} \left[e^{-a(r/\sigma-1)} - e^{-b(r/\sigma-1)} \right] \tag{15}$$

where a, b and c are positive parameters ($a > b$). In principle, the DY potential contains three arbitrary parameters which can be varied independently. But, if we keep the value of $V(r)$ at the minimum constant, this condition reduces equation 15 to a two-parameter potential. Instead of two of the three original parameters we will use two more intuitive ones: the position of the minimum, r_0, and $\delta = (r_1 - r_0)/r_0$ characterizing the range of the attractive part, with r_1 ($r_1 > r_0$) denoting the distance for which the potential has dropped to 1% of its value at the minimum. In Table II we gather the parameters r_0/σ and δ corresponding to six representative cases to be considered below, the range of the attractions being denoted by L (long-ranged), I (intermediate) and S (short-ranged). The value of the potential at the minimun has been chosen arbitrarily so that $V(r_0/\sigma) = -1$.

Table II. Parameters defining the DY potential

range	$r_0/\sigma - 1$	δ
L	0.12	1.34
L	0.10	0.79
I	5.0×10^{-3}	3.28×10^{-2}
I	2.5×10^{-3}	1.65×10^{-2}
S	2.0×10^{-3}	1.30×10^{-2}
S	1.5×10^{-3}	1.05×10^{-2}

For the reference fluid phase we consider a HS fluid since $\int dr g_{\text{HS}}(r) V_{\text{HS}}(r) = 0$, with $g_{\text{HS}}(|\mathbf{r}_1 - \mathbf{r}_2|) = \rho_{\text{HS}}(|\mathbf{r}_1 - \mathbf{r}_2|)/\rho^2$ denoting the pair-correlation function of the fluid and ρ the uniform density, whereas the combination of $g_{\text{HS}}(r)$ with the DY potential leads to an expression involving the Laplace transform of $r g_{\text{HS}}(r)$ which, within the PY approximation, is known analytically *(38)*. This allow us to evaluate exactly $\tilde{F}$ for the fluid phase which is then minimized with respect to the free parameter $\lambda = \sigma_{\text{HS}}/\sigma$, where σ_{HS} denotes the HS diameter.

For the reference solid phase we choose an Einstein solid of particles bounded harmonically to the sites of the fcc lattice by a spring constant $K/2$. For this ideal reference system we have $\rho_0(\mathbf{r}_1, \mathbf{r}_2) = \rho(\mathbf{r}_1)\rho(\mathbf{r}_2)$, with $\rho(\mathbf{r})$ denoting the local density which, for the Einstein solid, has the form of a sum of normalized Gaussians (equation 5) with $\alpha = \beta K/2$. It can be easily shown that for this simple case $\tilde{F}$ can also be determined exactly for the solid phase. The best estimate for the solid free energy can then be found by minimizing $\tilde{F}$ with respect to α.

van der Waals theory. A more general approach for studying the solid-solid transition consists of developing a van der Waals theory for the solid phase similar to the original van der Waals theory for fluids (*39*). The starting point is again the Gibbs-Bogoliubov inequality but now we consider the following approximations: 1) The pair potential $V(r)$ is split into a HS reference part $V_{HS}(r)$ and an attractive part $V_A(r)$. 2) Correlations are neglected within the domain of $V_A(r)$, and 3) The upper bound of the free energy is taken as the estimate of F. These simplifications reduce the Gibbs-Bogoliubov inequality to the equation:

$$F = F_{HS} + \frac{1}{2} \int d\mathbf{r}_1 \int d\mathbf{r}_2\, \rho_{HS}(\mathbf{r}_1)\rho_{HS}(\mathbf{r}_2)V_A(r_{12}). \tag{16}$$

where $\rho_{HS}(\mathbf{r})$ denotes the local density of the HS reference system. Note that equation 16 reproduces the well-known van der Waals equation for the fluid phase in the uniform limit $\rho_{HS}(\mathbf{r}) \rightarrow \rho$:

$$f = f_{HS} + \frac{1}{2}\rho \int d\mathbf{r}V_A(r) \tag{17}$$

where $f = F/N$ and $f_{HS} = F_{HS}/N$. The free energy of the HS fluid f_{HS} can be obtained by thermodynamic integration of the equation of state, which is taken here to have the simple form:

$$\frac{\beta P_{HS}}{\rho} = \left[1 - \frac{\rho}{\rho_0}\right]^{-1} \tag{18}$$

where P_{HS} is the pressure of the HS fluid and ρ_0 is the maximum density for which the HS fluid can exist.

For the solid phase we assume that $V_A(r)$ is continuous (the case where $V_A(r)$ is discontinuous is somewhat more complicated and will not be considered here (*39*)). Due to the strong localization of the particles in the HS crystal, we approximate the local density $\rho_{HS}(\mathbf{r})$ in equation 16 by a delta function, yielding:

$$f = f_{HS} + \frac{1}{2}\sum_j V_A(r_j) \tag{19}$$

where the sum runs over the lattice sites, r_j being the distance of site j to an arbitrary site taken as origin. In this case f_{HS} will be also prescribed by adopting a simple equation of state for the solid phase, which has the simple form of a free-distance approximation:

$$\frac{\beta P_{HS}}{\rho} = \left[1 - \left(\frac{\rho}{\rho_{cp}}\right)^{1/3}\right]^{-1} \tag{20}$$

where P_{HS} is the pressure of the HS solid and ρ_{cp} is the fcc close-packing density. Before constructing the phase diagrams in the van der Waals approach, the relative position of the HS free energies of the fluid and solid phases have to be fixed. This is done by imposing that the HS free energies change their stability at the density predicted by the GELA (*6*). This leads to a condition for the determination of ρ_0 while ρ_{cp} is determined by the given crystal structure.

Results. The two theoretical approaches we have just described give the same qualitative results for the phase diagrams of simple fluids. We will be limited here to the consideration of the phase diagrams obtained from the simple model. As expected, the nature of the phase diagram depends on the ratio of the range of the attractive interactions to the range of the repulsions, i.e. on δ. Let $t = 1/\beta\epsilon$ be the dimensionless temperature and $\rho\sigma^3$ the dimensionless density. In Figures 1,2, and 3 we plot in the $(t, \rho\sigma^3)$-plane the stable phase transitions corresponding to the six cases gathered in Table II. The following situations occur:

Long-ranged attractions (Figure 1). There are two stable phase transitions: a vapor-liquid phase transition ending in a critical point, and a fluid-solid phase transition. The solid-solid transition is therefore metastable. For $\delta = 1.34$ (full dots and continuous lines) the phase diagram corresponds to that of the well-known LJ potential. Note that by reducing the range of the attractions to $\delta = 0.79$ (open dots and dashed lines) the major effect is to lower the vapor-liquid critical point whereas the vapor-liquid-solid triple point changes very little. This situation corresponds to values typical of atomic systems.

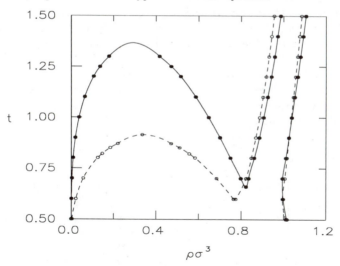

Figure 1. Phase diagram of the DY potential for long-ranged attractions.

Intermediate attractions (Figure 2). In this case both the vapor-liquid and the solid-solid transitions are metastable with respect to the fluid-solid transition. In such a diagram there are hence no critical or triple points. The main effect of reducing the range of the attractions from $\delta = 0.41$ (full dots and continuous lines) to $\delta = 3.28 \times 10^{-2}$ (open dots and dashed lines) is to lower the shoulder on the fluid side and to move the solid line of the coexistence to higher densities. This class of phase diagram is the one obtained experimentally by Ilett et al. (*10*) and theoretically in references (*31-32*) for colloidal dispersions.

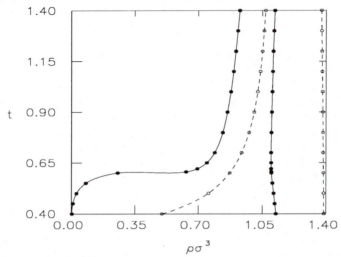

Figure 2. Phase diagram of the DY potential for intermediate attractions.

Short-ranged attractions (Figure 3).   There are two stable phase transitions: an isostructural solid-solid transition and a fluid-solid transition. The vapor-liquid transition is therefore metastable. By reducing the range of the attractions from $\delta = 1.65 \times 10^{-2}$ (full dots and continuous lines) to $\delta = 1.05 \times 10^{-2}$ (open dots and dashed lines) the major effect is the lowering of the fluid-solid-solid triple point while the solid-solid critical point remains practically constant. This situation corresponds to the MC simulations of Bolhuis et al. (*11,12*) for short-ranged SW potentials.

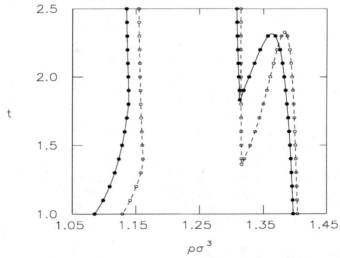

Figure 3. Phase diagram of the DY potential for short-ranged attractions.

Conclusions

We have reviewed the freezing of simple fluids and shown that the well-known phase diagram of atomic systems described in any elementary physics course is only one of the various types of phases diagrams which can be encountered in nature. It has been found from two simple theoretical approaches that when the range of the attractions relative to the range of the repulsions is reduced, new classes of phase diagrams develop. Such diagrams have been observed either experimentally or by MC simulations and could be of relevance to colloidal dispersions.

Acknowledgements

I would like to acknowledge M. Baus, H. N. W. Lekkerkerker and A. Daanoun for fruitful collaboration. I also thank J. M. R. Parrondo, R. Brito and M. S. Ripoll from the Universidad Complutense de Madrid for useful discussions and the Dirección General de Investigación Científica y Técnica (Spain) PB91-0378 for financial support.

Literature Cited

1. Ramakrishnan, T. V.; Yussouff, M. *Phys. Rev. B* **1979**, 19, 2775.

2. Haymet, A. D. J.; Oxtoby, D. W. *J. Chem. Phys* **1981**, 74, 2559.

3. Tarazona, P. *Mol. Phys* **1984**, 52, 81.

4. Curtin, W. A.; Ashcroft, N. W. *Phys. Rev. A* **1985**, 32, 2909.

5. Denton, A. R.; Ashcroft, N. W. *Phys. Rev. A* **1989**, 39, 4701.

6. Lutsko, J. F.; Baus, M. *Phys. Rev. A* **1990**, 41, 6647.

7. Barrat, J. L.; Hansen, J. P.; Pastore, G. *J. Chem. Phys* **1987**, 86, 6360.

8. Laird, B. B.; Kroll, D. M. *Phys. Rev. A* **1990**, 42, 4810.

9. Hansen, J. P.; Verlet, L. *Phys. Rev* **1969**, 184, 151.

10. Ilett, S. M.; Orrock, A.; Poon, W. C. K.; Pusey, P. N. *Phys. Rev. E* **1995**, 51, 1344.

11. Bolhuis, P.; Frenkel, D. *Phys. Rev. Lett.* **1994**, 72, 2211.

12. Bolhuis, P.; Hagen, M. H. J.; Frenkel, D. *Phys. Rev. E* **1994**, 50, 4880.

13. Baus, M. *J. Phys.: Condens. Matter* **1990**, 2, 2111.

14. Singh, Y. *Phys. Rep*, **1991**, 207, 351.

15. Evans, R. In *Fundamentals of Inhomogeneous Fluids*; Editor, D. Henderson. ; Marcel Dekker, Inc.: New York, 1992; pp 85-175.

16. Hansen J. P.; McDonald, I. R. *Theory of Simple Liquids*; Academic Press Inc.: London, 1986; pp 152-153, 532-536.

17. Hoover, W. G.; Ree, F. M. *J. Chem. Phys* **1968**, 49, 3609.

18. Hoover, W. G.; Gray, S. G.; Johnson, K. W. *J. Chem. Phys.* **1971**, 55, 1128.

19. Derjaguin, B.; Landau, L. *Acta Physicochim, URSS* **1941**, 14, 633.

20. Verwey, E. J. W.; Overbeek, J. G. *Theory of the Stability of the Lyophobic Colloids*; Elsevier: Amsterdam, 1948.

21. Robbins, M. O.; Kremer, K.; Grest, G. G. *J. Chem. Phys.* **1988**, 88, 3286.

22. Thimuralai, D. *J. Phys. Chem.* **1989**, 93, 5637.

23. Meijer, E. J.; Frenkel, D. *J. Chem. Phys.* **1991**, 94, 2269.

24 Lutsko, J. F.; Baus, M. *J. Phys. Condens. Matter* **1991**, 3, 6547.

25 Tejero, C. F.; Lutsko, J. F.; Colot, J. L.; Baus, M. *Phys. Rev. A* **1992**, 46, 3373.

26. Baxter, R. J. *J. Chem. Phys.* **1968**, 49, 2270.

27. Zeng, X. C.; Oxtoby, D. W. *J. Chem. Phys.* **1990**, 93, 2692.

28. Marr, D. W.; Gast, A. P. *J. Chem. Phys.* **1993**, 99, 2024.

29. Tejero, C. F.; Baus, M. *Phys. Rev. E* **1993**, 48, 3793.

30. Asakura, S.; Oosawa, F. *J. Polymer Sci.* **1958**, 33, 183.

31. Gast, A. P.; Hall, C. K.; Russel, W. B. *J. Colloid Interface Sci.* **1983** 96, 251.

32. Lekkerkerker, H. N. W.; Poon, W. C. K.; Pusey, P. N.; Stroobants, A.; Warren, P. B. *Europhys. Lett.* **1992**, 20, 559.

33. Hagen, M. H. J.; Frenkel, D. *J. Chem. Phys.* **1994**, 101, 4093.

34. Hagen, M. H. J.; Meijer, E. J.; Mooij, G. C. A. M.; Frenkel, D; Lekkerkerker, H. N. W. *Nature* **1993**, 365, 425.

35 Coussaert, T.; Baus, M. *Phys. Rev. E.* **1995**, 52, 862.

36. Tejero, C. F.; Daanoun, A.; Lekkerkerker, H. N. W.; Baus, M. *Phys. Rev. Lett.* **1994**, 73, 752.

37. Tejero, C. F.; Daanoun, A.; Lekkerkerker, H. N. W.; Baus, M. *Phys. Rev. E* **1995**, 51, 558.

38. Wertheim, M. S. *Phys. Rev. Lett.* **1963**,10, 321.

39. Daanoun, A.; Tejero C. F.; Baus, M. *Phys. Rev. E* **1994**, 50, 2913.

Chapter 22

Density-Functional Theory
from $\hbar$ = 0 to 1

Recent Classical and Quantum Applications to Aluminum Siting in Zeolites and the Freezing of Simple Fluids

Shepard Smithline

Cray Research, Inc., 655 E. Long Oak Drive, Eagan, MN 55121

Quantum and classical density functional theory have become important tools for describing many body phenomena in physics and chemistry. In this paper we review density functional theory, emphasizing the quantum and classical connections of the two theories. We show that both versions of the theory write the density functional as the sum of an ideal term and a term arising from inter-particle interactions. By differentiating the functional with respect to the density, it is straightforward to derive an equation that minimizes the functional. In the application of the quantum theory to electronic structure, the value of the functional at the minumum is the energy, while in the classical theory, the minimum value of the functional is the free energy or grand potential, depending on the thermodynamic conditions at hand. In addition, two examples of the theory are presented. The first, an electronic structure application, is to aluminum siting in zeolites, while the second, an application of the classical theory, is to the freezing of simple fluids. The electronic structure theory is able to predict the optimal location of aluminum in the zeolite cage, while the classical theory, despite some notable successes, is, on the whole, less successful in describing the liquid- solid transition. The reasons for the apparent shortcomings of the classical theory are discussed. Finally, it is speculated that the weighted density formalism, because it provides a means for constructing non-local functionals, might provide a framework for deriving improved functionals which is important for the further development of the theory.

Density functional theory has become ubiquitous in physics and chemistry. The theory has its origins in the Hohenberg-Kohn theorems which apply to quantum and classical systems (1). This allows the theory to be applied to a wide range of phenomena, ranging from nucleation in liquid-vapor systems (2) to the electronic structure of atoms and molecules (1). This chapter discusses density functional theory, emphasizing some of the parallels between the classical and quantum versions of the theory. In addition,

0097–6156/96/0629–0311$15.00/0

we present two applications of the theory, one to aluminum siting in zeolite structures (3) and the other to the freezing of simple liquids (4).

The paper is organized as follows. In section II, we discuss some of the general quantum-classical connections in density functional theory, paying particular attention to the Hohenberg-Kohn theorems which form the foundation of density functional theory (1). Section III reviews the Kohn-Sham equations of electronic structure theory[1] and applies them to a zeolite cluster, which models a heterogenous catalyst used in hydrocarbon separation (3). The fourth section shows how the classical theory can be applied to the freezing of simple fluids and is used to study the transition in a simple hard sphere fluid. Section V summarizes the weighted density functional formalism (5), describing how this idea can be used in quantum and classical systems to construct improved functionals. Finally, in section VI we conclude.

Quantum-Classical Connections

Two theorems proved by Hohenberg and Kohn provide the formal basis of density functional theorm (6),(7). The first theorem states that the external potential uniquely specifies the density of the system and vice versa: the density specifies the external potential of the system. Here density can refer to a quantum or classical density at constant particle number, temperature, and volume (canonical ensemble), or constant chemical potential, temperature and volume (grand canonical ensemble). The first Hohenberg-Kohn Theorem implies that once the density ρ is known everything that can be known is known about the system. For instance, in electronic structure theory, once the positions of the nucleii are given (the external potential in this case), the first Hohenberg-Kohn Theorem tells us that the electron density is determined. Since the density fixes the total number of electrons, we can write the complete N electron Hamiltonian and solve, in principle, for the many body wavefunction. Analogously, in the classical theory, fixing the external potential allows us to specify the canonical distribution function for any given system which has a particular interaction potential. The distribution function, in turn, allows us to calculate the equilibrium average of any observable.

The second Hohenberg-Kohn theorem provides the theoretical underpinning for determining the density. It states for a given interaction potential and external potential $V_{ext}(r)$ there exists an energy functional F[p] which is minimized when $\rho = \rho_{equil}$. Moreover, at the minimum, this functional equals the energy, free energy, or grand potential, depending on the thermodynamic conditions at hand.

The proofs of the Hohenberg-Kohn theorems are well known (1), so we do not present them here. Both the classical and quantum proofs rely on the functional

$$F[f] = \mathrm{Tr}\, f\left(A + \beta^{-1}\ln f\right) \qquad (1)$$

which has the property that when $f = f_o$, the equilibrium distribution function, F[f] takes on its lowest value. This can be proved simply by considering the difference $F[f] - F[f_o]$ and showing that this difference is greater than zero for all $f \neq f_o$ using a Gibbs inequality (8). Since the proof uses the standard properties of the trace, the trace may be a classical canonical trace, the classical grand canonical trace, or their quantum counterparts. Furthermore, the proofs utilize the fact that the equilibrium distribution function is the exponential of an energy divided by the trace of f, so f may refer to the classical or quantum distribution function.

The zero temperature quantum case, corresponding to the conditions under which most quantum chemistry calculations are performed, is a special case of 1. As

$\beta \to \infty$, F becomes the energy

$$F \to E = < \psi_{Ni} |\hat{H}| \psi_{Ni} > .$$ (2)

According to the variational theorem, the energy takes on its lowest value when $\psi_{Ni} = \psi_{Ni}^{0}$, the ground state wavefunction. Thus, the variational theorem play the role of the Gibbs function at zero temperature and we can make the following analogies:

1. Energy   < - - - >   Free energy
2. The expectation value   < - - - >   Trace
3. The ground state wavefunction ψ_{Ni} < - - - >   The distribution function f
4. The variational theorem   < - - - >   The Gibbs theorem

Clearly, the zero temperature Hohenberg-Kohn theorems are closely related to their finite temperature cousins (1).

Functionals are generally constructed by writing them as the sum of two parts, an ideal term, which ignores inter-particle interactions and is specified exactly, plus a term which accounts for the interaction and is known approximately. For instance, in electronic structure theory, the reference system is a collection of non-interacting electrons (1) and one writes

$$E[\rho] = T_s[\rho] + \text{"}corrections\text{"}$$ (3)

where $T_s[\rho]$ is the kinetic energy of a system of non-interacting electrons and the "correction" is $J[\rho] + E_{xc}[\rho]$, $J[\rho]$ being the classical coulomb potential and $E_{xc}[\rho]$ being the exchange correlation potential. This last term includes the quantum exchange terms left out of $J[\rho]$ and the kinetic energy corrections to $T_s[\rho]$. We show below that $J[\rho] + E_{xc}[\rho]$ plays the role of a mean field potential which generates the density.

Similarly, for the classical case one often writes (8)

$$F[\rho] = F_{Ideal}[\rho] + \varphi[\rho].$$ (4)

The expression F_{ideal} is the free energy of a non-interacting system. In applications to simple fluids, the reference system a monatomic gas. The second term in equation 4, $\varphi[\rho]$, is the contribution to the free energy due to interactions and like $J[\rho] + E_{xc}[\rho]$, gives rise to an effective one body potential which determines the density.

Interestingly, not all density functional theories construct functionals by writing the functional as an ideal plus interacting piece. The geometric measures theory of Rosenfeld is one such example. In effect, it writes the free energy functional by interpolating between low and high density limits. We describe this theory in connection to hard sphere freezing in section IV.

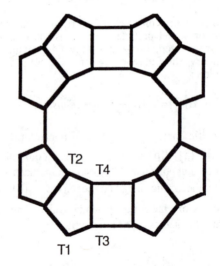

Figure 1.   Definition of Al substitution sites.

Electronic Structure Application

Kohn-Sham Theory. Almost all electronic structure calculations using density functional theory are carried out within the Kohn-Sham formalism (1). In this section we briefly review this approach and show how we can apply the formalism to an important problem in catalysis: the relative energetics of aluminum siting in zeolites (1).

The Kohn-Sham method starts with the functional

$$E[\rho] = T_s[\rho] + J[\rho] + Exc[\rho].$$ (5)

By writing the density as $\rho = \sum_i |\psi_i|^2$ and differentiating equation 5 with respect to the density, subject to the constraint that the Kohn-Sham orbitals ψ_i remain orthonormal, one can derive the Kohn-Sham equations

$$\left(-\frac{1}{2}\nabla_i^2 + V_{eff}\right)\psi_i = \varepsilon_i\psi_i$$ (6)

where ε_i is a Lagrange multiplier resulting from the constraints on the orbitals and V_{eff} is the effective one body potential that generates the density.

The exchange-correlation potential, which is included in the effective potential, is often computed using the local density approximation (LDA). LDA breaks space into small regions of nearly constant density. Since the exchange correlation energy for a uniform electron gas is known, the exchange correlation energy for the entire system can be computed by quadrature. The LDA approximation does surprisingly well, and is discussed below and by other authors in this volume.

Application to Zeolites. We illustrate quantum density functional theory by studying zeolites (3). Zeolites are important catalysts, particularly in the petroleum industry, where they play a critical role in the catalytic cracking of large hydrocarbon molecules. The catalytic activity of zeolites is known to depend largely on their acidic properties which arise when aluminum atoms are replaced by silicon atoms. Consequently, it is of interest to know the location of aluminum atoms in the zeolite structure. Ab initio computations have predicted geometries and energies of a large number of compounds, but catalytic compounds have generally been too complex to study by such methods. To overcome these computational limitations, semi-empirical orbital calculations can be used to calculate the preferred aluminum position; however these methods require the selection of various parameters whose accuracy has not been determined for these compounds. As a result, we were lead to investigate the distribution of Al in mordenite using density functional theory. Calculations on two different clusters were performed:

    (i) A 39 atom cluster with formula $Si_7O_{20}H_{12}$

    (ii) A 75 atom cluster, $Si_{14}O_{39}H_{22}$

in which the Si was replaced by Al at four different locations, known as T sites, and the relative energetics of the substitution was determined. Figure 1 shows the locations of these sites. T1 and T2 reside on rings with 5 tetrahedral atoms (eg. Al or Si) while T3 and T4 are four atom rings.

In addition to choosing the cluster geometry, a method must be chosen to truncate the cluster in such a way which preserves valency and charge distribution. This was accomplished by adding hydrogen atoms to the cluster to terminate the bonds.

Bonds lengths of .98 A for O-H and 1.43 for Si-H were used. The geometries were taken from the de-hydrated structures of Schlenker, Pluth and Smith (9). All calculations were performed with DGauss, a density functional program developed by Cray Research, as part of the UniChem quantum chemistry package.(9) Calculations were performed with a valence double-zeta basis set with polarization (DZVP); these basis sets are comparable in size to 6-31G*, but have been optimized for DFT. Becke-Perdew corrections to the local density approximation were added self-consistently. Table I below shows that T2 is predicted to be the most stable site (3).

Table. I   The relative energies of the different clusters for the four different T sites. All energies are measured in kcal/mole.

Cluster	T1 Site	T2 Site	T3 Site	T4 site
39 atom cluster	24.6	0.0	11.3	13.9
74 atom cluster	10.10	0.0	4.12	16.18

Our prediction that T2 is the preferred site is consistent with experiments (10),(11), though some ambiguity remains. Bodart et. al. found that their experimental results are consistent with assuming that T3 and T4 are preferentially occupied, a result which was confirmed by previous computational studies on smaller clusters using Hartree Fock theory (12). In contrast, Itabishi drew different predictions from adsorption and NMR studies on synthetic mordenites. They argued that T2 site is a plausible location of Al which is consistent with our results.   One effect that we have not completely investigated is the role that crystal relaxation plays in determining the relative ordering of the various sites. We are currently investigating this effect by allowing the aluminum and the nearest neighbor oxygen atoms to relax in the 74 atom cluster. Our preliminary results indicate that T2 remains the lowest energy site. We hope to report the results of this calculation in a future work.

Classical Statistical Mechanics Application

In contrast to quantum density functional theory, which is most often applied to questions of electronic structure, the classical theory is applied to a much wider variety of systems, ranging from liquid - vapor nucleation to polymeric systems. The intermolecular interactions in these systems are far more complex than in electronic structure problems, and because one usually does not know the exact Hamiltonian for these systems, relatively simple models are used to construct functionals. The results of these calculations are then compared to computer simulations. While the comparison to simulation may be quantitative, the real value of these theories is the qualitative understanding they often provide of actual many body systems. Here we apply the classical theory to the freezing of simple liquids.

Freezing Theory. The starting point for developing a theory of first order phase transitions, such as freezing, is the Legendre transform of equation 11, or the grand potential functional,

$$\Omega[\rho] = \int d\vec{r}\rho(\vec{r})\, V_{ext}(\vec{r}) + F_{ideal}[\rho(\vec{r})] - \phi[\rho(\vec{r})] - \mu \int d\vec{r}\rho(\vec{r}) \tag{7}$$

where μ is the chemical potential   Here the reference system is an ideal monatomic fluid whose free energy is given by a simple analytic expression. Note, that working in the grand ensemble allows us to conveniently equate the chemical potential of the liquid and solid, one of the conditions which must be satisfied at the transition point.

Differentiating equation 7 with respect to $\rho(\bar{r})$, and setting this functional derivative to zero yields

$$\rho(\bar{r}) = \lambda^{-3} \exp[\beta(\mu + V_{ext}(\bar{r}) + C[\rho(\bar{r})])] \tag{8}$$

where

$$C[\rho(\bar{r})] = \delta\phi/\delta\rho. \tag{9}$$

Note, that equation 20 shows explicitly that the density is determined by an effective one body potential, analogous to V_{eff} in Kohn-Sham theory.

It is straightforward to use this approach to study the freezing transition.(13) The derivation is given in reference 14. The final result is

$$\ln[\rho_s(\bar{r})/\rho_L] = \int d\bar{r}' \, C[\bar{r}; \mu_s, T_s] - C[\mu_L, T_L] - \varepsilon \tag{10}$$

where

$$C[\bar{r}; \mu_s, T_s] - C[\mu_L, T_L] = \int d\bar{r}' c^2_L(|\bar{r} - \bar{r}'|)[\rho_s(\bar{r}) - \rho_L][\rho_s(\bar{r}') - \rho_L]$$
$$+ \int d\bar{r}d'\bar{r}'' c^3_L(|\bar{r} - \bar{r}'|, |\bar{r} - \bar{r}''|)[\rho_s(\bar{r}) - \rho_L][\rho_s(\bar{r}') - \rho_L][\rho_s(\bar{r}'') - \rho_L] + \dots \tag{11}$$

Equation 9 is the fundamental equation of the freezing theory and admits non-trivial solutions for a set of chemical potentials and temperatures. The solution corresponding to the freezing point is identified as that solution where the pressures of the two phases, calculated to the same order in perturbation theory, are equal. When the equality of pressures is satisfied, the solutions of 9 are guaranteed to generate a thermodynamically consistent transition point, since 9 was derived assuming that the temperatures and chemical potentials of the two phases are equal.

Equation 9 is analogous to equation 6 in Kohn-Sham theory, as both arise by minimizing a functional with respect to a density subject to a constraint - the perfect crystal constraint in classical theory and the orthonormality of the Kohn-Sham orbitals in the quantum theory. Just as in Kohn-Sham theory, equation 9 is solved by expanding $\rho(\bar{r})$ in a basis, such as Gaussians centered on lattice sites at the solid positions, R_i.

$$\rho(\bar{r}) = \sum_i (\alpha/\pi)^{3/2} \exp[-\alpha(r - R_i)^2] \tag{12}$$

or trigonometric functions,

$$\rho(\bar{r}) = \sum_{n=0} \mu_n e^{i\bar{k}_n \cdot \bar{r}} \tag{13}$$

summed over reciprocal lattice vectors k_n. The particular lattice vectors or reciprocal lattice vectors are chosen to describe the symmetry of the particular solid one wishes to study (eg. fcc, hcp) . The order parameters (α or μ_n) are determined by substituting equations 12 or 13 into 9 and solving the resulting algebraic equation. Equation 12 restricts the solid to isotropic and harmonic oscillations about the lattice sites while 13 lifts this approximation, being a general representation for a periodic structure.

Expanding the solid density in a basis illustrates one of the central assumptions of the theory. One must assume the symmetry of the solid; the theory does not predict the symmetry of the solid, though it does predict the magnitude of the lattice vectors. Currently, there is no a priori way of addressing the issue of spontaneous symmetry breaking in classical density functional theory.

Another major assumption concerns expansion 11. Even if the expansion (4) is carried out to infinite order, a functional Taylor expansion like 11 suppresses fluctuations by prohibiting configurations in which some regions are liquid-like and others to be solid-like. This assumption is known as the homogeneity approximation and is central in all mean field theories of phase transitions. Like the assumption of spontaneous symmetry breaking, we know of no way of addressing this approximation.

Application to the Freezing of Hard Spheres. The freezing theory has been applied to a wide variety of problems, ranging from systems with purely repulsive forces, such as hard spheres and inverse power potentials, to systems with attractive forces, such as Lennard-Jones fluids and the one-component plasma, to more complex systems such as water and polymers (4). The results are some what inconsistent. For hard spheres the theory performs well, yielding a reasonably accurate freezing density and the density change on freezing, for instance (4). For moderately complex systems, such as inverse power potentials, the theory does less well, failing for instance to predict the bcc phase $\epsilon(\sigma/r^n)$, $n \le 6$ where ϵ and σ measure the strength and characteristic length of the interaction (14). For more complex fluids, the theory exhibits "re-entrant" behavior, successfully predicting, for instance, the freezing of water (15) and polyethylene (16).

Given this somewhat inconsistent performance, we re-examined the theory. One of the critical assumptions of the classical theory is the truncation of expansion 11 at second order. While the higher order terms are not thought to be small, it is often assumed the integral of them times the corresponding density difference is. This assumption is usually not checked because the evaluation of these terms is very difficult. Recently, however, a theory for constructing free energy functionals has been proposed and allows the higher order terms to be computed by functional differentiation of $F[\rho(r)]$ (17). When the triplet term is computed for hard spheres, it is found to agree quite well with computer simulations, and thus, it is natural to see how this term affects the results of the freezing theory. Before we present these results, we begin with a brief derivation of this Rosenfeld's free energy functional. Our discussion closely follows reference 18, and the reader is encouraged to consult it for more details.

Rosenfeld postulated the following general form for the excess free energy functional (that part over and above the ideal contribution) as,

$$F_{ex}[\{\rho_i(\vec{r})\}] = \int d^3\vec{r} \; \phi[\{n_m(\vec{r}), n_q(\vec{r})\}] = \int d^3\vec{r} \; \phi[\{n_\alpha(\vec{r})\}], \qquad (14)$$

where $n_\alpha(\vec{r})$ is a weighted density

$$n_\alpha(\vec{r}) = \int d^3\vec{r}' \; \rho_i(\vec{r}') \, \omega(\vec{r} - \vec{r}'), \qquad (15)$$

where $\rho_i(\vec{r})$ is the density of species i. Rosenfeld gives the explicit forms for $\omega_i^\alpha(\vec{r}-\vec{r}')$. They depend on the geometrical properties of the spheres and consist of vector and scalar functions. Equation 14 allows us to write the excess pressure function, $\Pi[\{n_\alpha\}]$, and chemical potentials, μ_{ex} and μ_{id}, as

$$\Pi = -\phi + \sum_\alpha n_\alpha \partial \phi / \partial n_\alpha \tag{16}$$

$$\mu_i^{ex} = \int d^3\vec{r}' \sum_\alpha \partial \phi / \partial n_\alpha [\{n_\gamma(\vec{r})\}] \omega_i^\alpha(\vec{r}-\vec{r}') \tag{17}$$

$$\mu_i^{id} = \ln[\rho_i(\vec{r})\lambda_i^3] \tag{18}$$

Although it is probably not readily apparent at this point, the assumption that the excess free energy density is a functional of only the fundamental-measures weighted density, restricts the final form of $\phi[\{n_\alpha\}]$ considerably. If we impose the exact relation for the uniform chemical potential $\mu_i \rightarrow PV_i$ for $R_i \rightarrow \infty$, on our fundamental measure description of the non-uniform fluid, the following differential equation must be satisfied,

$$\Pi + n_0 = \partial \phi / \partial n_3 \tag{19}$$

This relation can be derived by observing that the scalar and vector weights respectively satisfy

$$\int d^3\vec{r}' \omega_i^m(\vec{r}') = 1, \ R_i, \ S_i, \ V_i \tag{20}$$

$$\int d^3\vec{r}' \vec{\omega}_i^q(\vec{r}') = 0, \tag{21}$$

where R_i, S_i, and V_i are the radius, surface area, and volume of sphere i. Consequently for homogeneous fluids, equation 17 can be re-written as

$$\mu_i^{ex} = \delta\phi / \delta n_0 + \delta\phi / \delta n_1 R_i + \delta\phi / \delta n_2 S_i + \delta\phi / \delta n_3 V_i \tag{22}$$

Now since $P_{id} / kT = \sum_i \rho_i$, where P_{id} is the ideal pressure, then

$$V_i \left[\frac{P_{id}}{kT} \right] = V_i \left[\int d^3\vec{r} \sum_i \rho_i \ \delta(|\vec{r}|-R_i)/(4\pi R_i^2) \right] = V_i n_0 \tag{23}$$

since ω_i^0 is defined as $\omega_i^0(\vec{r}-\vec{r}') = \delta(|\vec{r}'|-r)/(4\pi r^2)$. Therefore,

$$PV_i = V_i \{n_0 + \Pi\} \tag{24}$$

and imposing $\mu_i \rightarrow PV_i$ yields,

$$V_i \Pi + V_i n_0 = \ln[\rho(\bar{r})\lambda^3] + \delta\phi/\delta n_0 + \delta\phi/\delta n_1 R_i + \delta\phi/\delta n_2 S_i + \delta\phi/\delta n_3 V_i \;. \tag{25}$$

Upon dividing by V_i and letting $R_i \to \infty$, only $\delta\phi/\delta n_3$ survives, giving the scaled particle differential equation 19. Using 19 we can write equation 16 as,

$$-\phi + \sum_\alpha n_\alpha \partial\phi/\partial n_\alpha + n_0 = \partial\phi/\partial n_3 \tag{26}$$

Furthermore, the only positive integer power combinations which have units of V^{-1} (as required by the virial theorem) are

$$n_0, \; n_1 n_2, \; \left(n_2\right)^2, \; \bar{n}_1.\bar{n}_2, \; n_2(\bar{n}_1.\bar{n}_2), \tag{27}$$

and therefore form a basis for ϕ,

$$\phi = f_0 n_0 + f_{12} n_1 n_2 + f_{222}\left(n_2\right)^2 + f_{12}\bar{n}_1.\bar{n}_2 + f_{222} n_2(\bar{n}_1.\bar{n}_2) \;. \tag{28}$$

The functions f are dimensionless, and for generality are allowed to depend on n_3, a dimensionless quantity. Substituting equation 28 into 26 results in five differential equations which allow the f's to be determined. The boundary conditions of the differential equations are chosen so that equation 28 reproduces the low density expression for the free energy density and the three particle diagram for $c^2(r)$. As a result it can be shown that,

$$\phi = \phi_s + \phi_v \tag{29}$$

where

$$\phi_s = -n_0 \ln(1-n_3) + n_1 n_2/(1-n_3) + 1/(24\pi)n_2^3/(1-n_3)^2 \tag{30}$$

and

$$\phi_v = \bar{n}_1.\bar{n}_2/(1-n_3) + 1/(8\pi)n_2(\bar{n}_2.\bar{n}_2).(1-n_3)^2 \tag{31}$$

Now the m-th order direct correlation functions are given by

$$c_{i_1\dots i_m}^m(\bar{r}_1, \bar{r}_2, \dots \bar{r}_m) = -\frac{\delta F_{ex}[\{\rho_i(\bar{r})\}]/kT}{\delta\rho_{i_1}(\bar{r})\dots\delta\rho_{i_m}(\bar{r})}$$

$$= \int d^3\bar{x} \sum_{\alpha_1\dots\alpha_m} \left\{\partial^m\phi/\partial n_{\alpha_1}\dots\partial n_{\alpha_m}\right\}\omega^{\alpha_1}(\bar{r}_1-\bar{x})\dots\omega^{\alpha_m}(\bar{r}_m-\bar{x}) \tag{32}$$

If we fourier transform $c_{i_1\dots i_m}^m$ and note that in the uniform liquid $\partial^m\phi/\partial n_{\alpha_1}\dots\partial n_{\alpha_m}$ are independent of position, then in k-space the direct correlation function is simply a linear combination of weight functions

$$\tilde{c}_{ij}^2(k) = -\sum_{\alpha\beta} \partial^2 \phi / \partial n_\alpha \partial n_\beta \, \tilde{\omega}_i^\alpha(k) \tilde{\omega}_j^\beta(k) \tag{33}$$

Equation 33 can be inverted to yield the real space representation which can be shown to be equivalent to the Percus-Yevick direct correlation function. Similarly for the triplet function we can write,

$$\tilde{c}^3(k_1, k_2, k_3) = -\sum_{\alpha,\beta,\gamma} \partial^3 \phi / \partial n_\alpha \partial n_\beta \partial n_\gamma \tilde{\omega}_i^\alpha(k_1) \tilde{\omega}_j^\beta(k_2) \tilde{\omega}_j^\gamma(k_3) \delta(k_1 + k_2 + k_3) \tag{34}$$

There are a total of 33 terms in expression 34, explicit expressions for which are given in Rosenfeld's paper. Our derivation of 34, which is in excellent agreement with computer simulation results and relies on the scaled particle differential equation, shows how one can construct a free energy functional that effectively interpolates between the low and high density results.

Now, returning to the third order contribution to the free energy, we note that the triplet function's contribution to the free energy can be evaluated (18) as

$$f_{ex}^3 = \sum_{\vec{k}_1, \vec{k}_2, \vec{k}_3} c^{(3)}(k_1, k_2, k_3) \mu_{k_1} \mu_{k_2} \mu_{k_3} \tag{35}$$

where the sum is over all reciprocal lattice vectors $\vec{k}_i$ of the solid subject to the triangle condition: $\vec{k}_3 = -\vec{k}_1 - \vec{k}_2$. The coefficients μ_k are the fourier coefficients of a Gaussian solid density. The results, shown in Table II, indicate that the third order term is nearly as large as the second order term, and as a result, there is no a priori way of justifying truncation of equation 11 at second order.

Table II. The second and third order contribution to the free energy functional divided by $kT\rho_L \sigma^3$ of fcc hard spheres (of diameter σ) of average solid density $\rho_s \sigma^3$ and gaussian width $\alpha\sigma^3$ as a function of liquid density, solid density and gaussian width:
a. $\rho_L \sigma^3 = .79$, $\rho_s \sigma^3 = .975$, $\alpha\sigma^3 = 180$; b. $\rho_L \sigma^3 = .975$, $\rho_s \sigma^3 = .975$, $\alpha\sigma^3 = 180$;
c. $\rho_L \sigma^3 = .79$, $\rho_s \sigma^3 = .79$, $\alpha\sigma^3 = 50$; d. $\rho_L \sigma^3 = .975$, $\rho_s \sigma^3 = .975$, $\alpha\sigma^3 = 50$
For comparison the results of Curtin-Ashcroft (reference 22) are also shown.

	Second-Order Terms	Third-Order Terms (us)	Third-Order Terms (Curtin-Ashcroft)
a	-3.8857	-1.6502	-1.841
b	-3.2576	-4.2856	
c	-1.5802	-1.5918	
d	-2.1094	-1.588	

These results strongly suggest that the apparent agreement between the second order theory freezing theory and computer simulation data is fortuitous in sense that one does not understand *why* truncating at second order should result in a good description of the liquid-solid transition. While we performed calculations only for hard spheres, we expect our results to be qualitatively correct for more realistic fluids which contain attractive as well as repulsive forces. The reason for the apparent failure of expansion equation 11 is explored in section V.

Non-local Functionals

One of the common assumptions of LDA theory and the freezing theory discussed above is that the free energy functional is a local functional of the density. That is, the density is evaluated at a point in space and from the value of the density at that point in space, the free energy is evaluated. However, we know quantum exchange is inherently non-local. For example within Hartree-Fock theory exchange operator $K_b(\vec{r}_1)$ is given by

$$K_b(\vec{r}_1)\chi_a(\vec{r}_1) = \left[\int d^3\vec{r} \, \chi_b(\vec{r}_2) r_{12}^{-1} \chi_a(\vec{r}_2) \right] \chi_b(\vec{r}_1) \tag{36}$$

which is clearly non-local. Similarly, we expect the free energy functional for the classical freezing theory to have non-local character. The origin of the classical non-locality can be understood from

$$F = \int d^3\vec{r}\rho(\vec{r})\Psi(\rho(\vec{r})) \tag{37}$$

where $\Psi(\rho(\vec{r}))$ is the free energy per-particle in a homogeneous system. This expression for the free energy implies that a particle at $\vec{r}$ is only affected by particles around it in a given range of interaction. If the range of the particle interactions is smaller than the length scales over which the density varies, then it is often reasonable to break the system up into small pieces, each one of nearly constant density, evaluate the free energy of each piece as if it were part of a homogeneous system, and add them up (integrate them) to get the total free energy. This prescription is followed in LDA theory. However, this approach is of limited use in the liquid -solid transition. Since the density of the solid reaches large peak values, the value of the free energy functional of the corresponding homogeneous system, is almost certain to be quite different the free energy functional of the real system. In fact, the density may even be impossible to achieve, as is the case for hard spheres, when the density exceeds the close-packing limit.

It is possible to imagine other schemes which employ a homogenous reference system to evaluate the free energy of an inhomogeneous system, but, if the density is not smooth on the relevant length scales, then a local density approximation is likely to break down. A fluid at a first order phase transition or up against a hard wall are clearly such cases. Thus, even if we could sum the series of equations 11 to beyond third order, it is not clear, a priori, that this would result in a good description of the liquid-solid transition, since expansion equation 11 is an inherently local approximation.

Still the notion of using a homogeneous system to describe inhomogeneous systems is appealing, provided one can introduce non-locality into the functional. One way to construct a non-local functional is to use a weighting functional. This idea was first used in electronic structure density functional theory,(19) and subsequently used by researchers in classical liquid theory (20). The essential idea is to replace $\rho(\vec{r})$ in equation 7 by a weighted density,

$$\overline{\rho}(\vec{r}) = \int d^3\vec{r}' \, w\left(\vec{r},\vec{r}', \overline{\rho}(r)\right)\rho(\vec{r}) \tag{38}$$

Note that Rosenfeld's scalar weights are a special case of 38 for which we set

$$w\left(\vec{r},\vec{r}',\ \overline{\rho}\ (r)\right)=\omega(\vec{r},\vec{r}') \tag{39}$$

There are two general ways of choosing the weight functions. One is to select them by a priori means, as done by Rosenfeld. The other is to "tailor" the weight function to reproduce some known properties of the homogeneous fluid. There are disadvantages to both techniques. If we adopt Rosenfeld's theory, then the weighted density cannot easily be interpreted as an effective local density due to the presence of the vector weight functions, as the weighted density now becomes a vector quantity not a simple scalarfunction. On the other hand, if we tailor the weight functions to mimic some property of the homogeneous fluid then it is possible to derive a relation for $w\left(\vec{r},\vec{r}',\rho\ (r)\right)$. For instance, if we use the full weight function then one can derive a differential equation which relates the weight function to the pair correlation function (21). Alternatively, if one is restricted to scalar weighted densities which are simple linear averages of the density (as in equation 15), one can derive weight functions, though they are non-trivial, as they involve derivatives of delta functions (22). The "tailored" functionals also have been applied to study the freezing of simple fluids (4) and while in some cases they give better answers, they still are not free from artifacts (4),(17). Interestingly enough, had we used the Rosenfeld weighting functions, and determined ϕ_{ex} by tailoring it to reproduce the Percus-Yevick direct correlation function instead of using the scaled particle differential equation, we would not have found Rosenfeld's vector component of ϕ_{ex} and for certain values of the k vectors, the vector part of the triplet function contributes about half the value of the value of the total function. As a result of the somewhat arbitrary nature of deriving weighted density functionals, it is clear that the idea needs to be further refined before it can be used reliably to describe the liquid-solid transition.

It should also be pointed out that the weighted density approximation was abandoned in electronic structure theory because other approaches proved to be more reliable (23). The original implementation of the weighted density idea used the random-phase approximation to compute the exchange-correlation functional for a homogeneous electron gas (21). However, if one could tabulate the exact pair function for a homogeneous electron gas, similar to the way one tabulates the exchange-correlation energy of the electron gas, then the RPA approximation could be lifted. As a result, the weighted density approximation, when combined with the various scaling relations used to construct the gradient corrected non-local functionals, might lead to new and more accurate models for the exchange-correlation energy.

Conclusion

Density functional theory has its origins in the Hohenberg-Kohn theorems. Since these theorems apply equally well to quantum and classical systems, density functional theory can be applied to problems in quantum and classical mechanics. Given the theories' common origin, it is not surprising that there are many similarities between the quantum and classical versions of the theories, and in this paper we discussed some of these similarities. We showed that both the quantum and classical versions of the theory often write the appropriate energy functionals the sum of an ideal and an interaction term. By minimizing the functional with respect to the density, it is straightforward to derive an equation for the density where the density is determined by a mean field potential arising from the inter-particle interactions.

Besides discussing quantum-classical analogies in density functional theory, we illustrated the theory by presenting two applications, one to aluminum siting in zeolites and the other to the freezing of classical fluids. The quantum version of the theory is able to predict the optimal location of aluminum in the zeolite cage while the classical theory, despite some notable successes, is less successful in describing the liquid- solid transition. The apparent shortcomings of the classical theory reflect the inherent difficulty of developing a first principles theory for first order phase transitions. Indeed, we argued that a density functional theory of freezing probably requires a sophisticated non-local free energy functional. Unfortunately, it is not at all clear how to construct such a functional. One possibility is to use a weighted density formalism. Besides being potentially useful in classical theory, this approach might also be helpful in deriving newnon-local functionals for quantum density functional theory, although additional work needs to be done to further develop this idea. Nevertheless, whatever the outcome of future research into weighted densities, we are optimistic that, given the past successes of density functional theory, the theory will continue to be an exciting and useful formalism for describing many body phenomena, be they quantum or classical systems.

Acknowledgements

The author is happy to acknowledge the many chemists at Cray Research, including George Fitzgerald, Rich Graham, Chengthe Lee, and Eric Stahlberg, who taught him quantum density functional theory. The author also thanks Cray Research, Inc. for providing the computer resources used to carry out some of the calculations reported here.

Literature Cited

1.　　Parr, R.; Yang, W.　*Density Functional Theory of Atoms and Molecules*; Oxford University Press: New York, NY 1989.

2.　　Oxtoby, D.W. ; this volume.

3.　　Carpenter, J.E., Fitzgerald, G., Eades, R.E. *Computer Aided Innovation of New Materials II* **1993**, 1035.

4.　　Singh, Y. *Physics Reports.* **1992**, *207*, 351 and references cited therein.

5.　　Denton, A.R.; Ashcroft, N.W. *Phys. Rev A.* **1989**, *39*, 4701 and references cited therein

6.　　Hohenberg, P.; Kohn, W. *Phys. Rev B.* **1964**, *136*, 864.

7.　　Mermin, N. D. *Phys. Rev A.* **1965**, *137,* 1441.

8.　　Evans, R.; *Adv. in Phys.* **1979**, *28*, 143.

9.　　Korminicki, A.; Fitzgerald, G. *J. Chem. Phys.* **1993**, *98*, 1398.

10.　Itabash, K.; Okada, T.; Iagawa, K. In *New Developments in Zeolite Science and Technology*; Murakami, Y.; Iijma, A.; Ward, J. W.; Eds.; Kodansha: Tokyo, 1969.

11.　Bodart, P.; Nagy, J.B.; Debras, G.; Gabelica, Z., Jacobs, P.A. *J. Phys. Chem.* **1986**, *90*, 5183.

12.　Derone, E. G.; Fripiat, J. G. In *Proceedings of the Sixth International Zeolite Conference*; Olson, D.; Bisio, A., Eds.; Butterworths: U.K., 1984.

13.　Haymet, A.D.J.; Oxtoby, D.W. *J. Chem Phys.* **1986**, *84*, 1769.

14.　Laird, B.B.; Kroll, D.M. *Phys Rev. A.* **1990**, *42*, 4810.

15. Ding, K.; Chandler, D.; Smithline, S.J.; Haymet, A.D.J. *Phys. Rev. Lett.* **1987**, *59*, 1698.

16. McCoy, J.D; Nath,S., this symposium.

17. Rosenfeld, Y; Levesque D.; Weis, J.J. *J. Chem. Phys,* **1990** *92*, 6818.

18. Smithline, S. J.; Rosenfeld, Y. Phys. Rev. A. **1990**, *42*, 2434.

19. Gunnarsson, O.; Jones, M.; Lundqvist, B.I. *Phys. Rev. B* **1979** *20*, 3136.

20. Hansen, J.P.; McDonald,I.R. *Theory of Simple Liquids;* Academic Press: New York, NY, 1986.

21. Curtin, W. A.; Ashcroft, N.W. *Phys. Rev A.* **1985** *32*, 2909.

22. Kierlik, M E.; Rosinberg, L. *Phys. Rev. A.* **1990**, *42*, 3382.

23. Becke, A. D. *J. Chem. Phys.* **1992**, *96*, 2157 and references cited therein.

FURTHER APPLICATIONS
AND THEORETICAL DEVELOPMENT
IN ELECTRONIC STRUCTURE
DENSITY-FUNCTIONAL THEORY

Chapter 23

The Calculation of NMR Parameters by Density-Functional Theory

An Approach Based on Gauge Including Atomic Orbitals

Georg Schreckenbach, Ross M. Dickson, Yosadara Ruiz-Morales, and Tom Ziegler[1]

Department of Chemistry, University of Calgary, 2500 University Drive Northwest, Calgary, Alberta T2N 1N4, Canada

Schemes for calculating nuclear magnetic resonance (NMR) shielding tensors and spin-spin coupling constants have been implemented in the Amsterdam Density Functional program system (ADF). The shielding tensors are calculated by the Gauge Including Atomic Orbitals (GIAO). This method and the calculation of the coupling constants are tested for a number of smaller molecules and it is shown that calculations of couplings to transition-metal nuclei and shielding tensors in metal complexes are feasible.

1. Introduction.

Nuclear magnetic resonance (NMR) is used extensively (1) as a practical tool in chemical research. Many of its applications can be carried out based on a simple effective Hamiltonian in which the observed shifts and spin-spin coupling constants are used as parameters without any further interpretation. However, an understanding of how electronic and geometrical effects influence these parameters has not been established in detail except for a few classes of compounds (1a-c), although such an understanding might enhance the amount of useful information obtained from NMR experiments.

Computational methods based on molecular orbital theory can in principle provide the required insight (1a-c), and the comparison between calculated and observed NMR spectra might further help in the identification of new species. With this in mind, several first principle methods capable of calculating NMR parameters have appeared over the last decade (1a-c).

Density functional theory (DFT) (2) forms the basis for some of the approaches used in computational studies of the shielding tensor $\vec{\vec{\sigma}}$. Recent advances in DFT have made it possible to use this approach for shielding calculations. Malkin *et al.* have published a series of pioneering papers on the calculation of NMR properties, including shielding (3a-g) and spin-spin coupling (3h). To calculate the shielding, they combine modern DFT with the "individual gauge for localized orbitals" (IGLO) method (4).

We have recently presented a method in which the NMR shielding tensor is calculated by combining the "Gauge Including Atomic Orbital" (GIAO) (5) approach

[1]Corresponding author

0097–6156/96/0629–0328$15.00/0

with DFT. Our implementation makes full use of the modern features of DFT in terms of accurate exchange-correlation (XC) energy functionals and large basis sets (6).

Of particular interest -- and still a challenge -- are applications in multi-nuclear NMR (1c, d), e.g., transition metal NMR. We present here results for carbonyl complexes $M(CO)_6$ (M= Cr, Mo, W).

In the present paper we address as well the calculation of nuclear spin-spin coupling constants. Spin-spin coupling constants are very difficult to compute. In the non-relativistic formulation of Ramsey (7) there are four terms, each of which has different requirements with respect to correlation and basis set. Malkin *et al.* (3a,g) provided the first practical implementation of these calculations in a density functional program, using a basis of Gaussian-type basis functions. In this paper we illustrate an implementation using a basis of Slater-type rather than Gaussian functions, and examine the feasibility of calculating spin-spin coupling constants in transition-metal systems.

2. The Calculation of Shielding Tensors.

Shielding tensors and the GIAO-DFT method. The details of the GIAO-DFT method have already been described previously (6a,b). However, we will have to stress a few points about NMR in general and the GIAO formalism in particular to facilitate the discussion in the next sections.

In NMR one considers the interaction energy between a nuclear magnetic moment $\vec{\mu}_N$ in an electronic system and an external homogeneous magnetic field $\vec{B}_o$. The presence of $\vec{B}_o$ will induce an internal magnetic field $\vec{B}_{\text{int}} = \vec{B}_d + \vec{B}_p$ in the electronic system so that the total interaction energy is given by

$$E = -\vec{\mu}_N \cdot (\vec{B}_d + \vec{B}_p + \vec{B}_o)$$
$$= -\vec{\mu}_N \cdot (1 + \vec{\vec{\sigma}}) \cdot \vec{B}_o \qquad (1)$$
$$= -\vec{\mu}_N \cdot (1 + \vec{\vec{\sigma}}^d + \vec{\vec{\sigma}}^p) \cdot \vec{B}_o$$

where

$$\vec{B}_d = -\vec{\vec{\sigma}}^d \cdot \vec{B}_o$$
$$\vec{B}p = -\vec{\vec{\sigma}}^p \cdot \vec{B}_o \qquad (2)$$

Here $\vec{\vec{\sigma}} = \vec{\vec{\sigma}}^p + \vec{\vec{\sigma}}^d$ are referred to as the shielding tensors; one third of the trace of $\vec{\vec{\sigma}}$ is the shielding constant. The vector $\vec{B}_d$ represents the diamagnetic component of the induced field. It is in most cases opposite to $\vec{B}_o$ with $\vec{B}_o \cdot \vec{B}_d < 0$ for any orientation of $\vec{B}_o$, and thus the diamagnetic shielding tensor $\vec{\vec{\sigma}}^d$ must according to eq. 2 have a positive trace and positive symmetrical diagonal components. The paramagnetic part of the induced field $\vec{B}_p$ points in the direction of $\vec{B}_o$ with $\vec{B}_o \cdot \vec{B}_p > 0$ for any orientation of $\vec{B}_o$. In this case the paramagnetic shielding tensor $\vec{\vec{\sigma}}^p$ must have a negative trace and negative symmetrical diagonal components according to eq. 2. The diamagnetic shielding $\vec{\vec{\sigma}}^d$ depends only on the unperturbed electron density, ρ^0, while the paramagnetic shielding, $\vec{\vec{\sigma}}^p$, contains the density up to first order with respect to the external magnetic field.

Then, the st-component of the diamagnetic tensor $\vec{\vec{\sigma}}^d$ is given in our GIAO formalism as (6)

$$\sigma_{st}^d = \alpha^2 \sum_i^{occ} n_i \left\{ \left\langle \Psi_i \left| \sum_v c_{vi}^o \frac{1}{2r_N^3} \left[\vec{r}_N \cdot \vec{r}_v \delta_{st} - \vec{r}_{N_s} \cdot \vec{r}_{v_t} \right] \chi_v \right\rangle \right. \right.$$

$$\left. + \alpha^2 \sum_{\lambda,v} c_{\lambda i}^o c_{vi}^o \left\langle \chi_\lambda \left| \left[\frac{\vec{r}_v}{2} \times \left(\vec{R}_v - \vec{R}_\lambda \right) \right]_s \left(\frac{\vec{r}_N}{r_N^3} \times \vec{\nabla} \right)_t \right| \chi_v \right\rangle \right\} \quad (3)$$

Here the occupied molecular orbital (MO) Ψ_i has been expanded in terms of atomic functions χ_v and the coefficients c_{vi}^o. The zero superscript indicates that Ψ_i is calculated with zero magnetic field strength, $\vec{B}_o = 0$, as an eigenfunction to the Kohn-Sham operator $h(o)$ with the eigenvalue ε_i^o. The vector $\vec{r}_v \equiv \vec{r} - \vec{R}_v$ denotes the position of the electron relative to the nucleus at which the atomic orbital χ_v is centered, and $\vec{r}_N$ is the position of the electron relative to the nuclear magnetic moment under consideration (relative to the NMR nucleus N). Further α is the dimensionless fine structure constant given as 1/137. The diamagnetic shielding tensor $\vec{\vec{\sigma}}^d$ of eq. 3 is gauge invariant and an expectation value of Hermitian operators. It depends only on the unperturbed occupied orbitals for which $n_i \neq 0$; this has been pointed out before.

The st-component of the paramagnetic tensor $\vec{\vec{\sigma}}^p$ is given according to the GIAO formalism as

$$\sigma_{st}^p = \alpha^2 \sum_i^{occ} n_i \left\{ \sum_{\lambda,v} c_{\lambda i}^o c_{vi}^o \left\langle \chi_\lambda \left| \frac{1}{2} \left(\vec{R}_\lambda \times \vec{R}_v \right)_s \left(\frac{\vec{r}_N}{r_N^3} \times \vec{\nabla} \right)_t \right| \chi_v \right\rangle \right.$$

$$\left. + \sum_j^{occ} \alpha S_{ij}^{(1,s)} \left\langle \Psi_i \left| \left[\frac{\vec{r}_N}{r_N^3} \times \vec{\nabla} \right]_t \right| \Psi_j \right\rangle + \sum_a^{unocc} \alpha u_{ai}^{(1,s)} \left\langle \Psi_i \left| \left[\frac{\vec{r}_N}{r_N^3} \times \vec{\nabla} \right]_t \right| \Psi_a \right\rangle \right\} \quad (4)$$

where

$$u_{ai}^{(1,s)} = \frac{F_{ai}^{(1,s)} - \varepsilon_i^{(0)} S_{ai}^{(1,s)}}{\varepsilon_i^{(0)} - \varepsilon_a^{(0)}} \quad \text{with } i \neq a \quad (5),$$

$$S_{pq}^{(1,s)} = \alpha \sum_{\lambda,v} c_{\lambda p}^{(0)} c_{vq}^{(0)} \left\langle \chi_\lambda \left| \left[\frac{\vec{r}}{2} \times \left(\vec{R}_v - \vec{R}_\lambda \right) \right]_s \right| \chi_v \right\rangle \quad (6),$$

and

$$F_{pq}^{(1,s)} = \alpha \sum_{\lambda,v} c_{\lambda p}^{(0)} c_{vq}^{(0)} \left\{ \left\langle \chi_\lambda \left| \left[\frac{-\vec{r}_v}{2} \times \vec{\nabla} \right]_s \right| \chi_v \right\rangle + \left\langle \chi_\lambda \left| \left[\frac{\vec{r}}{2} \times \left(\vec{R}_v - \vec{R}_\lambda \right) \right]_s h(0) \right| \chi_v \right\rangle \right\} \quad (7),$$

The index i runs over orbitals occupied in the field free ground state and the index a runs over the corresponding unoccupied orbitals. The paramagnetic shielding tensor $\vec{\vec{\sigma}}^p$ of eq. 4 is also gauge invariant by itself and an expectation value of Hermitian operators. The leading contribution to the paramagnetic shielding is the last term in eq. 4. It represents the first order magnetic coupling between an occupied molecular orbital, i, and a virtual orbital, a. This coupling is facilitated by way of the first order coefficient $u_{ai}^{(1,s)}$, which is inversely proportional to the difference of the eigenvalues, eq. 5.

It is worth noting that the shielding is completely formulated in terms of the (occupied and virtual) MO's, eqs. 3-7. This is an advantage of the GIAO formalism as it allows for the detailed analysis of the different orbital contributions to the shielding tensors.

Computational details. The above established formalism has been implemented into the Amsterdam Density Functional Package (ADF) (8). All properties are evaluated using the given numerical integration scheme of ADF. We will use non-local XC energy functionals (9), unless otherwise stated. We employ uncontracted Slater Type Orbitals as basis functions. Our basis sets are generally of triple ζ quality in the valence region, and of double ζ quality for core MO's. They are augmented by two sets of (p or d) polarization functions per atomic center. The experimental geometries are the basis for all calculations.

GIAO-DFT calculation of shielding constants for simple molecules. Calculated shielding constants for a representative set of small molecules are collected in Table I.

TABLE I. Calculated and Experimental Shielding Constants for a Number of Small Molecules

Molecule	Atom	Isotropic Shielding Constants (ppm)		
		DFT-GIAO[a]	DFT-IGLO[b]	Experiment
CH_4	C	191.2	187.7	195.1
	H	31.4	31.2	30.6
CH_3F	C	111.4	101.4	116.8
	F	462.3	450.7	471.6
	H	27.2	26.7	26.6
H_2O	O	331.5	324.3	344.0
	H	31.2	31.1	30.1
N_2	N	-72.9	-78.9	-61.6
C_2H_2	C	110.4	108.9	117.2
	H	30.4	30.0	29.3
Benzene	C	50.0	48.8[c]	57.2
H_2CO	C	-15.7	-26.6	-8.4
	O	-418.8	-455.6	-312.1
	H	20.7	20.8	18.3
F_2	F	-282.7	-250.6	-232.8

[a] Reference 6a. [b] Uncoupled DFT-IGLO: Ref. 3b. We cite the results for the same LDA/NL functional as in our DFT-GIAO method. [c] Ref. 3e.

We compare our results with those obtained by the "uncoupled" DFT-IGLO of Malkin and co-workers (3a-c) as well as with experiment. The agreement with experiment is generally satisfactory. The GIAO shieldings are of about the same quality as those obtained by the IGLO method, in most cases even slightly better, Table I. A direct comparison is however not possible since this would require to use exactly the same basis sets.

The agreement with experiment is not as good for non-hydrogen shifts in molecules like H_2CO or F_2 (this applies to both the GIAO and the uncoupled IGLO methods). These molecules are difficult cases for DFT in general.

Calculated GIAO-DFT shielding anisotropies for simple molecules. The shielding anisotropy, i.e., the individual tensor components of the shielding tensor, should be even more sensitive to the quality of the quantum-chemical method used than the (averaged) shielding constant. Here, we define the shielding anisotropy $\Delta\sigma$ as the difference between the parallel and the orthogonal principle components

$$\Delta\sigma = \sigma_\parallel - \sigma_\perp \qquad\qquad (8)$$

In Table II, we compare shielding constants and anisotropies for another representative set of small molecules with experimental results. We can see that the quality of the averaged shielding constant and the tensor components is comparable for any given molecule. Thus, we get in general good agreement between theory and experiment. However, we note large deviations between the calculated and experimental shielding anisotropies for those molecules (notably CO out of the given list) where the calculated shielding constant isn't reliable either. Note that the magnitude of the anisotropy can exceed the shielding constant considerably, Table II.

Table II. Calculated and Experimental Shielding Constants and Anisotropies

Molecule / Atom		Shielding (ppm)			
		Isotropic Shielding		Shielding Anisotropy	
		DFT-GIAO[a]	Experiment	DFT-GIAO[a]	Experiment
H_2	H	26.46[b]	26.26±1.5	1.64[b]	2.0[c]
HF	F	412.5	410	104.2	108
NH_3	N	262.0	264.5	-48.1	-40
CO_2	C	56.1	58.8	345.9	335
HCN	C	91.5	82.1	286.1	284.6±20
	N	8.4	-20.4	502.9	563.8±8
CO	C	-9.3	1.0	424.1	406
	O	-68.4	-42.3	718.7	676.1

[a] Reference 6a. [b] LDA. [c] Calculated with the Coupled Hartree-Fock method.

XC Functionals. In Table III we look at the influence of the XC functional on the calculated shielding constants. By comparing results of first and second generation DFT (LDA and LDA/NL, respectively) with experiment, we note a remarkable improvement for the latter method for all non-hydrogen nuclei (up to 57 ppm change for H_2CO). Malkin et al. had observed a similar strong influence for their DFT-IGLO method (3a,b), Table III. We have also included into Table III the results of the "coupled DFT-IGLO" approach of the same authors (3a,c,e). The idea of this approach is to model the current dependency of the XC potential (which is neglected in the "uncoupled" methods) by introducing a first order change into it. The authors do this in a somewhat *ad hoc* fashion. The influence of the new coupling is negligible in many cases. However, it leads to a significant reduction of the absolute value of the shielding and therefore to better agreement with experiment for some of the "difficult" molecules like F_2 or H_2CO, Table III. The influence of the XC functional can be traced back to well-known changes in the differences of the orbital energies, eq. 5, between occupied and low lying virtual MO's (6a). The influence of these small changes is therefore less pronounced in singly bound systems with large HOMO-LUMO separations, Table III.

The term "uncoupled" DFT is somewhat misleading, since even this level of theory accounts for correlation effects, and only contributions of the induced current to the correlation are neglected. Uncoupled Hartree-Fock theory on the other hand excludes correlation effects completely.

Table III.   The Influence of the Energy Functional Used

Molecule/ Atom		Isotropic Shieldings (ppm)					
		DFT-GIAO[a]		DFT-IGLO			Exp.
				Uncoupled[b]		Coupled[c,d]	
		LDA	LDA/NL	LDA	LDA/NL[c]	LDA/NL	
NH_3	N	267.2	262.0	259.3	253.4	253.3	264.5
	H	31.2	31.6	30.8	31.2	31.2	32.43
HF	F	415.1	412.5	412.7	409.0	409.6	410
	H	29.4	30.0	29.1	29.7	29.6	28.7
N_2	N	-83.2	-72.9	-86.7	-78.9	-69.3	-61.6
C_2H_2	C	102.9	110.4	102.5	108.9	109.6	117.2
F_2	F	-310.2	-293.7	-271.8	-250.6	-197.8	-232.8
H_2CO	C	-31.7	-15.7	-40.4	-26.6	-12.3	-8.4
	O	-475.8	-418.8	-504.7	-455.6	-362.6	-312.1

[a] Reference 6a. [b] Ref. 3a. [c] We cite the results for the same LDA/NL functional as in our DFT-GIAO method. [d] Ref. 3b.

Frozen core approximation. For heavier elements one might wonder to what degree it is possible to make use of the frozen core approximation in which orbitals of lower energy are taken from atomic calculations. The extension of the shielding calculations (eqs. 3 to 7) to include the frozen core approximation has been discussed in detail earlier (6b). Here, we address this question in connection with calculated ^{77}Se shieldings in Table IV. The core at the selenium atom contains the *1s*, *2s*, and *2p* shells while the core of the second period atoms carbon and fluorine contains the *1s* shell only. This choice was taken according to the discussion in reference 6b. The deviation in the calculated shielding between the frozen core results and the all electron calculations (numbers in brackets) is always smaller than 10 ppm, Table IV. We note also from Table IV that some of the deviation between the frozen core and all electron cases cancels when relative shifts are considered; the deviation does not exceed 5 ppm in this case.

Let us now compare the calculated results to experimentally obtained values. The deviation between theory and experiment is considerable for all the ^{77}Se shieldings, Table IV. However, we get a much better agreement between theory and experiment when we consider relative shifts instead of absolute shieldings, Table IV. The experimental accepted standard for ^{77}Se shifts is liquid Dimethyl Selenide, $(CH_3)_2Se$. We have therefore included this compound into our investigation.

The agreement between theory and experiment is good for ^{77}Se shifts for the few compounds that have been considered here. The experimental uncertainty is certainly large, as are gas-to-liquid shifts and solvation effects. The former are as big as 119 ppm for H_2Se. Solvent and counterion effects amount to a shift range of 25 ppm in the example of the cyclic tetra selenium dication, Table IV. Calculated shieldings refer of course to a single molecule at zero temperature whereas all the experimental data is obtained at finite temperatures and pressures; most of the experiments were carried out in solution or neat liquids. All these effects can yield considerable shifts, and make a direct comparison between theory and experiment difficult. On the theoretical side, it is likely that our basis sets are not yet completely saturated.

Table IV. Calculated and Experimental ^{77}Se Shieldings and Shifts for a few Molecules (Numbers in ppm)

molecule	absolute shielding		chemical shift	
	exp.	calculated[a]	experiment[b]	calculated[a]
$(CH_3)_2Se$, (C_{2V})	2069	1,666 (1,673)	0	0 (0)
H_2Se	2,401	2,093 (2,096)	-345 (g) -226 (l)	-427 (-423)
SeF_6	1,438	988 (992)	610 (g) 631 (l)	678 (681)
CSe_2	1,738	1,441 (1,448)	331 (g) 299 (sol)	225 (225)
Se_4^{2+}		-170	1,923-1,958 (sol)[c]	1,836

[a]Calculated shieldings from frozen core calculations and (in brackets) from all electron calculations. [b]G - gas phase; l - liquid; sol - solution. [c]Result depending on solvent and counterion.

A special case is the Se_4^{2+} ion. This is a highly correlated molecule, and traditional Hartree-Fock based methods are unable to predict the chemical shift for this ion. However, the DFT result compares well with experiment; DFT is indeed capable of handling such systems.

Let us now come back to the absolute shieldings. We note that the calculated absolute shieldings seem to be uniformly too small by about 300 ppm, Table IV. The experimental absolute shielding scale is based on the absolute shielding of SeF_6 that was found to be $1,438\pm64$ ppm. However, this value is based on a theoretically predicted (diamagnetic) shielding value of the free selenium atom. This theoretical value has been corrected explicitly for relativistic effects, in particular the relativistic contraction of the core density. Other relativistic effects are probably not yet important for ^{77}Se chemical shifts. The magnitude of the necessary correction is estimated at 300 ppm. Therefore, we find that our calculated shielding values are uniformly too small by about 300 ppm, Table IV. This uniform error of 300 ppm cancels of course when (relative) chemical shifts are calculated. This point illustrates the importance of absolute shielding scales for the test of theoretical methods.

Transition Metal Complexes. The Example of $Cr(CO)_6$, $Mo(CO)_6$, and $W(CO)_6$. Table V displays calculated (6c) and experimental absolute ^{13}C NMR shielding tensor components for the three hexacarbonyls $M(CO)_6$ (M=Cr, Mo, and W). A similar compilation is given in Table VI for the ^{17}O shielding. The calculated and observed tensor components σ_{ss} (s = x, y, z) compare well for both ^{13}C and ^{17}O, with differences within the experimental error limits. The σ_{ss} tensor components in

Tables V and VI correspond to an orientation in which the CO ligand probed by NMR has the $^{13}C^{17}O$ bond vector along the z-axis.

Table V. A Comparison between Experimental and Calculated[i] Absolute ^{13}C Chemical Shielding Tensor Components for the CO Molecule and Group 6 Metal Carbonyls

System	σ_{xx} ppm (expt)[a]	σ_{yy} ppm (expt)[a]	σ_{zz} ppm (expt)[a]	Anisotropy $\Delta\sigma$ ppm[b] (expt)[a]	Isotropic shielding[c] σ_i ppm (expt)[a]	Other work[f] (Absolute Scale[a])
CO	-149.4 [g]	-149.4	273.6	423.0	-8.4	
	(-132.3 [d])	(-132.3)	(273.4)	(406(s) ±1.4)	(1.0 (sr))	
Cr(CO)₆	-167.9 [g]	-167.9	264.2	432.3	-23.9	-20.4
	(-167.6±15 [e])	(-167.6±15)	(255.4±15)	(423±30)	(-26.6±15(s, l))	
Mo(CO)₆	-164.0 [g]	-164.0	266.4	430.4	-20.5	-19.4
	(-157.6±15 [e])	(-157.6±15)	(260.4±15)	(417±30)	(-17.6±15)	
W(CO)₆ Rel.[h]	-161.5 [g]	-161.5	272.1	433.6	-16.9	
	-149.6 [h]	-149.6	266.7	416.4	-10.9	-6.2
	(-138.6±15 [e])	(-138.6±15)	(256.4±15)	(395±30)	(-6.6±15(s, l))	

q

[a]The data originally reported in ppm relative to liquid TMS are converted to absolute shielding using the ^{13}C absolute shielding scale in which $\sigma(^{13}C$ in liquid TMS)= 185.4 ppm based on ^{13}C in CO(molecular beam)= -42.3±17.2 ppm, see Ref. 6c.   [b]$\Delta\sigma$ is defined as $\Delta\sigma = \sigma_{zz}-1/2(\sigma_{xx}+\sigma_{yy})$. Here the z-axis is pointing along the CO bond vector of the ligand probed by NMR.   [c]$\sigma_i = 1/3(\sigma_{xx} + \sigma_{yy} +\sigma_{zz})$.   [d,e]Ref. 6c.   [f]Ref. 3f.   [g]Non-relativistic NL-SCF calculation, see ref. 6d.   [h]Relativistic NL-SCF-QR calculation.   [i]Reference 6c.

Table VI. A Comparison between Experimental and Calculated[i] Absolute ^{17}O Chemical Shielding Tensor Components for the CO Molecule and Group 6 Metal Carbonyls

System	σ_{xx} ppm (expt)[a]	σ_{yy} ppm (expt)[a]	σ_{zz} ppm (expt)[a]	Anisotropy $\Delta\sigma$ ppm[b] (expt)[a]	Isotropic shielding σ_i ppm[c] (expt)[a]
CO	-307.3 [g]	-307.3	410.6	717.9	-67.9
	(-267.6±26(sr) [e])	(-267.6±26(sr))	(408.47±26)	(676.1±26(sr))	(-42.7±17.2)
Cr(CO)₆	-302.8 [g]	-302.8	373.9	676.7	-77.2
	(-307.1±10-20 [d])	(-271.1±10-20)	(401.9±10-20)	(691±10-20)	(-59.1±10-20)
Mo(CO)₆	-293.5 [g]	-293.4	362.5	655.9	-74.8
	(-277±10-20 [d])	(-248.1±10-20)	(386.9±10-20)	(650±10-20)	(-46.1±10-20)
W(CO)₆ Rel[h]	-291.5 [g]	-291.4	359.9	651.4	-74.4
	-268.3 [h]	-268.2	351.7	620.0	-61.6
	(-259.1±10-20 [d])	(-228.1±10-20)	(374.9±10-20)	(619±10-20)	(-40.1±10-20).

[a] The experimental data originally reported in ppm relative to liquid H₂O are converted to absolute shielding using the ^{17}O absolute shielding scale in which $\sigma(^{17}O$ in liquid H₂O)= 307.9 ppm, see Ref. 6c.   [b-i] See Table V.

It is worth pointing out that the experimental σ_{xx} and σ_{yy} components should be equal for a single $M(CO)_6$ molecule. That they are different must be attributed to crystal effects.

All hexacarbonyls have observed absolute isotropic shieldings, σ_i, that are negative for ^{13}C as well as ^{17}O. Further, σ_i decreases in absolute terms down the triad, with the largest drop at the $5d$ element tungsten. Carbon as well as oxygen are seen to be deshielded in $Cr(CO)_6$ and $Mo(CO)_6$ compared to free CO. In $W(CO)_6$ this deshielding is only marginal for carbon and changed to a shielding for oxygen. All the experimental trends are reproduced by the GIAO-DFT scheme after the inclusion of relativity (6d), Tables V and VI. Kaupp (3f) *et al.* have recently published calculated isotropic ^{13}C shielding constants for several transition metal complexes based on their IGLO-DFT scheme. It follows from Table V that their calculated isotropic carbon shieldings for the hexacarbonyls are in good agreement with experimental values and our estimates based on the GIAO-DFT method.

Outlook. DFT is particularly suited to deal with many-electron systems, e.g., transition metal complexes. We have given examples of this above. However, a proper treatment of heavier element compounds (e.g., $4d$ and $5d$ complexes) requires the inclusion of relativity. Work is in progress to introduce relativistic terms as well (6d). The frozen core approximation is an important step in this direction, see above. Once completed, the new program would for the first time allow to cover the complete range of multi-nuclear NMR by first-principle theoretical methods. Multinuclear NMR (1) is a field of tremendous and growing experimental significance.

3. The Calculation of Spin-Spin Coupling Constants

As the experimental field of multi-nuclear NMR increases in importance, the ability to provide theoretical calculations of NMR properties such as spin-spin coupling constants for a wide variety of systems is desirable. Semi-empirical theories are only reliable for analogues of already-understood systems, while traditional *ab initio* methods are constrained by the expense involved in treating electron correlation adequately. Density functional theory provides a third path between these two methods and is now often the method of choice for investigating inorganic systems, especially those containing transition metals.

The ability to calculate NMR spin-spin coupling constants from density functional theory for such systems would be valuable. Work on DFT calculations of chemical shielding tensors in transition metal systems has already been discussed in a previous section. In the present section we address the calculation of nuclear spin-spin coupling constants.

General theory of nuclear spin-spin coupling. The theory of nuclear spin-spin coupling is well established; we review it here for the reader's convenience and to establish symbols and terminology. The reader may wish to consult one of several excellent reviews on the subject for further information (10,11). Nuclear spin-spin coupling is the interaction energy of two nuclear magnetic moments. It can be shown that the reduced spin-spin coupling constant K is the second derivative of the energy of the system with respect to the nuclear magnetic moments $\vec{\mu}_A = \gamma_A \hbar \hat{I}_A$:

$$K_{xy}(A,B) = \frac{\partial^2 E}{\partial \mu_{Ax} \partial \mu_{By}} \qquad (9).$$

It is a second-rank tensor but we shall concern ourselves here only with the isotropic part, $K = (1/3)(K_{xx} + K_{yy} + K_{xx})$. The reduced coupling constant K is preferred in theoretical discussions over the ordinary spin-spin coupling constant J,

$$J(A,B) = h\gamma_A \gamma_B K(A,B) / 4\pi^2 \qquad (10),$$

because J is proportional to the product of the nuclear gyromagnetic ratios $\gamma_A \gamma_B$ while K is independent of the isotopic identities of the nuclei. Reduced coupling constants are expressed in SI units of 10^{19} kg m^{-2} s^{-2} A^{-2} throughout this paper.

The general theory of second-derivative properties (12) tells us that for a pair of perturbations μ_A and μ_B,

$$\hat{H} = \hat{H}_o + \mu_A \hat{H}^{(\mu_A)} + \mu_B \hat{H}^{(\mu_B)} \qquad (11)$$

then

$$\frac{\partial^2 E}{\partial \mu_{Ax} \partial \mu_{By}} = \left[\frac{\partial}{\partial \mu_B} \left\langle \Psi(\mu_B) \left| \hat{H}^{(\mu_A)} \right| \Psi(\mu_B) \right\rangle \right]_{\mu_A = \mu_B = 0} \qquad (12),$$

i.e., in order to calculate second derivatives of the energy we need the first derivative of the wave function. In density functional theory we avoid explicit reference to the wave function and reformulate this in terms of the Kohn-Sham molecular orbitals (MOs) and the density constructed therefrom. A similar formalism is also the basis of the shielding calculation as outlined in an earlier section.

There are four different terms in the non-relativistic Hamiltonian described by Ramsey (7): The diamagnetic and paramagnetic spin-orbit terms (DSO and PSO), the spin-dipolar term (SD), and the Fermi-contact term (FC).

$$\hat{H}^{(\mu_C)} = \hat{H}_{dso} + \hat{H}_{pso} + \hat{H}_{fc} + \hat{H}_{sd} \qquad (13)$$

Each of these terms gives a separate contribution to the isotropic spin-spin coupling, and in the present work we evaluate the first three of these. The spin-dipolar term, however, is both complicated to implement and generally very small (10), so it is not dealt with in the present work. There are also two anisotropic contributions, one from the cross-term of the spin-dipolar and Fermi-contact interactions, one from the classical dipolar interaction of the nuclei, both of which we also omit. The well known (7) analytical form of the terms in eq. 13 as well as the details of the calculations discussed below have been given elsewhere (13).

Spin-spin coupling constants of main group compounds. We have performed all-electron calculations of spin-spin coupling constants on a range of test systems, using the local spin-density approximation and the triple-zeta valence, double-zeta core, doubly polarized basis described in reference 13.

Coupling constants for the 10-electron hydrides are shown in Table VII. The calculated geminal hydrogen-hydrogen couplings do not show the experimental trend across the row. The correct trend *is* reproduced in coupled Hartree-Fock (CHF) calculations (10), although the absolute magnitudes are very poor unless correlation is added (10). The approximate cancellation of the diamagnetic and paramagnetic spin-orbit terms also appears in the CHF calculations, leading to the conclusion that the Fermi-contact term in these LDA calculations has the wrong behavior.

Table VII. Reduced Coupling Constants in some 10-electron Hydrides

Molecule	coupling	FC	DSO	PSO	total	exp
CH_4	$^2K(H,H)$	-.55	-.29	.31	-.53	-1.03
NH_3	$^2K(H,H)$	-.62	-.43	.50	-.69	-.87
H_2O	$^2K(H,H)$	-1.03	-.59	.72	-.90	-.60
CH_4	$^1K(C,H)$	39.6	.27	.8	40.5	41.3
NH_3	$^1K(N,H)$	50.6	.07	2.9	53.6	50
H_2O	$^1K(O,H)$	41.4	.05	8.0	49.4	48
HF	$^1K(F,H)$	16.3	.05	18.1	34.4	46.9

The one-bond couplings between heavy atoms and hydrogen show (proportionally) much better agreement with experiment, with the exception of hydrogen fluoride. Results for hydrogen fluoride can be substantially improved by going to a basis set with both triple-zeta core and polarization flexibility: The Fermi contact term then becomes 25.9, leading to a total coupling of 43.7, compared with an experimental value of 46.9.

Table VIII. Reduced Coupling Constants in some Second Row Hydrides

Molecule	coupling	FC	DSO	PSO	total	exp
SiH_4	$^2K(H,H)$	.08	-.20	.11	-.01	.23
PH_3	$^2K(H,H)$	-.77	-.12	.11	-.78	-1.12
H_2S	$^2K(H,H)$	-.85	-.15	.18	-.82	
SiH_4	$^1K(C,H)$	69.9	.0	-.1	69.8	84.9
PH_3	$^1K(N,H)$	25.3	.0	1.4	26.7	37.8
H_2S	$^1K(O,H)$	16.8	.0	5.2	22.0	
HCl	$^1K(F,H)$	3.7	.0	13.2	16.9	32

In the second-row hydrides shown in Table VIII the geminal proton-proton couplings seem to have the correct experimental trend, but the lack of experimental couplings for H_2S makes this uncertain. The one-bond couplings to hydrogen are affected by a shortfall in the Fermi contact term of some 10--20 units, which increases monotonically across the row. This is probably due to basis set inadequacies for the Fermi contact term, that are similar to those just described for hydrogen fluoride.

In Table IX we observe that the hydrogen-hydrogen couplings are also poor in the simple hydrocarbons, but the carbon-proton one-bond couplings are in quite good agreement with experiment. Two-bond carbon-proton couplings are somewhat worse, as might be expected due to their small magnitude. The carbon-carbon couplings in the series ethane--ethene--ethyne are qualitatively reproduced. The spin-dipolar term may be relevant to these couplings, particularly for C_2H_2.

In Table X N_2 and to a lesser degree CO are poor because the cancellation of different contributions is exceedingly sensitive to geometry. A similar geometry sensitivity may occur in CO_2. The results for H_2 are somewhat disappointing, but the overestimation of the Fermi contact term here is probably due to the local spin-density approximation.

The couplings to fluorine in CH_3F are unsatisfactory; this is probably again due to an inadequate basis for fluorine. The large discrepancy for the C-F coupling is also partly due to the neglect of the spin-dipolar mechanism. The other couplings follow the trends just noted for other simple hydrocarbons: The proton-proton coupling is poor, but the carbon-proton one-bond coupling is just a few percent smaller than experiment. In general we note that any small coupling involving a balance of competing terms is liable to be poor for the usual reason that small errors in large numbers lead to a large error in their difference. There are additional problems with proton-proton couplings that seem to be associated with their small magnitude as well, but one-bond couplings involving heavy atoms are generally much larger and therefore quite reliable. In these cases there is a common tendency for the calculation to underestimate the coupling.

Table IX. Reduced Coupling Constants for some Light Hydrocarbons

Molecule	coupling	FC	DSO	PSO	total	exp
CH_4	$^2K(H,H)$	-.55	-.29	.31	-.53	-1.03
C_2H_6	$^2K(H,H)$	-.66	-.24	.26	-.64	
C_2H_4	$^2K(H,H)$	.31	-.32	.32	.31	.21
C_2H_6	$^3K(H,H)$ trans	1.23	-.26	.24	1.21	.67
C_2H_6	$^3K(H,H)$ gauche	.22	-.08	.08	.22	.67
C_2H_4	$^3K(H,H)$ cis	.57	-.09	.0-6	.54	.97
C_2H_4	$^3K(H,H)$ trans	1.07	-.29	.23	1.01	1.59
C_2H_2	$^3K(H,H)$	.15	-.30	.36	.21	.80
CH_4	$^1K(C,H)$	39.6	.1	.8	40.5	41.3
C_2H_6	$^1K(C,H)$	40.2	.2	.6	41.0	41.3
C_2H_4	$^1K(C,H)$	47.6	.1	.3	48.1	51.8
C_2H_2	$^1K(C,H)$	79.5	.1	0.0	79.1	82.4
C_2H_6	$^2K(C,H)$	-.6	-.1	.1	-.6	-1.5
C_2H_4	$^2K(C,H)$	1.3	-.2	-.4	.6	-.8
C_2H_2	$^2K(C,H)$	14.3	-.4	1.8	15.7	16.3
C_2H_6	$^1K(C,C)$	39.6	.2	.1	39.8	45.6
C_2H_4	$^1K(C,C)$	103.5	.1	-13.3	90.3	89.1
C_2H_2	$^1K(C,C)$	262.2	.0	7.5	269.7	226.0

Table X. Reduced Coupling Constants in some other First-row Compounds

Molecule	coupling	FC	DSO	PSO	total	exp
H_2	$^1K(H,H)$	28.0	-.1	.3	28.2	23.3
N_2	$^1K(N,N)$	39.0	-.3	-39.0	-.3	-20.0
CO	$^1K(C,O)$	-29.1	-.2	-37.2	-66.6	-40.1
CO_2	$^1K(C,O)$	-41.4	.2	-11.7	-53.0	-39.4
CH_3F	$^2K(H,H)$	-.23	-.25	.24	-.23	-.80
CH_3F	$^1K(C,H)$	46.8	.2	.1	47.1	49.4
CH_3F	$^1K(C,F)$	-106.1	.2	13.8	-92.2	-57.0
CH_3F	$^2K(F,H)$	1.88	-.12	1.22	2.94	4.11

Transition metal carbonyls. We have carried out calculations on a series of three transition-metal complexes for which experimental couplings are known. One can see instantly from Table XI that a single contribution (the Fermi contact term) dominates the calculated couplings, and so there is reason to believe that the total couplings will reflect reality. This is confirmed by comparison with experiment.

Table XI. Reduced Coupling Constants in Three Transition Metal Carbonyls

Molecule	coupling	FC	DSO	PSO	total	exp
$[V(CO)_6]^-$	$^1K(V,C)$	133	.3	-.6	127	146
$FeCO_5$	$^1K(Fe,C_{ax})$	206	.4	-.28	178	239
$FeCO_5$	$^1K(Fe,C_{eq})$	246	.3	-9	237	239
$[Co(CO)_4]^-$	$^1K(Co,C)$	354	.3	-.1	353	400±20

The reduced coupling constant increases going across the periodic row, and the calculated couplings are approximately 15 % smaller than experiment. This suggests that further calculations of transition-metal coupling constants should be feasible. The remaining error could be due to the approximate density functional, geometry, vibrational averaging, neglect of the spin-dipolar mechanism, or some combination of these.

4. Conclusions

We presented a modern implementation of the GIAO method for the calculation of the shielding tensor using DFT. The calculated shielding constants and tensors agree well with experimental results; the quality is certainly comparable with the results of other theoretical methods. However, the results for some strongly correlated systems are not yet satisfactory. The paramagnetic shielding contribution of these systems is very large and is generally overestimated in magnitude. Our method has also been applied to heavier elements such as ^{77}Se, as well as to the ^{13}C and ^{17}O shifts in metal carbonyls. The DFT-GIAO scheme is capable of handling even highly correlated systems like Se_4^{2+}.

We conclude from the previous discussion that the current state of affairs does not yet allow to decide how to improve DFT methods for the calculation of shielding tensors: The reason of the observed errors for some systems might be that our XC functionals are inappropriate for a precise description of the orbital energies. If this is the case, then "next generation" XC potentials should be able to fix the problems.

We have demonstrated that density-functional calculations of NMR spin-spin coupling constants are feasible on transition metal compounds using Slater-type basis functions. Basis set requirements for the Fermi-contact term are stringent in some cases, as they are with Gaussian-type basis functions. Furthermore, contributions from core orbitals of the responding nucleus in the calculation of the Fermi-contact term can be significant, although use of the frozen core approximation on the perturbing nucleus may still be practical.

With a triple-zeta doubly polarized basis (13) set, we find that couplings between heavy atoms can be calculated with a typical error of less than 15 %. Larger errors may occur (i) when the coupling constant has contributions of differing sign from the various mechanisms, (ii) when basis set requirements are more stringent, as for fluorine and chlorine, (iii) when the Fermi-contact term is less than about 1×10^{19} kg m^{-2} s^{-2} A^{-2} as in most proton-proton couplings, and (iv) when the spin-dipolar mechanism (neglected here) is significant.

The local spin-density functional used here has been superseded in some applications by gradient-corrected density functionals, but preliminary calculations have revealed no meaningful improvement in results for the spin-spin coupling with the more expensive gradient-corrected exchange-correlation potentials V_{XC}. We speculate that this is connected with the unphysical behavior of the common gradient-corrected potentials near the nucleus. We anticipate that further studies of the exchange-correlation potential, as opposed to the exchange-correlation energy density, will yield significant improvements for properties such as the nuclear spin-spin coupling constant.

Acknowledgments

This work has been supported by the Natural Sciences and Engineering Research Council of Canada (NSERC), as well as by the donors of the Petroleum Research Fund, administered by the American Chemical Society (ACS-PRF No 20723-AC3). G.S. acknowledges the Graduate Faculty Council, University of Calgary, for a scholarship, Y.R.-M. acknowledges a scholarship from DGAPA-UNAM (Mexico), and R.M.D. is thankful for a post-doctoral fellowship from the University of Calgary.

References.

1 (a) Fukui, H. *Magn. Res. Rev.* **1987**, *11*, 205.
   (b) Chesnut, D. B. In *Annual Reports on NMR Spectroscopy*; Webb, G. A.; Ed.; Academic Press: New York, **1989**; Vol. 21.
   (c) Tossel, J. A. (Ed.), *Nuclear Magnetic Shielding and Molecular Structure; NATO ASI C386*; Kluwer Academic Publishers: Dordrecht, The Netherlands, **1993**.
   (d) Mason, J. (Ed.), *Multinuclear NMR*, Plenum Press: New York, **1987**.
2 (a) Parr, R. G.; Yang, W. *Density-Functional Theory of Atoms and Molecules*; Oxford University Press: New York, Oxford, **1989**.
   (b) Ziegler, T. *Chem. Rev.* **1991**, *91*, 651.
3 (a) Malkin, V. G.; Malkina, O. L.; Erikson, L. A.; Salahub, D. R. In *Modern Density Functional Theory: A Tool for Chemistry*; Vol. 2 of *Theoretical and Computational Chemistry*; Politzer, P.; Seminario, J. M. (Eds.); Elsevier: Amsterdam, The Netherlands, **1995**.
   (b) Malkin, V. G.; Malkina, O. L.; Salahub, D. R. *Chem. Phys. Lett.* **1993**, *204*, 80.
   (c) Malkin, V. G.; Malkina, O. L.; Salahub, D. R. *Chem. Phys. Lett.* **1993**, *204*, 87.
   (d) Woolf, T. B.; Malkin, V. G.; Malkina, O. L.; Salahub, D. R.; Roux, B. *Chem. Phys. Lett.* ; **1995**, *239*, 186
   (e) Malkin, V. G.; Malkina, O. L.; Casida, M. E.; Salahub, D. R. *J. Am. Chem. Soc.* **1994**, *116*, 5898.
   (f) Kaupp, M.; Malkin, V. G.; Malkina, O. L.; Salahub, D. R. *J. Am. Chem. Soc.* **1995**, *117*, 1851; erratum *ibid.* **1995**, *117*, 8492.
   (g) Kaupp, M.; Malkin, V. G.; Malkina, O. L.; Salahub, D. R. *Chem. Phys. Lett.* **1995**, *235*, 382.
   (h) Malkin, V.G.; Malkina, O.L.; Salahub, D.R. *Chem. Phys. Lett.* **1994**, *221*, 91.
4 (a) Kutzelnigg, W. *Israel J. Chem.* **1980**, *19*, 193.
   (b) Schindler, M.; Kutzelnigg, W. *J. Chem. Phys.* **1982**, *76*, 1919.
   (c) Kutzelnigg, W.; Fleischer, U.; Schindler, M. In *NMR - Basic Principles and Progress;* Springer-Verlag: Berlin, **1990**; Vol. 23, p. 165.
5 (a) Ditchfield, R. *Mol. Phys.* **1974**, *27*, 789.
   (b) Wolinski, K.; Hinton, J. F.; Pulay, P. *J. Am. Chem. Soc.* **1990**, *112*, 8251.
   (c) Friedrich, K.; Seifert, G.; Grossmann, G. *Z. Phys. D* **1990**, *17*, 45.
6 (a) Schreckenbach, G.; Ziegler, T. *J. Phys. Chem .* **1995**, *99*, 606.
   (b) Schreckenbach, G.; Ziegler,T. *Int. J. Quantum Chem.*, submitted.
   (c) Ruiz-Morales, Y.; Schreckenbach, G.; Ziegler,T. *J. Phys. Chem .*, submitted.
   (d) Schreckenbach, G.; Ziegler, T. *J. Phys. Chem.*, to be submitted.
7 Ramsey, N.F. *Phys. Rev.* **1953**, *91*, 303.
8 te Velde, G. *Amsterdam Density Functional (ADF), User's Guide, Release 1.1.3*; Department of Theoretical Chemistery, Free University: Amsterdam, **1994**.
9 (a) Becke, A. D. *Phys. Rev.* **1988**, *A38*, 3098.
   (b) Perdew, J. *Phys. Rev.* **1986**, *B33*, 8822; erratum *ibid.* **1986**, *B34*, 7406.
10 Kowalewski, J. *Ann. Rep. NMR Spectrosc.* **1982**, *12*, 81, and references therein.
11 Fukui, H. *Nuclear Magnetic Resonance (Specialist Periodical Reports)* **1994**, *23*, 124, and references therein.
12 McWeeny, R. *Methods of Molecular Quantum Mechanics*; Academic Press: London, 2nd ed., **1992**.
13 Dickson, R.M.; Ziegler,T. *J.Phys.Chem.*, submitted.

Chapter 24

Hybrid Hartree–Fock Density-Functional Theory Functionals: The Adiabatic Connection Method

Jon Baker, Max Muir, Jan Andzelm, and Andrew Scheiner

Biosym/Molecular Simulation, 9685 Scranton Road, San Diego, CA 92121–3752

Drawing upon a large body of recent work covering a wide range of chemistry and comparing results obtained using local and nonlocal density functionals, as well as comparisons with traditional *ab initio* techniques such as Hartree-Fock and MP2, we demonstrate that hybrid HF-DFT functionals - exemplified here by Becke's original 3-parameter ACM functional - are the best density functionals currently available, certainly for organic chemistry and possibly in other areas as well. Results with the ACM functional are typically of better quality than MP2 and only marginally more expensive computationally than standard Hartree-Fock, making ACM the method of choice for accurate, cost-effective *ab initio* computations.

The use of Density Functional Theory (DFT) [1,2] for the calculation of molecular structures, properties and energetics has exploded during the past few years. There have been major advances in both theory - the development of nonlocal functionals (to better model the effects of rapidly changing densities near atomic nuclei) [3-5], the advent of analytical gradients [6-10] and second derivatives [11-14] - and in the practical tools for applying the theory - computational codes such as DMOL [15], DGAUSS [16] and ADF [17] are now well established and are becoming increasingly more sophisticated; traditional *ab initio* packages such as GAUSSIAN [18], CADPAC [19] and TURBOMOLE [20] are including DFT capability as a major part of their functionality. In fact DFT has now become so prevalent and its applicability and the quality of its predictions so great that in the not too distant future it is likely to take over from Hartree-Fock (HF) as the "basic" *ab initio* technique. One has only to glance through the recent chemical literature to note the increasing number of publications with a DFT component and the growing number of groups doing work in this area.

There are two major reasons for the increasing use of density functional methods in quantum chemistry. One - already mentioned in the preceeding paragraph - is that the results *are* good. The second, and perhaps the most important reason, is that - certainly compared to the more accurate *ab initio* techniques - the method is fast. Depending on the implementation, DFT calculations range from being 2 or 3 times more costly than standard Hartree-Fock to an order of magnitude or more faster, with the relative speed advantage becoming even greater with increasing system size. Since DFT includes electron correlation, we have a method that is potentially much more accurate than traditional HF and yet is, at worst, no more expensive for large systems and can in fact be made many times faster.

0097–6156/96/0629–0342$16.50/0

Some of the early successes of DFT are documented in the excellent review article by Ziegler [21] and the book edited by Labanowski and Andzelm [22]. A key paper that helped considerably in popularizing DFT methods, particularly amongst *ab initio* quantum chemists, was the study by Johnson, Gill and Pople of the performance of a number of density functionals [10]. These authors looked at geometries, dipole moments, vibrational frequencies and atomization energies of 32 small neutral molecules (a subset of the G2 data set [23]) with the 6-31G* basis set [24] and compared DFT with HF, MP2 and QCISD results. The findings from this work were that DFT geometries were slightly worse, dipole moments were at least as good, vibrational frequencies were on the whole better, and for atomization energies two density functionals - BVWN and BLYP (combinations of Becke's nonlocal exchange [3] with Vosko, Wilk and Nusair's local correlation [25] and Lee, Yang and Parr's nonlocal correlation functional [5], respectively) - gave excellent agreement with experiment and were in fact the only acceptable theoretical methods.

A notable success for DFT methods was the highly correlated molecule, FOOF. This is a system which is very difficult to get "right" by traditional *ab initio* methods, geometrical predictions are often at considerable variance with experiment and high levels of theory and large basis sets are needed to give reliable results. Using local DFT and a modest numerical basis, Dixon et al. obtained excellent agreement with the experimental F-O and O-O bond lengths [26].

Despite the success with FOOF, local DFT has been more-or-less discredited, at least for organic chemistry, due to its marked tendency to overbind, i.e., to make a molecule more stable relative to its separated atoms than it really is. This is illustrated in Table 1 which gives theoretical and experimental atomization energies for a number of systems (taken from ref.10); as can be seen the local (SVWN) functional is consistently overbinding - denoted by the negative sign for the average and maximum errors - relative to experiment (although not as *underbinding* as the HF values, which are terrible). As noted by Johnson et al. [10], the nonlocal BLYP functional gives much superior results.

Conventional wisdom has it that the nonlocal functionals in common use - particularly BP [3,4] and BLYP [3,5] - typically give results of the quality of MP2 or better at a significantly reduced cost. However, let us not get too carried away. At the time of writing DFT methods cannot rival the most sophisticated *ab initio* methods - such as configuration-interaction or coupled cluster techniques - for accuracy. Unfortunately, these methods are enormously expensive and are typically only used for benchmark calculations on relatively small systems. Virtually all of the standard post-HF methods for including electron correlation involve systematically improving the HF wavefunction (by including more and more excitations from the HF reference determinant for example); such methods for *systematic* improvement are currently unavailable in DFT - everything depends on the quality of the initial density functional.

The design of increasingly accurate functionals has been - and will doubtless continue to be - an area of active research. We have already noted the development of nonlocal functionals (which involve the gradient of the density in their definition) which has been a major improvement. However, despite the quality of the results, more accuracy is needed in many cases to seriously rival the best *ab initio* methods and there are increasing signs that in certain areas even the best nonlocal functionals are seriously deficient. For example, Pople and coworkers have shown that conventional Kohn-Sham methods consistently and significantly underestimate the barrier for the reaction $H + H_2 \longrightarrow H_2 + H$; in particular, LDA in unmodified form fails completely, predicting H_3 to be a stable species [27]. *Negative* barrier heights have been reported for certain reactions, and although this mainly seems to be a problem with local DFT there does appear to be a general trend emerging for even nonlocal DFT barriers to be perhaps too low and this could be a major problem for radical reactions in particular [28]. There have also been quite dramatically different predictions of the energy ordering in larger systems between, e.g., MP2 and nonlocal DFT; for example in the relative stabilities of isomers of C_{20} calculated using MP2 and BLYP [29]. Despite the many successes, there is clearly a need for more accurate and reliable functionals.

Table 1
Atomization Energies for some small molecules (kcal/mol)
(theoretical values with the 6-31G* basis; taken from ref.10)

compound	HF	MP2	QCISD	SVWN[a]	BLYP	Expt.
H_2	75.9	86.6	91.2	107.5	103.2	103.3
CH_4	300.4	354.2	353.9	436.8	389.9	392.5
NH_3	170.2	232.4	230.7	306.0	270.1	276.7
H_2O	131.7	188.8	183.7	240.8	207.3	219.3
HF	82.1	118.2	114.0	146.2	124.4	135.2
C_2H_2	271.9	365.6	351.2	438.6	383.4	388.9
C_2H_4	394.2	489.4	481.7	600.9	528.1	531.9
C_2H_6	506.0	608.5	603.1	752.1	660.9	666.3
HCN	184.9	287.3	269.7	346.5	306.2	301.8
CO	168.3	254.3	237.4	293.4	257.4	256.2
H_2CO	237.8	335.5	321.9	417.6	361.8	357.2
CH_3OH	331.5	434.8	425.3	551.2	475.3	480.8
N_2	105.1	212.1	192.3	257.3	231.3	225.1
NO	46.4	134.8	124.8	193.8	162.8	150.1
O_2	28.9	117.6	99.0	174.6	136.8	118.0
H_2O_2	109.4	219.6	206.8	310.4	252.8	252.3
CO_2	234.7	381.0	347.8	464.3	392.9	381.9
F_2	-34.3	36.8	27.9	83.6	54.4	36.9
av.error	107.2	23.2	34.0	-47.1	7.2	
max.error	160.3	57.8	63.2	-85.8	18.8	

[a] SVWN consistently overbinding as shown by the negative value for the average
and maximum error; HF (especially), MP2 and QCISD are consistently underbinding.
BLYP is neither.

Very recently a new class of DFT functionals is becoming popular - hybrid HF-DFT functionals which mix part of the "exact" Hartree-Fock exchange in with the density functional. These functionals derive from ideas expounded by Becke based on a rigorous *ab initio* formulation of the Kohn-Sham exchange-correlation energy known as the "adiabatic connection" formula [30]. One of the strict "limits" of this formula is the pure exchange-only energy of the Kohn-Sham determinant with no dynamical correlation whatsoever; this is essentially identical to the conventional Hartree-Fock exchange energy.

One way of looking at DFT, which is particularly appropriate for our purposes, is to treat the Kohn-Sham equations in the same manner as in standard Hartree-Fock with a Fock matrix given by [10]

$$F = H + J + K^{xc} \qquad (1)$$

where H and J are the usual one- and two-electron Coulomb matrices and the exchange-only K matrix in HF theory has been replaced by a DFT exchange-correlation matrix K^{xc}.

Becke has argued that "traditional" (if we may call them that) density functionals, including gradient-corrected, in which the HF exchange energy is replaced *entirely* by the DFT exchange-correlation energy, no matter how sophisticated they become, will not be able to correctly reproduce the exchange-only "limit" of the adiabatic connection formula, and the way to rectify this is to mix in part of the "real" exchange, or - in other words - not replace all of K by K^{xc} but leave part of the original K behind. This may also be related to the well known problem in DFT of the so-called Coulomb "self-interaction" of the electrons, a problem that is eliminated in Hartree-Fock theory due to an exact cancellation of Coulomb and exchange integrals.

The hybrid (ACM) functional proposed by Becke in ref.30 is a linear combination of various density functionals together with a term representing the exact exchange

$$E^{xc} = E^{xc}_{svwn} + a_0(E^x_{hf} - E^x_{svwn}) + a_x \Delta E^x_{b88} + a_c \Delta E^c_{pw91} \qquad (2)$$

Here E^{xc}_{svwn} is the total (local) SVWN exchange-correlation energy [25], E^x_{hf} is the Hartree-Fock exchange energy, E^x_{svwn} is the (local) Slater exchange, ΔE^x_{b88} is Becke's 1988 (nonlocal) correction to the exchange energy [3] and ΔE^c_{pw91} is Perdew and Wang's 1991 (nonlocal) correction to the correlation energy [4]. The coefficients a_0, a_x and a_c, are empirical parameters determined by the best fit to the experimental data from Pople's G2 set [23]; they take values of 0.2, 0.72 and 0.81, respectively. Recently, Becke has developed another functional with only a single fit parameter (the amount of exact exchange included, which has increased to 0.28; see ref.31 and an independent article in this volume); however, all results reported herein are with the original ACM functional as given in Eq.2.

In this article we present a comparison of results obtained using local, nonlocal and the hybrid ACM functional, together with some HF, MP2 and other methodologies, focussing on geometries, energetics and barrier heights (including reaction profiles) for a number of chemical systems. Much of the work presented summarizes previous and pending publications from the (now disbanded) theory group at what was formerly Biosym; the work on hydrogen bonding is previously unpublished. Our contention - overwhelmingly supported by a large body of calculations - is that hybrid HF-DFT functionals - as represented here by the ACM functional - are overall the best density functionals currently available, certainly for organic chemistry and possibly in other areas too.

RESULTS

1. Organic Chemical Reactions

We start by summarizing the results of a study of twelve "typical" small organic reactions, six closed-shell and six radical [32]:

Closed-Shell (singlets)

vinyl alcohol ---> acetaldehyde	(keto-enol tautomerism)
butadiene + ethylene ---> cyclohexene	(parent Diels-Alder)
s-tetrazine ---> 2HCN + N_2	
trifluoromethanol ---> carbonyl fluoride + HF	
ring opening of cyclobutene	
rotation in butadiene	

Radical (doublets)

$FO + H_2$ ---> $FOH + H$
$OH + H_2$ ---> $H_2O + H$
ring opening of cyclopropyl radical
$CH_3 + C_2H_4$ ---> propyl radical
$CH_3 + H_2CO$ ---> ethoxy radical
$H + HCCH$ ---> $H_2 + CCH$

The study looked at geometries, heats of reaction and barrier heights using semiempirical (MNDO, AM1), traditional *ab initio* (HF, MP2) and density functional (BLYP, ACM) methods. No detail will be given here - we refer the reader to the original publication for a discussion of individual reactions [32] - instead we present a summary in the form of Table 2, which gives heats of reaction and barrier heights for all twelve reactions, and Table 3, giving average and maximum deviations from experiment for barrier heights, heats of reaction and bond lengths. All *ab initio* calculations used the 6-31G* basis set (with five pure as opposed to six Cartesian components in the d-polarization functions).

As can be seen from Table 3, results with the ACM functional are overall clearly the best; they are significantly better than MP2, especially for barrier heights. Other conclusions reached in the study (bearing in mind the limitations of the 6-31G* basis set) were [32]:

(1)   Barrier heights calculated with BLYP are almost always *lower* than barriers calculated with MP2, particularly for radical reactions. (For MP2 spin contamination in the underlying UHF wavefunction plays a large part in artificially raising the barrier; this problem is almost non-existent in DFT wavefunctions [33].)

(2)   BLYP barriers for radical reactions tend to be *too* low compared to experiment (Table 3). In particular, reactions with very low barriers may be erroneously predicted to have no barrier at all (e.g. $OH + H_2$ and $CCH + H_2$).

(3)   ACM barriers tend to be *higher* than the corresponding BLYP barrier, particularly for radical reactions, and are typically in better agreement with experiment. For radicals, ACM barriers are still lower than with MP2.

(4)   Although giving at least as good energetics on average as MP2, there are some disturbing instances (OH + H$_2$) where the BLYP PES shows "unusual" behaviour (such behaviour is not confined to the BLYP functional). The ACM functional, which includes a portion of the exact exchange, appears to remedy the situation.

(5)   The ACM functional gives excellent geometries which are often in better agreement with experiment than MP2. BLYP geometries are often poor with bond lengths that are typically too long.

The "unusual" behaviour mentioned on the BLYP OH + H$_2$ potential energy surface refers to the apparent negative barrier height for this reaction (Table 2). This anomalous result prompted an in-depth study of the reaction profile at various levels of theory, the results of which are presented below.

2. OH + H$_2$ ---> H$_2$O + H

The potential energy surface for this reaction was examined at H-O.....H-H distances between 1.0 and 2.1 A along a C$_s$ reaction path with a variety of density functionals, both local (Slater exchange only - Snull - and SVWN) and nonlocal (BP, BLYP and ACM), and compared to potential energy curves obtained at the HF, MP2 and CCSD(T) levels [34].
        Full details are given in the original work [34] but essentially the interatomic O.....H distance was varied from 1.0 to 2.1 A in 0.05 A steps, and at each fixed O.....H distance a full geometry optimization in the space of all remaining variables (within C$_s$ symmetry) was carried out. The resulting energy profiles, with the separated reactants OH + H$_2$ as the energy zero, are shown in fig.1. The 6-31G* (5d) basis was used throughout.
        The reaction profiles are given on two plots: fig.1a shows the profiles for HF, MP2, CCSD(T) and ACM while the remaining density functionals (Snull, SVWN, BP and BLYP) are shown on fig.1b. In both cases the far right hand side of the profile - at an interatomic distance of 2.1 A - essentially represents the reactants, while the far left - at an interatomic distance of 1.0 A - represents the products, H$_2$O + H. (This O.....H distance is more-or-less the correct O-H bond length in water; the optimized H-H distance at this point - not shown on the plots - is very large (typically in excess of 2.0 A) indicating we have essentially a separated hydrogen atom.)
        Within the limitations of the basis set used, we regard the CCSD(T) result to be "exact" and these calculations were done to give an *ab initio* "standard" against which the other methods could be compared. Thus we are looking for a barrier height of around 10 kcal/mol and a heat of reaction of around -10 kcal/mol (experimentally these values are ~4 and -14.6 kcal/mol, respectively).
        The two sets of profiles (figs.1a and 1b) are clearly different. The standard *ab initio* procedures, along with ACM, all show a definite barrier; it's too high with HF and too low with ACM - although fortuitously the ACM barrier is in excellent agreement with experiment (the heat of reaction is in excellent agreement with the CCSD(T) value) - but in all cases there is a clear barrier. None of the "pure" density functionals give any barrier at all. The local functionals (Snull and SVWN) are attractive over the entire range of interatomic distances studied and predict a very exothermic reaction - clear evidence of overbinding; the two nonlocal functionals (BP and BLYP) are essentially flat in to 1.3 A after which they dip down to form the products. There is *no* barrier with DFT.
        Closer examination of the BLYP profile shows there is a very slight "dip" around 1.9 A, with a corresponding "rise" around 1.4-1.5 A. This "rise" is still below the energy of the reactants and acts as a kind of "spurious" transition state - hence the predicted "negative barrier height" with this functional.

Table 2

Heats of reaction (ΔH_f) and barrier heights (EA) for 12 "typical" organic reactions (kcal/mol; see text and ref.32 for more details)

	MNDO	AM1	HF	MP2	BLYP	ACM	Expt.
1. Vinyl Alcohol ---> Acetaldehyde							
EA	91.2	73.6	70.0	55.4	48.7	52.3	39.4
ΔH_f	-7.4	-8.0	-17.8	-17.5	-16.1	-15.5	-9.8
2. Butadiene + Ethylene ---> Cyclohexene							
EA	49.7	23.7	46.2	32.3	21.3	20.3	25
ΔH_f	-34.2	-36.1	-38.6	-35.5	-32.5	-47.1	
3. Tetrazine ---> 2HCN + N$_2$							
EA	97.5	72.6	71.3	40.5	29.9	47.8	51.8
ΔH_f	11.3	-24.3	-71.2	-72.4	-39.6	-28.7	-46.4
4. CF$_3$OH ---> CF$_2$O + HF							
EA	91.2	66.0	59.5	43.9	34.0	41.0	45.1
ΔH_f	16.7	6.4	15.9	11.9	12.5	15.8	5.6
5. Cyclobutene ---> trans-Butadiene							
EA	49.8	35.5	45.2	35.9	29.8	36.1	32.9
ΔH_f	-11.4	-2.1	-19.0	-13.5	-8.7	-15.0	-8.2
6. cis-Butadiene ---> trans-Butadiene							
EA	0.2	1.1	2.7	3.0	4.0	3.5	3.9
ΔH_f	-0.3	-0.8	-3.0	-2.6	-3.8	-3.5	-2.5
7. FO + H$_2$ ---> FOH + H							
EA	41.3	21.0	36.1	22.0	14.7	16.3	17.4
ΔH_f	12.4	13.1	15.0	-1.5	16.9	12.9	3.9
8. OH + H$_2$ ---> H$_2$O + H							
EA	30.7	11.5	26.5	12.7	-0.8[a]	3.3	3.95
ΔH_f	-9.8	-2.6	2.0	-16.9	-5.2	-7.5	-14.6
9. Cyclopropyl radical ---> Allyl radical							
EA	23.1	24.5	24.1	35.5	18.1	24.7	22.0
ΔH_f	-25.2	-29.0	-34.6	-21.7	-32.6	-26.8	-22.8
10. Methyl radical + Ethylene ---> Propyl radical							
EA	13.7	1.7	11.3	18.1	5.0	6.2	7.9
ΔH_f	-34.9	-36.8	-20.7	-29.0	-20.8	-26.8	-25.5
11. Methyl radical + Formaldehyde ---> Ethoxyl radical							
EA	21.5	8.2	11.9	19.8	3.4	4.6	6.8
ΔH_f	-3.6	-14.4	-16.7	-5.1	-11.1	-16.4	-10.8
12. H + HCCH ---> H$_2$ + CCH							
EA	41.7	30.4[b]	34.0	51.0	25.7[b]	28.4	22.2
ΔH_f	40.0	30.4	22.3	46.0	25.7	27.8	

[a] *negative* barrier height with BLYP
[b] no TS could be located for AM1 and BLYP

Table 3
Average and maximum deviations from experiment for barrier heights,
heats of reaction and bond lengths for the compounds and reactions given
in Table 2

Barrier Heights (kcal/mol; 12 reactions)

	MNDO	AM1	HF	MP2	BLYP	ACM
Av.Error	23.4	9.3	13.6	9.9	5.9	3.7
Max.Error	51.8	34.2	30.6	28.8	21.9	12.9

Heats of Reaction (kcal/mol; 10 reactions)

	MNDO	AM1	HF	MP2	BLYP	ACM
Av.Error	10.9	7.5	10.5	6.3	5.9	6.8
Max.Error	57.7	22.1	24.8	26.0	13.0	17.7

Bond Lengths (Angstroms; 35 bonds)

	MNDO	AM1	HF	MP2	BLYP	ACM
Av.Error	0.020	0.020	0.014	0.010	0.015	0.008
Max.Error	0.165	0.091	0.065	0.041	0.034	0.022

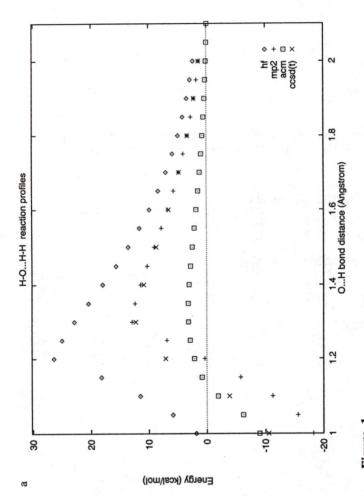

Figure 1
Reaction profiles along a C_s reaction path for OH + H_2 --> H_2O + H with the 6–31G* (5d) basis. Plots show the approaching H-O....H-H distance (Å) versus energy (kcal/mol) relative to the separated reactants. At each (fixed) O....H distance a full optimization of all remaining degrees of freedom was carried out (see text for more details). (a) HF, MP2, CCSD(T) and ACM; (b) Snull, SVWN, BP and BLYP. Reprinted with permission from ref.34.

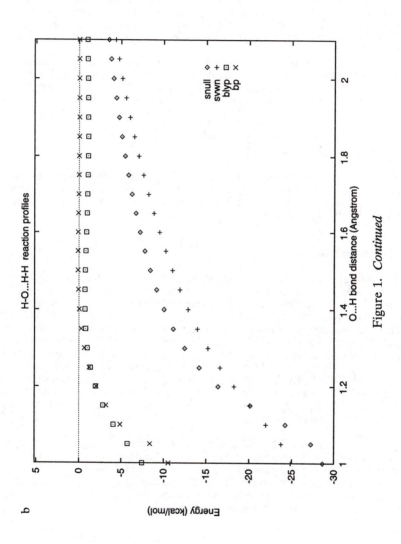

Figure 1. *Continued*

Comparing the four methods that predict a barrier - HF, MP2, CCSD(T) and ACM (fig.1a) - with the four that do not - Snull, SVWN, BP and BLYP (fig.1b) - then it is surely more than simply coincidence that *all of the methods that (correctly) predict a barrier include at least a portion of the exact exchange.* None of the "pure" density functionals include *any* exact exchange and in our opinion the incorrect description for this reaction is primarily due to deficiencies in the (local) Slater exchange term. In many cases - including this one - the ACM functional, because it incorporates a portion of the exact exchange, is able to correct somewhat for the deficiences in the Slater exchange and provide a much better description of reaction profiles.

The OH + H_2 reaction is just one example of the marked tendency of "pure" density functional methods to give too low, or non-existent, barrier heights, especially for low-barrier radical reactions. The following hydrogen abstraction reactions involving methane also fail to give a barrier with DFT-only functionals [35]

$$CH_4 + OH \quad ---> \quad CH_3 + H_2O$$
$$CH_4 + Cl \quad ---> \quad CH_3 + HCl$$
$$CH_4 + H \quad ---> \quad CH_3 + H_2$$

and there are likely to be many more. Problems are not confined to radical reactions only, for example Latajka et al. report that DFT-only methods give too low barriers for proton transfer reactions (e.g. the FHF anion); the results are much improved with hybrid HF-DFT schemes [36].

3. Fluoromethanes

We are currently in the closing stages of a systematic density functional study of fluorination in methane, ethane and ethylene to determine whether DFT can accurately reproduce all the structural and conformational changes that occur following repeated fluorine substitution in these compounds [37]. There are several fairly subtle conformational effects (of order ~1 kcal/mol) on fluorine substitution as well as some marked changes in geometrical parameters across a series. In the above-mentioned study we investigate geometries, dipole moments, rotational barriers and stabilization energies using semiempirical (MNDO, AM1, PM3), *ab initio* (HF, MP2) and density functional (SVWN, BP, BLYP, ACM) methods. Two basis sets were used in the study; 6-31G* and a much larger TZ2P basis [38]. We summarize below results for the fluoromethanes. An extensive series of mainly Hartree-Fock studies on fluorocarbons, including fluoromethanes, was published in the late 1980s by Dixon [39].

Table 4a shows C-F bond lengths for the fluoromethanes at all levels of theory studied along with experimental values (all taken from standard tables [40]). Table 4c gives calculated and experimental dipole moments and Table 4b gives incremental geminal stabilization energies (IGSTAB - the increase in thermodynamic stability of a geminal substituted system relative to the corresponding monosubstituted system); the latter can be derived from the isodesmic reactions:

$$\Delta H^o_f \text{ (kcal/mol)}$$

	ΔH^o_f (kcal/mol)	
2 CH_3F ---> CH_4 + CH_2F_2	-12.3 $\pm$ 4.3	(3a)
3 CH_3F ---> 2 CH_4 + CHF_3	-32.2 $\pm$ 6.5	(3b)
4 CH_3F ---> 3 CH_4 + CF_4	-49.6 $\pm$ 8.4	(3c)

which lead to IGSTABs for fluorine in CH_2F_2, CHF_3 and CF_4 of -6.2, -10.7 and -12.4 kcal/mol, respectively [41].

One of the most well known fluorine substituent effects is the decrease in C-F bond length (and corresponding increase in bond strength) with increasing fluorine

Table 4a
C-F bond lengths in the fluoromethanes (Angstroms)

compound	MNDO	AM1	PM3	6-31G* basis						TZ2P basis						Expt.
				HF	MP2	SVWN	BP	BLYP	ACM	HF	MP2	SVWN	BP	BLYP	ACM	
CH3F	1.347	1.375	1.351	1.365	1.391	1.364	1.389	1.397	1.377	1.363	1.388	1.374	1.400	1.412	1.383	1.382
CH2F2	1.352	1.372	1.349	1.338	1.365	1.348	1.370	1.377	1.356	1.334	1.359	1.351	1.374	1.383	1.357	1.357
CHF3	1.353	1.368	1.346	1.317	1.344	1.333	1.353	1.359	1.338	1.312	1.336	1.333	1.354	1.361	1.337	1.332
CF4	1.347	1.358	1.337	1.302	1.331	1.322	1.341	1.347	1.326	1.296	1.321	1.320	1.340	1.346	1.323	1.323
av.error	0.021	0.023	0.017	0.018	0.009	0.007	0.015	0.022	0.004	0.022	0.004	0.005	0.019	0.028	0.002	

Table 4b
Incremental geminal stabilization energies for fluorine in the fluoromethanes (kcal/mol)

compound	MNDO	AM1	PM3	6-31G* basis						TZ2P basis						Expt.
				HF	MP2	SVWN	BP	BLYP	ACM	HF	MP2	SVWN	BP	BLYP	ACM	
CH2F2	-0.8	-1.5	-4.6	-7.4	-8.2	-9.5	-8.3	-8.0	-8.2	-6.1	-6.9	-7.8	-6.6	-6.3	-6.6	-6.2
CHF3	-1.6	-2.4	-8.9	-13.0	-14.3	-16.6	-14.3	-13.7	-14.2	-9.9	-11.1	-12.6	-10.3	-9.7	-10.4	-10.7
CF4	-1.5	-2.1	-12.2	-15.7	-17.1	-20.0	-17.0	-16.2	-16.9	-12.4	-13.6	-15.4	-12.5	-11.6	-12.8	-12.4
av.error	8.5	7.8	1.2	2.3	3.8	5.6	3.8	2.9	3.7	0.3	0.8	2.2	0.3	0.6	0.4	

Table 4c
Calculated dipole moments in the fluoromethanes (debye)

compound	MNDO	AM1	PM3	6-31G* basis						TZ2P basis						Expt.
				HF	MP2	SVWN	BP	BLYP	ACM	HF	MP2	SVWN	BP	BLYP	ACM	
CH3F	1.76	1.62	1.44	1.99	1.88	1.56	1.63	1.62	1.72	2.00	1.92	1.76	1.81	1.88	1.85	1.86
CH2F2	2.21	2.04	1.81	2.05	1.90	1.53	1.61	1.60	1.72	2.11	1.99	1.80	1.85	1.92	1.91	1.97
CHF3	2.23	2.08	1.88	1.70	1.56	1.28	1.34	1.33	1.43	1.76	1.64	1.51	1.53	1.59	1.58	1.65
av.error	0.31	0.25	0.27	0.09	0.06	0.37	0.30	0.31	0.20	0.13	0.03	0.14	0.13	0.04	0.05	

substitution in fluorinated methanes, a trend that can clearly be observed in the experimental C-F bond lengths in Table 4a and the IGSTABs in Table 4b.

Theoretically, the semiempirical methods do a very poor job of reproducing this trend. C-F bond lengths do decrease with AM1 and PM3 but only very slightly and they actually *increase* with MNDO (except for CF_4); both MNDO and AM1 fail to reproduce the experimental IGSTAB trend although the PM3 results are good with an average error of only 1.2 kcal/mol, a value that is better than any of the *ab initio* calculations with the 6-31G* basis.

All of the *ab initio* methods reliably reproduce both the C-F bond shrinkage and the increasing geminal stabilization, at least relatively. In absolute terms, HF bond lengths are too short and BP and BLYP values too long. The best results by far for C-F bond lengths are ACM/TZ2P with an average error of only 0.002 A; next best are MP2/TZ2P and ACM/6-31G*, both with average errors of 0.004 A. The worst results in absolute magnitude are BLYP/TZ2P which are systematically too long (by 0.028 A on average) although the bond length contraction across the series is well reproduced. For the stabilization energies, results with the TZ2P basis are clearly superior to those with the smaller 6-31G* basis; apart from SVWN/TZ2P the average error is less than 1 kcal/mol in all cases. Perhaps surprisingly, Hartree-Fock gives the best overall results, although BP and ACM are very close (average errors of 0.3, 0.3 and 0.4 kcal/mol with the TZ2P basis, respectively). However, some caution is in order regarding the exact ordering, since the error bars on the heats of reaction (3a-3c, above) are rather large.

For the dipole moments, once again the semiempirical methods are not especially successful at duplicating the experimental values or trend; all three methods predict that CHF_3 has a larger dipole moment than CH_2F_2 whereas the reverse is the case experimentally. The calculated DFT/6-31G* dipole moments are not much better; except for ACM the other three density functionals show CH_3F with a larger dipole moment than CH_2F_2 which is again the reverse of the experimental observation. All DFT dipole moments are consistently too low with this basis set. Both HF and, especially, MP2 give much better results.

The DFT results improve significantly with the larger TZ2P basis. Both BLYP and ACM now give good agreement with experiment, with average errors of 0.04 and 0.05 debye, respectively. Best of all is MP2/TZ2P, with an average error of just 0.03 debye. Hartree Fock dipole moments actually worsen with the better basis, all values are now too high.

Overall, the best theoretical method for reproducing the observed physical properties of the fluoromethanes that we have examined here is clearly ACM/TZ2P. Its nearest rival is MP2/TZ2P.

4. Oxides of Nitrogen

Another study currently in progress is an investigation of the geometries, energetics and properties of various nitrogen oxides [42]. This study was undertaken following calculation of the heats of formation of NO, N_2O, NO_2, N_2O_3, N_2O_4, N_2O_5 and NO_3 using local DFT; apart from NO all calculated values were exothermic, and dramatically so, in marked contrast to experiment (all the nitrogen oxides have endothermic heats of formation). This is shown in Table 5 which gives calculated heats of formation with various density functionals using a TZ2P basis along with experimental values and spin-projected MP4 (based on HF geometries) with a split-valence + polarization basis.

As can been seen, for the six oxides of nitrogen shown, heats of formation with local DFT (SVWN) are all exothermic and bear no relation at all to experiment (the average error is a staggering 184.3 kJ/mol). There is a steady improvement on progressing to the nonlocal BP and BLYP functionals, but even here the calculated values are always too exothermic; however results with the hybrid ACM functional are

Table 5
Heats of formation of some nitrogen oxides (kJ/mol)
(DFT calculations used a TZ2P basis; PMP4 used a split-valence plus
polarization basis. All calculated values include ZPVE corrections)

compound	SVWN	BP	BLYP	ACM	PMP4	Expt.
N_2O	-19.5	38.5	55.5	61.9	82.8	85.5
NO_2	-55.8	-2.7	10.9	21.4	35.7	35.9
N_2O_3	-89.9	30.5	45.5	82.1	112.8	82.8
N_2O_4	-254.7	-66.7	-35.7	-3.3	34.1	18.7
N_2O_5	-288.0	-50.3	-15.2	11.2	65.0	11.3
NO_3	-86.0	20.9	39.1	71.9	98.6	77.5
av.error	184.3	56.9	35.4	11.1	20.5	

generally good, with a lower average error even than MP4 (which is enormously more expensive, even with the smaller basis set).

5. Hydrogen Bonding

A reliable description of hydrogen bonding is a challenge for theory due to the weakness of the interactions involved compared to, e.g., those in a conventional covalent bond. Effects such as basis set superposition error (BSSE), which can normally be neglected, may need to be addressed, especially if the basis used is not sufficiently large. A nice discussion of the various theoretical considerations involved in order to obtain accurate structures and energetics for hydrogen-bonded and other van der Waals systems is given in the review article by Chalasinski and Szczesniak [43].

There have now been several DFT studies of hydrogen bonding and the general conclusion appears to be that - with a reliable basis set - nonlocal density functionals can provide an adequate description of hydrogen bonding (e.g., see ref.44). However, a very recent article from Del Bene and coworkers [45] does suggest that, although results are generally good, care is required in the selection of a basis set, and MP2 results are often better, even when compared against hybrid functionals. For very weak van der Waals dimers, Pulay has shown that current functionals fail to describe the dispersion interaction properly near the van der Waals minimum and has concluded that "present DFT theories are probably not useful for the investigation of weakly interacting systems" [46].

We present here a study of three hydrogen-bonded systems: HF..HF, CO..HF and OC..HF. We concentrate of the HF dimer as this has been extensively studied theoretically [47], including a recent DFT study [48], and there is reliable experimental data available for comparison. To the best of our knowledge there have been no previous DFT studies on CO..HF and OC..HF. Comparisons are made between HF, MP2, SVWN, BP, BLYP and ACM.

We have used the same TZ2P basis that was used in our studies on the fluoromethanes and nitrogen oxides (subsections 3 and 4, above). Tests have shown that BSSE effects are small with this basis; consequently we have made no corrections for BSSE. Additional calculations on HF..HF with a diffuse s function on hydrogen and diffuse s and p functions on fluorine had minimal effect on optimized geometries, so no extra diffuse functions were included in the basis.

Table 6a presents optimized monomer and dimer geometries, tables 6b and 6c show dipole moments and vibrational frequencies, respectively, and table 6d gives calculated dimer binding energies. Note that the latter are D_e values, uncorrected for ZPVE (the harmonic approximation is often inaccurate for weakly bound complexes). The experimental binding energy reported for the HF dimer is the D_0 value obtained by Dayton et al. [49] corrected using the zero-point energy difference proposed by Racine and Davidson [46,47].

Looking first at the calculated quantities for the monomers (CO and HF) we see that once again ACM geometries are in excellent agreement with experiment (table 6a). Hartree-Fock bond lengths are too short and BP and BLYP are too long.

As is well known the Hartree-Fock dipole moment of CO has the wrong sign (the dipole vector points away from the O atom); all the correlated methods give the correct dipole direction with ACM giving the best agreement with experiment (table 6b). The MP2 dipole moment is too high. For HF, all methods give reasonable dipole moments although the Hartree-Fock and SVWN values are a bit too high.

It is well established that DFT, both local and nonlocal, gives fairly good agreement with observed vibrational frequencies [10,50]. However, this is somewhat fortuitous since what is typically compared are calculated harmonic and experimental anharmonic frequencies; what should be compared is clearly harmonic with harmonic. If this is done, then for both CO and HF, ACM gives by far the best results (table 6c).

For the HF..HF dimer, most of the methods studied give fair to good agreement with the experimentally determined interatomic distances; a notable exception is SVWN which predicts a far too short intermolecular F-F distance and a correspondingly too large increase in the H-F distances compared to that in the monomer (table 6a). The calculated dimer binding energy is also too high (table 6d). This is a further manifestation of the overbinding tendency of local DFT. The inability of LDA to deal with hydrogen bonding has already been noted by Salahub and coworkers [51]. For the bond angles, although the FFH[d] angle is well predicted by almost all methods, the calculated FFH[a] angle is too low in all cases except Hartree-Fock. Overall, MP2 gives perhaps the best agreement with the experimental geometry; however ACM is also very close and gives very similar geometrical parameters to MP2.

The best agreement with the *observed* vibrational frequencies for HF..HF is with the BP nonlocal functional, closely followed by BLYP (table 6c). However, we again reiterate that we are comparing calculated harmonic with experimental anharmonic frequencies. Harmonic frequencies should always be higher than anharmonic and we note that the only methods for which the calculated frequencies are consistently higher than experiment are MP2 and ACM; these two methods are in fact in fairly good agreement with one another. We would conjecture that if experimental harmonic frequencies were available for the HF dimer, the best agreement would be provided by ACM. Although the unscaled Hartree-Fock frequencies give the worst agreement with experiment, if the usual scaling factor of 0.89 is employed then the average (scaled) error falls to 30 cm^{-1} due to the much improved agreement for the two highest modes.

There is very little experimental data available for the CO..HF dimers. A microwave study in the early 1980s led Legon and coworkers to conclude that the dominant species present in the gas phase was OC..HF, which was linear with an average C-F distance of 3.047 A [52].

All theoretical methods predict OC..HF to be more stable than CO..HF, although the difference is not particularly marked at the Hartree-Fock level. Overall, the best agreement with the limited experimental data is given by MP2 with again the ACM results being very similar. As with the HF dimer, SVWN is too binding.

One cannot draw any firm conclusions from such a limited study, but the ACM functional does appear to give results very comparable to MP2 and generally better than the other density functionals examined here.

Table 6a
Monomer and dimer geometries for dimers of HF and CO
(Angstroms and degrees; all calculations used a TZ2P basis set)

compound	HF	MP2	SVWN	BP	BLYP	ACM	Expt.
rCO	1.103	1.135	1.128	1.136	1.138	1.124	1.128
rHF	0.898	0.918	0.932	0.929	0.933	0.920	0.917
CO..HF							
rCO	1.105	1.135	1.130	1.137	1.138	1.126	
rHF	0.900	0.919	0.937	0.931	0.935	0.921	
rOH	2.150	2.165	1.820	2.160	2.157	2.169	
OC..HF							
rCO	1.100	1.132	1.123	1.132	1.133	1.121	
rHF	0.901	0.925	0.950	0.941	0.943	0.929	
rCH	2.277	2.057	1.797	1.996	1.994	2.022	
av. rCF	3.178	2.982	2.747	2.937	2.937	2.951	3.047

$$F \overset{\displaystyle H^d}{\underset{\displaystyle H^a}{\cdots\cdots\cdots F}}$$

HF..HF							
rFF	2.810	2.719	2.548	2.740	2.741	2.720	2.72-2.79
rFH^d	0.902	0.924	0.948	0.938	0.942	0.927	0.921
rFH^a	0.901	0.922	0.937	0.933	0.937	0.923	0.919
$aFFH^a$	119.8	108.9	99.2	105.5	105.5	108.8	117 ± 6
$aFFH^d$	6.9	7.9	9.4	8.1	8.1	7.6	7-9

Table 6b
Calculated dipole moments for HF, CO monomers and dimers (debye)

compound	HF	MP2	SVWN	BP	BLYP	ACM	Expt.
CO	-0.15	0.28	0.22	0.19	0.15	0.12	0.11
HF	1.94	1.88	1.91	1.85	1.86	1.88	1.82
CO..HF	2.48	1.99	2.37	2.08	2.14	2.15	
OC..HF	2.23	2.79	3.05	2.78	2.75	2.69	
HF..HF	3.58	3.34	3.36	3.28	3.30	3.37	

Continued on next page

Table 6. *Continued*

Table 6c
Calculated harmonic vibrational frequencies for CO, HF and HF..HF (cm^{-1})

compound	HF	MP2	SVWN	BP	BLYP	ACM	obs.	har.
CO	2422	2119	2373	2121	2108	2215	2143	2170
HF	4470	4160	3990	3986	3930	4127	3963	4139
HF..HF	4434	4113	3918	3940	3886	4087	3930	
	4385	4034	3677	3769	3758	3963	3868	
	511	597	758	610	607	599	510	
	438	495	584	493	481	490	400	
	198	221	296	234	222	223	216	
	139	168	187	162	167	164	150	
av. error	177	92	125	55	59	75		

Table 6d
Calculated dimer binding energies (kcal/mol)

compound	HF	MP2	SVWN	BP	BLYP	ACM	Expt.
CO..HF	1.6	1.5	3.9	0.8	1.6	1.1	
OC..HF	2.3	4.1	8.4	3.7	4.2	3.6	
HF..HF	4.1	5.0	9.3	4.3	5.3	4.5	4.8 ± 0.2

6. DFT Validation Study

Finally we present some preliminary results from a systematic density functional study of the geometries, ionization potentials and dipole moments of over 100 small to medium-sized molecules (ranging in size from 2 to 9 atoms) and the energetics of over 300 reactions involving these molecules (corrected for ZPVE) which we have recently completed [53]. The study encompasses both local, nonlocal and hybrid (ACM) functionals with a range of Gaussian basis sets from split-valence + polarization to large uncontracted correlation-consistent, containing f functions on first and second row atoms and d functions on hydrogen [54]. Calculated values are compared with experiment and with traditional HF and MP2 techniques. Some of these results have been presented elsewhere [55], but only for a subset of our total data base.

A full list of the molecules in our data set is given in Table 7. Again, no detail of individual molecules and reactions are given; instead we present a summary of the average errors for all bond lengths and the average error in the calculated heats of reaction for the over 300 reactions examined (including atomization energies, hydrogenations, oxidations and miscellaneous) as compared to experimental values. For more details see refs. [53,55].

Table 7
Molecules in DFT validation study

LiH	PH	HCOH	S$_2$
BeH	PH$_2$	HCOOH	Cl$_2$
BF$_2$H	PH$_3$	CH$_3$OH	ClO$_2$
BH$_2$	SH	CH$_3$NH$_2$	Cl$_2$O
BN	H$_2$S	CH$_3$PH$_2$	NSF
BO	HCl	N$_2$	(CH$_3$)$_2$O
BO$_2$	Li$_2$	N$_2$H$_2$ (trans)	PO
BS	LiF	N$_2$H$_2$ (cis)	HPO
B$_2$H$_6$	C$_2$H$_2$	N$_2$H$_4$	HNO
CH	C$_2$H$_4$	NH$_2$CN	HNO$_3$
CH$_2$(1)	C$_2$H$_6$	CH$_2$N$_2$	CS
CH$_2$(3)	CH$_3$CCH	HNCNH	SO
CH$_3$	H$_2$CCCH$_2$	NO$_2$	SO$_2$
CH$_4$	CH$_3$CHO	NF	SO$_3$
NH	CH$_2$CHOH	O$_2$	NaCl
NH$_2$	CH$_3$CHCH$_2$	H$_2$O$_2$	ClF
NH$_3$	CN	F$_2$	Si$_2$H$_6$
OH	HCN	FCN	CH$_3$F
H$_2$O	HNC	FOOF	CH$_3$Cl
HF	HCP	FOH	CH$_3$SH
SiH	HCNO	F$_2$CO	CH$_2$F$_2$
SiH$_2$(1)	HNCO	CO$_2$	CH$_2$NH
SiH$_2$(3)	CO	Na$_2$	HOCl
SiH$_3$	HCO	Si$_2$	NOCl
SiH$_4$	H$_2$CO	P$_2$	NOF

systems in the last column are cyclic

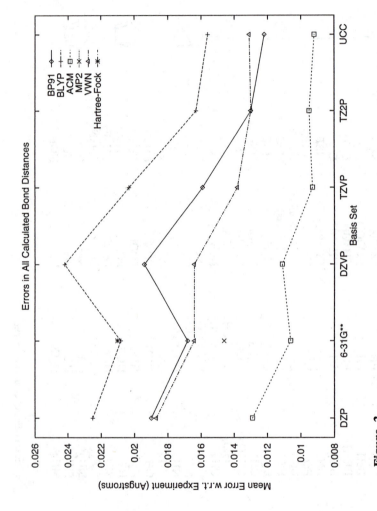

Figure 2
Average error in calculated bond lengths versus basis set for the set of molecules in Table 7. Basis size generally increases from left to right; UCC is the large (uncontracted) correlation-consistent basis of Dunning. (See text and ref.53 for more details).

Fig.2 shows the average error in bond length plotted against (increasing) basis set size. The figure is self-explanatory and clearly shows that the ACM functional consistently provides better average geometries than all the other functionals examined (SVWN, BP and BLYP) for all basis sets. For the larger triple-zeta quality bases, the average error in the calculated bond lengths is less than 0.01 A. The greater relative accuracy of ACM is particularly noticeable with the smaller double-zeta basis sets. Note that local SVWN geometries are usually better than nonlocal and BLYP geometries are clearly and consistently worst of all, with an average error greater than 0.02 A for almost all basis sets. For the 6-31G** basis, both HF and MP2 averages are given; MP2 geometries are on the whole better than "pure" DFT geometries but clearly inferior to ACM.

Average errors in calculated reaction energies are plotted in fig.3a and, on an expanded scale for the nonlocal functionals, fig.3b. Again ACM is clearly superior to the other methods; the average error with the TZ2P basis approaches 6 kcal/mol for ACM, compared to over 7 kcal/mol for BP and almost 8 kcal/mol for BLYP (fig.3b). All the nonlocal functionals are markedly better than (local) SVWN for which the average error is around 25 kcal/mol and shows almost no improvement as basis size increases. With the 6-31G** basis, all the nonlocal functionals show better average energetics than MP2; HF is clearly worst of all with an average error of over 50 kcal/mol! Note that the DZVP and TZVP bases were specifically optimized for DFT [56] (unlike all other basis sets which were optimized for HF or HF-based wavefunctions) and the fact that the BP and BLYP energetics show a clear relative improvement in these cases suggests that effort directed towards basis set optimization for the ACM functional ought to prove rewarding.

7. Organometallics

Almost all of the work presented above involves typical small to medium-sized "organic" systems. Although we have done only limited work to date on systems involving transition metals, there is increasing indication in the recent literature that hybrid HF-DFT schemes provide improved accuracy relative to standard nonlocal functionals for organometallics as well. We quote just a few examples below.

In a study of the successive binding energies of $Fe(CO)_5^+$, Ricca and Bauschlicher [57] found that although the BLYP functional alone gave very poor results, the hybrid B3LYP functional (this is similar to the ACM functional as defined in Eq.2, except that the Perdew 91 correlation functional (PW91) is replaced by the LYP functional) gave excellent results, better than those obtained from a modified coupled-pair approach (MCPF); additionally B3LYP gave better geometries and frequencies than MP2 and these authors opined that B3LYP might be the "method of choice for all but highly accurate calculations on small systems containing transition metals". Further studies on $Co(H_2)_n^+$ [58] and $V(H_2)_n^+$ [59] confirmed the reliability of the hybrid functional.

Similarly in a study of cationic transition-metal methyl complexes MCH_3^+ (M=Sc-Cu,La,Hf-Au) which specifically examined the performance of hybrid HF-DFT functionals, Holthausen et al. concluded that such methods were "a promising alternative to rigorous high level *ab initio* theory, at least for a description of the electronic structure of singly bonded open-shell transition metal complexes" [60]. An essentially similar conclusion was drawn by the same group in a related study of the bonding in cationic first-row transition metal methylene complexes MCH_2^+ [61].

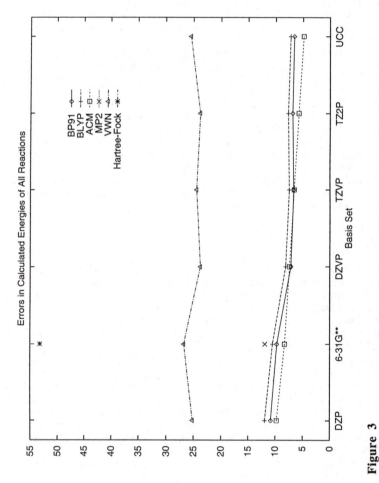

Figure 3
Average error in calculated heats of reaction (including ZPVE corrections) versus basis set for over 300 reactions derived from the set of molecules in Table 7. Basis size generally increases from left to right; UCC is the large (uncontracted) correlation-consistent basis of Dunning. (See text and ref.53 for more details). (a) All density functionals studied; (b) Nonlocal functionals (including ACM) on an expanded scale.

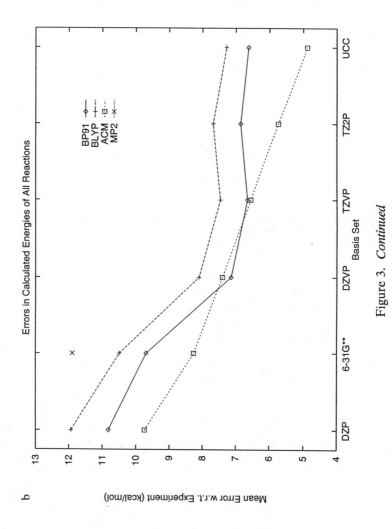

Figure 3. *Continued*

SUMMARY

In this article we have attempted to show that, for a wide range of chemical systems, better predictions on average for molecular structures and energetics (and also certain molecular properties) are obtained with hybrid HF-DFT functionals - such as ACM - than with "DFT-only" functionals, both local and nonlocal, or with standard *ab initio* procedures, such as HF and MP2. Our summary of the current state-of-the-art in density functional applications is:

1.   Local density functionals give reasonable geometries and
      vibrational frequencies but energetics (heats of reaction,
      barrier heights) are generally poor. Molecular stability is
      exaggerated due to overbinding. Weakly bound species in
      particular are poorly described by LDA.

2.   Nonlocal functionals usually give much improved energetics
      (typically of the same quality or better than MP2) but geometries
      for standard covalent systems are often worse, particularly for
      BLYP which gives bond lengths that are systematically too long.

3.   DFT-only barrier heights for low-barrier reactions are usually
      too low, or non-existent, especially for radical reactions.

4.   Functionals which include a portion of the true HF exchange
      (such as ACM) provide *significantly* better results than standard
      nonlocal functionals. In particular

      *   geometries are better than MP2

      *   heats of reaction are better than MP2

      *   barrier heights are better than MP2

      *   computational cost is *much* less than standard
           implementations of MP2

In conclusion, we would argue that hybrid HF-DFT functionals are overall the best density functionals currently available, certainly for organic chemistry and potentially in other areas of chemistry as well.

Despite our glowing recommendation of the ACM functional, it is only fair to point out that currently it requires calculation of all the two-electron integrals in order to determine the HF exchange. This requirement means that, in terms of computational speed, it can never be faster than a standard Hartree-Fock calculation, whereas other non-hybrid DFT implementations - which do not need any HF exchange - can, by suitable approximations to the Coulomb term, be made much faster. For example, DGAUSS [16] uses a "resolution of the identity" approach, along with an auxiliary basis, to replace the four-centre two-electron coulomb integrals by products of three-centre integrals which can be computed much more efficiently. DMOL [15] uses a purely numerical basis and solves the Poisson equations to approximate the Coulomb term; again this is theoretically much faster. Of course, for the ACM functional - and in similar implementations for other density functionals - since the two-electron integrals are calculated for the exchange term, then they are available "for free" so to speak for the Coulomb term, which can be determined with much greater accuracy. It remains to be seen if a functional can be developed which can include the effects of "exact exchange" without having to "include" the two-electron integrals as well.

ACKNOWLEDGMENTS

Much of the work presented here was supported in part by a grant to Biosym Technologies from the National Institute of Standards and Technology/Advanced Technology Program.

LITERATURE CITED

[1]   P.Hohenberg and W.Kohn
        Phys.Rev.B. **136** (1964) 864;
[2]   W.Kohn and L.J.Sham
        Phys.Rev.A. **140** (1965) 1133
[3]   A.D.Becke
        Phys.Rev.A **38** (1988) 3098
[4]   J.P.Perdew
        in "Electronic Structure of Solids", ed. P.Ziesche and H.Eschrig,
        (Akademie Verlag, Berlin, 1991)
[5]   C.Lee, W.Yang and R.G.Parr
        Phys.Rev.B **37** (1988) 785
[6]   L.Fan, L.Verluis, T.Zeigler, E.J.Baerends and W.Ravenek
        Int.J.Quant.Chem.Symp. **22** (1988) 173
[7]   R.Fournier, J.Andzelm and D.R.Salahub
        J.Chem.Phys. **90** (1989) 6371;
[8]   B.I.Dunlap, J.Andzelm and J.W.Mintmire
        Phys.Rev.A **42** (1990) 6354
[9]   B.Delley
        J.Chem.Phys. **94** (1991) 7245
[10]   B.G.Johnson, P.M.W.Gill and J.A.Pople
        J.Chem.Phys. **98** (1993) 5612
[11]   R.Fournier
        J.Chem.Phys. **92** (1990) 5422;
[12]   B.I.Dunlap and J.Andzelm
        Phys.Rev.A **45** (1992) 81
[13]   A.Komornicki and G.J.Fitzgerald
        J.Chem.Phys. **98** (1993) 1399
[14]   N.C.Handy, D.J.Tozer, G.J.Laming, C.W.Murray and R.D.Amos
        Israel J.Chem. **33** (1993) 331
[15]   DMOL, v2.3.5, Biosym Technologies, San Diego, CA (1994)
[16]   DGAUSS, Cray Research Inc., Eagan, MN
[17]   ADF, Department of Theoretical Chemistry, Vrije University,
        Amsterdam
[18]   M.J.Frisch, G.W.Trucks, H.B.Schlegel, P.M.W.Gill, B.G.Johnson,
        M.W.Wong, J.B.Foresman, M.A.Robb, M.Head-Gordon, E.S.Replogle,
        R.Gomperts, J.L.Andres, K.Raghavachari, J.S.Binkley, C.Gonzalez,
        R.L.Martin, D.J.Fox, D.J.Defrees, J.Baker, J.J.P.Stewart and J.A.Pople
        GAUSSIAN 92/DFT, Gaussian Inc., Pittsburgh, PA (1993)
[19]   R.D.Amos, I.L.Alberts, J.S.Andrews, S.M.Colwell, N.C.Handy,
        D.Jayatilaka, P.J.Knowles, R.Kohayashi, N.Koga, K.E.Laidig,
        P.E.Maslen, C.W.Murray, J.E.Rice, J.Sanz, E.D.Simandiras
        A.J.Stone and and M.D.Su
        CADPAC 5, Department of Chemistry, University of Cambridge,
        Cambridge, UK (1992)
[20]   R.Ahlrichs, M.Bar, M.Ehrig, M.Haser, H.Horn and C.Kolmel
        TURBOMOLE, v.2.5, Biosym Technologies, San Diego, CA (1995)

[21] T.Ziegler
      Chem.Rev. **91** (1991) 651 and references therein
[22] J.Labanowski and J.Andzelm, Eds.
      Density Functional Methods in Chemistry
      (Springer-Verlag, New York, 1991)
[23] L.A.Curtiss, K.Raghavachari, G.W.Trucks and J.A.Pople
      J.Chem.Phys. **94** (1991) 7221
[24] P.C.Hariharan and J.A.Pople
      Theor.Chim.Acta. **28** (1973) 213
[25] We use the parametrization by S.H.Vosko, L.Wilk and M.Nusair,
      Can.J.Phys. **58** (1980) 1200 of the exact uniform gas results of
      D.M.Ceperley and B.J.Alder, Phys.Rev.Letts. **45** (1980) 566
[26] D.A.Dixon, J.Andzelm, G.Fitzgerald and E.Wimmer
      J.Phys.Chem. **95** (1991) 9197
[27] B.G.Johnson, C.A.Gonzales, P.M.W.Gill and J.A.Pople
      Chem.Phys.Letts. **221** (1994) 100
[28] C.A.Gonzales and B.G.Johnson
      Poster at satellite symposium of 8th International Congress of
      Quantum Chemistry, Cracow, Poland, June 13-16, 1994
[29] K.Raghavachari, D.L.Strout, G.K.Odom, G.E.Scuseria, J.A.Pople,
      B.G.Johnson and P.M.W.Gill
      Chem.Phys.Letts. **214** (1993) 357
[30] A.D.Becke
      J.Chem.Phys. **98** (1993) 5648
[31] A.D.Becke
      J.Chem.Phys. to be published
[32] J.Baker, M.Muir and J.Andzelm
      J.Chem.Phys. **102** (1995) 2063
[33] J.Baker, A.Scheiner and J.Andzelm
      Chem.Phys.Letts. **216** (1993) 380
[34] J.Baker, J.Andzelm, M.Muir and P.R.Taylor
      Chem.Phys.Letts. **237** (1995) 53
[35] D.A.Dixon, Dupont, private communication
[36] Z.Latajka, Y.Bouteiller and S.Scheiner
      Chem.Phys.Letts. **234** (1995) 159
[37] M.Muir and J.Baker
      to be published
[38] A.Schafer, H.Horn and R.Ahlrichs
      J.Chem.Phys. **97** (1992) 2571
[39] D.A.Dixon
      J.Phys.Chem. **92** (1988) 86
[40] D.R.Lide, Editor, Handbook of Chemistry and Physics
      (74th edition, CRC Press, Boca Raton, FL, 1993)
      reprinted from: Kagaku Benran, 3rd edition, vol.II (1984) 649
[41] B.E.Smart
      "Fluorinated Organic Molecules", chap.4 in "Molecular Structure and
      Energetics", vol.3, ed. J.F.Liebman and A.Greenberg (VCH, 1986)
[42] H.Munakata, T.Kakumoto and J.Baker
      to be published
[43] G.Chalasinski and M.M.Szczesniak
      Chem.Rev. **94** (1994) 1723
[44] K.Kim and K.D.Jordan
      J.Phys.Chem. **98** (1994) 10089
[45] J.E.Del Bene, W.B.Person and K.Szczepaniak
      J.Phys.Chem. **99** (1995) 10705
[46] S.Kristyan and P.Pulay
      Chem. Phys. Letts. **229** (1994) 175

[47] S.C.Racine and E.R.Davidson
      J.Phys.Chem. **97** (1993) 6367
[48] Z.Latajka and Y.Bouteiller
      J.Chem.Phys. **101** (1994) 9793
[49] D.C.Dayton, K.W.Jucks and R.E.Miller
      J.Chem.Phys. **90** (1989) 2631
[50] J.Andzelm and E.Wimmer
      J.Chem.Phys. **96** (1992) 1280
[51] F.Sim, A.St.-Amant, I.Papai and D.R.Salahub
      J.Am.Chem.Soc. **114** (1992) 4391
[52] A.C.Legon, P.D.Soper and W.H.Flygare
      J.Chem.Phys. **74** (1981) 4944
[53] A.Scheiner, J.Baker and J.Andzelm
      to be published
[54] T.H.Dunning
      J.Chem.Phys. **90** (1989) 1007
[55] J.Andzelm, J.Baker, A.Scheiner and M.Wrinn
      Int.J.Quant.Chem. in press
[56] N.Godbout, D.R.Salahub, J.Andzelm and E.Wimmer
      Can.J.Chem. **70** (1992) 560
[57] A.Ricca and C.W.Bauschlicher
      J.Phys.Chem. **98** (1994) 12899
[58] C.W.Bauschlicher and P.Maitre
      J.Phys.Chem. **99** (1995) 3444
[59] P.Maitre and C.W.Bauschlicher
      J.Phys.Chem. **99** (1995) 6836
[60] M.C.Holthausen, C.Heinemann, H.H.Cornehl, W.Koch and H.Schwarz
      J.Chem.Phys. **102** (1995) 4931
[61] M.C.Holthausen, M.Mohr and W.Koch
      Chem.Phys.Letts. **240** (1995) 245

Chapter 25

Copper Corrosion Mechanisms of Organopolysulfides

Anne M. Chaka[1], John Harris[2], and Xiao-Ping Li[2]

[1]Lubrizol Corporation, 29400 Lakeland Boulevard, Wickliffe, OH 44092–2298
[2]Biosym/Molecular Simulation, 9685 Scranton Road, San Diego, CA 92121–3752

Organopolysulfide lubricant additives are effective antiwear agents which protect ferrous metal components, but also cause corrosion of copper-based alloys such as bronze and brass in the same mechanical systems. In commercial organopolysulfides of the type $R\text{-}(S)_n\text{-}R$, the corrosive behavior of polysulfides dramatically increases when $n \geq 4$. Three possible reasons for this behavior are examined using local and nonlocal density functional theory as well as post-Hartree-Fock theory at the MP2 level. In addition we present some of the first results using a new density functional program, Fast_Structure, based on the Harris functional.

Organopolysulfide lubricant additives which effectively passivate and protect ferrous metals often corrode copper-containing metal alloys such as bronze and brass. This is a serious limitation as ferrous and non-ferrous metals are commonly used to fashion different parts of the same mechanical system. The ultimate goal is to minimize the corrosive behavior of the additives while preserving their effective wear protection performance. In commercial organopolysulfides of the type $R\text{-}S_n\text{-}R$, where $n=2\text{-}6$, the longer sulfide chains are known empirically to be much more corrosive with respect to copper than the shorter chains where $n \leq 3$. Currently little is known about the corrosion mechanisms involved which can explain this difference in reactivity. We propose three hypotheses in an effort to determine why the corrosive behavior of polysulfides dramatically increases when $n \geq 4$. The first hypothesis proposes that the S-S bonds are weaker and hence more reactive in the longer polysulfides. The second hypothesis suggests that the longer polysulfides are more corrosive because they are capable of removing copper atoms from the surface via a chelation mechanism. The third hypothesis considered is that the steric bulk of the hydrocarbon side chains can inhibit corrosion by limiting the contact of the sulfur with the surface in the shorter polysulfide chains, but not for the inner sulfur atoms in the longer chains.

In this study we present some of the first results utilizing a new density functional program Fast_Structure (FS) (1), based on the Harris functional with trial densities constructed from spherically symmetric site-densities. These results are compared with Kohn-Sham (2) density functional theory with and without nonlocal gradient corrections, as well as post-Hartree-Fock results at the MP2 level (3).

Methodology

Polysulfides are linear molecules with dihedral angles of approximately 90°. Disulfides can exist in either *d* or *l* conformations, and polysulfides can form either right or left-handed helices. Chains with at least five atoms can be described as either *cis* or *trans* depending on whether the two terminal atoms are on the same or opposite side of the plane formed by the three central atoms. Fibrous sulfur S_∞ exists in the all-*trans* conformation, forming a helical structure (*4*). In this study we use polysulfides in the all-*trans* conformation, analogous to S_∞, and with twofold symmetry. Starting geometries for all neutral species in this study were obtained using Fast_Structure. Additional refinement was performed using Hartree-Fock self consistent field (SCF) theory and MP2 methodology as implemented in HONDO 94.8,(*5*) and Kohn-Sham DFT in DMol 2.3.5 (*6*). The functional used for the local density calculations is that of Vosko, Wilk, and Nusair (*7*). For the nonlocal density calculations we used Becke's gradient corrections to the exchange (*8*) and that of Lee, Yang, and Parr (*9*) for the correlation. All basis sets used are of double zeta plus polarization (DZP) quality, with Delley's numerical DNP for the DFT work (*6*), and Dunning and Hay's DZP (*10*) for the Hartree-Fock and MP2. All calculations on open-shelled systems were performed with spin unrestricted.

Fast_Structure, a new program based on the Harris functional, numerical basis sets, and a novel optimization technique, can be very useful as a fast initial geometry optimization method without the difficulties associated with parameterization. For a detailed description of the methodology behind Fast_Structure and benchmark results, see the chapter by Li, *et al.*, in this volume.

To determine the utility of Fast_Structure for the polysulfide molecules in this study, a detailed examination of the geometry and potential surface, including planar rotational transition state structures, of H_2S_2 and H_2S_4 was performed. The results are presented in Tables I-V and Figure 1. For the equilibrium geometries of H_2S_2 and H_2S_4, Fast_Structure gives an H-S bond distance and dihedral angles which are indistinguishable from experiment. S-S bond distances are from 0.02 to 0.05 Å too long, but in between the values obtained by DMol with and without nonlocal gradient corrections. There is good agreement with the SSS bond angle in H_2S_4, but a 5° difference with the HSS bond angle in H_2S_4. All of the other methods differ by more than 6° from the experimental value for the HSS angle, so the latter may be in some doubt.

Table I. Ground state geometry of H_2S_2

	FS (VWN)	DMol (VWN)	DMol (BLYP)	SCF	MP2	Expt.
r(HS) Å	1.348	1.371	1.387	1.338	1.337	1.35[a]
r(SS) Å	2.097	2.062	2.128	2.077	2.066	2.055[b]
<(HSS)	97.1°	98.2°	98.2°	98.4°	98.6°	91.95°
<(HSSH)	90.4°	90.0°	90.7°	90.2°	90.3°	90.6°[c]

[a]Ref. 11. [b]Ref. 12. [c]Ref. 13.

Examination of the planar rotational transition states in Tables II and IV and Figure 1, reveals that Fast_Structure predicts the transition state structure geometry with the same accuracy as the ground state geometry relative to the other methods. Experimental values for the *anti* rotational barrier height about the unhindered S-S bond in H_2S_2 range from 1.9 to 6.9 kcal/mol (*15-18*), and Fast_Structure gives a

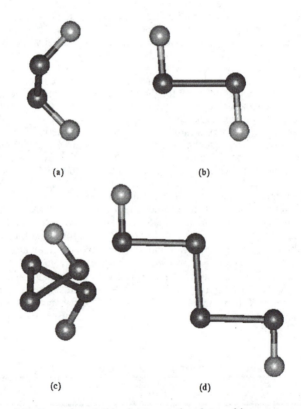

Figure 1. (a). Ground and (b) planar rotational transition state geometry of H_2S_2. (c). Ground and (d) planar triple rotational transition state geometry of H_2S_4.

Table II.   Geometry of H_2S_2 Planar Transition State

	FS (VWN)	DMol (VWN)	DMol (BLYP)	SCF	MP2
r(HS) Å	1.348	1.365	1.365	1.330	1.333
r(SS) Å	2.141	2.116	2.117	2.106	2.110
<(HSS)	91.2°	92.1°	92.2°	94.4°	93.2°

Table III.   Ground State Geometry of H_2S_4

	FS (VWN)	DMol (VWN)	DMol (BLYP)	SCF	MP2	Expt [a]
r(H-S1) Å	1.351	1.370	1.386	1.338	1.337	
r(S1-S2) Å	2.084	2.068	2.139	2.071	2.064	2.03
r(S2-S3) Å	2.091	2.059	2.122	2.071	2.065	2.07
<(HSS)	95.6°	97.4°	97.3°	98.1°	97.9°	
<(SSS)	105.7°	108.7°	111.0°	105.9°	106.6°	105°
<(HSSS)	85.3°	83.3°	84.4°	84.0°	84.3°	
<(SSSS)	77.4°	82.9°	87.8°	79.5°	78.9°	78°

[a]Ref. 14.

Table IV.   Geometry of H_2S_4 Planar Transition State

	FS (VWN)	DMol (VWN)	DMol (BLYP)	SCF	MP2
r(H-S1)	1.350	1.368	1.348	1.331	1.336
r(S1-S2)	2.146	2.112	2.145	2.108	2.114
r(S2-S3)	2.137	2.107	2.137	2.106	2.111
<(HSS)	89.8°	91.1°	90.4°	93.4°	92.2°
<(SSS)	92.7°	92.4°	93.1°	95.7°	94.0°

Table V. Rotational barrier heights for planar H_2S_2 and H_2S_4(kcal/mol)

	SCF	MP2	FS (VWN)	DMol (VWN)	DMol (BLYP)	Exp't
H_2S_2	5.8	5.6	5.4	6.7	5.8	1.9-6.9[a]
H_2S_4	23.1	25.3	25.9	30.3	27.9	-

[a]Experimental rotational barrier heights in kcal/mol: 1.9 (*15*), 2.7(*16*), 5.99(*17*), 6.9(*18*).

barrier height within this range. As can be seen from the calculated rotational barriers in Table V, the barrier height of 5.4 kcal/mol obtained by Fast_Structure is the lowest of all the theoretical methods used, but differs from the MP2 and BLYP results by only 0.3 and 0.5 kcal/mol, respectively. The Fast_Structure result is, however, 1.3 kcal/mol lower than the DMol result calculated using the same LDA functional VWN. The values for H_2S_4 in Table V are for the completely planar *anti, anti, anti* conformation. The value of 25.9 kcal/mol calculated by Fast_Structure for this physically unrealistic triple rotational transition state is within the range of 25.4 - 30.3 kcal/mol obtained by the other correlated methods.

Overall, the equilibrium and rotational transition state geometries as well as the barrier heights obtained by Fast_Structure indicate that the method gives a sufficiently accurate description of the potential energy surface to serve as a fast geometry optimization tool for the molecules investigated in this study.

Results

Hypothesis #1 *The sulfur-sulfur bonds in the longer polysulfides are weaker and more reactive, and hence more corrosive.*

In this section we will examine the lability of the sulfur-sulfur bond with respect to homolytic cleavage

$$RS\text{-}SR' \rightarrow RS\bullet + R'S\bullet$$

and copper assisted reductive cleavage

$$RS\text{-}SR' + Cu(I)OX \rightarrow RSCu(II)OX + R'S\bullet$$

where X is a proton or cluster representing the surface.

Bond lengths and the energies of the HS_nH series are shown in Tables VI and VIII. SCF, MP2 and VWN all give results in very good agreement with experiment. BYLP, as in many other systems (*19*), overestimates bond lengths. Fast_Structure, with a relatively loose gradient criterion for geometry optimizations, yields bond lengths between VWN and BLYP, thus providing excellent starting points for additional refinement with the more rigorous methods. An examination of Table VI shows that no matter which method is used, there is no significant difference in S-S bond lengths as one goes down the HS_nH series to longer polysulfides. Bond indices calculated at the SCF level shown in Table VII, however, do indicate that there is a weakening in bond strength in the longer polysulfides.

Examination of bond dissociation energy (DE), in which the stability of the thiyl free radical products plays a role, is consistent with the results of the bond index analysis. Bond dissociation energy is defined as the energy of the reaction A-B -> A• + B• in the gas phase at 0K. There is some experimental indication that S-S bonds are weaker in the longer polysulfides, but values for the bond dissociation energies show considerable variation. For H_2S_2 the reported experimental values for the S-S dissociation energy range from 59.6 to 80.4 kcal/mol (*20-26*), with most recent values converging on 64-65 kcal/mol with a reported uncertainty of ±6 kcal/mol (*22*). Part of the variability is due to the use of different experimental techniques performed in different laboratories, making comparison difficult. Franklin and Lumpkin (*25*), however, have reported the S-S DE for both the H_2S_2 and H_2S_3 using the same experimental technique, and found a decrease in DE from 80.4 kcal/mole for the disulfide to 64 kcal/mol for the trisulfide. Despite the experimental variability, all reported results for the di- and trisulfides are higher than the 33 kcal/mol obtained for elemental sulfur, an S_8 'polysulfide' with a cyclic structure which undergoes homolytic cleavage to form the linear biradical $\bullet S\text{-}S_6\text{-}S\bullet$ (*27*), supporting the concept of weaker bonds in the longer chains.

Calculated values for the polysulfide dissociation energies reported in Table IX and Reference 26 reflect the trend of decreasing bond strength with increasing chain length despite differences in the absolute values obtained for the correlated methods. BLYP gives values closest to the range of experimental results, followed by MP2. The consistency of the trends observed with MP2 and BLYP for all the polysulfides suggest that the conclusions drawn on the results as calculated would be valid. LDA overestimates bond strength. Calculations at the SCF level severely underestimate

Table VI. Bond lengths of HS$_n$H polysulfides (Å)

	SCF	MP2	FS (VWN)	DMol (VWN)	DMol (BLYP)	Expt.
HSSH						
H-S	1.338	1.337	1.347	1.371	1.387	1.35[a]
S-S	2.077	2.066	2.097	2.062	2.128	2.055[b]
HSSSH						
H-S	1.332	1.337	1.35	1.372	1.386	
S-S	2.064	2.065	2.083	2.064	2.133	
HSSSSH						
H-S	1.332	1.337	1.351	1.37	1.386	
HS-S	2.061	2.063	2.084	2.068	2.139	2.03[c]
HSS-S	2.064	2.065	2.091	2.059	2.122	2.07[c]
HSSSSSH						
H-S	1.332	1.337	1.353	1.371	1.387	
HS-S	2.065	2.064	2.085	2.063	2.133	
HSS-S	2.063	2.064	2.087	2.068	2.132	
HSSSSSSH						
H-S	1.332	1.338	1.350	1.371	1.387	
HS-S	2.061	2.062	2.082	2.064	2.132	
HSS-S	2.064	2.066	2.088	2.065	2.128	
HSSS-S	2.063	2.065	2.097	2.072	2.145	

[a]Ref. 11. [b]Ref. 12. [c]Ref. 14.

Table VII. Bond Indices of HS$_n$H polysulfides

	SCF
HSSH	
H-S	0.95
S-S	1.00
HSSSH	
H-S	0.95
S-S	0.98
HSSSSH	
H-S	0.95
HS-S	0.99
HSS-S	0.96
HSSSSSH	
H-S	0.95
HS-S	0.99
HSS-S	0.96
HSSSSSSH	
H-S	0.95
HS-S	0.99
HSS-S	0.96
HSSS-S	0.96

Table VIII. Total Energies of HS_nH polysulfides (Hartrees)

	SCF	MP2/DZP	DMol (VWN)	DMol (BLYP)
HSSH	-796.1749	-796.5819	-794.8866	-797.6153
HSSSH	-1193.6827	-1194.2883	-1191.7635	-1195.8397
HSSSSH	-1591.1906	-1591.9955	-1588.6400	-1594.0634
HSSSSSH	-1988.6983	-1989.7026	-1985.5158	-1992.2867
HSSSSSSH	-2386.2062	-2387.4103	-2382.3924	-2390.5105
Thiyl Radicals				
HS•	-398.0643	-398.2462	-397.3770	-398.7577
HSS•	-795.5845	-795.9677	-794.2674	-797.0052
HSSS•	-1193.0900	-1193.6718	-1191.1528	-1195.2322
HSSSS•	-1590.5990	-1591.3802	-1588.0340	-1593.4563
HSSSSS•	-1988.1060	-1989.0867	-1984.9116	-1991.6787

Table IX. Bond dissociation energies of HS_nH (kcal/mol)

	SCF	MP2/DZP	DMol (VWN)	DMol (BLYP)	Expt.
HS-SH	29.1	56.1	83.2	62.7	59.6 - 80.4 [a]
HS-SSH	21.3	46.7	74.8	48.1	64 [b]
HS-SSSH	22.8	48.6	69.2	48.1	
HS-SSSSH	22.0	47.8	65.8	45.6	
HS-SSSSSH	22.5	48.5	65.2	46.5	
HSS-SSH	13.6	37.7	66.0	33.2	
HSS-SSSH	15.0	39.6	60.0	30.9	
HSS-SSSSH	14.3	39.2	57.1	30.7	
HSSS-SSSH	16.5	41.8	54.5	28.9	

[a]Experimental bond dissociation energies in kcal/mol for H_2S_2: 59.6 (20), 64.1 (21), 65±6 (22), 66 (23), 72 (24), 80.4 (25).
[b]Ref. 25.

bond strength, as would be expected from correlation errors in a nonisodeismic reaction.

This decrease bond dissociation energies with increasing chain length can be attributed to differences in stabilization of the unpaired electron in the fragments resulting from homolytic cleavage. The HS• fragment is the most unstable of all and is responsible for the large energy required to disrupt the S-S bond in H_2S_2. The results using the BLYP functional indicate that the second sulfur atom in HSS• provides an additional stabilization energy of 15 kcal/mol over the HS• radical. These results are consistent with the reported stability of the disulfur thiyl radical from the experimental work on dimethyl polysulfides of Pickering et al.(27). Examination of the atomic spin populations indicates clearly that delocalization of the unpaired electron is responsible for the lower energy. In HS• the spin population is entirely on the sulfur atom, whereas in H-S1-S2• it is 0.31 on S1 and 0.69 on the terminal S2 atom. The stabilization of the radical by a three-electron bond between the two sulfur atoms at the end of the chain was postulated in the fifties (28) and is supported by these nonlocal DFT results. For MP2, the average stabilization energy going from HS• to HSS• is less than BLYP at 9 kcal/mol. Spin population analyses were not performed at the MP2 level, but results obtained at the SCF level indicate less delocalization occurs than in nonlocal DFT with spin populations of 0.11 and 0.89 on S1 and S2, respectively, increasing the bond order between them from 0.95 in the intact polysulfide to 1.23 in the radical.

Increasing the length of the thiyl radical from two to three sulfurs atoms has a much less dramatic stabilization effect. The delocalization of the unpaired electron is primarily limited to the two terminal sulfur atoms, and the addition of the third sulfur atom has a negligible effect. The spin population becomes 0.38 on S2 and 0.56 on S3 in the H-S1-S2-S3• fragment using the BLYP functional, and the stabilization energy increases an average of 2 kcal/mol over the HSS• fragment. For LDA the stabilization energy is 6 kcal/mol. At the HF level the spin populations of 0.11 and 0.87 on S2 and S3, respectively, represent a negligible change from S1 and S2 in HSS• above. The MP2 energy on the DZP optimized geometry even shows a slight increase in DE of 1.9 kcal/mol and hence a slight destabilization of HSSS• compared to HSS•.

The magnitude of the stabilization effect of adding additional sulfurs to the thiyl radical fragment diminishes even further on going from three to four sulfur atoms, being less than 1 kcal/mol for BLYP and MP2, and approximately 3 kcal/mol for VWN. Change in spin populations is negligible, as the addition of the fourth sulfur does not contribute to delocalization.

Some of the variability and errors in treating correlation when calculating bond dissociation energies can be minimized if the energies are calculated relative to H_2 and the isodeismic character of the reaction can be maintained:

$$RS\text{-}SR' + H_2 \; \text{->} \; RS\text{-}H + R'S\text{-}H$$

The isodeismic results for the HS_nH series are reported in Table X. The reaction energies closest to experiment were again achieved with the BLYP functional. The differences in the isodeismic reaction energies can be related to the S-S bond dissociation energies to obtain relative values. All correlated methods used are in close agreement indicating a decrease in the S-S bond dissociation energy of 13.0, 13.7, 12.0 kcal/mol (compared to 11.6 kcal/mol experimentally (*21*)) between the S1 and S2 atoms in the longer polysulfides relative to the bond in H_2S_2, as well as a general weakening between S2 and S3 relative to S1 and S2 in all molecules. The MP2 results show a smooth decrease in the S1-S2 and S2-S3 dissociation energies as the polysulfide chain lengthens, whereas both local and nonlocal DFT show some irregularities of less than a kcal/mol. For the central S3-S4 bond in H_2S_6, both local and nonlocal DFT indicated a slight strengthening of 1 kcal/mol of the sulfide bond relative to the S2-S3 bond of H_2S_4, whereas MP2 showed a weakening of 1.2 kcal/mol.

The results for the dimethylpolysulfide series reported in Tables XI and XII below show a similar trend of decreasing DE as the polysulfide chains increase in length, with a difference between the disulfide and the central S3-S4 bond in the hexasulfide being 19.9, 24.9, 39.0, and 23.7 kcal/mol for MP2, DMol VWN, and DMol BLYP, respectively. Several points are worth noting in comparing this series to the HS_nH series, the first being the effect of the methyl groups on DE. For dimethyl disulfide the S-S DE decreases 2.3 kcal/mol compared to H_2S_2 using the BLYP functional, but increases 3.5 kcal/mol using MP2. The actual values of DE for BLYP and MP2, however, converge on the same value of 60 kcal/mol, which is 7-13 kcal/mol higher than the range of experimental values. For $CH_3SS\text{-}SSCH_3$ the DE calculated by BLYP (37.8 kcal/mol) and MP2 (34.8 kcal/mol) are in much better agreement with the experimental value of 36.6 kcal/mol obtained by the relatively accurate free-radical scavenger technique (*31*).

Table X. Reaction Energies RS-SR' + H$_2$ -> RS-H + R'S-H (kcal/mol)

	SCF	MP2/ DZP	MP2	DMol (VWN)	DMol (BLYP)	Expt [a]
HS-SH	-21.7	-17.4	-17.4	-26.1	-11.9	-13.7
HS-SSH	-19.6	-12.1	-12.1	-19.1	-5.7	-5.1
HS-SSSH	-19.6	-11.7	-11.7	-19.3	-6.1	-6.9
HS-SSSSH	-19.7	-11.7	-11.7	-19.8	-6.4	
HS-SSSSSH	-19.6	-11.4	-11.4	-19.3	-6.1	
HSS-SSH	-17.4	-6.4	-6.4	-12.3	0.1	-2.1
HSS-SSSH	-17.6	-6.0	-6.0	-13.0	-0.6	
HSS-SSSSH	-17.5	-5.7	-5.6	-13.0	-0.6	
HSSS-SSSH	-17.6	-5.2	-5.2	-13.2	-1.0	

[a]Ref. 21.

Table XI. Bond lengths for CH$_3$S$_n$CH$_3$ (Å)

	SCF	MP2	FS (VWN)	DMol (VWN)	DMol (BLYP)	Expt.[a]
CH$_3$SSCH$_3$						
C-S	1.816	1.814	1.820	1.806	1.867	1.810
S-S	2.065	2.056	2.054	2.039	2.103	2.03
CH$_3$SSSCH$_3$						
C-S	1.813	1.811	1.811	1.802	1.876	
S-S	2.070	2.065	2.069	2.103	2.122	
CH$_3$SSSSCH$_3$						
C-S	1.814	1.812	1.813	1.802	1.863	
CS-S	2.068	2.056	2.059	2.052	2.107	
CSS-S	2.080	2.074	2.118	2.083	2.165	
CH$_3$SSSSSCH$_3$						
C-S	1.809		1.815	1.803	1.864	
CS-S	2.011		2.059	2.045	2.109	
CSS-S	2.052		2.106	2.085	2.153	
CH$_3$SSSSSSCH$_3$						
C-S	1.817		1.823	1.817	1.863	
CS-S	2.065		2.063	2.065	2.103	
CSS-S	2.075		2.095	2.075	2.156	
CSSS-SSSC	2.074		2.080	2.074	2.131	

[a]Ref. 32.

How do the energetics of sulfide bond cleavage change in the presence of the copper surface? XPS and low angle Xray diffraction indicate that Cu$_2$O is the predominant species on the surface of the copper strip used in the corrosion test (*33*). As the surface being corroded is actually Cu$_2$O, the formal +1 oxidation state of copper is used in this study. The simplest Cu(I) species which can be used to approximate the copper oxide surface is Cu(I)OH in the following reaction:

$$RS-SR' + CuOH \rightarrow RS-CuOH + R'S\bullet$$

Table XII.   Bond Dissociation Energies for $CH_3S_nCH_3$ (kcal/mol)

	SCF	MP2/ DZP	DMol (VWN)	DMol (BLYP)	Expt.
CH_3S-SCH_3	29.4	59.6	89.7	60.4	67-73.2[a]
CH_3S-$SSCH_3$	21.6	48.0	68.9	49.7	46[b]
CH_3S-$SSSCH_3$	22.5	50.3	71.1	49.2	
CH_3S-$SSSSCH_3$	22.8	50.6	64.8	47.4	
CH_3S-$SSSSSCH_3$	22.4	50.5	69.1	45.3	
CH_3SS-$SSCH_3$	11.8	34.8	46.1	37.8	36.6[c]
CH_3SS-$SSSCH_3$	13.5	37.4	48.4	37.2	
CH_3SSS-$SSSCH_3$	14.4	39.7	50.7	36.7	

[a]Experimental S-S bond dissociation energies for CH3S-SCH3 in kcal/mol: 67 (29), 69 (24), 70.7 (21) 73.2 (29).
[b]Ref. 30.
[c]Ref. 31.

For this and subsequent reactions energies in this work involving metal containing clusters, we report nonlocal DFT-based results due to greater computational efficiency, although both MP2 and BLYP have exhibited trends consistent with each other and with the majority of experimental data. As can be seen from the results in Table XIII, considerable energy from the bond dissociation is recovered from the interaction of the radical fragments with CuOH. For DMol VWN, this amount ranges between 55 and 77 kcal/mol depending upon which fragment interacts with the surface, making the reaction very exothermic even if only one of the two fragments is stabilized by the surface. For both fragments, the amount of energy recovered would be twice that amount. For DMol BLYP, the amount of energy recovered ranges between 37 and 52 kcal/mol. If the uncaptured fragment is the methylsulfide radical CH3S• the bond dissociation reaction is still endothermic by 5.4 to 10.4 kcal/mol, but becomes exothermic when the uncaptured fragment is the disulfide radical CH3SS•. If both fragments are 'recovered' by the 'surface', then the bond dissociation becomes spontaneous in all cases in DFT.

To design a finite cluster to include more than one copper atom, several considerations are involved. Ideally one would like a cluster which models the reactivity, electron density, and the symmetry of the periodic surface, yet is sufficiently small to be computationally feasible. In addition, one would like to eliminate dangling bonds, except at the surface, yet maintain electronic neutrality, as electrostatic forces are very long range and can easily distort the charge density of the cluster.

In the crystal structure of Cu_2O shown in Figure 2, one can see that each oxygen atom is tetrahedrally coordinated in a diamond-like structure, with linearly coordinated copper atoms serving as spacers between each pair of oxygen atoms. Hence, a logical cluster would be one shown in Figure 3a, with hydrogen atoms being used to replace the terminal shell of copper atoms. Assuming each copper atom forms one covalent and one Lewis acid bond, and each oxygen forms two covalent and two Lewis base bonds, the cluster will have a net +5 charge. A neutral cluster can be constructed by using only four terminating hydrogen atoms instead of nine as shown in Figure 3b. The positions of the terminating hydrogens are optimized, as are the

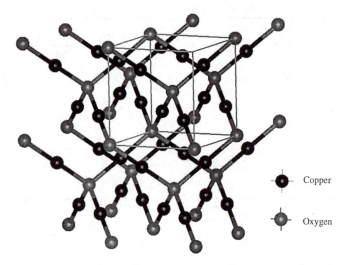

Figure 2. Crystal structure of Cu_2O showing tetrahedrally coordinated oxygen atoms and linear coordinated copper.

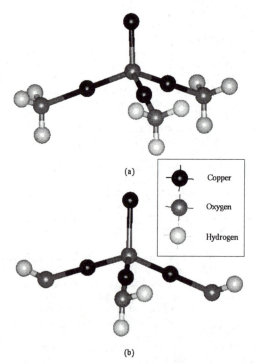

Figure 3. Copper oxide clusters: (a). $[Cu_4O_4H_9]^{+5}$ (b). $Cu_4O_4H_4$

Table XIII. Reaction Energy for RS-SR' + CuOH -> RS-CuOH + R'S•
(kcal/mol)

Reaction Product	DMol (VWN)	DMol (BLYP)
$CH_3S•$ CuOH + •SCH_3	12.3	7.8
$CH_3SS•$ CuOH + •SCH_3	8.8	5.4
$CH_3SSS•$ CuOH + •SCH_3	4.6	6.8
$CH_3SSSS•$ CuOH + •SCH_3	10.1	10.4
$CH_3SSSSS•$ CuOH + •SCH_3	2.6	7.0
$CH_3S•$ CuOH + •$SSCH_3$	-8.6	-2.8
$CH_3SS•$ CuOH + •$SSCH_3$	-13.9	-6.5
$CH_3SSS•$ CuOH + •$SSCH_3$	-18.1	-5.2
$CH_3SSSS•$ CuOH + •$SSCH_3$	-12.7	-1.5
$CH_3SSS•$ CuOH + •$SSSCH_3$	-15.8	-5.8

central copper and its adjacent oxygen atom, while the remainder of the atoms in the cluster are fixed.

Mulliken population analyses performed with DSolid (*34*) on Cu_2O with three-dimensional periodic boundary conditions shows a charge of +0.41 on copper and -0.81 on oxygen in the crystal. This charge on copper is reproduced almost exactly in CuOH (0.41) and very closely in the neutral $Cu_4O_4H_4$ cluster (0.38). The effect of the +5 charge on the cluster in Figure 3a is apparent when compared to the electronically neutral CuOH and $Cu_4O_4H_4$ clusters when each forms a complex with •SH. From the geometries in Table XIV and the Mulliken populations in Table XV, it can be clearly seen that the +5 charge results in a slight lengthening of the Cu-S bond (0.05Å), a shortening of the Cu-O bond (0.05Å), and considerable shifting of electron density away from the reactive site. In the +5 charged cluster, the sulfur atom has a charge of +0.31, whereas in the neutral clusters CuOH and $Cu_4O_4H_4$, it has charges of -0.33 and -0.44, respectively, indicating an additional transfer of approximately 2/3 of an electron from the sulfur atom to the copper oxide cluster. The additional charge resides almost exclusively on the terminal hydrogen and peripheral oxygen atoms. The charge on the central copper atom in all three SH complexes is relatively unaffected by the total charge, with populations ranging from 0.26 to 0.31.

For the neutral clusters, a small amount of charge is actually transferred from the sulfur atom, 0.08*e* for the CuOH cluster and 0.18*e* for the $Cu_4O_4H_4$ cluster. There is some reference in the literature (*35*) to reductive cleavage of polysulfides by metal surfaces in which an electron is transferred from the metal to the sulfur to produce the sulfide anion:

$$R-S-S-R' \xrightarrow{metal} R-S• + R'S:-$$

For a passivated metal surface such as copper oxide where the copper is in a formal +1 oxidation state, such reductive cleavage is clearly not the case. Examination of the electronic structure shows that the very stable d^{10} electronic configuration of the copper atom remains intact.

Surprisingly, whether or not the cluster is neutral has a negligible effect on the binding energy despite the rather considerable effect on the charge distribution. The binding energy of HS• to the charged cluster is -83.7 kcal/mol, only 0.5 kcal/mol lower than the -83.2 kcal/mol obtained for the neutral $Cu_4O_4H_4$ cluster. The -108.9

Table XIV. Geometry of HSCuOX complexes

	HS•CuOH	HS•Cu$_4$O$_4$H$_4$	[HS•Cu$_4$O$_4$H$_9$]$^{+5}$
r(HS) Å	1.371	1.384	1.388
r(SCu) Å	2.072	2.161	2.109
r(CuO) Å	1.753	1.930	1.980
<(HSCu)	95.7°	94.9°	97.9°
<(SCuO)	174.2°	177.9°	176.9°

Table XV. Mulliken populations in Copper Oxide Clusters (VWN)

	Free	Complexed
Neutral CuOH		
S	-0.25	-0.33
H	0.25	0.29
Cu	0.41	0.31
O	-0.84	-0.80
H	0.44	0.53
Neutral Cu$_4$O$_4$H$_4$		
S	-0.25	0.14
H	0.25	0.44
Cu	0.38	0.26
O	-0.86	-0.87
2 Cu	0.34	0.40
1 Cu	0.33	0.34
2 O	-0.85	-0.80
1 O	-0.99	-0.99
2 H	0.63	0.63
2 H	0.43	0.49
[Cu$_4$O$_4$H$_9$]$^{+5}$		
Cu	0.82	0.31
O	-0.95	-0.91
3 Cu	0.52	0.50
3 O	-1.26	-1.26
9 H	0.81	0.82
S	-0.25	-0.44
H	0.25	0.27

kcal/mol binding energy for the CuOH cluster was 25 kcal/mol lower than for the larger clusters, probably because free CuOH is intrinsically higher in energy than a larger cluster.

Calculations of the binding energy of the CH3S$_n$• series with the larger clusters using the BLYP functional are currently in progress.

Hypothesis #2: *The longer polysulfides are more corrosive because they can chelate the copper atoms and remove them from the metal surface.*

The optimized geometries obtained with the BLYP functional of the H_2S_4, H_2S_5, and H_2S_6 chelation complexes with Cu(I) are shown in Figure 4. To determine the strain energy in the three complexes, the energy of the HS_nH chain is calculated in the conformation shown in Figure 4, minus the copper atom, and compared to the energy of the equilibrium geometry. As can be seen from the results in Table XVI, the potential energy surface for the distortion of the polysulfide chains is relative flat, and the strain energy for the polysulfide chain is fairly low at less than 15 kcal/mol.

Table XVI. Strain Energies of $[Cu \cdot H_2S_n]^{+1}$ chelation complexes (kcal/mol)

	$[Cu \cdot H_2S_4]^{+1}$	$[Cu \cdot H_2S_5]^{+1}$	$[Cu \cdot H_2S_6]^{+1}$
DE (reaction)[a]	-72.5	-79.5	-83.4
DE HS_nH strain	10.8	8.3	14.6
DE Cu(I) bend	102.7	93.9	90.9
<SCuS[b]	121.7°	149.2°	158.9°

[a]Energy of the reaction $Cu(I) + H_2S_n \rightarrow [Cu \cdot H_2S_n]^{+1}$
[b]Optimum <SCuS is 174.2°, shown in Figure 5.

Quite a different picture emerges, however, when the strain energy for bending the S-Cu-S bond is considered. A $[H_2S \cdot Cu \cdot SH_2]^{+1}$ complex is used as a benchmark compound to determine the minimum energy of the S-Cu-S bond. Three possible molecular symmetries of $[H_2S \cdot Cu \cdot SH_2]^{+1}$, C_2, D_{2d}, and D_{2h}, shown in Figure 5, were examined to determine the optimum S-Cu-S angle. The lowest energy structure exhibits C_2 symmetry and has a S-Cu-S bond angle of 174.2°. The D_{2d}, and D_{2h} structures were 27.2 and 28.8 kcal/mol, respectively, higher in energy than the C_2 geometry. The strain energy of the bent S-Cu-S bond in the chelation complexes was determined by fixing the copper and sulfur atoms at the positions they occupy in the complexes, optimizing the positions of the two hydrogens added to each sulfur, and comparing the energy to the C_2 structure. As can be seen from this comparison in Table XVI, the strain energies are exceedingly large: 102.7, 93.9, and 90.9 kcal/mol, for the H_2S_4, H_2S_5, and H_2S_6 Cu(I) complexes respectively. Therefore, it can be concluded that polysulfide chains containing 4 to 6 sulfurs are not long enough to accommodate the linear geometry required by a Cu(I) ion, making it extremely unlikely that such a chelation mechanism is primarily responsible for the corrosion of a Cu_2O surface. Once in solution, it is possible that two chains or four fragments of sulfide chains may form a tetrahedral complex with the Cu(I) ion, but steric considerations make this unlikely to occur at the surface.

Hypothesis #3 *The steric bulk of the side chains can inhibit corrosion by limiting the contact of the sulfur with the surface in the shorter polysulfide chains.*
To study the effect of bulky hydrocarbon side chains, *t*-butyl groups are chosen as model compounds. Geometries are obtained using Fast_Structure. The Connolly surfaces of the di-*t*-butyl sulfides were calculated using InsightII (*36*) with a 5.0 Å probe radius to approximate an ideal, relatively flat solid surface. As can be seen from the results in Figures 6 and 7, the di- and trisulfides are well-protected with only 19.7 and 33.8 Å[2], respectively, of exposed surface area, whereas the tetra-, penta- and hexasulfides have 48.5, 66.9 and 74.8 Å[2], respectively, and hence are relatively exposed.

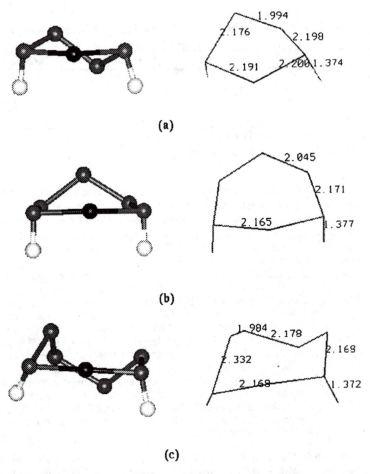

Figure 4. Cu(I) chelation complexes: (a). $[Cu \cdot H_2S_4]^{+1}$ (b). $[Cu \cdot H_2S_5]^{+1}$ (c). $[Cu \cdot H_2S_6]^{+1}$

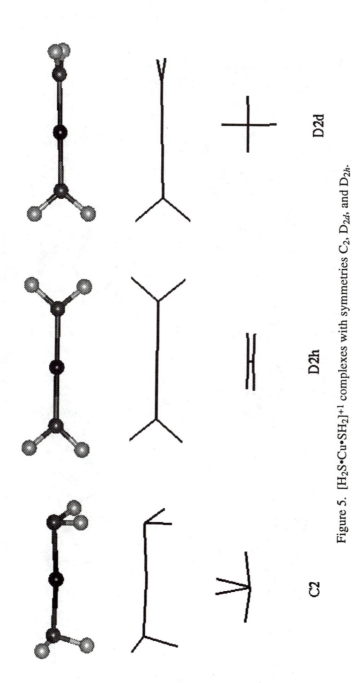

Figure 5. $[H_2S \cdot Cu \cdot SH_2]^{+1}$ complexes with symmetries C_2, D_{2d}, and D_{2h}.

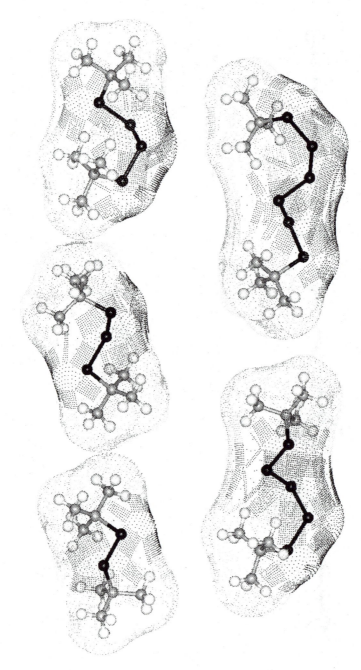

Figure 6. Solvent-accessible surfaces of di-t-butylpolysulfides calculated using a 5 Å probe radius. Dark areas indicate regions of exposed sulfur.

Lowest Frequencies of the Bending Vibrational Modes (cm⁻¹)

H_2S_2	H_2S_3	H_2S_4	H_2S_5	H_2S_6
329	180	76	35	19

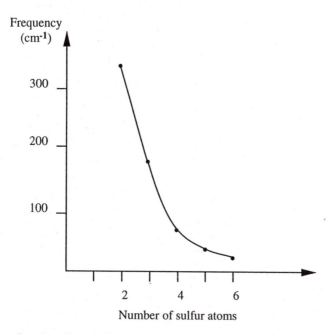

Figure 7.   Results of force constant calculations on HS_nH showing the exponential decrease of the lowest frequency bending mode as the number of sulfur atoms increases.

Accessibility of the sulfur atoms to the surface is not determined solely by the surface area exposed in the static lowest energy conformation, but also by the sulfide chain's flexibility. Force constant calculations on HS_nH compounds were performed using Fast_Structure. The results shown in Figure 7 indicate that the lowest frequency of the bending vibrational modes decreases exponentially as the number of sulfur atoms increases. This behavior implies that the sulfide chains are much easier to bend for $n \geq 4$, thus increasing the possibility of exposing unhindered sulfur to the surface and increasing corrosive behavior.

Conclusions

After investigation of the three hypotheses to explain greater corrosiveness of the longer polysulfides - weaker, more reactive bonds in the longer polysulfides, chelation, and steric protection by hydrocarbon side chains - we conclude that the most important factor is the greater reactivity of the longer chains. This greater reactivity is due to the greater stabilization of the radical product of the homolytic cleavage in the longer chains, resulting in a lower bond dissociation energy. The difference in bond dissociation energy using BLYP is 34 kcal/mol higher for the central sulfide bond in the hexasulfide than the disulfide, and 19.2 kcal/mol than the trisulfide for the HS_nH series, and 24 and 13 kcal/mol, respectively, for the analogous DE for the dimethylpolysulfide series.

All the sulfide bonds, however, are labile in the presence of copper oxide, as the bond dissociation energy is more than compensated for by the exothermic adsorption of both sulfide radicals onto exposed copper sites on the surface. Hence we conclude that the steric protection provided by bulky hydrocarbon sidechains such as t-butyl groups can be useful in the protection of the copper surface from the di- and trisulfides.

Chelation by the longer polysulfide chains was shown not to be a viable mechanism of corrosion, at least for copper species in the +1 formal oxidation state found in Cu_2O. The distortion of the S-Cu-S bond required to chelate a linearly coordinated Cu(I) atom requires too high an energy penalty to be feasible. It is much more likely that the polysulfide chain would be cleaved into fragments upon contact with the surface rather than chelating a copper atom, and that it is these fragments which are removing the copper atoms, irrespective of whether they initially came from the same polysulfide molecule or not. Longer polysulfide chains may increase the probability of two fragmented sulfide chains being in close proximity to a single site on the surface, but this is probably not as important a factor as their greater reactivity.

Of the theoretical methods used in this study, DMol with the BLYP functional and MP2 were found to give bond dissociation energies in closest agreement with the available experimental data, although greater accuracy would be desirable. We are currently investigating in greater detail the errors due to correlation and spin contamination. Bond lengths calculated with BLYP were systematically too long. Fast_Structure, based on the Harris functional with trial densities constructed from spherically symmetric site-densities, was found to give ground state and rotational transition state structures which agreed well with fully self-consistent methods and with experiment, thus demonstrating its value as a fast method to obtain reasonable starting geometries without parameterization .

Literature Cited

1. Li, X.P.; Andzelm, J.; Harris, J.; Chaka, A.M. "A Fast Density Functional Method For Chemistry". This volume. Fast_Structure is a precursor to the FastStruct/SimmAnn module marketed by Biosym/MSI, San Diego, CA 92121.
2. Kohn, W.; Sham, L.J. *Phys. Rev.* **1965**, *140*, 1133.
3. Pople, J.A.; Binkley, J.S.; Seeger, R. *Int. J. Quant. Chem.* **1984**, *5*, 280.
4. Tuinstra, F. *Acta Crystallogr.* **1966**, *20*, 341.
5. Dupuis, M.; Marquez, A.; Davidson, E.R. HONDO 94.8,**1994**, IBM Corporation, Neighborhood Road, Kingston, NY 12401.
6. Delley, B. *J. Chem. Phys.* **1990**, *92*, 508; DMol Version 2.3.5 (1993), Biosym/MSI, San Diego., CA 92121
7. Vosko, S.H.; Wilk, L.; Nusair, M. *Can. J. Chem.* **1980**, *5*, 1629.
8. Becke, A.D. *Phys. Rev. A* **1988**, *38*, 864.
9. Lee, C.; Yang, W.; Parr, R.G. *Phys. Rev. B* **1988**, *37*, 785.
10. Dunning, T.H.; Hay, P.J. In *Modern Theoretical Chemistry*; Shaefer, H.F., Ed.; Plenum Press: New York, 1975; Vol. 2.
11. Wimmewisser, M.; Haase, J. *Z.Naturforsch* **1968**, *23a*, 56.
12. Wimmewisser, G.; Wimmemisser, M.; Gordy, W. *J. Chem. Phys.* **1968**, *49*, 3465-3478

13. Wilson, M.K.; Badger, R.M. *J. Chem. Phys.* **1949**, *17*, 1232.
14. Abrahams, S.C. *Quart. Revs. (London)* **1956**, *10*, 424.
15. Schwartz, M.E. *J. Chem. Phys.* **1969**, *51*, 348.
16. Feher, F.; Schulze-Rettmer, R. *Z. anorg. u. allgem. Chem.* **1956**, *295*, 123.
17. Veillard, A.; Demuynck, J. *Chem. Phys. Lett.* **1969-1970**, *4*, 476.
18. Redington, R.L. *J. Mol. Spectrosc.* **1962**, *9*, 469.
19. Delley, B.; Wrinn, M.; Luthi, H.P. *J. Chem. Phys.* **1994**, *100*, 5785-5791.
20. Sehon, A.H. *J. Am. Chem. Soc.* **1953**, *74*, 4722-4733.
21. Benson, S.W. *Thermochemical Data*; Wiley & Sons: New York, NY, 1968.
22. Mackle, H. *Tetrahedron* **1963**, *19*, 1159-1170.
23. Luft, N.W. *Monatsh.* **1955**, *86*, 474.
24. Cottrell, T.L. *The Strength of Chemical Bonds*; Academic Press, Inc.: New York, NY, 1954.
25. Franklin, J.L.; Lumpkin, H.E. *J. Am. Chem. Soc.* **1952**, *74*, 1023.
26. We gratefully acknowledge additional data provided by a reviewer who obtained obtained 56, 54, 55, 55, and 53 kcal/mol with B3LYP/6-311++G*, G1, G2MP2, and G2, respectively, which brings into question the higher values reported in the literature for the H_2S_2 bond dissociation energies.
27. Pickering, T.L.; Saunders, K.J.; Tobolsky, A.V. In *The Chemistry of Sulfides*; Tobolsky, A.V, Ed.; Interscience Publishers; John Wiley & Sons: New York, NY, 1968, pp. 61-72
28. Gee,G.; Fairbrother, F.; Merrall, G.T. *J. Polymer Sci.* **1955**, *16*, 459.
29. Guryanova, E.N. *Quarterly Reports on Sulfur Chemistry* **1970**, *5*, 113-123.
30. Pickering, T.L. Ph.D. Thesis, Princeton 1966.
31. Kende, I.; Pickering, T.L.; Tobolsky, A.V. *J. Am. Chem. Soc.* **1965**, *87*, 5582.
32. Meyer, M. *J. Mol. Struct.* **1992**, *273*, 99-121.
33. Experiments performed in our laboratories.
34. DSolid, Version 1.0, Biosym/MSI, CA 92121.
35. Kajdas, C. *Lubrication Science* **1994**, *6-3*, 203-228.
36. InsightII, Version 2.3.5, Biosym/MSI, San Diego, CA 92121.

Chapter 26

A Fast Density-Functional Method for Chemistry

Xiao-Ping Li[1], Jan Andzelm[1], John Harris[1], and Anne M. Chaka[2]

[1]Biosym/Molecular Simulation, 9685 Scranton Road,
San Diego, CA 92121–3752
[2]Lubrizol Corporation, 29400 Lakeland Boulevard,
Wickliffe, OH 44092–2298

Standard methods for performing Kohn-Sham calculations involve self-consistency cycling and require either the solution of Poisson's equation for a density given as a quadratic sum over the orbital basis, or the use of a separate density basis and a fitting procedure. An alternate approach using a functional different from but closely related to that of Kohn and Sham avoids both self-consistency cycling and density fitting and allows accurate results with comparatively less effort. The main elements of this approach and some applications that illustrate its potential in chemistry are discussed.

Over the past few years, the usefulness of the Kohn-Sham density functional/local density approximation (DFT-LDA) approach [1] in chemistry has been widely recognized. Recently, the method was validated in systematic calculations for a set of transition-metal compounds involving first- and second-row transition metals [2,3]. Determining structural information for transition-metal compounds has presented a significant challenge for theoretical methods. Hartree-Fock theory [4], so successful when applied to organic systems, fails to predict accurately the structures of organometallics and more accurate methods that include electron correlation explicitly are computationally demanding. Recently, a successful parametrization of a semi-empirical method has been accomplished for several transition metals [5]. The semi-empirical approach is very fast, but it is limited to systems for which parameters are available. There are many possible hybridization schemes and oxidation states of transition metals bonded to various ligands and this makes determination of semi-empirical or force-field parameters difficult and not easily transferable. It was found [2,3] that the local density approximation provides accurate structures involving covalent metal-ligand bonds. In the case of dative bonds such as occur in metal carbonyls,

0097–6156/96/0629–0388$15.00/0

the LDA gives somewhat shortened metal-ligand bond lengths. This is due to the tendency of the LDA to overestimate attractive nonbonded interactions and is corrected when gradient corrections are included.

The availability of a first principles method that predicts accurate structures at a lesser computational cost than correlated ab initio methods is therefore rather important. Nevertheless, the standard implementation of the DFT method, via self-consistent solution of the Kohn Sham equations, remains a major computational problem and alternative methods of implementation that can reduce this cost are highly desirable. The main purpose of this paper is to outline the elements of such a method and to validate it via explicit calculation, particularly for organometallic molecules.

The crucial advantage of the Kohn-Sham DFT approach over conventional quantum chemical methods is the replacement of the many-electron wavefunction by the density as the quantity that is varied, and the resulting reduction to a "one-particle" problem. In carrying out Kohn-Sham calculations some computational difficulties arise that are due ultimately to the definition of the Kohn-Sham density functional, E_{ks}, in terms of V-representable densities. Each trial density must be generated by a Schrödinger equation and cannot be chosen freely. In practice, this means that E_{ks} cannot be minimized directly by optimizing within a restricted density basis. Furthermore, if atomic orbitals are used to solve the one-particle problem, the evaluation of E_{ks} requires, in principle, four-center integrals. To circumvent this, density fit (or density-projection) procedures are commonly used in conjunction with self-consistency cycling (see, eg., Ref. [6]).

An alternate procedure is to exploit the properties of a different density functional, E, that has come to be known as the Harris Functional [7-9]. This possesses the same stationary points as the Kohn-Sham functional but is defined on function space. The crucial advantage of E as distinct from E_{ks} is that it can be evaluated using a density basis chosen for calculational convenience, for example, a sum over spherically symmetric site densities. The disadvantage is that, unlike E_{ks}, E does not obey a minimum principle. Finnis first pointed out that E displays a maximum for "reasonable" density variations [10]. Robinson and Farid showed that, within the local density approximation (LDA), E displays a saddle point, but with positive curvature occurring only for density fluctuations composed solely of spatially rapidly varying components [11]. Since it is quite straightforward to eliminate this possibility, a maximum property of E can be guaranteed. Maximizing E is then the equivalent in the new approach to minimizing E_{ks} (commonly achieved via "self-consistency cycling").

Other authors have exploited the favourable properties of E, notably Sankey and Niklewski [12], who restricted the trial density to a sum of overlapped neutral atom densities and made a series of approximations that resulted in an extremely fast tight-binding-like scheme. A series of applications, notably to large carbon systems, has shown that this approach can work rather well [13]. The scheme we describe here is more conservative and focuses in the first instance on universality (i.e. full coverage of the periodic table) rather than on efficiency for a limited class of elements. The scheme is a further development of a method suggested previously [14] and described in detail by Lin and Harris [15]. The most important innovation is provision for charge transfer which is crucial in general applications. Recently, the Sankey-Niklewski scheme has also been modified to allow for charge transfer [16].

The organization of this paper is as follows: In Section 2, we give a brief description of the scheme and detail those calculational features that differ from the proposal of Lin and Harris [15]. These refer in the main to the orbital and density bases used, the manner of performing integrals and the provision for charge transfer. In Section 3, we then quote results on the structure of a series of molecules, which illustrate that the present scheme gives an excellent description of structural properties over a wide class of bonding environments.

Theoretical Discussion

The theoretical elements underlying the basic scheme are described in detail in Ref. [15] and we restrict ourselves here to a brief summary. The energy functional, for N electrons in external (usually nuclear) field $V_{ext}(x)$ is:

$$E[n] = \sum_n a_n \varepsilon_n - \int d\bar{x}\, n(\bar{x}) \left\{ \frac{1}{2}\Phi(\bar{x}) - \varepsilon_{xc}(\bar{x}) + \mu_{xc}(\bar{x}) \right\} \qquad \text{Eq. 1}$$

where ε_n are the eigenvalues of the one-particle Schrödinger equation

$$\left\{ -\frac{\nabla^2}{2m} + V_{\text{eff}}(\bar{x}) - \varepsilon_n \right\} \Psi_n(\bar{x}) = 0 \qquad \text{Eq. 2}$$

with potential construction

$$V_{\text{eff}}(\bar{x}) = V_{\text{ext}}(\bar{x}) + \Phi(\bar{x}) + \mu_{xc}(\bar{x}) \qquad \text{Eq. 3}$$

Here, Φ is the Coulomb potential corresponding to $n(\bar{x})$ and

$$\mu_{xc}(\bar{x}) \equiv \frac{d}{dn(\bar{x})} \left\{ n(\bar{x})\varepsilon_{xc}(\bar{x}) \right\} \qquad \text{Eq. 4}$$

We assume the local density approximation (LDA) for the exchange-correlation energy density

$$\varepsilon_{xc}(\bar{x}) = \varepsilon_{xc}^{h}(n(\bar{x}))$$ Eq. 5

with $\varepsilon_{xc}^{h}(n)$ the energy of a homogeneous electron gas having density n. A notable feature of the energy functional, Eq. 1, is that its evaluation does not involve explicitly the "out" density from the Schrödinger equation,

$$n(\bar{x}) = \sum_{n} a_{n} |\Psi_{n}(\bar{x})|^{2}$$ Eq. 6

where the a_n are appropriate occupation numbers. The "in-density" appearing in E[n] in Eq. 1 is independent of the orbitals that are generated via the solution of Eq. 2 and it is this freedom that gives Eq. 1 its flexibility with regard to approximation.

The evaluation of Eq. 1 and its derivatives is achieved by first choosing a convenient density basis. This defines the potential appearing in Eq. 2, which is then solved using an orbital basis. As an appropriate trial density we use the form

$$n(\bar{x}) = \sum_{i} \sum_{\upsilon} Z_{i}^{\upsilon} n_{i}^{\upsilon}(|\bar{x} - \bar{x}_{i}|)$$ Eq. 7

where $n_{i}^{\upsilon}(|\bar{x} - \bar{x}_{i}|)$ is an atom centered, spherically-symmetric site density and Z_{i}^{υ} is a parameter that may be fixed or can vary. In Eq. 7, the sum over "i" refers to a sum over sites, and the sum over "nu" allows for two or more density basis functions per site. Since only spherically symmetric site densities are used, the solution of Poisson's equation for any density of the form in Eq. 7 is trivial. If the Z_{i}^{υ} are fixed, the energy and forces are determined by a single evaluation of Eq. 1 and its derivatives with respect to the nuclear positions (formulae for which are given in Ref. [15]). The time limiting step in the calculation is the solution of the one-particle Schrödinger equation as in Eq. 2, but only one solution is needed for a given set of nuclear locations so the speed up as compared with a traditional DFT code (other things being equal) is roughly the number of iterations the latter requires to become "self-consistent." This can be as small as 3 and as large as 50 depending on the application.

The individual density basis functions in Eq. 7 are numerical and are generated using an atom program. For light atoms, two functions per site, one representing the core and one the valence, are often sufficient. This is because s- and p-orbitals in the first row have similar extents. Supplementary functions calculated using ionic configurations to improve accuracy can be added. For heavier atoms, in particular for transition elements it is important to allow for weighting of individual shells. This is because transfer between shells that accompanies chemical bonding can cause quite drastic changes in the density. Thus, in the third row separate functions are used to represent the s- and d- shells.

In general, it will be necessary to allow for some variation in the multiplicative parameters, Z_i^v, that influence the charge distribution. The variation may need to describe a charge transfer between sites or an effective change of intra-site electron configuration (eg., s => d transfer). This can be done conveniently using the maximum principle obeyed by E with respect to density fluctuations. The "forces" acting on the charge variables Z_i^v are given by Eq. 9 of Ref. 15 and all that is necessary is to maximize E subject to the constraint on the overall norm

$$\sum_i \sum_v Z_i^v = N \qquad\qquad \text{Eq. 8}$$

The maximum principle can be guaranteed since all density fluctuations corresponding to a given Z_i^v have the same spatial form. It is therefore easy to ensure that the electrostatic contribution to the curvature outweighs the exchange-correlation component, which is the condition for a maximum. (Another way of saying this is that all density fluctuations corresponding to a given site density have the same decomposition in Fourier space. It is necessary only to ensure that each function of the density basis has sufficient weight in low-k components.) In practice, this will be true for any atom-derived density of the type considered. A potential problem would arise here only if fine tuning of the density were to be attempted by, eg., including density basis functions with strong angular variations. Since the Harris functional is quite flat near to its saddle point, it is likely that acceptable accuracy can invariably be achieved without such fine tuning.

The one-particle Schrödinger equation, Eq. 2, is solved approximately by expanding the one-electron wavefunctons, Ψ_n, in a basis of atomic orbitals (AO's), χ_i, in the usual way,

$$\Psi_n(\bar{x}) = \sum_i c_i \chi_i(\bar{x} - \bar{x}_i) \qquad\qquad \text{Eq. 9}$$

and solving the resulting secular problem

$$\left[H_{ij} - \varepsilon_n S_{ij} \right] c_j^n = 0 \qquad\qquad \text{Eq. 10}$$

Here H_{ij} and S_{ij} are Hamiltonian and overlap matrices, respectively.

In a departure from the method in Ref. [15], the orbital basis is numerical rather than analytic, comprising AOs generated using an atom program with finite range boundary conditions, so that the orbitals vanish continuously at a cutoff radius, r_i^c. These radii can be chosen to combine accuracy with efficiency and numerical stability. The r_i^c determine the range beyond which overlap of AO's centered on different sites vanishes and, therefore, control the number of integrals that contribute to the Hamiltonian and overlap matrices. As Sankey and Niklewski, in particular, have noted [12], finite range orbitals have a critical advantage over conventional AOs when calculations are carried out for extended systems. All AOs are generated using an atom program with neutral atomic, and ionic configurations. The latter include "polarization functions" or "atom-unoccupied orbitals" that are important whenever the corresponding levels are low-lying. (Occasionally, truncated Slater orbitals are preferred for polarization functions.)

The overall time-limiting feature of the scheme is the evaluation of the three-center integrals that contribute to the matrix elements in Eq. 10. Each of these is performed using the efficient weight-function method of Delley [6]. Energy derivatives are calculated as detailed in Ref, [15] (cf. Eq. 10. This is equivalent to the corresponding expression, also Eq. 10, in Ref. [17]), with additional contributions that arise because of the dependence of the numerical grids on the nuclear positions. The energy gradients and energy are then consistent to high accuracy even when the numerical meshes are sparse. In contrast to Sankey and Niklewski, we have not invoked approximating procedures for the three-center integrals. This is because, initially, our goal is to retain universality at the same level as conventional DFT-LDA schemes (some of the Sankey/Niklewski approximations breakdown for transition elements, for example). We believe, nevertheless, that there are many possibilities for approximation that can reduce the CPU requirement of the code very significantly.

Geometry optimization can be carried out either directly, using a BFGS scheme [18], with or without optimization of the charge parameters, or by dynamical methods. In the former case, it should be possible in principle to optimize the nuclear positions and the charge density variables simultaneously. However, we have not yet found a procedure for this that was sufficiently stable. Currently, the charge variables are optimized separately for each set of nuclear positions. This is a stable, but not efficient procedure.

Dynamical methods (simulated annealing) employ MD algorithms and can, in principle, propagate the charge variables along with the nuclear positions. As detailed elsewhere [10], the density parameters can be assigned a (negative) mass such that they are continuously and adiabatically driven to the maximum of E[n]. This method, which is a real-space version of the propagation method of Car and Parrinello for plane-wave coefficients [19], achieves continuous updating of the trial density so this is optimized at all times at very little additional computational cost. This is certainly the method of choice for simulated annealing and molecular dynamics if it is relatively straightforward to find appropriate masses for the density variables. This is a matter of building up experience.

Results

An important aspect of the calculational scheme described above is the limited nature of the density basis. Since the actual density of a system displaying covalent bonding cannot be represented in the form given in Eq. 7, it might seem that the use of this ansatz may involve unacceptable error whatever basis functions are used. In this section, we will quote results for a series of molecules which demonstrate, so far as structural information is concerned, that the errors made are actually quite small.

Table 1 shows structural parameters for a number of dimers and small organic molecules. These all-electron calculations were performed using the local density approximation with the VWN parameterization, and the "standard basis" option for density and orbitals of the FastStruct/SimAnn module (see below). The final column in Table 1 shows the percentage error as compared with experiment. For the most part, the errors are of the order of typical LDA errors. That is, the use of the functional in Eq. 1 in conjunction with a trial density constructed from spherically symmetric site densities does not involve much loss of significance in the results.

Next we consider some transition-metal compounds involving first- and second-row transition metals. It has previously been demonstrated [2,3] that DFT-LDA calculations (using DMol) give rather accurate structures for these compounds. In Table 2, we compared these earlier data with corresponding results obtained using the current approach (as implemented in the FastStruct/SimAnn module) and with experimental data. The comparison does not yield a definitive measure of the difference between the simplified calculation and the Kohn-Sham limit because slightly different exchange correlation functionals were used (von Barth-Hedin for DMol and VWN for FS/SA), and the orbital basis functions were also somewhat different (the "standard" option of the respective modules). Nevertheless, the close agreement between the two sets of calculations allows the conclusion that the relative simplicity of the trial density in Eq. 7 does not obviate rather accurate calculation of structural parameters.

Finally, we consider applications to larger organometallic molecules. The structure of bis(N-methyl-5-nitrosalicylideneaminato) nickel(II) was determined recently by Kamenar et al [20]. This is a square planar Ni compound with 41 atoms, as displayed in Figure 1. The FS/SA optimized structure of this molecule reproduces all the qualitative features of the experimental structure. To give a more quantitative comparison, selected bond distances and bond angles are presented in Table 3. As previously, the FS/SA results are close to the DMol results, with a maximum difference of about 2% (in the C-N bond of the nitro

Table 1. Structures of dimers and small molecules

System	Calculated Error		Exp. [4]	Percent
Li$_2$	r(LiLi)	2.638	2.673	-1.3%
Be$_2$	r(BeBe)	2.535	2.49	1.8%
B$_2$	r(BB)	1.630	1.60	1.9%
C$_2$	r(CC)	1.232	1.24	-0.7%
N$_2$	r(NN)	1.117	1.098	1.8%
O$_2$	r(OO)	1.249	1.208	3.4%
F$_2$	r(FF)	1.409	1.412	-0.2%
Na$_2$	r(NaNa)	2.957	3.078	-4.1%
Al$_2$	r(AlAl)	2.478	2.466	0.5%
Si$_2$	r(SiSi)	2.301	2.245	2.5%
P$_2$	r(PP)	1.910	1.893	0.9%
S$_2$	r(SS)	1.982	1.889	4.9%
Cl$_2$	r(ClCl)	2.089	1.988	5.1%
Cu$_2$	r(CuCu)	2.178	2.220	-1.9%
LiF	r(LiF)	1.594	1.564	1.9%
NaCl	r(NaCl)	2.346	2.361	-0.6%
CuF	r(CuF)	1.760	1.745	0.8%
CuCl	r(CuCl)	2.059	2.051	0.4%
TiCl$_2$	r(TiCl)	2.148	2.170	-1.0%
CH	r(CH)	1.156	1.120	3.2%
CH$_2$	r(CH)	1.059	1.029, 1.078	2.9%,-1.8%
	<(HCH)	135.05	144.7, 136.0	
CH$_3$	r(CH)	1.064	1.079	-1.4%
CH$_4$	r(CH)	1.087	1.094	-0.6%
C$_2$H$_2$	r(CC)	1.185	1.203	-1.2%
	r(CH)	1.029	1.060	-3.0%
C$_2$H$_4$	r(CC)	1.323	1.339	-1.2%
	r(CH)	1.054	1.086	-3.0%
	<(HCC)	122.71	121.2	
C$_2$H$_6$	r(CC)	1.514	1.536, 1.531	-1.5%, -1.1%
	r(CH)	1.066	1.091, 1.096	-2.3%, -2.8%
	r(CH)	1.066	1.091, 1.096	-2.3%, -2.8%
	<(HCC)	113.58	110.9	
C$_6$H$_6$	r(CC)	1.372	1.399	-2.0%
	r(CH)	1.040	1.084	-4.2%
CO	r(CO)	1.143	1.128	1.3%
CH$_2$O	r(CO)	1.219	1.208	0.9%
	r(CH)	1.067	1.116	-4.6%
	<(HCO)	122.90	121.8	
CH$_3$OH	r(CO)	1.420	1.425, 1.421	-0.4%,-0.1%
	r(CH)	1.069	1.094, 1.094	-2.3%,-2.3%
	r(OH)	0.967	0.945, 0.963	2.3%, 0.4%
	<(HCO)	109.16	108.5, 107.2	
	<(COH)	114.42	108.0, 108.0	
CH$_2$CO	r(CO)	1.178	1.161	1.5%
	r(CC)	1.305	1.314	-0.7%
	r(CH)	1.066	1.083	-1.6%
	<(CCH)	120.32	118.7	
CO$_2$	r(CO)	1.183	1.162	1.8%

Continued on next page

Table 1. *Continued*

System	Calculated Error		Exp. [4]	Percent
CN	r(CN)	1.169	1.175	-0.5%
HCN	r(CN)	1.148	1.154	-0.5%
	r(CH)	1.028	1.063	-3.4%
CH₃CN	r(CN)	1.387	1.424	-2.7%
	r(CH)	1.066	1.101	-3.3%
	r(CN)	1.157	1.166	-0.8%
	<(HCN)	113.88	109.1	
HF	r(HF)	0.967	0.917	5.5%
CH₃F	r(CH)	1.064	1.098	-3.2%
	r(CF)	1.404	1.382	1.6%
	<(FCH)	111.59	108.5	
CHF₃	r(CH)	1.067	1.098	-2.9%
	r(CF)	1.383	1.333	3.8%
	<(FCH)	110.73	110.3	
CF₄	r(CF)	1.379	1.321	4.4%
H₂O	r(OH)	0.984	0.957	2.8%
	<(HOH)	111.00	104.5	
H₂O₂	r(OO)	1.430	1.475, 1.452	3.1%, 1.5%
	r(OH)	0.986	0.950, 0.965	3.8%, 2.2%
	<(HOO)	105.25	94.8, 100.0	
	<(HOOH)	120.19	119.8, 119.1	
NH₃	r(NH)	1.005	1.012	-0.7%
	<(HNH)	115.71	106.7	
NO	r(NO)	1.189	1.151	3.3%
HNO₃	r(N=O)	1.253	1.206	3.9%
	r(N-O)	1.454	1.405	3.5%
	r(OH)	0.999	0.960	4.1%
	<(O=N=O)	130.51	130.0	
	<(NOH)	106.56	102.0	
NOF	r(NF)	1.501	1.520	-1.3%
	r(NO)	1.186	1.130	5.0%
	<(FNO)	110.96	110.2	
H₂S₂	r(SS)	2.126	2.055	3.5%
	r(SH)	1.375	1.327	3.6%
	<(SSH)	95.62	91.3	
	<(HSSH)	90.38	90.6	
SiH₂	r(SiH)	1.561	1.516	3.0%
	<(HSiH)	92.23	92.1	
C₄SH₄	r(CS)	1.746		
	r(C-C)	1.408		
	r(C=C)	1.354		
	r(CH1)	1.407		
	r(CH2)	1.409		
	<(CSC)	90.17		
	<(SCC)	111.77		
	<(CCC)	113.14		
Si₃	r(SiSi)	2.275		
	<(SiSiSi)	90.64		

Table 2. Geometries for selected First-Row Transition-Metal Compounds

Molecule	Parameter	DMol [3]	FS	expt [3]
ScF	r(ScF)	1.795	1.775	1.787
ScF$_3$	r(ScF)	1.833	1.815	1.91
TiF$_4$	r(TiF)	1.747	1.735	1.745
Cl$_3$TiCH$_3$	r(TiC)	2.012	2.004	2.047
	r(TiCl)	2.174	2.150	2.185
	r(CH)	1.105	1.084	1.098
	<(ClTiC)	106.3	109.6	105.6
	<(HCTi)	108.1	109.1	109.0
VOF$_3$	r(VO)	1.574	1.561	1.569
	r(VF)	1.721	1.718	1.729
	<(FVF)	110.7	108.5	111.2
VOCL$_3$	r(VO)	1.573	1.548	1.570
	r(VCl)	2.131	2.117	2.142
	<(ClVCl)	110.8	111.5	111.3
VF$_5$	r(VFax)	1.747	1.731	1.734
	r(VFeq)	1.712	1.715	1.703
CrO$_2$F$_2$	r(CrO)	1.567	1.572	1.575
	r(CrF)	1.702	1.708	1.720
	<(OCrO)	108.3	109.1	107.8
	<(FCrF)	110.4	115.7	111.9
CrO$_2$Cl$_2$	r(CrO)	1.568	1.573	1.581
	r(CrCl)	2.099	2.088	2.126
	<(OCrO)	109.2	110.7	108.5
	<(ClCrCl)	110.2	111.8	113.3
[CrO$_4$]-2	r(CrO)	1.661	1.662	1.66
[MnO$_4$]-1	r(MnO)	1.611	1.599	1.629
Cr(CO)$_6$	r(CrC)	1.869	1.866	1.909
	r(CO)	1.157	1.138	1.137
Fe(CO)$_5$	r(FeCax)	1.774	1.772	1.807
	r(FeCeq)	1.772	1.773	1.827
	r(CaxO)	1.152	1.136	1.152
	r(CeqO)	1.159	1.140	1.152
Ni(CO)$_4$	r(NiC)	1.785	1.777	1.838
	r(CO)	1.154	1.136	1.141

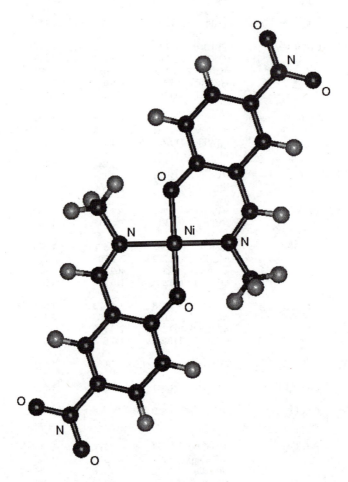

Figure 1. Bis(N-methyl-5-nitrosalicylideneaminato)Nickel(II)

Table 3.   Selected bond distances and angles of bis(N-methyl-5-nitrosalicylideneaminato) nickel(II)

Parameter	DMol	FS/SA	expt [20]
r(Ni-N)	1.88	1.88	1.925
r(Ni-O)	1.82	1.82	1.826
r(N-CH₃)	1.45	1.45	1.475
r(N-C)	1.30	1.27	1.296
r(O₂N-C)	1.43	1.40	1.449
r(O-C)	1.30	1.27	1.299
r(N-O)	1.24	1.24	1.225
r(C-Cben)	1.41	1.40	1.431
r(Cben-Cben)	1.42	1.37	1.411
a(O-Ni-N)	93.4	91.2	93.2
a(Ni-N-C)	124.8	123.1	124.8
a(Ni-O-C)	131.0	130.8	131.0
a(Ni-N-CH₃)	121.3	120.0	120.4

group and the O-Ni-N bond angle), and both are in satisfactory agreement with the experimental values.

4. Discussion and Conclusions

Both the DMol and the FS/SA codes use numerical basis functions, but in the standard options of the codes, used in the above calculations, there are differences.   FS/SA uses finite range functions that vanish identically beyond a given radius.   This feature is not very important for open molecular systems, but allows a very substantial speed-up when very many neighbors must be taken into account (as when periodic boundary conditions are invoked, for instance).   In addition, polarization functions are constructed differently and are sometimes absent in the FS/SA code where they are present in DMol (for the carbon atom, for instance). Whereas preliminary tests have shown that the FS/SA standard option is reasonably accurate, more careful work is needed to establish the consequences of basis error.   The present calculations do suggest, nevertheless, that residual errors are very unlikely to be larger than a few percent and serve to establish clearly that the overall scheme yields structures that are close to the Kohn-Sham limit (i.e., the structures that would be obtained in a fully self-consistent, orbital basis converged Kohn-Sham calculation).

There are several factors that are important in understanding why the ansatz in Eq. 7 does not cause larger errors than are evident in these results. First, the Harris functional is relatively flat away from its stationary point. The quadratic error term for E and E_{ks} has been analyzed in detail by Zaremba [21] and also by Finnis [10], who presented a direct comparison of the behaviour of E_{ks} and E away from a common stationary point. The larger curvature of E_{ks} is related to the overshoot of the electrostatic potential that results when a density is made more, or less, compact, which itself is connected with the inherent instability of the Kohn-Sham self-consistency cycle.

The flatness of E explains to some extent why a simple ansatz like Eq. 7, which must be quite far from the true density (at least locally), gives a reasonable energy. However, it is a general observation that the absolute total energy given by the scheme is less accurate than the structural parameters. The reason for this is that structural information requires accurate tracking of changes that are local (in the sense that the energy differences refer to sets of nuclear coordinates that are close). Terms in the energy that are constant within the particular geometry do not contribute to the forces and have no bearing on the structural information. Such terms may have to do with, for example, intra-site adjustments that have little influence on the variation of the chemical bonds that have formed. Such terms will in general differ, however, when a global change such as the breaking of a strong chemical bond occurs. For this reason, the accurate calculation of absolute reaction energies may require a more careful optimization of the energy functional (whether this is E or E_{ks}) than a determination of the structure of local minima, or even of the reaction path joining these minima. The energy surface so determined can then be relatively easily adjusted by performing detailed Kohn-Sham minimisations for selected points along the path.

The most obvious way of improving the total energy within the present scheme is to extend the orbital and density bases and there are a number of ways in which this could be done. Alternatively, the energy can be improved via perturbation theory (by adding an estimate of the quadratic correction term[10]). This has the advantage that the simplicity and speed of the present method, which is more than adequate for most structural studies in molecules and periodic solids, would be retained in full.

Acknowledgments

The calculations reported in this paper were performed using the FastStruct/SimAnn module that is a product of Biosym/MSI Inc.

Literature Cited

1. W. Kohn and L. J. Sham, Phys. Rev. 140 A 1133 (1965).
2. T. Ziegler, Chem. Rev. 91, 651 (1991) and references therein
   K. Labanowski and J. Andzelm, Eds. "Density Functional Methods
   in Chemistry" (Springer-Verlag, New York,1991) and references
   therein.
3. C. Sosa, J. Andzelm, B. Elkin, E. Wimmer, K. Dobbs and D. A.
   Dixon, J. Phys. Chem. 96 (1992) 6630.
4. W. J. Hehre, L. Radom, P. Schleyer and J. P. Pople, Ab Initio
   Molecular Orbital Theory (Wiley, New York, 1986).
5. W. Thiel, A. A. Voityuk, Int. J. Quant. Chem. 44 (1993) 807.
6. B. Delley, J. Chem. Phys. 92, 508 (1990).
7. B. Delley, D. E. Ellis, A.J . Freeman, E. J. Baerends and E. J. Post,
   Phys. Rev. B27, 2132 (1983) .
8. J. Harris, Phys. Rev. B31 1770 (1985).
9. W. M. C. Foulkes and R. Haydock, Phys. Rev. B39, 12520 (1989).
10. M. W. Finnis, J. Phys: Condens Matter 2 331 (1990).
11. I. J. Robertson and B. Farid, Phys. Rev. Lett. 66, 3265 (1991).
12. O. Sankey and J. Niklewski, Phys. Rev. B40 3979 (1989).
13. M. O'Keeffe, G. B. Adams and O. Sankey, Phys. Rev. Lett. 68 2325
    (1992).
14. J. Harris and D. Hohl, J. Phys. Condens. Matter 2, 5161 (1990).
15. Z. Lin and J. Harris, J. Phys. Condens. Matter 5, 1055 (1992).
16. A. A. Demkov, J. Ortega, O. F. Sankey, M. P. Grumbach, (preprint).
17. B. Delley, J. Chem. Phys. 94, 7245 (1991).
18. J. Baker, J. Comput. Chem. 7 385 (1986).
19. R. Car and M. Parrinello, Phys. Rev. Lett. 55, 2471 (1985).
20. B. Kamenar, B. Kaitner, A. Stefanovic, N. Waters, Acta, Cryst.
    C46 (1990) 1923.
21. E. Zaremba, J. Phys: Condens Matter 2 2479 (1990).

Chapter 27

Density-Functional Calculations of Radicals and Diradicals

Myong H. Lim, Sharon E. Worthington, Frederic J. Dulles, and Christopher J. Cramer[1]

Department of Chemistry and Supercomputer Institute, University of Minnesota, Minneapolis, MN 55455–0431

Open-shell molecules highlight some of the greatest strengths and weaknesses of density functional theory (DFT). When accurate geometries are used, doublet hyperfine couplings are in general well-predicted by spin-polarized DFT calculations. However, for a series of 25 phosphorus-containing radicals, when geometries are optimized at the same level of theory as is used for the prediction of hyperfine couplings, no pure DFT functional does as well as either the UHF or UMP2 levels of theory with identical basis sets. However, hybrid HF/DFT functionals do perform almost as well as MP2. In diradicals, on the other hand, the DFT formalism appears to account well for nondynamical and dynamical correlation effects that are not included in single-determinant Hartree-Fock theory. Singlet-triplet gaps are predicted well for a number of carbenes and nitrenium ions. Finally, a spin-annihilation procedure that involves the construction of a Slater determinant from DFT orbitals is shown to permit the accurate calculation of open-shell singlet energies.

1. Introduction

Open-shell molecules pose particular challenges for electronic structure methods. Within the context of Hartree-Fock (HF) theory, one may either pursue a restricted open-shell approach (ROHF) or alternatively an unrestricted approach (UHF) (1). ROHF theory has the virtue that it produces wavefunctions that are eigenfunctions of the total spin operator S^2—this is usually not the case for UHF theory. On the other hand, UHF theory permits spin polarization of the doubly-occupied orbitals, where ROHF theory does not.

The latter property is important in certain cases. For instance, if an atom with a nucleus that has a magnetic moment lies in a nodal plane of the singly-occupied molecular orbital (SOMO) of a doublet, ROHF theory predicts the atom to have a zero

[1]Corresponding author

0097–6156/96/0629–0402$15.25/0

electron spin resonance (ESR) hyperfine coupling constant (e.g., all of the atoms in the planar $^2A_2''$ methyl radical). However, experiment verifies that spin polarization of the doubly occupied orbitals can cause a net spin density at nuclei in the nodal plane of the SOMO, i.e., non-zero hyperfine couplings are observed *(2)*.

Regrettably, by allowing for spin polarization, UHF theory also allows spin contamination, i.e., differentiation between the spatial parts of the alpha and beta orbitals causes electronic states of higher multiplicity to appear as artifacts in UHF wavefunctions. In cases where spin contamination is particularly severe, one-electron properties, e.g., hyperfine coupling constants, will be inaccurately predicted because of the poor quality of the wavefunction. Calculations on open-shell electronic structures thus manifest something of the truism that one cannot have one's cake and eat it too.

However, an interesting feature of density functional theory that has been noted in the recent literature is that it seems to suffer minimally from spin contamination in its spin polarized (i.e., unrestricted) form *(3-6)*. It is perhaps not surprising then that many groups have examined the performance of various density functionals for the prediction of hyperfine coupling constants *(7-16)* and other magnetic properties *(16-18)*. Although Suter et al. identified limitations of DFT in the calculation of hyperfine coupling constants for the NH radical cation *(15)*, its performance for a variety of other small molecules has been in general equal in quality to post-HF ab initio calculations.

We recently examined a series of 25 phosphorus containing radicals with the goal of comparing the utility of UHF, projected UHF (PUHF), and second-order perturbation theory (UMP2) for the prediction of isotropic hyperfine coupling constants *(19)*. Within this set of molecules, experimental measurements are available for 20 different ^{31}P couplings, 8 ^{19}F couplings, 7 ^{35}Cl couplings, and 5 ^1H couplings. High-level theoretical studies have also been reported for various members of the set. The first topic of this chapter will be an analysis of the performance of a variety of DFT functionals applied to this data set.

A second difficulty in working with open-shell systems can occur when the open-shell system is an excited state above a closed-shell ground state. In that case, it can be difficult to calculate accurately the excited state energy relative to that of the ground state. This is particularly the case in many one-center diradicals, e.g., methylene *(20-24)*, where accurate calculation of singlet-triplet gaps has proven challenging. In particular, when there are low-lying excited states, the theory must adequately account for non-dynamical correlation and must do so in a consistent way for both the open- and closed-shell multiplets; naturally, it is also important to account consistently for dynamical correlation effects as well. These requirements render essentially useless any comparison between restricted Hartree-Fock (RHF) and UHF energies. Instead, more time-consuming approaches, e.g., multiconfiguration self-consistent-field (MCSCF) or single-reference coupled cluster calculations including single, double, and perturbative triple excitations (CCSD(T)), must typically be employed.

Since DFT functionals include correlation directly into the self-consistent-field (SCF) equations, they can *in principle* handle the multiplet splitting problem more efficiently that conventional molecular orbital techniques. That is to say, DFT accounts for non-dynamical correlation not by the linear combination of two or more configurations (as in MCSCF approaches) but rather by construction of the functional for what is formally a single-configuration representation. We *(5,25,26)* and others *(4,27-33)* have examined this issue in a variety of molecular systems, and the second part of this chapter is devoted to this endeavor.

Finally, DFT is a single-determinant theory in the sense that it is not possible to solve the DFT SCF equations for an *intrinsically* multiconfigurational electronic state. For example, open-shell singlet states must be represented by a linear combination of (at least) two configuration state functions. As originally described by Slater *(34)*, and developed more fully by Ziegler et al. *(35)*, the energies of such states may be calculated via the approximate "sum method". An alternative approach is to consider the construction of a Slater-determinant-type wavefunction from the DFT orbitals—with such a plan it might be preferable to refer to the subject orbitals as being "approximate Hartree-Fock molecular orbitals" *(36)*. Once such a determinant is constructed, it is possible to apply spin annihilation operators to mixed-state wavefunctions in order to calculate excited states having the same spatial symmetry as the mixed-state wavefunction. An example of this approach as applied to *E*-diazene (HN=NH) concludes the present chapter.

2. Methodology

Prediction of ESR Isotropic Hyperfine Coupling Constants. The calculation of ESR hyperfine couplings (hfs) has received considerable attention *(10,37-51)*. The *isotropic* hyperfine coupling to nucleus X, a_X, is calculated from

$$a_X = (8\pi/3)gg_X\beta\beta_X\rho(X),\tag{1}$$

where g is the electronic g factor, β is the Bohr magneton, g_X and β_X are the corresponding values for nucleus X, and $\rho(X)$ is the Fermi contact integral which measures the unpaired spin density at the nuclear position. When the spin density is expanded in a finite basis set, the Fermi contact integral is evaluated from

$$\rho(X) = \sum_{\mu\nu} P^{\alpha-\beta}_{\mu\nu} \phi_\mu(R_X) \phi_\nu(R_X),\tag{2}$$

where $P^{\alpha-\beta}$ is the one-electron spin density matrix, the summation runs over all basis functions ϕ, and the evaluation of the overlap between elements μ and ν is only at the nuclear position, R_X. The one-electron spin-density matrix is available for a wide variety of methodologies, including semiempirical theory *(44,48,52,53)*, UHF and ROHF theory, post-Hartree Fock treatments using many-body perturbation theory (MP2) *(40,41,49,54-56)* and configuration interaction (CI) *(38,39,42,43,46,54,57-67)*; more recent work has examined DFT *(7-14)*. This chapter examines ROHF and several hybrid and pure DFT methods for a data set of 25 phosphorus-containing molecules, as described further below.

Spin Annihilation of DFT Wavefunctions. In spin-polarized DFT calculations, open-shell singlet energies can be estimated by a statistical approximation under certain circumstances ("statistical exchange approximation" was the term used by Slater for Xα calculations—the generalization of Ziegler et al. is more properly referred to as the "sum method") *(34,35)*. This procedure makes use of the 50:50 determinant

$$^{50:50}\Psi = \left[\varphi_a(1)\alpha(1)\varphi_b(2)\beta(2) - \varphi_a(2)\alpha(2)\varphi_b(1)\beta(1) \right], \tag{3}$$

where 1 and 2 are the indices of the unpaired electrons occupying orbitals φ_a and φ_b. The 50:50 superscript emphasizes that this configuration is an equal combination of the open-shell singlet and the $S_z = 0$ triplet, provided the two pure spin states are characterized by the same spatial orbitals.

When the Hamiltonian contains no spin-dependent terms, the separation

$$\left\langle ^{50:50}\Psi \mid H \mid ^{50:50}\Psi \right\rangle = \frac{1}{2}\left(\left\langle ^3\Psi_0 \mid H \mid ^3\Psi_0 \right\rangle + \left\langle ^1\Psi \mid H \mid ^1\Psi \right\rangle \right) \tag{4}$$

may be accomplished. One may thus estimate the energy of the open-shell singlet from the readily calculated expectation value of the Hamiltonian operating on the 50:50 and triplet configurations.

An alternative method for calculating the open-shell singlet energy involves application of the annihilation operator A_{s+1} defined by *(68)*

$$A_{s+1} = \frac{s^2 - \{(s+1)[(s+1)+1]\}}{[s(s+1)] - \{(s+1)[(s+1)+1]\}}, \tag{5}$$

where S is the total spin operator, s is the desired total spin, and $s+1$ is the contaminant (next highest) spin state to be annihilated. Although initially employed for the calculation of UHF energies *(69,70)*, the operator may also be applied to a determinant formed from Kohn-Sham orbitals, in which case the spin-annihilated energy is calculated as

$$E_{\text{PDFT}} = {}^{50:50}E_{\text{DFT}} + \frac{\sum_i \langle \Psi^0_{\text{KS}} | H | \Psi^i_{\text{KS}} \rangle \langle \Psi^i_{\text{KS}} | A_{s+1} | \Psi^0_{\text{KS}} \rangle}{\langle \Psi^0_{\text{KS}} | A_{s+1} | \Psi^0_{\text{KS}} \rangle}. \tag{6}$$

That is, the energy of the 50:50 state is calculated in the conventional way, i.e., by plugging in α and β densities to the DFT energy expression. The spin-annihilation correction, however, is calculated using matrix elements of the Schrödinger Hamiltonian and the spin-annihilation operator expressed in the KS MO basis *(36)*.

General. For the phosphorus-containing radicals, molecular geometries were fully optimized at the UHF, ROHF, UMP2, and several DFT levels of theory employing the 6-31G** basis set *(71-74)*. The nature of all stationary points was verified by analytic calculation of harmonic frequencies *(75)*. Calculation of isotropic hyperfine splittings (hfs) values was accomplished with equations 1 and 2 above, using the 6-311G** basis set *(76)* with one-electron spin-density matrix elements taken from ROHF and DFT wavefunctions. For anions, the 6-311++G* basis set *(77)* was also employed, but the effects on calculated hfs values were insignificant, and they are not reported here. The hfs calculations were carried out for geometries optimized at several different levels of theory. Details of the error analysis and a more complete description of the experimental data are available in a prior publication *(19)*. We have examined elsewhere issues associated with other choices of basis set, and note here only that the 6-311G** basis has been found to optimally balance efficiency and accuracy in our studies *(41)*.

Methylcarbene and -nitrenium geometries were optimized at the HF (restricted for singlet, unrestricted for triplet) level of theory with the cc-pVDZ basis set *(78)*—C_s methylnitrenium singlet spontaneously fragments to H_2 and $HCNH^+$ at all other levels of theory examined. Since our focus is on S-T gaps, the HF geometries were used for single-point calculations with other levels of theory. Dimethylcarbene and -nitrenium geometries were optimized at the DFT and complete-active-space self-consistent-field *(79,80)* (CASSCF) levels of theory with the cc-pVDZ basis set *(78)*. Construction of CASSCF active spaces is described in Section 4. Additional DFT single point calculations were performed either using the cc-pVTZ basis set *(78)* or the cc-pVQZ basis set *(78)* with g functions removed; multireference second order perturbation theory *(81-83)* (CASPT2N) and coupled-cluster calculations including all single, double, and perturbative triple excitations (CCSD(T)) were also carried out *(84-87)*.

E-Diazene geometries (C_{2h}) were optimized and energies calculated at the DFT level using the cc-pVDZ and cc-pVTZ basis sets.

Several DFT functionals were employed, combining either Xα local exchange or Becke's (B) non-local exchange functionals *(88)* with either the local correlation functionals of Vosko, Wilks, and Nusair *(89)* (VWN or VWN5), or the non-local alternatives developed by Lee, Yang, and Parr *(90)* (LYP), or Perdew *(91)* (P). Reference is also made to another nonlocal exchange method of Perdew and co-workers, PW *(92)*. Adiabatic connection method functionals combining Becke's non-local functional with some HF exchange (B3) were also employed *(93)*. Calculations were carried out using the MOLCAS *(94)*, GAMESS *(95)*, and GAUSSIAN 92/DFT *(96)* program suites.

3. Phosphorus-Containing Radicals

The 25 phosphorus-containing radicals examined are illustrated in Figure 1. Table 1 summarizes errors and linear regressions of the predicted isotropic hyperfine coupling constants from different levels of theory on the experimental data; this table includes analyses of data for hfs values calculated at the UHF and UMP2 geometries as well (said hfs values together with all geometrical data are available from the authors on request but, for reasons of space, they are not presented here).

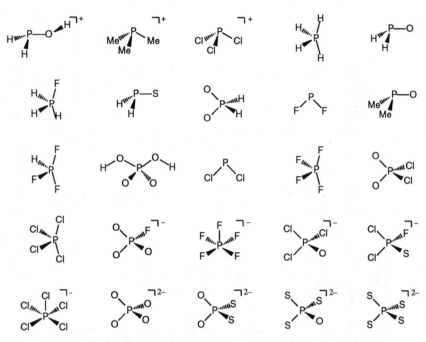

Figure 1. 25 molecules for isotropic hyperfine coupling constant calculations.

Figure 2 plots calculated data against experimental data for the BLYP/6-311G**//BLYP/6-31G** level of theory.

Several trends in Table 1 are worthy of note. To begin, the predicted hyperfine couplings that most closely match the experimental data are obtained from the hybrid functionals B3P and B3LYP. The mean unsigned error with these functionals is minimized for calculations employing the MP2 geometries. However, the predicted data are better correlated with the experimental data by linear regression when UHF geometries are employed. This situation arises because errors in calculated values using the UHF geometries are more consistently under-estimations (and are hence correctable) than is the case when MP2 geometries are employed. For self-consistently optimized geometries, the mean unsigned error for these functionals remains roughly the same as for UHF geometries; the linear regressions are somewhat degraded, however. For comparison purposes, we note that for this data set these two functionals predict isotropic hyperfine coupling constants about as well as MP2//MP2 calculations, but considerably less well than MP2//UHF calculations (the latter having a mean error of −10.5 G, a mean unsigned error of 16.5 G, a correlation coefficient R = 0.9986, and a standard error in the linear fit of 18.2 G) *(19)*.

Slightly lower quality predictions are obtained from ROHF calculations, which are not particularly sensitive to the geometry employed. Of course, in many instances the ROHF method predicts zero hfs because the atom(s) in question lie(s) in

Table 1. Error and Regression Analyses for Prediction of Isotropic Hyperfine Couplings[a]

| Level of Theory | | Error Analysis | | | Correlation | Regression Analysis | | |
Optimized Geometry	$P^{\alpha-\beta}_{\mu\nu}$	Mean Error, G	Std Dev, G	Unsigned Error, G	Const. R^2	Slope	Intercept, G	Standard Error, G
UHF	ROHF	27.7	41.4	39.2	0.9937	1.037	17.4	38.3
UHF	B3P	26.5	40.4	35.0	0.9949	0.950	-11.1	33.0
UHF	B3LYP	23.3	38.8	32.9	0.9948	1.038	12.4	35.1
UHF	BLYP	27.3	47.7	38.5	0.9924	1.051	12.7	42.2
UHF	BVWN	24.7	47.2	37.1	0.9918	1.041	13.1	43.9
UHF	XαP	27.9	47.6	38.6	0.9913	0.957	-14.5	43.5
MP2	ROHF	15.6	45.9	35.8	0.9912	0.980	21.5	45.5
MP2	B3P	14.8	36.4	29.7	0.9944	1.006	-16.8	36.8
MP2	B3LYP	11.4	37.8	28.9	0.9942	0.979	17.6	36.9
MP2	BLYP	15.5	41.6	32.6	0.9925	0.992	17.8	41.9
MP2	BVWN	12.9	43.4	32.6	0.9921	0.983	18.0	43.2
MP2	XαP	16.2	45.5	34.0	0.9915	1.013	-20.4	45.6
ROHF	ROHF	28.9	41.5	40.3	0.9939	1.039	17.9	37.9
B3P	B3P	11.5	42.0	30.9	0.9928	1.016	-16.3	41.9
B3LYP	B3LYP	8.8	48.8	34.3	0.9905	0.971	17.6	47.3
BLYP	BLYP	2.7	107.9	54.4	0.9571	0.916	28.5	100.4
BVWN	BVWN	2.6	112.8	60.5	0.9535	0.911	29.9	104.5
XαP	XαP	15.7	55.7	37.6	0.9869	1.007	-17.8	56.3

[a] All calculations employed the 6-31G** basis set for geometry optimization and the 6-311G** basis set for hfs values.

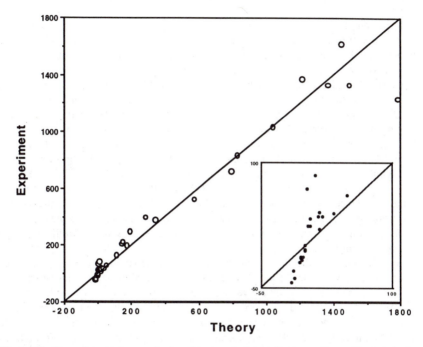

Figure 2. Experimental vs. calculated isotropic hyperfine couplings (Gauss) at the BLYP/6-311G**/BLYP/6-31G** level. The ideal line of unit slope and zero intercept is also shown. The inset expands the region from −50 to 100 Gauss.

the nodal plane of the SOMO (e.g., PF_2, H_2PO_4, PCl_2, PCl_2O_2, etc.)—this can cause significant disagreement when spin polarization gives rise to sizable couplings for these atoms (e.g., $a_P = 68$ G in PCl_2). Nevertheless, when this artifact does not pose a problem, the performance of this method may make it attractive given that the ROHF wavefunction is an eigenfunction of the total spin operator (i.e., a pure doublet).

A further increase of about 20% in the standard error for the linear regressions is observed when pure DFT functionals are used for hfs prediction at the UHF and UMP2 geometries. A *very* large increase in the error is observed for pure DFT calculations carried out at the DFT-optimized geometries—this error is especially severe for the BLYP and BVWN functionals. For the BLYP functional, we observe most P-X bond lengths to be considerably lengthened compared to MP2 values: on average, P-H bonds are 0.031 Å longer (11 data), P-F bonds are 0.030 Å longer (10 data), P-Cl bonds are 0.061 Å longer (10 data), P-O bonds are 0.024 Å longer (14 data) and P-S bonds are 0.043 Å longer (7 data).

The largest source of error, however, is associated with the failure of the BLYP and BVWN functionals to accurately predict even gross features of the molecular geometries for several of the radicals. PCl_4 offers perhaps the best example. Experiment indicates this molecule to have C_{2v} symmetry, with two chlorine atoms axially disposed in a trigonal bipyramid, and the remaining two chlorine atoms equatorially disposed *(97)*. The unpaired electron is localized predominantly in the remaining equatorial position. Calculations at all other levels of theory are consistent with this observation, predicting an angle at phosphorus of between 140 and 150 degrees between the two apical P–C bonds (i.e., there is some compression into the space occupied by the unpaired electron). Within the constraints of C_{2v} symmetry, however, the BLYP and BVWN geometries relax to structures having T_d symmetry. This causes an error of over 500 G in the predicted ^{31}P hyperfine coupling constant (expt. 1233 G). Moreover, if symmetry constraints are completely relaxed, the molecule dissociates into a van der Waals complex of PCl_3 and a chlorine atom! Similar distortions are observed for FPO_3^-, PCl_2FS^-, PO_4^{2-}, POS_3^{2-}, and PS_4^{2-}, although in several of these instances the situation is rendered more complicated because some functionals predict changes in the symmetries of the electronic ground states. These qualitative inaccuracies in describing the electronic structures are reminiscent of those observed by Ruiz et. al. *(98)* for Mulliken charge transfer complexes.

The errors in the PCl_4 geometry vary widely depending on which functional is employed. When Becke's non-local exchange functional (B) is employed, the Cl_{ax}-P-Cl_{ax} bond angle is always reduced (i.e., more in error) compared to Slater or Xα exchange. Even for those combinations of functionals that do not predict the T_d structure to be the lowest energy available within C_{2v} symmetry, the energy separation between the optimized geometry and the T_d geometry ranges from 0.5 to 3.7 kcal/mol at the DFT/cc-pVDZ level. At the MP2/cc-pVDZ level, this energy difference is 6.3 kcal/mol. The best DFT results occur when the non-local PWP functional is used. Here we note the work of Eriksson *(99)* whose PWP calculations predict a Cl_{ax}-P-Cl_{ax} bond angle of 140.4 degrees (slight compression compared to UHF = 149.2 degrees); however, at this geometry, the PWP ^{31}P hfs value is still significantly overestimated (1488.2 G) compared to that calculated for the UHF geometry (1319.7 G).

We highlight these results because Eriksson et al. have compared the PWP functional to others *(11)* (including some of those used here) and in particular have done so for PF_2 and PCl_2 *(100)*. Based on analysis of the spatial distribution of unpaired spin density, they have emphasized that, relative to the local density approximation (LDA), the non-local PWP functional redistributes spin from the inner-valence region to both the core and the outer valence regions—the former effect generally leads to an increase in calculated hyperfine couplings which improves their agreement with experiment. Laidig has emphasized that the LDA, VWN, B, P and LYP functionals all fail to adequately "push" electron density into the outer regions of atoms and molecules *(101)*. The PW and/or P functionals appear to correct this problem to some extent, although it is not clear that this has any significant impact on the predicted hfs values. Further work with this functional in the area of prediction of isotropic hyperfine coupling constants will be particularly interesting.

To conclude this section on a practical note, we note that the B3LYP and B3P methods predict isotropic hyperfine coupling constants only slightly less accurately than MP2 calculations when UHF geometries are employed. Accuracy is degraded when geometries optimized at these DFT levels are employed, but remains improved over any other fully DFT approach. This is due in part, presumably, because inclusion of some Hartree-Fock exchange does cause the geometries to more closely resemble those calculated at the UHF level. Since the DFT calculations scale more efficiently than MP2, these methods will be the most cost-effective for large molecules. Barone *(14)* has come to a similar conclusion for diatomics containing H, B, C, N, O, and F atoms, which suggests that the utility of these hybrid functionals for hfs prediction is not limited to the present phosphorus-based data set.

4. Singlet-Triplet Gaps in Carbenes and Nitrenium Ions

Methylene (CH_2) has a long history as a subject of theoretical studies *(20-24,102)*. Isoelectronic nitrenium (NH_2^+) has also been examined *(103-105)*. Of key interest is the accurate prediction of the S-T gaps in these systems. We *(5,25,26,106)* and others *(4,27-33)* have explored the efficacy of DFT in this regard, both for these simple parent systems, and for substituted versions. In general, modern DFT functionals appear to be remarkably accurate in accounting for the differential non-dynamic correlation in the two multiplets, and give excellent agreement with either experimental values (when available) or very high level ab initio calculations.

One question of fundamental interest is how substitution affects the S-T gap in these systems *(21,107-109)*. We have focused in earlier work on phenyl substitution *(5,106)* and incorporation of the hypovalent center into a three-membered ring *(26)*. Here we examine the more subtle effects of mono- and dimethyl substitution, with the additional goal of comparing DFT predictions against other trustworthy levels of theory. Prior studies of these two carbenes *(21,107,108,110-119)* and two nitrenium ions have appeared *(104,106)*, although none have employed DFT.

In order to facilitate comparisons between different theoretical methods, the same geometries were used for all levels of theory. Key geometrical parameters are listed in Table 2. For methylcarbene and methylnitrenium, HF geometries were used because at all other levels of theory singlet methylnitrenium (C_s) spontaneously fragments to H_2 and $HCNH^+$—for methylcarbene, the effect on the S-T gap of geometry optimization at higher levels of theory is expected to be fairly small, but it is important to recognize that the results discussed below should *not* be regarded as quantitative. For the dimethyl-substituted cases, MCSCF(2,2) geometries were used (the active space correlates the two nonbonding orbitals on the hypovalent center)— the triplets have C_{2v} symmetry and the singlets have C_2 symmetry *(118,119)*. Optimized DFT geometries (not reported here) were similar overall to those obtained at the CAS level, except that the triplet bond angle is predicted to be 4 to 5 degrees larger. This apparent overestimation has been noted *(25)* for methylene and nitrenium; the effect of angle widening is to stabilize the triplet over the singlet by 1 to 1.5 kcal/mol more than is found for the MCSCF geometries. One geometric trend worthy of note is that the bond angle at the hypovalent center is always larger for the nitrenium ion than for the corresponding carbene. This may be attributed to the tendency for the more electronegative (cationic) nitrogen to mix more s character into hybrid orbitals used to make σ bonds to its substituents than does carbon.

Table 2. Geometries employed for carbenes and nitrenium ions

Molecule	State	Level of Theory[a]	rXC, Å	rXH, Å	∠CXY, deg
CH3CH	$^1A'$	HF	1.487	1.106	105.5
(X=C,Y=H)					
	$^3A''$	HF	1.485	1.083	131.4
CH3NH+	$^1A'$	HF	1.364	1.027	112.1
(X=N,Y=H)					
	$^3A''$	HF	1.442	1.024	150.4
CH3CCH3	$^1A_1$	CAS(2,2)	1.493		111.1
(X=C,Y=C)					
	$^3B_1$	CAS(2,2)	1.490		129.0
CH3NCH3+	$^1A_1$	CAS(2,2)	1.396		119.7
(X=N,Y=C)					
	$^3B_1$	CAS(2,2)	1.445		144.3

[a] All calculations employed the cc-pVDZ basis set.

S-T gaps calculated at various levels of theory are collected in Table 3. The measured S-T gaps for methylene *(120,121)* and nitrenium *(122)* are 9.1 and 30.1 kcal/mol, respectively. For both the mono- and dimethyl-substituted systems, the S-T gap remains larger for the nitrenium than for the carbene. In addition, the singlet becomes the ground state for dimethylcarbene at the DFT levels of theory. We now compare the predictions from different levels of theory, and then discuss the effects of methyl substitution.

The S-T gap in methylcarbene changes by only 1 kcal/mol when the MCSCF active space is increased from a simple (2,2) calculation (correlating the two nonbonding orbitals on the carbene carbon) to a (12,12) calculation (correlating the full valence space). At the CASPT2 level, this difference is reduced to 0.5 kcal/mol, indicating that in this instance perturbation theory successfully corrects for some of the inadequacies in the (2,2) space in addition to accounting for some dynamic correlation. The situation for methylnitrenium is less well behaved—increasing the active space decreases the S-T gap by 3.5 kcal/mol at the MCSCF level and by 8.0 kcal/mol at the CASPT2 level. The cationic charge has the effect of reducing the total energy spanned by the occupied (and virtual) valence orbitals, and this presumably makes it increasingly important to more completely optimize them. For both methylcarbene and methylnitrenium, the best CASPT2 gaps are about 3 to 5 kcal/mol larger than those calculated at the CCSD(T) level. This is quite consistent with the situation observed for methylene, where CASPT2 overstabilizes the triplet by about 3

Table 3. Singlet-triplet gaps for carbenes and nitrenium ions [a]

Molecule	Level of Theory[b]	S-T gap, kcal/mol	
		X=C	X=N+
CH_3XH	CAS(2,2)	7.9	20.2
	CAS(12,12)	8.9	16.7
	CASPT2N(2,2)	11.3	24.4
	CASPT2N(12,12)	10.8	16.4
	CCSD(T)	7.3	13.2
	BVWN5	4.9	7.4
	BLYP	5.1	7.8
CH_3XCH_3	CAS(2,2)	3.8	15.9
	CASPT2N(2,2)	6.1	10.9
	CCSD(T)	1.8	8.6
	BVWN5	−0.3	2.9
	BLYP	−1.0	2.3

[a] Using the geometries found in Table 2. [b] All calculations employed the cc-pVDZ basis set.

kcal/mol *(82,83,123)*, and where CCSD(T) agrees well with experiment. The S-T gaps predicted by DFT, on the other hand, are considerably lower than the CCSD(T) values. For methylnitrenium, in the absence of experimental data, there are insufficient high-quality computational results with which to compare in order to evaluate the relative accuracies. For methylcarbene, however, several other studies exist *(21,107,108,110-115,118)*. Based on an analysis of other calculations, and on their own high-quality generalized valence bond (GVB) CI calculations, Khodabandeh and Carter *(115)* conclude that the S-T gap of methylcarbene is 3 ± 2 kcal/mol, which provides more support for the DFT predictions than for the CCSD(T) value. This agreement improves for DFT when the basis set is enlarged: the BVWN5 and BLYP predicted S-T gaps are 4.1 and 3.9 kcal/mol, respectively, with the cc-pVQZ basis set (these numbers are converged based on comparison to cc-pVTZ values). Moreover, as noted above, complete geometry optimization further stabilizes the triplet by roughly 1 kcal/mol. The larger gaps predicted by the other levels of theory may be associated with one or more of the following (i) greater sensitivity to an incomplete basis than is observed for DFT, (ii) less adequate accounting for dynamic correlation, and (iii) greater sensitivity to geometric inaccuracies. The second explanation seems likely to be associated with the bulk of the difference, given the results of Khodabandeh and Carter *(115)*.

In the dimethylated cases, the situation is similar in many respects. Although it would be interesting to compare the (2,2) active space to the full-valence CAS, this calculation is presently impractical. Attempts to use a (6,6) active space including the heavy atom sigma bonds proved problematic and were not pursued extensively. For dimethylcarbene compared to methylcarbene, there is a slight increase in the disagreement between the best CASPT2 prediction and that obtained at the CCSD(T) level. This difference is near the maximum in the range for the systematic error in S-T gaps discussed by Andersson et al. *(82,83,123)*. Once more, the DFT predicted S-T gaps for both molecules are smaller than those found at the CCSD(T) level; this difference is sufficient to cause DFT to predict dimethylcarbene to have a singlet ground state. The predicted DFT gaps agree very well with the MRCI+Q results of Matzinger and Fülscher *(118)* and the CCSD(T)/TZ2P+f results of Richards et al. *(119)*; the latter calculations predict a singlet ground state with a gap of –0.9 and –0.8 kcal/mol, respectively. Agreement between the two DFT functionals is again good, and the S-T gaps are very nearly converged with respect to basis set size: changes no larger than 0.4 kcal/mol are observed on going to the cc-pVTZ basis set. This is in contrast to the CCSD(T) level of theory. Our cc-pVDZ results mirror polarized double-ζ results of Richards et al. *(119)*, but the latter workers observed the gap to change by 2.7 kcal/mol (in favor of the singlet) on going to a doubly-polarized triple-ζ basis set.

As for the chemical effects of substitution, it has been argued that the S-T gap of methylcarbene is reduced by comparison to carbene because the singlet is stabilized by hyperconjugative interactions (i.e., π donation) with the methyl group *(21,107,108)*. Destabilization of the singlet, on the other hand, may compete because of steric interactions between the two groups on the hypovalent center: as discussed by Carter and Goddard *(109)*, bond-angle widening decreases the amount of p character in the associated σ bonds, thus reducing the s character available to the doubly-occupied nonbonding orbital. Khodabandeh and Carter *(115)*, based on a detailed analysis of their GVB-CI results, discount the importance of these two effects for methylcarbene, however, and emphasize that the methyl group destabilizes the *triplet* by (i) decreasing the p character in the in-plane SOMO, (ii) reducing electron donation to the electrophilic hypovalent center, and (iii) raising the energy of the π-like SOMO by repulsive orbital interactions. While these effects probably also influence the methylnitrenium S-T gap, it is apparent that hyperconjugation plays a more significant role in this instance. In the singlet, the N–C bond length is reduced by 0.08 Å compared to the triplet, consistent with considerable π character in this bond (note that in methylcarbene, the C–C bond *lengthens* on going from triplet to singlet, consistent with decreased s character in the σ bond). Moreover, the effect of methyl substitution on the S-T gap is considerably larger in the nitrenium case than in the carbene: the nitrenium singlet is stabilized by roughly 23 kcal/mol and the carbene singlet by about 5 kcal/mol (based on comparison of the DFT predictions for MeXH to experimental values for XH_2). This observation is consistent with the nitrenium nitrogen being a much more aggressive π acceptor than the carbene carbon. Such differential π acceptance has also been noted for phenyl substitution *(5)*.

On attachment of a second methyl group, both the carbene and nitrenium S-T gaps are further reduced by about 5 kcal/mol. Hence, for the carbene case, there appears to be little if any steric destabilization of the singlet *(109)*. The effect of the second methyl substitution on the nitrenium S-T gap, on the other hand, is much smaller than that of the first. One interpretation would be that the accepting ability of the methylated nitrenium ion singlet is no better than the methylated carbene (i.e., one methyl group "saturates" the difference associated with the positive charge in the nitrenium). However, the observed angle widening and the crowding imposed by shorter heavy-atom bonds in the nitrenium singlet (compared to the carbene) *may* indicate that there is here some competition between sterics and hyperconjugation in the singlet.

In conclusion, for identical geometries, MCSCF and CASPT2 methods predict S-T gaps 3 to 5 kcal/mol larger than those found at the CCSD(T) level. The latter theory itself predicts gaps that are 1 to 6 kcal/mol larger than those found at the DFT level. For the case of methylcarbene, where the highest-quality calculations at other levels of theory are available, DFT appears to provide the more reliable results. Although the separate effects of hyperconjugation and sterics on the S-T gap of methylcarbene have been assigned to be minimal *(115)*, hyperconjugation plays a dominant role in the more aggressively electrophilic nitrenium ions, and steric destabilization of the singlet state is observed for dimethylnitrenium.

5. Spin Annihilation of Kohn-Sham-orbital-based Slater Determinants

Back et al. have measured the near-UV spectrum for the symmetry-forbidden $1^1B_g \leftarrow 1^1A_g$ transition in *E*-diazene *(124,125)*. CI calculations have been performed for several excited states of this system, both at the semiempirical *(126)* and ab initio *(127-132)* levels. We here compare to the latter studies by examining the relative energies of the ground state singlet, triplet (^3B_g), and first-excited-state singlet as calculated at the UHF and BLYP levels. The excited-state (open-shell) singlet energies are calculated using the spin annihilation technique described in Section 2 and also using the sum method of Ziegler et al. *(35)*.

Optimized geometries for the various states using the cc-pVDZ basis set within the constraints of C_{2h} symmetry are provided in Table 4. Experimental bond lengths have been determined from N_2 matrix IR data *(133)*, and the HNN angle

Table 4. Optimized geometries for *E*-diazene

Level of Theory	Electronic State	rNN, Å	rNH, Å	$\angle$HNN, deg
UHF	^1A_g	1.215	1.020	107.3
	^3B_g	1.254	1.010	120.3
	$^{50:50}B_g$	1.253	1.007	120.3
BLYP	^1A_g	1.262	1.058	105.2
	^3B_g	1.289	1.033	118.4
	$^{50:50}B_g$	1.289	1.035	118.6
Experiment	^1A_g	1.252	1.028	106.9

Table 5. Excited state energies for E-diazene (kcal/mol)

Electronic State	C_{2h} Geometry	UHF	BLYP
^3B_g	^1A_g (vertical)	45.1	49.7
	^3B_g (relaxed)	34.9	41.0
sum method ^1B_g	^1A_g (vertical)	75.3	64.9
	$^{50:50}B_g$ (relaxed)	65.3	56.1
annihilated ^1B_g	^1A_g (vertical)	55.4	64.8
	$^{50:50}B_g$ (relaxed)	44.9	55.7

from gas-phase IR data *(134)*. The BLYP level provides the better prediction for rNN, while the UHF level is somewhat better for the remaining two parameters. Although the ^1B_g state has not been optimized (a manual optimization would be required) it is clear from the similarity of the 50:50 mixed singlet/triplet structure to the triplet structure that the pure open-shell singlet should be very similar in geometry. Geometrical parameters calculated at the UHF level for the $^{50:50}B_g$ state agree closely with those reported by Del Bene et al. *(132)* from UHF/6-31+G** calculations.

Vertical and relaxed (C_{2h}) excitation energies from single-point calculations using the cc-pVTZ basis set are compiled in Table 5. For this system, the excited-state singlet energies at the BLYP level differ only very slightly for the two methods of calculation—this is not the case at the UHF level, where large differences are observed, and these UHF results are not discussed further. Experimental data are not available for the triplet, but the predicted value for the vertical excitation at the BLYP level may be compared to those calculated at the MRCI levels by Vasudevan et al. *(127)* and Kim et al. *(135)*, namely 47.7 and 60.7 kcal/mol, respectively. The discrepancy between the two MRCI results is surprisingly large. Although our BLYP results are in much better agreement with the predictions of Vasudevan et al., this excitation awaits experimental resolution.

The absorption maxima in the near-UV spectrum occur over a range of 78.4 to 83.0 kcal/mol *(124,125)*. Vasudevan et al. *(127)* suggested this region to be blue-shifted relative to the "usual" vertical transition since in the absence of an inducing mode the transition moment for this forbidden excitation would be expected to be zero. Peric et al. *(128)* performed a complete vibrational analysis at the MRCI level and concluded that one quantum (3120 cm^{-1}, 8.9 kcal/mol) of excitation in the asymmetric N-H stretching mode (v_5) afforded good agreement with the various vibrational progressions in the experimental spectrum. Back et al. *(125)* refined and reinterpreted their spectroscopic data on the basis of the theoretical calculations and agreed that the vertical separation between the two singlet potential energy curves should be 71.8 ± 2.3 kcal/mol. The present BLYP results underestimate this value by about 10%. This may be compared with the MRCI results of Vasudevan et al. *(127)* and Kim et al. *(135)*, in this case 68.5 and 81.8 kcal/mol, respectively. Again there is significant disagreement between the two MRCI studies, although it is perhaps noteworthy that they predict a very similar energy separation between the two B_g multiplets, suggesting that one or the other set of calculations may be unbalanced in its description of the ground state.

Finally, the experimental value for the $1^1B_g \longleftarrow 1^1A_g$ T_0 transition (i.e., from lowest vibrational state to lowest vibrational state) is 59.4 ± 3.5 kcal/mol *(125)* (again including one vibrational quantum in v_5)—accounting for differential zero-point energies *(128,135)* leads to a T_e value of 61.8 kcal/mol. This number may be compared to the relaxed calculations in Table 5, and it is seen that the BLYP calculations are in fair agreement (the BLYP comparison ignores torsional relaxation of the excited-state singlet, which lowers the predicted T_e by about 1.5 kcal/mol *(128,131,132)*). The MRCI calculations of Peric et al. *(128)* and Kim et al. *(135)* predict T_e values of 58.9 and 70.3 kcal/mol, respectively.

The BLYP functional was originally chosen for this study because of its impressive performance in calculating the $1^1B_u \longleftarrow 1^1A_g$ vertical transition for *s-trans* 1,3-butadiene *(36)* (expt. 5.9 eV, predicted 6.1 eV), an excitation that has proven quite challenging for other levels of theory (and which furthermore provides an example of an excitation where the spin-annihilation technique provides much more reliable results than the statistical approximation). Performance of the BLYP functional in the case of E-diazene is slightly less satisfactory, and it may be that a different functional would be more efficacious. We are continuing to examine the issue of functional utility in this and other chemical systems with respect to the calculation of excited-state-singlet energies. However, it is worth emphasizing in closing that the spin-annihilated DFT calculations are considerably less costly than the MRCI calculations to which they have been compared, so from an efficiency standpoint, somewhat increased errors may be acceptable depending on the chemical questions under investigation.

6. Conclusions

Accurately accounting for unpaired spin density and its effect on one-electron properties can be challenging for any theoretical method. In doublets, one measure of the quality of electronic structure is the accuracy of predicted hyperfine splittings. For accurate molecular geometries, we and others have noted the efficacy of spin-polarized DFT calculations. However, for the series of 25 phosphorus-containing radicals presented here, inadequacies in DFT-optimized geometries lead to a significant degradation in the quality of predicted hyperfine splittings compared to the UHF or UMP2 levels of theory; hybrid HF/DFT functionals, however, perform about as well as MP2 under these circumstances.

Moving from one unpaired electron to two, DFT appears to capture significant nondynamical and dynamical correlation effects that are not included in single-determinant Hartree-Fock theory, and that are difficult to account for consistently with single-reference or multireference correlation approaches. For methyl- and dimethylcarbene and -nitrenium, the performance of two DFT functionals appears to be superior to MCSCF, CASPT2, and CCSD(T) calculations. However, experimental data are not yet available for these systems. Until such data are available, additional calculations using multireference CI approaches would clearly be desirable in order to further evaluate this contention.

Finally, the calculation of open-shell singlet energies by application of a spin-annihilation procedure to a Slater determinant formed from DFT orbitals appears to be promising as an efficient alternative to more computationally demanding procedures.

Acknowledgments. We gratefully acknowledge the support of the U.S. Army Research Office (DAAH04-93-G-0036) and the National Science Foundation. We further thank Profs. Jan Almlöf, Wes Borden, Emily Carter, Dan Falvey, and Robert McMahon, Dr. Susan Gustafson, and Mr. Bradley Smith for stimulating discussions, and Prof. Leif Eriksson for calculations on PCl4 using the PWP functional.

Literature Cited

(1)   Szabo, A.; Ostlund, N. S. *Modern Quantum Chemistry*; Macmillan: New York, 1982.
(2)   This is true even after accounting for vibrational averaging—see, for instance, reference 55.
(3)   Baker, J.; Scheiner, A.; Andzelm, J. *Chem. Phys. Lett.* **1993**, *216*, 380.
(4)   Murray, C. W.; Handy, N. C.; Amos, R. D. *J. Chem. Phys.* **1993**, *98*, 7145.
(5)   Cramer, C. J.; Dulles, F. J.; Falvey, D. E. *J. Am. Chem. Soc.* **1994**, *116*, 9787.
(6)   Pople, J.; Gill, P.; Handy, N. *Int. J. Quant. Chem.* in press.
(7)   Eriksson, L. A.; Wang, J.; Boyd, R. J. *Chem. Phys. Lett.* **1993**, *211*, 5.
(8)   Eriksson, L. A.; Malkin, V. G.; Malkina, O. L.; Salahub, D. R. *J. Chem. Phys.* **1993**, *99*, 9756.
(9)   Ishii, N.; Shimizu, T. *Phys. Rev. A* **1993**, *48*, 1691.
(10) Barone, V.; Adamo, C.; Russo, N. *Chem. Phys. Lett.* **1993**, *212*, 5.
(11) Eriksson, L. A.; Malkina, O. L.; Malkin, V. G.; Salahub, D. R. *J. Chem. Phys.* **1994**, *100*, 5066.
(12) Kong, J.; Eriksson, L. A.; Boyd, R. J. *Chem. Phys. Lett.* **1994**, *217*, 24.
(13) Ishii, N.; Shimizu, T. *Chem. Phys. Lett.* **1994**, *225*, 462.
(14) Barone, V. *Chem. Phys. Lett.* **1994**, *226*, 392.
(15) Suter, H. U.; Pless, V.; Ernzerhof, M.; Engels, B. *Chem. Phys. Lett.* **1994**, *230*, 398.
(16) Malkin, V. G.; Malkina, O. L.; Eriksson, L. A.; Salahub, D. R. In *Modern Density Functional Theory; A Tool for Chemistry, Theoretical and Computational Chemistry*; P. Politzer and J. Seminario, Eds.; Elsevier: Amsterdam, 1995; Vol. 2; p. 273.
(17) Malkin, V. G.; Malkina, O. L.; Salahub, D. R. *Chem. Phys. Lett.* **1994**, *221*, 91.
(18) Schreckenbach, G.; Ziegler, T. In *Abstracts of Papers of the 209th National Meeting of the American Chemical Society*; American Chemical Society: Washington D.C., 1995; PHYS #114.
(19) Cramer, C. J.; Lim, M. H. *J. Phys. Chem.* **1994**, *98*, 5024.
(20) Roos, B. O.; Siegbahn, P. M. *J. Am. Chem. Soc.* **1977**, *99*, 7716.

(21) Davidson, E. R. In *Diradicals*; W. T. Borden, Ed.; Wiley-Interscience: New York, 1982; p. 73.
(22) Goddard, W. A. *Science* **1985**, *227*, 917.
(23) Shavitt, I. *Tetrahedron* **1985**, *41*, 1531.
(24) Schaefer, H. F., III *Science* **1986**, *231*, 1100.
(25) Cramer, C. J.; Dulles, F. J.; Storer, J. W.; Worthington, S. E. *Chem. Phys. Lett.* **1994**, *218*, 387.
(26) Cramer, C. J.; Worthington, S. E. *J. Phys. Chem.* **1995**, *99*, 1462.
(27) Jones, R. O. *J. Chem. Phys.* **1985**, *82*, 325.
(28) Radzio, E.; Salahub, D. R. *Int. J. Quant. Chem.* **1986**, *29*, 241.
(29) *Density Functional Methods in Chemistry*; Labanowski, J.; Andzelm, J., Eds.; Springer-Verlag: New York, 1991.
(30) Gutsev, G. L.; Ziegler, T. *J. Phys. Chem.* **1991**, *95*, 7220.
(31) Jacobsen, H.; Ziegler, T. *J. Am. Chem. Soc.* **1994**, *116*, 3667.
(32) Arduengo, A. J.; Rasika Dias, H. V.; Dixon, D. A.; Harlow, R. L.; Klooster, W. T.; Koetzle, T. F. *J. Am. Chem. Soc.* **1994**, *116*, 6812.
(33) Arduengo, A. J.; Bock, H.; Chen, H.; Denk, M.; Dixon, D. A.; Green, J. C.; Herrmann, W. A.; Jones, N. L.; Wagner, M.; West, R. *J. Am. Chem. Soc.* **1994**, *116*, 6641.
(34) Slater, J. C. *Phys. Rev.* **1951**, *81*, 385.
(35) Ziegler, T.; Rauk, A.; Baerends, E. J. *Theor. Chim. Acta* **1977**, *43*, 261.
(36) Cramer, C. J.; Dulles, F. J.; Giesen, D. J.; Almlöf, J. *Chem. Phys. Lett.* **1995**, *245*, 165.
(37) Aagaard, O. M. Ph.D. Thesis, Technische Universiteit Eindhoven, 1991.
(38) Carmichael, I. *Chem. Phys.* **1987**, *116*, 351.
(39) Chipman, D. M. *J. Chem. Phys.* **1983**, *78*, 4785.
(40) Cramer, C. J. *J. Am. Chem. Soc.* **1991**, *113*, 2439.
(41) Cramer, C. J. *J. Mol. Struct. (Theochem)* **1991**, *235*, 243.
(42) Ellinger, Y.; Subra, R.; Berthier, G. *Nouv. J. Chim.* **1983**, *7*, 375.
(43) Feller, D.; Davidson, E. *J. Chem. Phys.* **1983**, *80*, 1006.
(44) Glidewell, C. *Inorg. Chim. Acta* **1984**, *83*, L81.
(45) Janssen, R. A. J.; Buck, H. M. *J. Mol. Struct. (Theochem)* **1984**, *110*, 139.
(46) Knight, L. B.; Earl, E.; Ligon, A. R.; Cobranchi, D. P.; Woodward, J. R.; Bostick, J. M.; Davidson, E. R.; Feller, D. *J. Am. Chem. Soc.* **1986**, *108*, 5065.
(47) Nakatsuji, H.; Izawa, M. *J. Chem. Phys.* **1989**, *91*, 6205.
(48) O'Malley, P. J.; MacFarlane, A. J. *J. Mol. Struct. (Theochem)* **1992**, *277*, 293.
(49) Sekino, H.; Bartlett, R. J. *J. Chem. Phys.* **1985**, *82*, 4225.
(50) Sieiro, C.; de la Vega, J. M. G. *J. Mol. Struct. (Theochem)* **1985**, *120*, 383.
(51) Smith, P.; Donovan, W. H. *J. Mol. Struct. (Theochem)* **1990**, *204*, 21.
(52) Knight, L. B.; Arrington, C. A.; Gregory, B. W.; Cobranchi, S. T.; Liang, S.; Paquette, L. *J. Am. Chem. Soc.* **1987**, *109*, 5521.
(53) Gorlov, Y. I.; Penkovsky, V. V. *Chem. Phys. Lett.* **1975**, *35*, 25.
(54) Barone, V.; Minichino, C.; Faucher, H.; Subra, R.; Grand, A. *Chem. Phys. Lett.* **1993**, *205*, 324.
(55) Cramer, C. J. *J. Org. Chem.* **1991**, *56*, 5229.

(56) Cramer, C. J. *Chem. Phys. Lett.* **1993**, *202*, 297.
(57) Lunell, S.; Eriksson, L. A.; Worstbrock, L. *J. Am. Chem. Soc.* **1991**, *113*, 7508.
(58) Knight, L. B.; Steadman, J.; Feller, D.; Davidson, E. *J. Am. Chem. Soc.* **1984**, *106*, 3700.
(59) Knight, L. B.; Earl, E.; Ligon, A. R.; Cobranchi, D. P. *J. Chem. Phys.* **1986**, *85*, 1228.
(60) Knight, L. B.; Winiski, M.; Kudelko, P.; Arrington, C. A. *J. Chem. Phys.* **1989**, *91*, 3368.
(61) Knight, L. B.; Ligon, A.; Cobranchi, S. T.; Cobranchi, D. P.; Earl, E.; Feller, D.; Davidson, E. *J. Chem. Phys.* **1986**, *85*, 5437.
(62) Knight, L. B.; Wise, M. B.; Childers, A. G.; Davidson, E. R.; Daasch, W. R. *J. Chem. Phys.* **1980**, *73*, 4198.
(63) Carmichael, I. *J. Phys. Chem.* **1987**, *91*, 6443.
(64) Barone, V.; Grand, A.; Minichino, C.; Subra, R. *J. Chem. Phys.* **1993**, *99*, 6787.
(65) Barone, V.; Grand, A.; Minichino, C.; Subra, R. *J. Phys. Chem.* **1993**, *97*, 6355.
(66) Buckingham, A. D.; Olegário, R. M. *Chem. Phys. Lett.* **1993**, *212*, 253.
(67) Knight, L. B.; Cobranchi, S. T.; Gregory, B. W.; Earl, E. *J. Chem. Phys.* **1987**, *86*, 3143.
(68) Löwdin, P.-O. *Phys. Rev.* **1955**, *97*, 1509.
(69) Amos, T.; Hall, G. G. *Proc. R. Soc. London* **1961**, *A263*, 483.
(70) Mayer, I. *Adv. Quantum Chem.* **1980**, *12*, 189.
(71) Ditchfield, R.; Hehre, W. J.; Pople, J. A. *J. Chem. Phys.* **1971**, *54*, 724.
(72) Hehre, W. J.; Ditchfield, R.; Pople, J. A. *J. Chem. Phys.* **1972**, *56*, 2257.
(73) Hariharan, P. C.; Pople, J. A. *Theor. Chim. Acta* **1973**, *28*, 213.
(74) Francl, M. M.; Pietro, W. J.; Hehre, W. J.; Binkley, J. S.; Gordon, M. S.; DeFrees, D. J.; Pople, J. A. *J. Chem. Phys.* **1982**, *77*, 3654.
(75) Hehre, W. J.; Radom, L.; Schleyer, P. v. R.; Pople, J. A. *Ab Initio Molecular Orbital Theory*; Wiley: New York, 1986.
(76) Krishnan, R.; Binkley, J. S.; Seeger, R.; Pople, J. A. *J. Chem. Phys.* **1980**, *72*, 650.
(77) Frisch, M. J.; Pople, J. A.; Binkley, J. S. *J. Chem. Phys.* **1984**, *80*, 3265.
(78) Dunning, T. H. *J. Chem. Phys.* **1989**, *90*, 1007.
(79) Roos, B. O.; Taylor, P. R.; Siegbahn, P. E. M. *Chem. Phys.* **1980**, *48*, 157.
(80) Pulay, P.; Hamilton, T. P. *J. Chem. Phys.* **1988**, *88*, 4926.
(81) Andersson, K.; Malmqvist, P.-Å.; Roos, B. O.; Sadlej, A. J.; Wolinski, K. *J. Phys. Chem.* **1990**, *94*, 5483.
(82) Andersson, K.; Malmqvist, P.-Å.; Roos, B. O. *J. Chem. Phys.* **1992**, *96*, 1218.
(83) Andersson, K.; Roos, B. O. *Int. J. Quant. Chem.* **1993**, *45*, 591.
(84) Cizek, J. *Adv. Chem. Phys.* **1969**, *14*, 35.
(85) Purvis, G. D.; Bartlett, R. J. *J. Chem. Phys.* **1982**, *76*, 1910.
(86) Scuseria, G. E.; Schaefer, H. F., III *J. Chem. Phys.* **1989**, *90*, 3700.
(87) Raghavachari, K.; Trucks, G. W.; Pople, J. A.; Head-Gordon, M. *Chem. Phys. Lett.* **1989**, *157*, 479.

(88) Becke, A. D. *Phys. Rev. A* **1988**, *38*, 3098.
(89) Vosko, S. H.; Wilks, L.; Nussair, M. *Can. J. Phys.* **1980**, *58*, 1200.
(90) Lee, C.; Yang, W.; Parr, R. G. *Phys. Rev. B* **1988**, *37*, 785.
(91) Perdew, J. P. *Phys. Rev. B* **1986**, *33*, 8822.
(92) Perdew, J. P.; Wang, Y. *Phys. Rev. B* **1986**, *33*, 8800.
(93) Becke, A. D. *J. Chem. Phys.* **1993**, *98*, 5648.
(94) Andersson, K.; Blomberg, M. R. A.; Fülscher, M. P.; Karlström, G.; Kellö, V.; Lindh, R.; Malmqvist, P.-Å.; Noga, J.; Olsen, J.; Roos, B. O.; Sadlej, A. J.; Siegbahn, P. E. M.; Urban, M.; Widmark, P.-O. *MOLCAS-3*; University of Lund: Sweden, 1994.
(95) Schmidt, M. W.; Baldridge, K. K.; Boatz, J. A.; Elbert, S. T.; Gordon, M. S.; Jensen, J. H.; Koseki, S.; Matsunaga, N.; Nguyen, K. A.; Su, S.; Windus, T. L.; Dupuis, M.; Montgomery, J. A. *J. Comp. Chem.* **1993**, *14*, 1347.
(96) Frisch, M. J.; Trucks, G. W.; Schlegel, H. B.; Gill, P. M. W.; Johnson, B. G.; Wong, M. W.; Foresman, J. B.; Robb, M. A.; Head-Gordon, M.; Replogle, E. S.; Gomperts, R.; Andres, J. L.; Raghavachari, K.; Binkley, J. S.; Gonzalez, C.; Martin, R. L.; Fox, D. J.; Defrees, D. J.; Baker, J.; Stewart, J. J. P.; Pople, J. A. *Gaussian 92/DFT, Revision G.1*; Gaussian, Inc.: Pittsburgh, PA, 1993.
(97) Mishra, S. P.; Symons, M. C. R. *J. Chem. Soc., Dalton Trans.* **1976**, 139.
(98) Ruiz, E.; Salahub, D.; Vela, A. *J. Am. Chem. Soc.* **1995**, *117*, 1141.
(99) L. A. Eriksson, unpublished calculations.
(100) Austen, M. A.; Eriksson, L. A.; Boyd, R. J. *Can. J. Chem.* **1994**, *72*, 695.
(101) Laidig, K. E. *Chem. Phys. Lett.* **1994**, *225*, 285.
(102) Bauschlicher, C. W.; Langhoff, S. R.; Taylor, P. R. *J. Chem. Phys.* **1987**, *87*, 387.
(103) Lee, S. T.; Morokuma, K. *J. Am. Chem. Soc.* **1971**, *93*, 6863.
(104) Ford, G. P.; Herman, P. S. *J. Am. Chem. Soc.* **1989**, *111*, 3987.
(105) Glover, S. A.; Scott, A. P. *Tetrahedron* **1989**, *45*, 1763.
(106) Falvey, D. E.; Cramer, C. J. *Tetrahedron Lett.* **1992**, *33*, 1705.
(107) Baird, N. C.; Taylor, K. F. *J. Am. Chem. Soc.* **1978**, *100*, 1333.
(108) Mueller, P. H.; Rondan, N. G.; Houk, K. N.; Harrison, J. F.; Hooper, D.; Willen, B. H.; Liebman, J. F. *J. Am. Chem. Soc.* **1981**, *103*, 5049.
(109) Carter, E. A.; Goddard, W. A. *J. Chem. Phys.* **1988**, *88*, 1752.
(110) Ha, T. K.; Nguyen, M. T.; Vanquickenborne, L. G. *Chem. Phys. Lett.* **1982**, *92*, 459.
(111) Kohler, H. J.; Lischka, H. *J. Am. Chem. Soc.* **1982**, *104*, 5884.
(112) Luke, B. T.; Pople, J. A.; Krog-Jesperson, M. B.; Apeloig, Y.; Karni, M.; Chandrasekhar, J.; Schleyer, P. R. *J. Am. Chem. Soc.* **1986**, *108*, 270.
(113) Evanseck, J. D.; Houk, K. N. *J. Phys. Chem.* **1990**, *94*, 5518.
(114) Gallo, M. M.; Schaefer, H. F., III *J. Phys. Chem.* **1992**, *96*, 1515.
(115) Khodabandeh, S.; Carter, E. A. *J. Phys. Chem.* **1993**, *97*, 4360.
(116) Modarelli, D. A.; Platz, M. S. *J. Am. Chem. Soc.* **1993**, *115*, 470.
(117) Ma, B.; Schaefer, H. F., III *J. Am. Chem. Soc.* **1994**, *116*, 3539.
(118) Matzinger, S.; Fülscher, M. P. *J. Phys. Chem.* **1995**, *99*, 10747.

(119) Richards, C. A.; Kim, S.-J.; Yamaguchi, Y.; Schaefer, H. F., III *J. Am. Chem. Soc.* **1995**, *117*, 10104.

(120) Bunker, P. R.; Jensen, P.; Kraemer, W. P.; Beardsorth, R. *J. Chem. Phys.* **1986**, *85*, 3724.

(121) Jensen, P.; Bunker, P. R. *J. Chem. Phys.* **1988**, *89*, 1327.

(122) Gibson, S. T.; Greene, P. J.; Berkowitz, J. *J. Chem. Phys.* **1985**, *83*, 4319.

(123) Andersson, K. *Theor. Chim. Acta* **1995**, *91*, 31.

(124) Back, R. A.; Willis, C.; Ramsay, D. A. *Can. J. Chem.* **1974**, *52*, 1006.

(125) Back, R. A.; Willis, C.; Ramsay, D. A. *Can. J. Chem.* **1978**, *56*, 1575.

(126) Ertl, P.; Leska, J. *J. Mol. Struct. (Theochem)* **1988**, *165*, 1.

(127) Vasudevan, K.; Peyerimhoff, S. D.; Buenker, R. J.; Kammer, W. E.; Hsu, H.-l. *Chem. Phys.* **1975**, *7*, 187.

(128) Peric, M.; Buenker, R. J.; Peyerimhoff, S. D. *Can. J. Chem.* **1977**, *55*, 1533.

(129) Peric, M.; Buenker, R. J.; Peyerimhoff, S. D. *Mol. Phys.* **1978**, *35*, 1495.

(130) Groenenboom, G. C.; van Lenthe, J. H.; Buck, H. M. *J. Chem. Phys.* **1989**, *91*, 3027.

(131) Michl, J.; Bonacic-Koutecky, V. *Electronic Aspects of Organic Photochemistry*; John Wiley & Sons: New York, 1990, p. 367.

(132) Del Bene, J. E.; Kim, K.; Shavitt, I. *Can. J. Chem.* **1991**, *69*, 246.

(133) Bondybey, V. E.; Nibler, J. W. *J. Chem. Phys.* **1973**, *58*, 2125.

(134) Carlotti, M.; Johns, J. W. C.; Trombetti, A. *Can. J. Phys.* **1974**, *52*, 340.

(135) Kim, K.; Shavitt, I.; Del Bene, J. E. *J. Chem. Phys.* **1992**, *96*, 7573.

Chapter 28

Density-Functional Approaches for Molecular and Materials Design

E. Wimmer

**Biosym/Molecular Simulation, Pore Club Orsay Université,
20 rue Jean Rostand, 91893 Orsay Cedex, France**

Density functional theory has become a practical and increasingly more widely used tool in molecular and materials design. This is due to three characteristics of this approach: (i) it enables accurate quantum mechanical calculations on fairly large molecules and unit cells; (ii) the theory is universal in the sense that it can be applied not only to solid state systems, but also to organic molecules, organometallic compounds, and surfaces, and (iii) the computational effort is reasonable. The practical implementations of density functional theory are based on the local density approximation, sometimes including generalized gradient corrections to improve the absolute values of binding energies. The capabilities of this approach are illustrated by the bonding of a silane molecule to a reconstructed Si(001) 2x1 surface, where cluster calculations show a rather localized nature of the molecule-surface interactions. Calculations of the adsorption geometry and the energetics of an ammonia molecule on a CuO(111) surface reveals that the N-Cu bond on a CuO surface is about one order of magnitude weaker compared with that in the Cu-tetrammine complex. This has significant consequences for the design of protective coatings for Cu surfaces. The final example shows the capability of density functional theory to predict equilibrium structures of fairly complex solids such as LaNi$_5$.

The discovery of molecules and materials with novel functional properties continues to be one of the most fascinating areas of scientific research and it is essential for meeting the growing demands of our industrialized societies in an era of changing economic conditions and increasing environmental concerns. The complexity of modern materials, the constraints on their production and processing, and the rising cost of experimental research make it mandatory to rationalize the research and development strategies by rapid pre-screening of all possible molecular and materials design options and by focusing as quickly as possible on the most promising candidates. The targeted performance characteristic of a molecule such as its binding strength to a surface or the ability of a compound to store hydrogen is just one of the conditions that need to be met for a successful molecular and materials design. In fact,

0097–6156/96/0629–0423$15.00/0

a whole range of other conditions need to be fulfilled such as low toxicity, low cost and assured availability of raw materials, low energy requirements and high safety of the synthesis and processing, environmental acceptability and recyclability.

A detailed understanding of the molecules and materials on the atomistic level greatly facilitates this task. To this end, theoretical and computational approaches have become a powerful tool in the research and development process. In order to be useful in molecular and materials design, three essential criteria have to be met by any theoretical and computational tool: (i) it has to be applicable to fairly large systems containing any type of atom in any bonding situation, (ii) its accuracy has to be predictable, and (iii), the computational effort has to be reasonable.

The first requirement comes from the great variety and heterogeneity of modern materials. While in the past the molecular and materials sciences were fairly well separated disciplines, today the boundary between these two areas is rapidly disappearing. For example, in the 1960's, materials science was almost synonymous with metallurgy and the prevailing structural materials for mechanical applications were steel and other metal alloys, whereas today the importance of synthetic polymers, ceramics, and composites as structural engineering materials has reached an unprecedented level and continues to grow *(1)*. Functional materials for electrical, optical, and magnetic applications have long been the domain of inorganic materials, yet molecular systems are rapidly gaining in importance in these areas. For example, liquid crystal displays are replacing cathode ray tubes and organic light emitting diodes are promising alternatives to III-V or II-VI semiconductors. At the same time, advances in silicon-based technologies are enabling the creation of microelectronic devices with unprecedented miniaturization in feature sizes.

The requirements of generality, system size, accuracy and computational effort are very hard to reconcile and for many decades it seemed nearly impossible to meet all of these simultaneously. The development of density functional theory, its expansion from solid state physics into the molecular sciences together with the astonishing progress in computer hardware have enabled a major step towards this goal *(2)*. However, in view of the tantalizing complexity of real systems and the enormous difficulties arising from the vast differences in length and time scales between atomistic processes and macroscopic behavior, it is clear that we are far from having a complete solution to accomplish the task of molecular and materials design through computer simulations. Nevertheless, theoretical and computational tools are able to make a significant contribution and their importance is increasing. It is the aim of the present contribution to discuss and illustrate the present capabilities of density functional methods as a tool in molecular and materials design, to point to current limitations, and to indicate development trends.

Density Functional Theory

Within the scope of this contribution, only the most relevant aspects of density functional theory are highlighted. For comprehensive treatments and reviews the reader is referred to the literature such as those given in references *(3)* and *(4)*.

Density functional theory offers a rigorous framework for the quantum mechanical description of any ensemble of atoms such as molecules, molecular aggregates, three-dimensional periodic solids, and surfaces containing any atom from the periodic table. Perhaps the most important practical problem that can be solved with density functional theory is the determination of the changes in the total energy of a system as a function of the positions of the atoms. This knowledge of the energy hypersurface is fundamental to the understanding of any chemical system. It allows the prediction of ground state structures, relative stability of conformations, the prediction of vibrational spectra and, at least in principle, the determination of transition state geometries and barrier heights in any chemical reaction.

In density functional theory *(5,6)* the total energy is expressed as a functional of the electron density, ρ, for given positions, $\mathbf{R}_\alpha$, of all atomic nuclei.

$$E = E[\rho(\mathbf{r}), \mathbf{R}_\alpha] \tag{1}$$

This functional has a minimum for the ground state electron density.

$$\left. \frac{\partial E[\rho]}{\partial \rho} \right|_{\rho = \rho_o} = 0 \tag{2}$$

The total energy is decomposed into three contributions, a kinetic energy, T_o, a Coulomb energy, U, due to classical electrostatic interactions among all charged particles in the system, and a term called the exchange-correlation energy, E_{xc}, that captures all many-body interactions.

$$E = T_o + U + E_{xc} \tag{3}$$

This decomposition is formally exact, but only the relationship between the Coulomb energy and the total charge distribution can be given directly. For any practical purposes, the calculation of the kinetic energy term requires good one-particle wave functions (molecular orbitals) and the exchange-correlation term needs to be approximated.

In wave function based quantum chemical methods such as Hartree-Fock theory with second order perturbation theory, coupled-cluster methods and configuration interaction (CI) methods the high and often prohibitive computational effort comes from the specific description of electron correlation. In fact, these methods implicitly assume that these many-body effects of the interacting electron system are of a long-range delocalized nature. The local density approximation (LDA) of density functional theory and also the more recent generalized gradient approximations (GGA) rest on the assumption that electrons are essentially "nearsighted". It turns out that this assumption is justified especially in regions of reasonably high electron density as found, for example, in the interior of a solid or a molecule. Therefore, as an approximation, the exchange-correlation energy is taken from known results of an interacting electron system of constant density ("homogeneous electron gas") and it is assumed that the exchange and correlation effects are not strongly dependent on inhomogeneities of the electron density away from a reference point, $\mathbf{r}$. Thus, the local electron density can be used to evaluated the exchange and correlation effects of a volume element around $\mathbf{r}$

$$E_{xc}[\rho] \approx \int \rho(\mathbf{r}) \, \varepsilon_{xc}^o[\rho(\mathbf{r})] d\mathbf{r} \tag{4}$$

The exchange-correlation energy per electron, ε_{xc}^o, in a system of interacting electrons of constant density is very well known *(7)*. It turns out that the LDA is an astonishingly good approximation and many structural properties and relative energy changes of a great variety of compounds are very well described. However, it has been found for many systems *(8-10)* that atomization energies are overestimated by the LDA. Thus, the calculation of absolute values for binding energies, but also dissociation energies and weak interactions such as hydrogen bonds or carbonyl bonds require methods beyond the local density approximation. One possibility is offered by gradient corrections such as those suggested by Becke *(11,12)*, Perdew *(13)* and Lee et al. *(14)*. In these approaches, terms depending on the gradient of the electron density are included in the expressions for the exchange and correlation energy. While these approaches definitely improve the values of binding energies of most compounds as well as the bond distances in many weakly bound systems, some of these gradient corrections contain parameters which are fitted in a somewhat ad-hoc manner. This clouds their predictive capabilities. In contrast, the LDA is exactly defined and rests only on fundamental physical constants. Thus, DFT-LDA calculations have an ab initio character.

The accuracy of density functional calculations is intimately related to the use of one-particle wave functions, ψ_i, which define the total electron density through

$$\rho(\mathbf{r}) = \sum_i f_i |\psi_i(\mathbf{r})|^2 \tag{5}$$

with f_i, being the occupation of level i. Using this decomposition of the total electron density into one-particle densities, the variational properties given by eqs. (1) and (2) lead to effective one-particle Schrödinger equations *(6)*, which are referred to as Kohn-Sham equations

$$\left[-\frac{1}{2}\nabla^2 + V_c(\mathbf{r}) + \mu_{xc}(\mathbf{r}) \right] \psi_i(\mathbf{r}) = \varepsilon_i \psi_i(\mathbf{r}) \tag{6}$$

Corresponding to the three terms in the total energy expression (3), namely the kinetic energy, the Coulomb energy, and the exchange-correlation energy, the effective one-particle Hamiltonian of the Kohn-Sham equations contains a kinetic energy operator, a Coulomb potential operator, and an exchange-correlation operator. The latter is related to the exchange-correlation energy by

$$\mu_{xc} = \frac{\partial E_{xc}[\rho]}{\partial \rho} \tag{7}$$

The theory has been generalized to spin-polarized systems *(15,16)* in which the density of the spin-up electrons and that of the spin-down electrons are different, as is the case in open-shell molecules and in magnetic systems. In practice this means that the Kohn-Sham equations have to be solved for the spin-up and spin-down electrons individually, which essentially corresponds to a spin-unrestricted formalism.

The fundamental quantities of density functional theory are the total electron density (and spin density for open-shell molecules and magnetic systems) and the corresponding total energy. The one-particle wave functions (i.e. molecular orbitals in the case of molecules) and the associated one-particle eigenvalues are, strictly speaking, only auxiliary quantities. In practice, however, these one-particle quantities are extremely useful in order to explain the frontier orbitals of a molecule, to distinguish between metallic and insulating behavior of a solid, and to interpret optical excitation energies and photoemission experiments or even subtle effects such as the magneto-optic Kerr effect and magnetic anisotropy energies. In fact, for metallic and delocalized systems, the eigenvalues actually correspond almost quantitatively to excitation energies and the highest occupied level is the negative of the ionization energy or work function. In semiconductors, where localization can be important, the LDA gap is typically 30-50% too small compared with the experimental optical gap. For very localized systems such as f-electrons in rare-earth compounds, these one-particle eigenvalues are far from excitation energies. This does not mean that in these cases the LDA is inappropriate. It simply means that one has to go one step further in the theory in order to describe localized excitation phenomena which are not metallically screened.

At present, there is not a single density functional implementation which would be equally applicable and efficient for all systems ranging from molecules to magnetic transition metal heterostructures. Therefore, a number of different approaches have emerged as discussed, for example, by Wimmer *(17)*.

Illustrative Examples

Chemisorption of silane on a Si(001)2x1 surface. The first example is related to the reaction of molecules with surfaces as they occur, for instance, in chemical vapor deposition processes in the fabrication of microelectronic components *(18)*. The technologically most important semiconductor surface is probably Si(001). This surface is known to reconstruct in the form of a (2x1) structure involving a dimerization of the surface silicon atoms *(19)*. It is not clear if the bonding of a silyl group to one of the Si dimer atoms and the attachment of the remaining hydrogen atom to the other Si dimer atom breaks the dimer bond. If one dimer bond would open in this way, how much would the neighboring Si dimer change its geometry? In other words, how local is the bonding of a silane molecule on a Si surface?

In the present study *(20)*, the chemisorption of silane, SiH_4, was modeled by a finite cluster containing 26 Si atoms as shown in Fig. 1. The bonds towards the bulk are saturated with H bonds with a Si-H distance of 1.496 Å, which correspsonds to the optimized Si-H distance in silane obtained on the same level of theory as used in the cluster calculations. The electronic structure, total energies and forces were calculated by an all-electron local density functional approach with the DMol program *(21,22)*. In this method, the density functional equations are solved with an expansion of the one-particle wave functions in a linear combination of numerical atom orbitals. For all Si atoms, d-polarization functions were added to the double-numerical basis set and found to be important to get correct bond lengths. The $1s$ and $2s$ core electrons of the Si atoms were treated by a frozen core approximation. This results in a total of 480 valence orbitals. The numerical integration were carried out with a "medium" grid *(22)* corresponding to a total of 109178 integration points. The geometry optimization was performed on the local density functional level by using the exchange-correlation terms given by Vosko, Wilk, and Nusair *(23)*. In the geometry optimization the top-most 11 Si atoms with their hydrogen atoms are relaxed and the remaining Si atoms, which would be attached to the bulk atoms, are kept at their bulk positions. In the present study the geometry optimization was terminated with a maximum gradient of 0.001 Ha/Bohr. Starting from nearly symmetric dimers, the geometry optimization leads to an asymmetric arrangement of the dimers as shown on the left-hand side of Fig. 1. A dimer bond distance of 2.33 Å is obtained for this asymmetric arrangement, which is larger than the previously calculated value of 2.21 Å for a symmetric arrangement *(24)* and the experimental value of 2.21 ± 0.04 Å obtained from PEXAFS measurements *(24)*. Upon chemisorption of -SiH_3 and -H on the two atoms of a surface dimer, these atoms move into the same plane parallel to the surface ("symmetric" arrangement) thereby restoring the tetrahedral coordination of the dimer atoms (cf. Fig. 1). The dimer bond length increases from 2.33 Å to 2.44 Å. Interestingly, the geometry of the neighboring asymmetric dimer is slightly reduced from 2.33 Å to 2.30 Å (cf. right-hand side of Fig. 1). This implies that the interaction of a silyl radical with a reconstructed Si(001)2x1 surface is of a fairly local nature in the sense that it does not affect significantly the geometry of neighboring dimers.

The chemisorption energy of a silane molecule on a Si(001)2x1 surface is calculated by subtracting the total energy of the clean Si(001)2x1 surface and that of an isolated silane molecule from the total energy of the cluster representing the

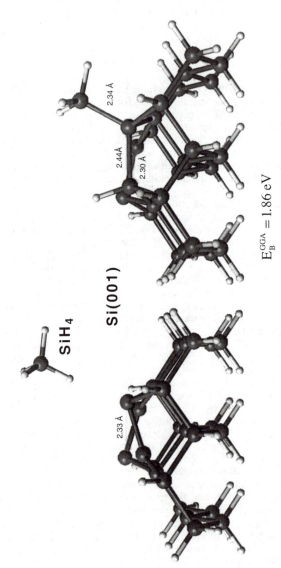

Figure 1. Cluster model of the reconstructed Si(001)2x1 surface. The bonds towards the bulk are saturated with H atoms. The cluster shown on the left-hand side represents the clean, reconstructed surface. The chemisorbed system is displayed on the right-hand side, where a silyl group is attached to one atom of a surface dimer and the remaining H atom from the silane molecule is bonded to the other dimer atom.

chemisorbed system as shown in Fig. 1. To this end, the geometry of a cluster representing the clean surface and that of an isolated silane molecule were optimized at the same level of theory. It is known that the local density approximation leads to good geometries, but tends to overestimate dissociation energies. Therefore, the total energies of all three systems, i.e. the clean cluster, the isolated silane molecule, and the chemisorbed system are recalculated by using a gradient-corrected form for the exchange-correlation potential as given by Becke *(12)* for the exchange and Lee, Yang, and Parr *(14)* for the correlation energies. The dissociative binding energy (at T=0) of a silane molecule on a reconstructed Si(001) surface is thus calculated to be 2.38 eV on the LDA level and 1.86 eV by using the B-LYP gradient corrections.

Binding of ammonia to a CuO(111) surface. In the design of protective coatings of copper surfaces it is critical to have a clear understanding of the interactions of various molecular functional groups with the metal surface. In the case of copper, one design strategy could start from the idea to exploit the Cu-N bonding. In fact, the $[Cu(NH_3)_4]^{2+}$ complex is known to be quite stable and it is not unreasonable to assume that there is also a significant bonding between NH_3 and the atoms on a copper surface. At ambient conditions, copper surfaces are likely to be covered by an oxide film which could be a Cu(I) or a Cu(II) oxide. Here we consider the case of the Cu(II) oxide. Inspection of the CuO crystal, which crystallizes in space group C2/c, reveals that the ($\bar{1}11$) surface might be a low-energy surface because it is fairly dense and contains both cations and anions (cf. Fig. 2). In the present calculations, The CuO(111) is modeled by a $Cu_{14}O_{14}$ cluster as shown in Fig. 2. An ammonia molecule is placed on top of this cluster and an energy minimization is carried out using the DMol approach as described earlier. In this case, a double-numerical basis has been employed and a "medium" grid is used for the numerical integrations of the matrix elements *(22)*. From Fig. 2 it can be seen that the ammonia molecule is bonded on top of a copper atom, as one might expect. The molecule is slightly tilted, indicating an attractive interaction between one of the hydrogen atoms of the ammonia molecule and an oxygen atom of the surface. For comparison, the geometry and binding energies are calculated for an isolated $[Cu(NH_3)_4]^{2+}$ complex using the same level of theory. In the tetrammine complex, the calculated Cu-N distance is found to be 2.03 Å, compared with a distance of 2.19 Å in the case of the CuO surface. Surprisingly, the bonding energy per Cu-N bond in the tetrammine complex is about one order of magnitude larger than the Cu-N interaction of the ammonia molecule on the CuO(111) surface. This means that the initial guideline for the design of adhesive molecules may not be justified since there is a surprisingly large difference of the behavior of Cu ions in a free complex and on a CuO surface.

Structure of LaNi₅. In the third example, the capabilities of density functional theory are illustrated for the calculation of structural properties of solids. One of the problems in the design of metal-hydride batteries is the expansion of materials such as LaNi₅ upon loading with hydrogen. It would be desirable to find alloys which maintain the electrochemical properties and hydrogen storage capabilities, yet show a smaller mechanical expansion and contraction within each battery cycle. As a first step it is necessary to demonstrate that reliable structural information can be obtained for the pure alloys.

Figure 3 shows the crystal structure of LaNi₅ which consists of a hexagonal lattice of the La atoms with the Ni atoms forming an intriguing pattern of Ni₄ units. The dense structure and metallic nature of LaNi₅ requires a band structure approach *(25)*. Furthermore, inclusion of relativistic effects are required for the adequate description of La. To this end, the augmented spherical wave method *(26)* with the atomic-sphere-approximation (ASA) as implemented in the ESOCS program *(27)* was

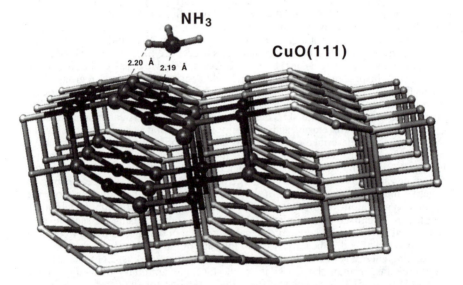

Figure 2. Section of the CuO(111) surface. The atoms displayed in darker shades are used for the quantum mechanical cluster calculations. The ammonia molecule is shown in its equilibrium position as obtained from a geometry optimization using the DMol program.

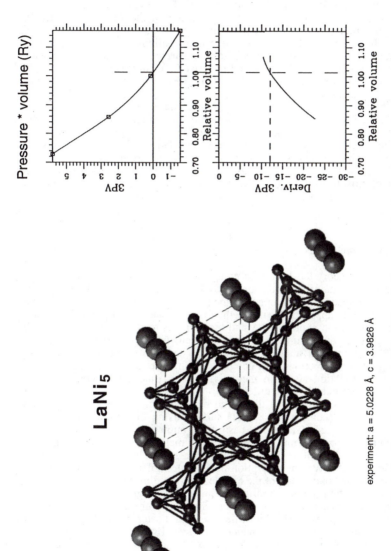

experiment: a = 5.0228 Å, c = 3.9826 Å

Figure 3. Crystal structure of LaNi5 and changes of the internal pressure as a function of isotropic volume changes calculated with the augmented spherical wave method as implemented in the ESOCS program. The volume on the x-axes of the two graphs on the right-hand side are relative to the experimental volume.

chosen. In this approach, the local density functional problem is solved by expanding the wave functions in a variational basis set which is constructed from atomic-like functions inside the atomic spheres which are matched to scattering functions. Relativistic effects are taken into account in the form of a semi-relativistic approach given by Koelling and Harmon (28). A major feature of the ASA approach is the use of a shape approximation to the effective potential in the form of spherically symmetric potentials around each atom. This implies that only isotropic volume changes can be studied, whereas anisotropic deformations would require a full-potential approach such as the full-potential linearized augmented plane wave (FLAPW) method (29) or the full potential linearized muffin-tin orbital (FP-LMTO) approach (30). However, these approaches would be computationally significantly more demanding than ESOCS.

On the right-hand side of Fig. 3 the pressure due to the electronic degrees of freedom of the system are shown as a function of the volume. It can be seen that vanishing pressure, i.e. equilibrium conditions are found for a volume of the unit cell which is within 2% of the experimental results (31). The derivative of the pressure with respect to volume changes, shown in the lower panel on the right hand side of Fig. 3, is proportional to the bulk modulus. These preliminary results demonstrate the capabilities of this first-principles approach to predict structural and energetic properties of fairly complex intermetallic phases.

Summary and Conclusions

Density functional theory has become a practical and useful tool for molecular and materials design. Its major strengths are (i) fairly large molecules and unit cells of about 100 atoms containing any element of the periodic table and more are accessible by this first-principles approach (ii) the method is quite accurate: it gives equilibrium bond distances to within about 0.05 Å and bond angles to within about 1-2°(8-10). Binding energies are overestimated by the local density approximation (LDA), but can be brought to within about 5 kcal/mol or better of experiment by using generalized gradient corrections. (iii) compared with Hartree-Fock calculations and correlated wave function ab initio methods, density functional calculations are computationally significantly more efficient.

These capabilities have been illustrated by the example of a silane molecule chemisorbed onto a reconstructed Si(001) surface using the DMol approach, i.e. numerical atomic orbitals as variational basis and a numerical integration to evaluate the Hamiltonian matrix elements, and a projection method of the charge density to solve Poisson's equation within the self-consistence cycles. The cluster calculations reveal that the dissociative chemisorption of a silane molecule on a reconstructed Si(001) increases the bond length of the affected surface dimer by about 0.1 Å while the dimer bond-length at the neighboring Si-dimer is not significantly altered. This indicates that the chemical interaction of the silyl radical with a silicon surface is of a local nature. This justifies, a posteriori, the use of a cluster model to describe the interactions of molecules with semiconductor surfaces. However, the cluster used in this study is still relatively small and more calculations will be necessary to investigate the convergence of the results with the cluster size. Given the capability of density functional theory (especially with the gradient corrections included) to predict fairly accurate binding energies, density functional theory could prove to be a valuable tool for the prediction of thermochemical data of molecule/surface interactions, thus enriching the experimental data base of this important class of chemical systems.

The case of an ammonia molecule adsorbed on a CuO(111) surface is another example of the use of density functional theory for obtaining data on binding energies, which then can be used in molecular design strategies. The present preliminary study indicates that the Cu-N bonding strength in a free tetrammine complex might be about one order of magnitude larger than in the case of an ammonia molecule adsorbed on top of a Cu atom on the CuO(111) surface. This surprisingly large difference has significant implications for the strategy of designing molecules that bind well to this oxide surface.

The third example demonstrates the use of density functional theory in a very different chemical environment, namely the LaNi$_5$ crystal. In this case, the chemical bonding between the La and Ni atoms involves localized *d*-electrons as well as itinerant *s* and *p* electrons which contribute to the metallic bonding in this system. While it is computationally more efficient to use a band structure method such as the augmented spherical wave (ASW) approach, one should keep in mind that the basic theory is the same as for the calculations on molecules and clusters. This universality is a remarkable feature of density functional theory.

Given the trend in materials science towards more complex materials involving combinations of organic and inorganic aspects, for example in organometallic compounds or in heterogeneous catalysis, density functional theory takes a unique place as a unified and universal tool. As stated earlier, the prediction of ground state geometries is perhaps one of the strongest practical aspects of density functional theory. Furthermore, the theory can also treat open-shell systems, for example radicals such as diphenyl-picryl-hydrazyl (2) as well as magnetic systems such as magnetic multilayer structures and systems with reduced dimensionality (32). In fact, in the area of magnetism density functional theory is making already a major contribution in the industrial design and development of magnetic materials for applications such as magneto-optic recording and magnetic reading heads.

Despite these successes, there are still a number of limitations and challenges, which are related to accuracy and computational speed. In contrast to wave function based ab initio methods, there is no systematic way to improve the accuracy beyond the local density approximation. The generalized gradient corrections are a step forward, but it seems that the current generation of these methods is reaching some fundamental limitations and a qualitative step forward is called for (33). The fundamental problem seems to be the "short-sightedness" underlying the local density approximation and any gradient expansion. In many cases, especially in solids, this approximation works surprisingly well. However, in situations of weak intermolecular interactions such as dative bonds in carbonyls, in the hydrogen bond, but also in the transition state of a chemical reactions, the low density of electrons in critical regions of the system leads to long-range interactions (delocalization of the exchange hole) which will have to be included in order to improve the accuracy.

Given the fact that a large majority of quantum chemists have only recently started to use density functional theory, the progress in the theory can be expected to accelerate. From the increasing body of density functional results for molecules and solids, which is made possible by the wider availability of density functional computer codes and powerful computer hardware, it can be expected that density functional methods will become an indispensable tool for molecular and materials design, hopefully helping to solve the many challenging technological problems of our industrialized societies.

Acknowledgments. The author thanks his colleague Catalina Guerra for her help and Prof. Jürgen Kübler for the discussions of the LaNi$_5$ system. All calculations reported here were carried out on IBM RS/6000 workstations.

References

1. Ashby, M. F. *MRS Bulletin* **1993**, 18, 43.
2. Wimmer, E.; Freeman, A. J.; Fu, C.-L.; Cao, P.-L.; Chou, S.-H.; and Delley, B. ACS Symposium Series 353, Jensen K. F. and Truhlar, D. G., editors, American Chemical Society: Washington, DC, 1987, Chapter 4, pp 49-68.
3. Parr, R. G.; and Yang, W.*Density-Functional Theory of Atoms and Molecules*, Oxford University Press: New York, NY, 1989
4. Ziegler, T. *Chem. Rev.* **1991**, 91, 651.
5. Hohenberg, P.; and Kohn, W. *Phys. Rev.* **1964**, 136, B 864.
6. Kohn, W; and Sham, L. J. *Phys. Rev.* **1965**, 140, A1133.
7. Ceperley, D. M.; and Alder, B. J. *Phys. Rev. Lett.* **1980**, 45, 566.
8. Weinert, M.; Wimmer, E.; and Freeman, A. J. *Phys. Rev. B* **1982**, 26, 4571.
9. Delley, B. *J. Chem. Phys.* **1991**, 94, 7245.
10. Andzelm, J.; and Wimmer, E. *J. Chem. Phys.* **1992**, 96, 1280.
11. Becke, A. *J. Chem. Phys.* **1986**, 84, 4524.
12. Becke, A. *Phys. Rev. A* **1988**, 38, 3098.
13. Perdew, J. P. *Phys. Rev. B* **1986**, 33, 8822.
14. Lee, C.; Yang, W.; and Parr, R. G. *Phys. Rev. B* **1988**, 37, 786.
15. von Barth, U.; and Hedin, L. *J. Phys. C* **1972**, 5, 1629.
16. Gunnarsson, O.; and Lundqvist, B. I. *Phys. Rev. B* **1976**, 13, 4274.
17. Wimmer, E. *J. Computer-Aided Materials Design* **1993**, 1, 215.
18. Jensen, K. F.; and Kern, W. in *Thin Film Processes - II*, Vossen, J. L.; and Kern, W., editors Academic Press:San Diego, 1991, pp 283 - 368.
19. Tang, S.; Freeman, A. J.; and Delley, B. *Phys. Rev. B* **1992**, 45, 1776 and references therein.
20. Guerra, C.; and Wimmer, E. unpublished.
21. Delley, B. *J. Chem. Phys.* **1990**, 92, 508.
22. *DMol User Guide,* version 2.3.5, Biosym Technologies: San Diego, CA, 1993.
23. Vosko, S, H.; Wilk, L.; and Nusair, M. *Can. J. Phys.* **1980**, 58, 1200.
24. Spiess, L.; Mangat.; P.S.; Tang; S.-P.; Schirm; K. M., Freeman, A. J.; and Soukiassian, P. *Surf. Sci. Lett.* **1993**, 289, L631.
25. Cabus, S.; Gloss, K.; Gottwig, U.; Horn, F.; Klemm, M.; Kübler, J.; Steglich, F.; and Parks, R.D. *Solid State Comm.* **1984** , 51, 909.
26. Williams, A. R.; Kübler, J.; and Gelatt, J. R. *Phys. Rev. B* **1979**, 19, 6094.
27. *ESOCS User Guide*, version 2.0, Biosym Technologies: San Diego, CA, Feb. 1995.
28. Koelling, D. D.; and Harmon, B. N. *J. Phys. C* **1977**, 10, 3107.
29. Wimmer, E.; Krakauer, H.; and Freeman, A. J. in *Advances in Electronics and Electron Physics,* ed. by Hawkes, P. W. Academic Press: Orlando, 1985, Vol. 65, p. 357 - 434 and references therein.
30. Methfessel, M. *Phys. Rev. B* **1988**, 38, 1537.
31. Thompson, P.; Reilly, J. J.; and Hastings, J. M. *J. of Less-Common Metals* **1987**, 129, 105.
32. Wang, D.-S.; Freeman, A. J.; and Krakauer, H. *Phys. Rev. B* **1982**, 26, 1340 and references therein.
33. Perdew, J., private communication

Chapter 29

Density-Functional Theory Studies on Beryllium Metal Fragments of 81, 87, and 93 Atoms

Richard B. Ross[1], C. William Kern[2], Shaoping Tang[3,4], and Arthur J. Freeman[3]

[1]PPG Industries, P.O. Box 9, Allison Park, PA 15101
[2]Department of Chemistry and [3]Department of Physics and Astronomy, Northwestern University, Evanston, IL 60208

Density functional molecular orbital theory has been applied to fragments of bulk beryllium consisting of 81, 87, and 93 atoms. Mulliken net charges for all atoms are found to be close to the bulk value of zero. ΔSCF ionization potentials are found to be 1.7 to 1.9 eV larger than the bulk workfunction (3.92 eV). Valence electron density plots provide greater detail than electron density plots from previous Hartree-Fock-Roothan studies. Valence electron deformation plots indicate increased electron density along the principal axis of symmetry (z) (perpendicular to the xy basal plane) in comparison to the x and y axial directions. In addition to comparisons with experimental bulk properties, the results are compared to previous Hartree-Fock-Roothan studies and density functional models of other beryllium clusters. It is possible to conclude that the calculated binding energies converge to near the bulk experimental value at ~70-80 atoms. The accuracy of the current model to determining binding energies and atomic populations for fragments of bulk beryllium as small as 81 atoms provides additional evidence that density functional methods are a useful tool to characterize the electronic structure of bulk metal systems.

The structure and dynamics of metal clusters are being characterized in increasing detail with the advent of modern instrumentation and sophisticated spectroscopic techniques. The experimental studies are providing increased understanding into the nature of intermetallic binding in clusters. For theoretical studies on clusters composed of more than a few atoms, two possible approaches to characterize the many-body interactions within the clusters are the Hartree-Fock-Roothan (HFR) and density functional theories (DFT). The methodologies reduce the N^7 dependence of highly correlated N-orbital schemes to a more tractable dependence of N^3-N^4. The Hartree-Fock-Roothan method leads to the best single determinant

---

[4]Current address: Materials Science Laboratory, Texas Instruments, Inc., Mail Stop 147, Dallas, TX 75243

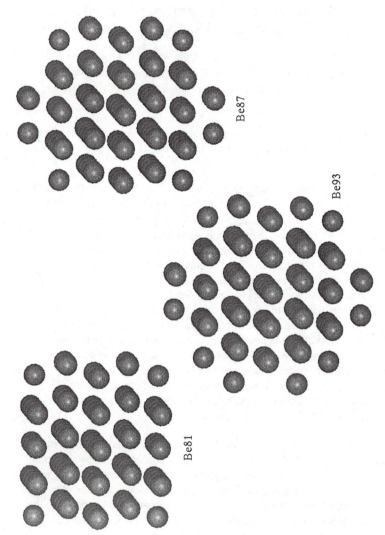

Figure 1. Three-dimensional perspective views of the 81, 87, and 93 atom clusters. The atoms in the clusters occupy positions as in the bulk hcp lattice.

uncorrelated wave function while the density functional theory method replaces the full Hamiltonian with an approximate correlated potential.

Previously, HFR studies have been carried out by Ross, Kern, Ermler, Pitzer and coworkers on clusters of 13 (*1*), 19 (*2*), 21 (*2*), 33 (*2*), 39 (*2*), 51 (*3*), 57 (*3*), 69 (*4*), 81 (*5*), 87 (*5*), 93 (*6*), 105 (*6*), 111 (*6*), 123 (*6*), and 135 (*7*) beryllium atoms. In these studies, evidence is presented for the convergence of bulk properties such as binding energy and net charges with cluster size. In addition, the size of the cluster required for convergence has been seen to depend on the property of interest.

A HFR study has also been carried out by Pettersson and Bauschlicher on a cluster of 55 beryllium atoms (*8*). A Hartree-Fock band structure calculation (*9*), density functional band theory studies (*10,11*), and orthogonalized plane wave studies (*12,13*) have also been performed in an examination of the electronic properties of bulk beryllium.

Recently, a density functional molecular orbital theory study has been performed by Tang and coworkers on a cluster of 135 beryllium atoms (*14*). In the present paper, density functional theory studies are reported for clusters of 81, 87, and 93 beryllium atoms. Calculated binding energies, ionization potentials, net charges, and electron densities are given and compared to previous work and experimental data where available.

Calculations

Three dimensional views of the metal fragments can be seen in Figure 1. The atoms occupy the positions as they would in the bulk hcp beryllium lattice. As discussed in detail previously (*2*), the fragments are derived by adding sets of atoms as they are found on successive coordination spheres about a central beryllium atom.

The calculations employ the local density functional approach (LDA) for molecules as implemented (*15a*) in the program DMol. The exchange and correlation potential employed is the explicit form of $V_{xc,\sigma}$ given by Hedin and coworkers (*16*). A detailed discussion of the formalism of the method has been presented previously by Delley (*15b*).

The basis functions in DMOL are generated numerically from the local density functional solutions for free atoms and for positively charged atoms. Five basis functions are used for beryllium which contain neutral Be-2s, Be^{+2}-2s, and hydrogenic C-2p orbitals. The C-2p orbitals are used as polarization functions.

Ground state neutral self-consistent field (SCF) studies have been carried out for all fragments. SCF studies have also been performed on the +1 ion. ΔSCF ionization potentials have been computed then as the difference of converged neutral and first ionized states.

The studies for the closed shell neutral and first ionized states have been carried out in spin-restricted and spin-unrestricted manners, respectively. The degree of convergence of the self-consistent iterations, measured by root mean square (rms) changes in the charge density, is set at 10^{-6} which allows the total energy to converge to 10^{-6} Ry.

The ground state neutral calculations on Be_{81}, Be_{87}, and Be_{93} have been carried out on a Cray Y-MP supercomputer requiring 66, 85, and 64 minutes of CPU time, respectively. The $+1$ ion SCF studies have been carried out on a Silicon Graphics R8000 workstation and required 351, 820, and 681 minutes for the same respective atom clusters.

Discussion for Selected Properties

Binding Energy. Calculated binding energies for the fragments are shown in Table I and plotted in Figure 2. Also shown for comparison are binding energies calculated from LDA studies on a fragment of 135 atoms (*14*) as well as on a series of smaller fragments (*17*). The experimental binding energy (*18*) and binding energies calculated for a series of HFR clusters (*1-7*) are also included for comparison.

Table I. Calculated Binding Energies (kcal/mol) for Beryllium Clusters

Fragment	DFT[a]	HFR[b]
Be13	50.6	12.0
Be19	59.8	17.2
Be21	59.4	17.9
Be33	59.4	16.5
Be39	65.0	21.7
Be51	70.1	25.8
Be57	70.8	25.4
Be69	73.0	28.2
Be81	73.2	28.3
Be87	73.4	28.5
Be93	73.2	28.0
Be105	-	29.9
Be111	-	30.5
Be123	-	29.6
Be135	77.5	31.7
Bulk Exp.[c]	75.3	

[a]DFT studies for clusters of 13 to 69 from Ref. 17. Binding energy for Be_{135} from Ref. 14.

[b]HFR binding energies from Ref. 1-7.

[c]Ref. 18.

As can be seen in the figure and table, the DFT binding energies converge to near the bulk value of 75.3 kcal/mol at ~70-80 atoms. Specifically, the binding energies for 69, 81, 87, and 93 atom fragments are found to be 73.0, 73.2, 73.4, and 73.2 kcal/mol, respectively. As the cluster size is increased to 135 atoms, the binding energy is overestimated slightly (77.5 kcal/mol) compared to the experimental value (75.3 kcal/mol (*18*)). It should be noted, however, that the binding energy will likely increase somewhat more if d-polarization functions are included in the basis set.

As can also be seen in Figure 2, the HFR binding energies begin to converge with cluster size at a point similar to the DFT results (~70-80 atoms). However, the calculated values of the binding energies are 28-32 kcal/mol for the largest clusters which is well below the bulk value of 75.3 (*18*). The underestimation of binding energy for the HFR clusters is likely due in largest part to the omission of electron correlation effects which are approximated in the DFT methodology through the exchange-correlation potential.

As can also be seen in Figure 2, the trends for increasing binding energy with cluster size for both the DFT and HFR methodologies are similar. For example, there is an increase between 13 and 19 atoms followed by a leveling off to 33 atoms. The binding energy then increases between 33 and 51 atoms. The binding energy then increases sharply to 57 atoms followed by a more gradual increase to 135 atoms in both cases. The similarity of trends between the two methodologies also increases confidence that the lowest states have been found in the HFR clusters which require a search amongst low-lying electron configurations.

An estimate of the contribution of electron correlation to the binding energy can be obtained by subtracting the Be_{13} HFR and DFT binding energies which yields a difference of 38.6 kcal/mol. Calculation of this difference for clusters from 19 through 135 atoms produces a rather narrow range of values from 42.6 to 45.8 kcal/mol which are only about 15% greater than the Be_{13} difference. Adding the Be_{19} difference of 42.6 kcal/mol to the HFR Be_{135} binding energy yields a binding energy of 74.3 kcal/mol which is to within 1 kcal/mol of the bulk binding energy (*18*). This analysis suggests that well over 95% of the electron correlation contribution to the binding energy of bulk beryllium metal is localized in the Be_{19} cluster.

Ionization Potential. Calculated ΔSCF ionization potentials from the studies are shown in Table II. As with binding energies, calculated values from previous HFR studies (*1-7*) and a DFT study on Be_{135} (*14*) are included for comparison. In this case, for lack of any other measurements, the calculated ionization potentials are compared to the experimental workfunction of the bulk metal (3.92 eV) (*19*). As can be seen from the calculated values in the table, the DFT ΔSCF ionization potentials range from 1.7 to 1.9 eV greater than the experimental workfunction. In contrast, the HFR ΔSCF ionization potentials agree to within 0.4 eV.

Atomic Charge. Net atomic charges calculated from Mulliken population analyses (*20*) are summarized in Table III. They are also shown in Figure 3 as a function of radius R for successive coordination shells of the symmetry-distinct

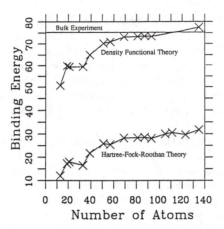

Figure 2. Calculated binding energies (kcal/mol) as a function of cluster size. Previously calculated HFR values (1-7) are included for comparison as are values from experiment (18) and DFT studies on a series of smaller clusters (17) and on Be$_{135}$ (14).

groups of atoms. Positive numbers represent a gain of charge and negative numbers a loss of charge.

As can be seen in the table and figure, calculated net charges for the central atom, BeO, are -0.038, -0.009, and +0.007 e for the 81, 87, and 93 atom clusters, respectively. These values are close to the bulk beryllium metal value of zero. The net charges for the remaining symmetry groups of atoms oscillate around the zero line with similar but not completely identical patterns. The absolute values of net charge range from 0.002 to 0.14 e indicating again the approximately neutral character of the atoms in the fragments. Lack of atomic relaxation may be partially responsible for the charges not being calculated to be exactly 0.0 e.

Similar observations have been seen for net atomic charges in a DFT study by Tang et al. on Be$_{135}$ (14). The net charge on the central beryllium atom was found to be 0.03 electrons. The net charges for the other symmetry-distinct groups of atoms also oscillate around the zero line without showing a clear trend. The absolute value of net charges are found to be small and range from 0.0-0.08 e. The similarity of the net charges for the 81, 87, 93, and 135 atom clusters indicates that they are either independent of cluster-size or converge to bulk limits in the DFT methodology at clusters of 81 atoms or less.

In contrast, HFR net charges calculated previously (5,6) for 81, 87, and 93 atom fragments range in absolute value from 0.02-1.0, 0.01-1.19, and 0.02-1.12 e, respectively. The charges on the central atom for the lowest states in the 81, 87, and 93 atom clusters were found to be 0.17, 0.18, and -0.06 e, respectively. The HFR average net charge for the central atom for the lowest states of 105 (6), 111 (6), 123 (6), and 135 (7) atom clusters have been found to be -0.09, -0.06, -0.06,

Table II. ΔSCF Ionization Potentials (eV) from the DFT studies and Lowest States of HFR Studies

Fragment	Δ SCF Ionization Potential (eV)	
	DFT	HFR[a]
Be13	-	4.85
Be81	5.90	4.28
Be87	5.70	3.78
Be93	5.81	4.20
Be105	-	4.31
Be111	-	4.48
Be123	-	4.03
Be135	5.85[b]	4.29

[a]HFR ionization potentials from Ref. 1 (Be$_{13}$), Ref. 5 (Be$_{81}$, Be$_{87}$), Ref. 6 (Be$_{93}$ through Be$_{123}$), and Ref. 7 (Be$_{135}$).
[b]Ref. 14.

and 0.02 e. The net charge on the central atom trends towards the bulk value of zero with increasing cluster size and hence is cluster-size dependent in the HFR methodology. This is in contrast to the DFT studies for which all net charges are close to zero for 81, 87, 93, and 135 atom clusters. This suggests that the DFT valence electron density is distributed more evenly over the atom cores compared to the HFR density which omits electron correlation.

Charge Density. Contour plots of the valence and deformation charge densities for the 81, 87, and 93 atom clusters are shown in Figures 4, 5, and 6 for the XY, XZ, and YZ planes, respectively. The deformation charge density is computed by subtracting the sum of atomic charges from the total charge for each respective cluster so that the charge redistributions resulting from DFT binding are seen more clearly. The dark and light lines represent charge gained or lost, respectively.

In the XY plane (Figure 4), there are two distinct triangular charge environments. The distinct environments are due to the presence of atoms in planes above and below one of the triangular regions. Examination of both the valence and deformation charge distributions for all three clusters display this characteristic pattern. The two different types of triangular regions alternate in six triangular segments around the central atom. In all three clusters, one region consists of a circular contour. In Be$_{81}$, the second region consists of 2 contours which are beginning to form a distorted hexagon shape. In Be$_{87}$ and Be$_{93}$, in the second region the three contours have transformed even more to a distorted hexagon. Examination of additional atoms in the next shell around the central atom also shows two different triangular regions of similar shapes. The increased electron density in the hexagon region, in comparison to the circular region, is indicative of the presence of atoms in planes above and below this region.

Table III. Net Atomic Charges (e) in the Density Functional Theory Model

Cluster	BeO	BeA	BeB	BeC	BeD	BeE	BeF	BeG	BeH	BeI	BeJ	BeK	BeL	BeM	BeN
Be$_{81}$	-0.038	0.014	0.036	0.042	0.008	-0.022	-0.014	0.07	-0.031	-0.065	-0.01	-0.004	0.081		
Be$_{87}$	-0.009	-0.002	0.04	0.017	0.02	-0.024	-0.003	0.063	-0.05	-0.064	-0.008	0.004	0.117	0.015	
Be$_{93}$	0.007	-0.006	0.02	0.009	0.009	0.027	-0.14	0.074	-0.067	-0.057	0.011	0.013	0.017	0.022	0.07

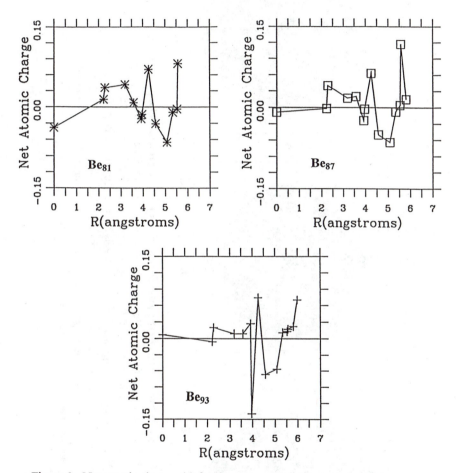

Figure 3. Net atomic charges (e) for the symmetry-distinct groups of atoms as a function of radius of the coordination shells for Be_{81}, Be_{87}, and Be_{93}. Points at identical radii arise to groups of atoms that are in the same coordination shell but nonequivalent from the symmetry point of view.

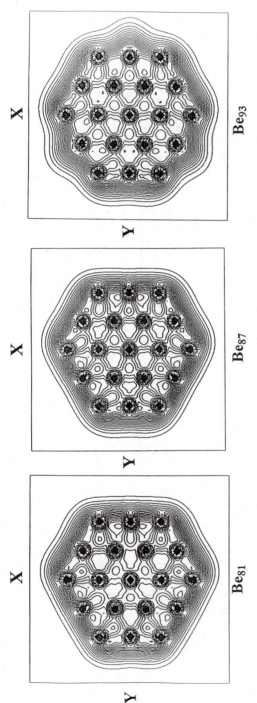

Figure 4. Contour plots of the valence charge (top) and deformation charge (bottom) densities for Be$_{81}$, Be$_{87}$, and Be$_{93}$ in the XY plane. The valence charge contour levels begin at 0.002 e/(a.u.)^3 with a spacing of 0.002 e/(a.u.)^3. The deformation charge density contour levels begin at ±0.001 e/(a.u.)^3 for positive and negative values, respectively, with a contour spacing of ±0.002 e/(a.u.)^3 for positive and negative values, respectively. The dark and light lines represent electronic charge gained and lost, respectively.

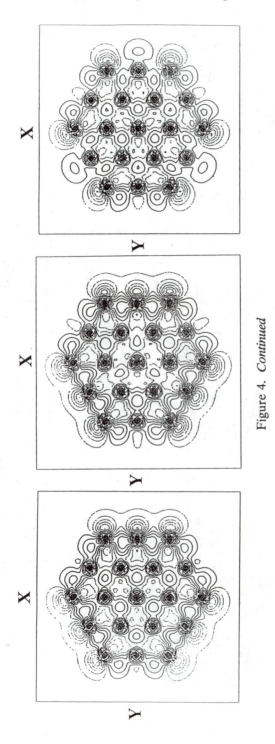

Figure 4. *Continued*

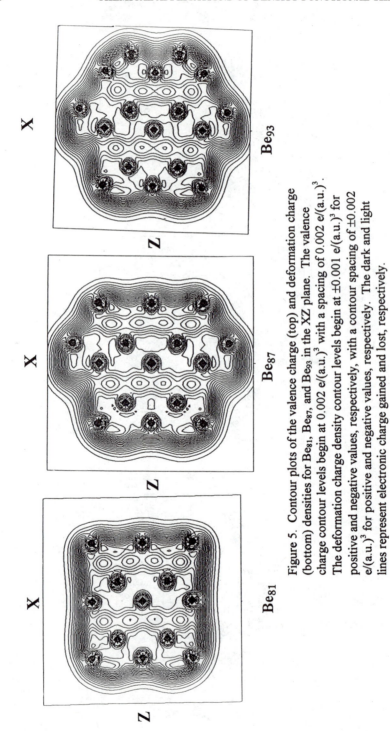

Figure 5. Contour plots of the valence charge (top) and deformation charge (bottom) densities for Be_{81}, Be_{87}, and Be_{93} in the XZ plane. The valence charge contour levels begin at 0.002 $e/(a.u.)^3$ with a spacing of 0.002 $e/(a.u.)^3$. The deformation charge density contour levels begin at ±0.001 $e/(a.u.)^3$ for positive and negative values, respectively, with a contour spacing of ±0.002 $e/(a.u.)^3$ for positive and negative values, respectively. The dark and light lines represent electronic charge gained and lost, respectively.

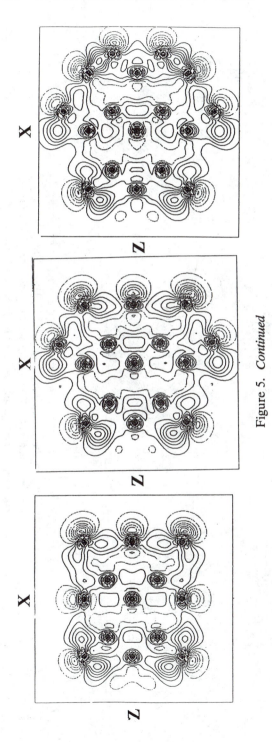

Figure 5. *Continued*

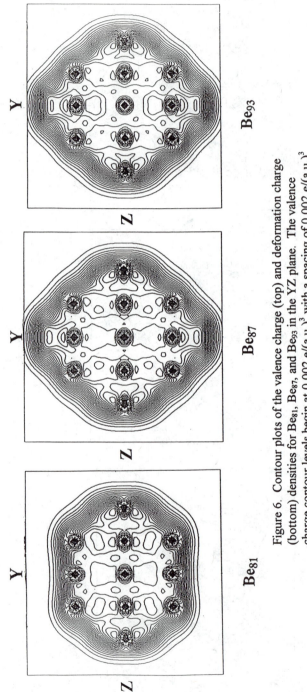

Figure 6. Contour plots of the valence charge (top) and deformation charge (bottom) densities for Be_{81}, Be_{87}, and Be_{93} in the YZ plane. The valence charge contour levels begin at 0.002 $e/(a.u.)^3$ with a spacing of 0.002 $e/(a.u.)^3$. The deformation charge density contour levels begin at ±0.001 $e/(a.u.)^3$ for positive and negative values, respectively, with a contour spacing of ±0.002 $e/(a.u.)^3$ for positive and negative values, respectively. The dark and light lines represent electronic charge gained and lost, respectively.

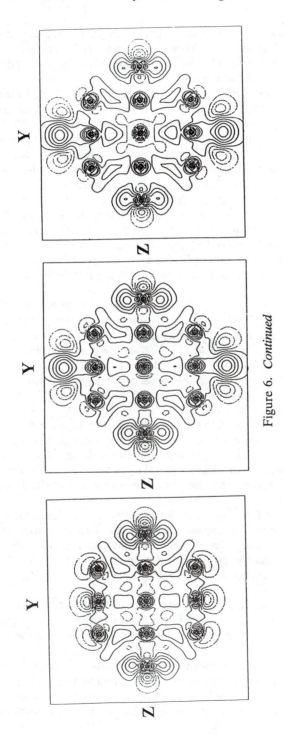

Figure 6. *Continued*

The presence of two distinct types of triangular charge regions around atoms in the XY plane is seen also in the DFT study (*14*) on a cluster of 135 beryllium atoms, thereby indicating that the onset of convergence of the charge distribution in the XY plane begins at 80-90 atoms.

A comparison of HFR charge densities for 81 (*5*), 87 (*5*), 93 (*6*), 105 (*6*), 111 (*6*), 123 (*6*), and 135 (*7*) atom clusters in the XY plane indicates again charge density convergence around the central beryllium atom starting at 80-90 atoms (or smaller). However in comparison to the DFT valence electron densities, a primary difference lies in the more homogenous nature of the charge distribution in the HFR cases. The increased sensitivity of the DFT density to nearby atoms results in sharper contours. In the HFR model, the lack of electron correlation tends to homogenize the valence electron density.

The HFR charge densities in the XY plane do not show any differences in the two triangular regions. The HFR charge densities were plotted at different contour levels than those in the current DFT studies. However, HFR Be_{135} charge densities (*14*) at the same contour levels as the current DFT studies also show a homogeneous charge density and no indications that two distinct triangular regions are present.

Examination of the valence and deformation charge densities in the XZ and YZ planes (Figures 5 and 6) indicates anisotropy in the charge distribution which is indicative of p orbital participation in bonding. Anisotropy is seen also in DFT studies on Be_{135} in the YZ plane and in XZ and YZ planes in HFR clusters of 81 (*5*), 87 (*5*), 93 (*6*), 105 (*6*), 111 (*6*), 123 (*6*), and 135 (*7*) atoms. The p orbital contribution to bonding agrees with a local density approximation (LDA) pseudopotential study (*11*) and an X-ray diffraction pattern (*21*) which exhibits sp^3 type bonding in beryllium metal.

In comparing the XZ valence charge densities (Figure 5) for the 81, 87, and 93 atom clusters, two recurrent "ribbons" of valence charge density are seen extending from the -Z to + Z direction. The charge density in these regions can be traced to the presence of atoms above and below the plane. The valence charge densities show this structural feature through the presence of circular contours within the ribbons.

The XZ deformation charge densities (Figure 5) indicate that charge density is decreased in the ribbon regions in comparison to a set of overlapping isolated atoms. The deformation charge density plots also show that more charge accumulates between atoms along the Z direction than in the X direction. This implies that the bonding along the Z direction may be stronger than in the X direction.

It can be seen also that the electron density changes around the central beryllium atom in the XZ plane as the cluster size increases. This suggests that the XZ charge density around the central atom has not converged in contrast to the XY plane density.

Analysis of the YZ plane valence charge density (Figure 6) shows variations around the central beryllium atom as well. The contours along the Y axis consist of distorted rectangles lying between the central atom and next nearest neighbors for the 81 and 87 atom fragments while for Be_{93} the first contours along the Y axis are further from the central atom and surround the next nearest neighbors. In the Z direction, the nearest contours to the central atom are distorted circular contours above and below the + Z and - Z directions for Be_{81}, triangular contours along the + Z and - Z axes for Be_{87}, and for Be_{93}, a pattern similar to Be_{81}.

The YZ plane deformation charge densities (Figures 6) show differences in the central Be charge environment as the cluster increases from 81 to 93 atoms. In addition, as in the XZ plane, more charge accumulates between atoms along the Z direction than in the Y direction which implies that the bonding along the Z direction may be stronger than in the Y direction. In fact, all four clusters (81, 87, 93, and 135 (*14*)) consistently show more charge accumulation along the Z axis than along the Y axis.

The differences observed in the XZ or YZ plane charge densities for all clusters is clearly not the case for the XY plane densities. Thus, charge densities in different planes may converge to bulk limits at different cluster sizes. Since bulk beryllium is not symmetrically equivalent in all directions, atoms are not added equivalently in all directions. The increased rate of convergence of the XY basal plane density may be enhanced since it is a reflection plane. Consequently, the atoms and electron density in this plane effectively see the bulk develop twice as fast in comparison to the case without a symmetry plane.

Conclusions

Density functional theory (DFT) calculations have been carried out on metal fragments of 81, 87, and 93 beryllium atoms. Calculated net charges for all atoms are nearly zero as in bulk beryllium. This is in sharp contrast to previous HFR studies in which net charge varies over a range of up to 1.12 e for clusters of similar size. In addition, the average net charge in the HFR model is cluster-size dependent, converging to the bulk limit around the central beryllium atom at a cluster of 135 atoms.

Calculated binding energies (73.2, 73.4, and 73.2 kcal/mol for the 81, 87, and 93 atom clusters) are found to be close to the bulk value of 75.3 kcal/mol (*18*). Comparable Hartree-Fock-Roothan studies underestimate the binding energy by 28-32 kcal/mol. This error is likely due in large part to the exclusion of electron correlation.

Plots of DFT valence electron charge densities exhibit sharper contours than corresponding Hartree-Fock-Roothan densities. Additional detail seen in the DFT model arises from atoms in planes above and below the basal plane. The electron density converges in the XY plane at 80-90 atoms unlike the XZ and YZ planes. Thus, electron densities can converge at different rates in different planes.

The accuracy of the current model to determining binding energies and atomic populations for fragments of bulk beryllium as small as 81 atoms provides additional evidence that density functional methods are a useful tool to characterize the electronic structure of bulk metal systems.

Acknowledgments

Computational results obtained using software programs from Biosym Technologies/ Molecular Simulations Inc. of San Diego - ab initio calculations were done with the *DMol* program and graphical displays were printed out from the *Insight® II* molecular modeling system.

Literature Cited

1. Ermler, W.C.; Kern, C.W.; Pitzer, R.M.; Winter, N.W. *J. Chem. Phys.* **1986,** *84*, 3937.
2. Ermler, W.C.; Ross, R.B.; Kern, C.W.; Pitzer, R.M.; Winter, N.W. *J. Phys. Chem.* **1988,** *92*, 3042.
3. Ross, R.B.; Ermler, W.C.; Pitzer, R.M.; Kern, C.W. *Chem. Phys. Lett.* **1987,** *134*, 115.
4. Ross, R.B.; Kern, C.W.; Pitzer, R.M.; Ermler, W.C.; Winter, N.W. *J. Phys. Chem.* **1990,** *94*, 7771.
5. Ross, R.B.; Ermler, W.C.; Luana, V.; Pitzer, R.M.; Kern, C.W. *Int. J. Quant. Chem: Quant. Chem. Symp.* **1990,** *24*, 225.
6. Ross, R.B.; Kern, C.W.; Pitzer, R.M.; Ermler, W.C. *Int. J. Quant. Chem.* **1995,** *55*, 393.
7. Ross, R.B.; Ermler, W.C.; Kern, C.W.; Pitzer, R.M. *Int. J. Quant. Chem.* **1992,** *41*, 733.
8. Pettersson, L.G.M.; Bauschlicher, C.W. *Chem. Phys. Lett.* **1982,** *130*, 111.
9. Dovesi, R.; Pisani, C.; Ricca, F.; Roetti, C. *Phys. Rev. B* **1982,** *25*, 3731.
10. Blaha, P.; Schwartz, K. *J. Phys. F* **1987,** *17*, 899.
11. Chou, M.Y.; Lam, P.K.; Cohen, M.L. *Phys. Rev. B* **1983,** *28*, 4179.
12. Loucks, T.L.; Cutler, P.H. *Phys. Rev.* **1964,** *133*, A819.
13. Herring, C.; Hill, A.G. *Phys. Rev.* **1940,** *86*, 132.
14. Tang, S.; Freeman, A.J.; Ross, R.B.; Kern C.W. *J. Chem. Phys.* **1995,** *103*, 2555.
15.(a) Delley, B. *J. Chem. Phys.* **1990,** *92*, 508; (b) ibid. **1991,** *94*, 7245.
16.(a) von Barth, U.; Hedin, L. *J. Phys. C* **1972,** *5*, 1629; (b) Hedin, L.; Lundqvist, B.I. *J. Phys. C* **1971,** *4*, 2064.
17. Tang, S.; Freeman, A.J.; Kern, C.W.; Ross, R.B.(unpublished).
18. Blaha P.; Schwartz, K. *J. Phys. F.* **1987,** *17*, 899.
19. Tompa, G.S.; Seidl, M.; Ermler, W.C.; Carr, W.E. *Surf. Sci.* **1987,** *L453*, 185.
20. Mulliken, R.S. *J. Chem. Phys.* **1955,** *23*, 1833.
21. Larsen, F.K.; Hansen, N.K. *Acta Cryst.* **1984,** *B40*, 169.

Chapter 30

Local and Gradient-Corrected Density Functionals

John P. Perdew, Kieron Burke[1], and Matthias Ernzerhof

Department of Physics and Quantum Theory Group, Tulane University,
New Orleans, LA 70118

The generalized gradient approximation (GGA) corrects many of the shortcomings of the local spin-density (LSD) approximation. The accuracy of GGA for ground-state properties of molecules is comparable to or better than the accuracy of conventional quantum chemical methods such as second-order Møller-Plesset perturbation theory. By studying various decompositions of the exchange-correlation energy E_{XC}, we show that the real-space decomposition of E_{XC} facilitates the most detailed understanding of how the local spin-density approximation and the Perdew-Wang 1991 GGA work. The real-space decomposition shows that the near universality of the on-top value for the exchange-correlation hole connects the homogeneous electron gas to inhomogeneous systems such as atoms and molecules. The coupling-constant decomposition shows that the exchange-correlation energy at full coupling strength $E_{XC,\lambda=1}$ is approximated more accurately by local and semi-local functionals than is the coupling-constant average E_{XC}. We use this insight both to critique popular hybrid functionals and to extract accurate energies from exact electron densities by using functionals for the exchange-correlation energy at full coupling strength. Finally, we show how a reinterpreted spin density functional theory can be applied to systems with static correlation.

Density functionals in quantum chemistry

The main goal of quantum chemistry is the reliable prediction of molecular properties [1]. The development of generalized gradient approximations (GGA's) [2–10] has made density functional theory [11–13] a serious competitor to conventional quantum chemistry methods for ground-state properties. The latter methods include Configuration Interaction techniques, Coupled-Cluster methods, and the

---

[1]Address after July 1, 1996: Department of Chemistry, Rutgers University, Camden, NJ 08102

0097–6156/96/0629–0453$15.00/0
© 1996 American Chemical Society

Møller-Plesset perturbation expansion [14]. They account for electron correlation, but suffer in general from basis set problems. These basis set problems are much less severe in current density functional methods, since (as we show below) the pair density is not expanded in a basis of one-particle functions.

The accuracy of the GGA is usually comparable to conventional quantum chemistry methods, at much lower computational cost. GGA's offer significant improvements in the calculation of molecular properties compared to their ancestor, the local spin-density (LSD) approximation [15]. The local spin-density approximation has not been popular amongst chemists, mainly because of its tendency to overestimate the binding energy of molecules.

In practical electronic structure calculations based on density functional theory [12], a set of independent-particle equations (the Kohn-Sham equations [15]) is solved. These equations require as input an approximation to the exchange-correlation energy E_{XC} as a functional of the electron density. The GGA approximations to E_{XC} depend on both the local spin-density $n_\sigma(\mathbf{r})$ and the gradient of the local spin-density. Among the popular GGA's, the Perdew-Wang (PW91) [7–10] functional allows the most detailed understanding of how GGA's work and why they work, since this approximation contains no empirical parameter and is constructed from first principles. Results of calculations with this form [9] show that it typically reduces exchange energy errors from 10% in LSD to 1%, and correlation energy errors from 100% to about 10%. PW91 corrects the LSD overestimate of atomization energies for molecules and solids in almost all cases, it enlarges equilibrium bond lengths and lattice spacings, usually correctly, and reduces vibrational frequencies, again usually correctly [10]. PW91 also generally improves activation barriers [16]. For recent results with PW91, see Refs. [17–27].

As indicated above, the exchange-correlation energy E_{XC} as a functional of the electron density is the crucial quantity in Kohn-Sham calculations. In this article we discuss various decompositions of the exchange-correlation energy, and we show which of these decompositions is accurately approximated by LSD and by the PW91 approximation. This analysis makes it possible to understand how and why local and semilocal functionals work even for highly inhomogeneous electron systems, such as atoms and molecules.

Decompositions of E_{XC}

The basic formula which serves as the starting point for various decompositions of E_{XC} is [11–13]

$$E_{XC} = \int_0^1 d\lambda \int_0^\infty du \, 2\pi u \int \frac{d\Omega_u}{4\pi} \int d^3r \, n(\mathbf{r}) \, n_{XC,\lambda}(\mathbf{r}, \mathbf{r} + \mathbf{u}). \qquad (1)$$

The exchange-correlation hole $n_{XC,\lambda}(\mathbf{r}, \mathbf{r} + \mathbf{u})$ at coupling strength λ is given in terms of the pair density $P_\lambda(\mathbf{r}\sigma, (\mathbf{r} + \mathbf{u})\sigma')$ by

$$n(\mathbf{r}) \left[n(\mathbf{r} + \mathbf{u}) + n_{XC,\lambda}(\mathbf{r}, \mathbf{r} + \mathbf{u}) \right] = \sum_{\sigma,\sigma'} P_\lambda(\mathbf{r}\sigma, (\mathbf{r} + \mathbf{u})\sigma'). \qquad (2)$$

σ and σ' are the spin variables of the electrons. $P_\lambda(\mathbf{r}\sigma, \mathbf{r}'\sigma')$ (where $\mathbf{r}' = \mathbf{r} + \mathbf{u}$) gives the probability density to find an electron with spin σ at $\mathbf{r}$ and an electron with spin σ' at $\mathbf{r}'$. The pair density is related to the many-electron wave function $\Psi_\lambda(\mathbf{r}, \sigma, \mathbf{r}', \sigma', \ldots, \mathbf{r}_N, \sigma_N)$ by

$$P_\lambda(\mathbf{r}\sigma, \mathbf{r}'\sigma') = N(N-1) \sum_{\sigma_3, \ldots, \sigma_N} \int d^3 r_3 \ldots \int d^3 r_N$$
$$\times \left| \Psi_\lambda(\mathbf{r}, \sigma, \mathbf{r}', \sigma', \ldots, \mathbf{r}_N, \sigma_N) \right|^2. \tag{3}$$

Ψ_λ is the ground-state wave function of a system in which the electron-electron repulsion operator is multiplied by λ and the external potential is varied with λ so that the electron density is equal to the physical ground-state density for all values of λ. $\Psi_{\lambda=1}$ is the interacting wavefunction found by traditional correlated methods, while $\Psi_{\lambda=0}$ is the exact exchange wavefunction (which is similar to that of Hartree-Fock [28]). Eqs. 1 and 2 suggest a number of possibilities to decompose E_{XC} by simply permuting the sequence of integrations and summations. The decomposed approximate exchange-correlation energy can then be compared with the corresponding exact quantity.

Energy-density decomposition of $\mathbf{E}_{\text{XC},\lambda}$. We examine the integrand of the expression

$$E_{\text{XC},\lambda} = \int d^3 r \, n(\mathbf{r}) \, \epsilon_{\text{XC},\lambda}(\mathbf{r}), \tag{4}$$

where

$$\epsilon_{\text{XC},\lambda}(\mathbf{r}) = \int_0^\infty du \, 2\pi u \int \frac{d\Omega_u}{4\pi} \, n_{\text{XC},\lambda}(\mathbf{r}, \mathbf{r} + \mathbf{u}). \tag{5}$$

(E_{XC} is related to $E_{\text{XC},\lambda}$ by $E_{\text{XC}} = \int_0^1 d\lambda E_{\text{XC},\lambda}$.) In the local spin-density approximation the energy per particle $\epsilon_{\text{XC},\lambda}(\mathbf{r})$ of an arbitrary inhomogeneous system is approximated by that of a homogeneous electron gas with spin density $n_\sigma(\mathbf{r})$, i.e., $\epsilon_{\text{XC}}(\mathbf{r}) = \epsilon_{\text{XC}}^{\text{unif}}(n_\uparrow(\mathbf{r}), n_\downarrow(\mathbf{r}))$. The error in E_{XC} resulting from this approximation is typically about 10%. The energy per particle close to a nucleus is usually overestimated, and that in the tail region of the electron density is underestimated [29]. However, the region close to the nuclei has a small volume and the contribution from the tail region is very small, since the energy per particle gets weighted by the electron density in Eq. 4. In the valence region of atoms and molecules, $\epsilon_{\text{XC}}^{unif}(n_\uparrow(\mathbf{r}), n_\downarrow(\mathbf{r}))$ shows a semiquantitative agreement with the exact energy per particle. However, the PW91 energy density is not so useful for understanding how the PW91 functional works [30], since the energy density $n\epsilon_{\text{XC},\lambda}^{PW91}$ has been simplified by an integration by parts, which leaves E_{XC} unchanged but leads to an ill-defined $n(\mathbf{r})\epsilon_{\text{XC},\lambda}^{GGA}(\mathbf{r})$.

Real-space decomposition of $\mathbf{E}_{\text{XC},\lambda}$. The real-space decomposition of the $E_{\text{XC},\lambda}$, which is defined by

$$E_{\text{XC},\lambda} = N \int_0^\infty du \, 2\pi u \, \langle n_{\text{XC},\lambda}(u) \rangle \tag{6}$$

where the system- and spherically averaged hole is

$$\langle n_{\text{XC},\lambda}(u) \rangle = \int \frac{d\Omega_u}{4\pi} \frac{1}{N} \int d^3r \, n(\mathbf{r}) \, n_{\text{XC},\lambda}(\mathbf{r}, \mathbf{r} + \mathbf{u}),$$

$$(7)$$

offers the most detailed insight into the LSD and PW91 functionals. In fact the PW91 exchange-correlation functional is based on a model for the system-averaged exchange-correlation hole $\int d^3r \, n(\mathbf{r}) \, n_{\text{XC},\lambda}(\mathbf{r}, \mathbf{r} + \mathbf{u})/N$. Detailed studies [31] of $\langle n_{\text{XC},\lambda}(u) \rangle$ have shown that even the LSD approximation to this quantity is remarkably accurate. Many exact conditions on the exact $\langle n_{\text{XC},\lambda}(u) \rangle$, such as the normalization condition $\int_0^\infty du \, 4\pi u^2 \, \langle n_{\text{XC},\lambda}(u) \rangle = -1$, are satisfied by LSD, since the LSD exchange-correlation hole is the hole of a possible physical system. The normalization condition on the hole together with the on-top ($u = 0$) value for $n_{\text{XC},\lambda}(\mathbf{r}, \mathbf{r} + \mathbf{u})$ set the scale for the exchange-correlation hole and therefore the scale for $\langle n_{\text{XC},\lambda}(u) \rangle$ and $E_{\text{XC},\lambda}$. Thus the on-top value of $n_{\text{XC},\lambda}(\mathbf{r}, \mathbf{r} + \mathbf{u})$ plays a crucial role in density functional theory. Investigations [32] on a number of systems show that the on-top value of the hole as a function of the local density is almost universal among Coulomb systems. Thus the LSD approximation to this quantity is very accurate, especially in the valence and tail regions of the electron density [32]. As a consequence, any approximate density functional should reproduce the correct LSD on-top value of the exchange-correlation hole in the limit of slowly varying electron densities. The PW91 functional has the LSD on-top value built in. It is based on a systematic expansion of the exchange-correlation hole in terms of the local density and the gradient of the local density. This gradient expansion approximation (GEA) to second order in ∇n improves the description of the hole at intermediate u values, but its spurious large-u behavior [33] violates a number of important constraints on the exact hole, such as the normalization condition. By restoring these conditions via the real-space cutoff procedure [8], we obtain the PW91 model for the exchange-correlation hole. Since the r integration in Eq. 7 involves an integration by parts which changes the local hole $n_{\text{XC},\lambda}(\mathbf{r}, \mathbf{r} + \mathbf{u})$, only the system-averaged hole is a well-defined quantity in the PW91 construction. Detailed studies [29,31] of the spherical- and system-averaged hole for molecules and atoms show that the PW91 approximation to this quantity significantly improves the LSD model. Other popular GGA's [5,6] do not provide models of the exchange-correlation hole and thus do not allow a detailed analysis of correlation effects on molecular bond formation.

Spin decomposition of $E_{\text{XC},\lambda}$. Another decomposition of the exchange-correlation hole and therefore of the exchange-correlation energy distinguishes between electrons with parallel and anti-parallel spins:

$$E_{\text{XC},\lambda} = E_{\text{XC},\lambda}^{\uparrow\uparrow} + E_{\text{XC},\lambda}^{\downarrow\downarrow} + E_{\text{XC},\lambda}^{\downarrow\uparrow},$$

$$(8)$$

where

$$E_{\text{XC},\lambda}^{\sigma,\sigma} = \frac{1}{2} \int d^3r \int d^3r' \frac{P_\lambda(\mathbf{r}\sigma, \mathbf{r}'\sigma) - n_\sigma(\mathbf{r})n_\sigma(\mathbf{r}')}{|\mathbf{r} - \mathbf{r}'|},$$

$$E_{\text{XC},\lambda}^{\uparrow\downarrow} = \int d^3r \int d^3r' \frac{P_\lambda(\mathbf{r}\uparrow, \mathbf{r}'\downarrow) - n_\uparrow(\mathbf{r})n_\downarrow(\mathbf{r}')}{|\mathbf{r} - \mathbf{r}'|}.$$

$$(9)$$

The Pauli principle prevents two parallel-spin electrons from coming close to each other, i.e., $P(\mathbf{r}\sigma, \mathbf{r}\sigma) = 0$. Electron-electron repulsion cannot deepen the corresponding hole at $u = 0$, so the spatial extent of the hole is not significantly reduced. Note that the deeper the hole at $u = 0$, the shorter-ranged it must be to satisfy the normalization condition. On the other hand, the on-top value of the exchange-correlation hole for two electrons with antiparallel-spin orientation is significantly lowered by electron-electron repulsion, and the normalization therefore assures that the spatial extent of the hole gets reduced by correlation. Local and semilocal approximations work best if the exchange-correlation hole is confined to a small region of space around the reference electron. In this case the information about the local density and the gradient of the density is sufficient to capture the important features of the exchange-correlation hole. Thus it is not surprising that local and semilocal functionals work better for the correlation energy between antiparallel-spin electron than they do for the correlation energy between parallel-spin electrons [29]. However, attempts to construct a hybrid scheme which uses a GGA for antiparallel-spin correlation and wave function methods for parallel-spin correlation are of limited use [29], since the correlation effects between parallel-spin electrons are as difficult to describe within a finite basis set approach as are correlation effects between antiparallel-spin electrons.

Finally we note that the approximate GGA for antiparallel-spin [34] predicts $E_c^{\downarrow\downarrow} + E_c^{\uparrow\uparrow}$ to be 20% of the total correlation energy of Ne, in good agreement with sophisticated wave function calculations which give a value [29] of 24%.

Coupling-constant decomposition of $\mathbf{E}_{\text{xc}}$. The kinetic correlation energy contribution T_c to the total energy need not be explicitly approximated as a functional of the electron density. T_c is implicitly accounted for in E_{xc} of Eq. 1 via the coupling-constant integration over the λ-dependent exchange-correlation hole [35]. This coupling-constant integration leads to another decomposition of the exchange-correlation energy:

$$E_{\text{xc}} = \int_0^1 d\lambda \, E_{\text{xc},\lambda}, \tag{10}$$

where

$$E_{\text{xc},\lambda} = \int_0^\infty du \, 2\pi u \int \frac{d\Omega_u}{4\pi} \int d^3r \, n(\mathbf{r}) \, n_{\text{xc},\lambda}(\mathbf{r}, \mathbf{r} + \mathbf{u}). \tag{11}$$

This decomposition has become a popular tool in density functional theory [36–38]. For $\lambda = 0$, the electrons are not Coulomb-correlated, so $E_{\text{xc},\lambda=0} = E_{\text{x}}$ accounts for the self-interaction correction and for the Pauli exclusion principle. Compared to the hole at finite values of λ, the hole at $\lambda = 0$ is shallower and therefore more long-ranged. At full coupling-strength ($\lambda = 1$), $E_{\text{xc},\lambda=1} = E_{\text{xc}} - T_c$. Electrons close to the reference electron get pushed away at small u-values and pile up at large u-values, making the hole deeper at $u = 0$ and more short-ranged. Local and semilocal approximations usually work best for small u. Thus they are least suitable for $\lambda = 0$ (the exchange-only limit). These expectations about the range of the hole are confirmed by a study [31] of the λ-dependent exchange-correlation hole, and are probably true for all systems.

The popular hybrid schemes [36–38] can be viewed as attempts to exploit this observation. We consider only hybrid schemes which recover the slowly-varying electron gas limit. In such a scheme, the coupling-constant integral is replaced by the weighted sum of the integrands at the endpoints of the coupling-constant integration and the density functional approximation to the exchange energy $E_{\text{xc},\lambda=0}$ is replaced by the Hartree-Fock exchange energy [29,38]. In formulas,

$$E_{\text{xc}} = aE_{\text{x}}^{HF} + (1-a)E_{\text{x}}^{GGA} + E_{\text{c}}^{GGA}. \tag{12}$$

The parameter a is usually adjusted to minimize the root-mean-square errors of various molecular properties. However, it has been demonstrated [32,29] that the parameter a is far from universal for molecular systems. This can be seen by considering the stretched H_2 molecule: The restricted Hartree-Fock hole is always distributed equally over both H-atoms, whereas the exact hole and its GGA model are localized on the H atom at which the reference electron is located. Thus no finite amount of exact exchange should contribute to E_{xc} in the limit of infinite stretching.

Exchange-correlation potential. It has been shown [39,40] that the PW91 approximation to the exchange-correlation potential $v_{\text{xc}} = \delta E_{\text{xc}}[n]/\delta n$, which appears in the Kohn-Sham Hamiltonian, deviates considerably from the exact potential, especially in the core and tail regions of atoms. However, we have argued that approximate local and semilocal density functionals give good approximations only to system-averaged quantities such as the system-averaged exchange-correlation hole. It is therefore more appropriate to study quantities which involve system-averages of the exchange-correlation potential v_{xc} and its components v_{x} and v_{c}, rather than the potential itself. The virial theorem shows that [41]

$$E_{\text{x}} = -\int d^3r \; n(\mathbf{r}) \; \mathbf{r} \cdot \nabla v_{\text{x}}(\mathbf{r}) \tag{13}$$

and

$$E_{\text{c}} + T_{\text{c}} = -\int d^3r \; n(\mathbf{r}) \; \mathbf{r} \cdot \nabla v_{\text{c}}(\mathbf{r}). \tag{14}$$

In this system average, the region close to the nuclei and the tail region of the electron density, where v_{xc} is not well approximated by PW91, get little weight. Due to error cancellation within the integral of Eqs. 14, the system-averaged quantities are far better approximated by PW91 than by LSD [31].

Exchange-correlation energies from exact densities. An approach which makes use of the improvement in accuracy of the LSD and GGA approximations at full coupling strength ($\lambda = 1$) relative to the coupling-constant average can be formulated based on knowledge of the exact density $n(\mathbf{r})$ corresponding to a known external potential $v(\mathbf{r})$ [42]. Several methods are known for calculating the exact Kohn-Sham potential $v_s(\mathbf{r})$ from a given electron density [43–47]. The exact exchange-correlation potential can then be obtained according to

$$v_{\text{xc}}(\mathbf{r}) = v_s(\mathbf{r}) - v(\mathbf{r}) - \int d^3r' \; \frac{n(\mathbf{r}')}{|\mathbf{r} - \mathbf{r}'|}. \tag{15}$$

The virial relations Eqs. 13 and 14 show that the virial of the exchange-correlation potential is given by the sum of $T_C + E_{XC}$. Since $E_{XC,\lambda=1} = E_{XC} - T_C$, we obtain the exact relation

$$E_{XC} = \frac{1}{2}[E_{XC,\lambda=1} - \int d^3r\ n(\mathbf{r})\ \mathbf{r} \cdot \nabla v_{XC}(\mathbf{r})]. \tag{16}$$

$E_{XC,\lambda=1}$ on the right hand side of this equation will now be replaced by its local or semilocal approximation, a replacement typically involving less error than the functional approximation of E_{XC}. As shown in Table 1, the resulting expressions for E_{XC} give a significant improvement compared to the local and semi-local approximations of E_{XC} itself.

Table 1: Exchange-correlation energies in Hartrees for several atoms [42] We compare exact values with those of LSD and PW91, and with improvements thereof (marked by a prime) using Eq. (16). Unless otherwise indicated, the exact values are from Ref. [48], while the approximate functionals are evaluated on the Hartree-Fock densities of Ref. [49].

atom	exact	LSD	LSD′	PW91	PW91′
H^a	-0.312	-0.290	-0.307	-0.314	-0.316
$H^{-\ a,b}$	-0.423	-0.409	-0.420	-0.425	-0.422
He	-1.068	-0.997	-1.048	-1.063	-1.066
Li	-1.827	-1.689	-1.786	-1.821	-1.829
$Be^{++\ b}$	-2.321	-2.107	-2.243	-2.298	-2.312
$Ne^{8+\ b}$	-6.073	-5.376	-5.776	-5.989	-6.036
Be	-2.772	-2.536	-2.686	-2.739	-2.755
Be^a	-2.772	-2.545	-2.691	-2.748	-2.760
N	-6.78	-6.32	-6.61	-6.77	-6.78
Ne^c	-12.48	-11.78	-12.20	-12.50	-12.47

[a] Approximate functionals evaluated on exact densities.
[b] Exact results from Ref. [39].
[c] Exact results from Ref. [43].

Static correlation in density functional theory. In the context of density functional theory, systems which show only dynamical correlation effects are called normal systems [50,51]. For small values of the coupling-constant λ, normal systems are well described by a single Slater determinant, and the on-top value of the hole is well reproduced by LSD and PW91. Systems with static correlation, such as H_2 stretched beyond the Coulson-Fisher point, are called abnormal systems. The exact wavefunction of stretched H_2 (with bond length $R \to \infty$) does not reduce to a single determinant as $\lambda \to 0$. Instead, it remains a Heitler-London type wavefunction, keeping one electron localized on each atom. But in a restricted single-determinant approximation, the electrons cannot localize on individual atoms. As a consequence, the on-top value of the exchange-correlation hole that results from a restricted calculation (Hartree-Fock, LSD, or GGA) is quite incorrect. To see this, note first that for a single Slater determinant, the on-top pair density is simply $P(\mathbf{r}, \mathbf{r}) = 2n_\uparrow(\mathbf{r})n_\downarrow(\mathbf{r})$. In a restricted calculation, in which $n_\uparrow(\mathbf{r}) = n_\downarrow(\mathbf{r}) = n(\mathbf{r})/2$, we find $P_{\lambda=1}(\mathbf{r}, \mathbf{r}) = P_{\lambda=1}^{\text{unif}}(n(\mathbf{r})/2, n(\mathbf{r})/2; u = 0)$, instead of the correct result, $P_{\lambda=1}(\mathbf{r}, \mathbf{r}) = 0$ for all $\mathbf{r}$. A cure for this problem is provided by an alternative interpretation of spin-density functional theory [50,51]. In this alternative interpretation, the quantities predicted are not the individual spin-densities $n_\uparrow(\mathbf{r})$ and $n_\downarrow(\mathbf{r})$, but the total density $n(\mathbf{r}) = n_\uparrow + n_\downarrow$ and the full coupling strength on-top pair density $P_{\lambda=1}(\mathbf{r}, \mathbf{r}) = P_{\lambda=1}^{\text{unif}}(n_\uparrow, n_\downarrow; u = 0)$. In abnormal systems, the spin symmetry can then be broken with impunity. In the stretched H_2 molecule, an electron with up-spin localizes on one hydrogen atom and an electron with down-spin localizes on the other. The spin-density in such an unrestricted calculation is obviously no longer accurately reproduced, but the total density is. Furthermore, $P_{\lambda=1}(\mathbf{r}, \mathbf{r}) = P_{\lambda=0}(\mathbf{r}, \mathbf{r})$ correctly vanishes for all values of $\mathbf{r}$, since either $n_\uparrow(\mathbf{r})$ or $n_\downarrow(\mathbf{r})$ is zero everywhere. This behavior of the on-top value of the pair density ensures that the unrestricted Kohn-Sham calculation gives an accurate dissociation energy.

Acknowledgments

This work has been supported in part by NSF grant DMR95-21353 and in part by the Deutsche Forschungsgemeinschaft.

References

[1] A. Szabo and N.S. Ostlund, *Modern Quantum Chemistry* (MacMillan, New York, 1982).

[2] D.C. Langreth and M.J. Mehl, Phys. Rev. B **28**, 1809 (1983).

[3] J.P. Perdew, Phys. Rev. B **33**, 8822 (1986); **34**, 7406 (1986) (E).

[4] J.P. Perdew and Y. Wang, Phys. Rev. B **33**, 8800 (1986); **40**, 3399 (1989) (E).

[5] A.D. Becke, Phys. Rev. A **38**, 3098 (1988).

[6] C. Lee, W. Yang, and R.G. Parr, Phys. Rev. B **37**, 785 (1988).

[7] J.P. Perdew, in *Electronic Structure of Solids '91*, edited by P. Ziesche and H. Eschrig (Akademie Verlag, Berlin, 1991), page 11.

[8] J. P. Perdew, K. Burke, and Y. Wang, *Real-space cutoff construction of a generalized gradient approximation: The PW91 density functional*, submitted to Phys. Rev. B, Feb. 1996.

[9] J. P. Perdew, J. A. Chevary, S. H. Vosko, K. A. Jackson, M. R. Pederson, D.J. Singh, and C. Fiolhais, Phys. Rev. B **46**, 6671 (1992); **48**, 4978 (1993) (E).

[10] K. Burke, J. P. Perdew, and M. Levy, in *Modern Density Functional Theory: A Tool for Chemistry*, edited by J. M. Seminario and P. Politzer (Elsevier, Amsterdam, 1995).

[11] R.G. Parr and W. Yang, *Density Functional Theory of Atoms and Molecules* (Oxford, New York, 1989).

[12] R.O. Jones and O. Gunnarsson, Rev. Mod. Phys. **61**, 689 (1989).

[13] R.M. Dreizler and E.K.U. Gross, *Density Functional Theory* (Springer-Verlag, Berlin, 1990).

[14] C. Møller and M. S. Plesset, Phys. Rev. **46**, 618 (1934).

[15] W. Kohn and L.J. Sham, Phys. Rev. **140**, A 1133 (1965).

[16] B. Hammer, K. W. Jacobsen, and J. K. Nørskov, Phys. Rev. Lett. **70**, 3971 (1993).

[17] A.D. Becke, J. Chem. Phys. **97**, 9173 (1992).

[18] R. Merkle, A. Savin, and H. Preuss, Chem. Phys. Lett. **194**, 32 (1992).

[19] D.J. Lacks and R.G. Gordon, Phys. Rev. A **47**, 4681 (1993).

[20] J. M. Seminario, Chem. Phys. Lett. **206**, 547 (1993).

[21] D. Porezag and M.R. Pederson, J. Chem. Phys. **102**, 9345 (1995).

[22] J.C. Grossman, L. Mitas, and K. Raghavachari, Phys. Rev. Lett. **75**, 3870 (1995); **76**, 1006 (1996) (E).

[23] E.I. Proynov, E. Ruiz, A. Vela, and D.R. Salahub, Int. J. Quantum Chem. S **29**, 61 (1995).

[24] A. Dal Corso, A. Pasquarello, A. Balderschi, and R. Car, Phys. Rev. B **53**, 1180 (1996).

[25] B. Hammer and J.K. Nørskov, Nature **376**, 238 (1995).

[26] A. Gross, B. Hammer, M. Scheffler, and W. Brenig, Phys. Rev. Lett. **73**, 3121 (1994).

[27] N. Moll, M. Bockstedte, M. Fuchs, E. Pehlke, and M. Scheffler, Phys. Rev. B **52**, 2550 (1995).

[28] A. Görling and M. Ernzerhof, Phys. Rev. A **51**, 4501 (1995).

[29] M. Ernzerhof, J.P. Perdew, and K. Burke, *Density functionals: Where do they come from, why do they work?*, a chapter to appear in *Density Functional Theory*, ed. R. Nalewajski, Spinger-Verlag, Berlin, 1996.

[30] P. Süle, O.V. Gritsenko, A. Nagy, E.J. Baerends, J. Chem. Phys. **103**, 10085 (1995).

[31] K. Burke, J.P. Perdew, and M. Ernzerhof, *System-averaged exchange-correlation holes*, work in progress.

[32] K. Burke, M. Ernzerhof, and J.P. Perdew, *Why semi-local functionals work: Accuracy of the on-top hole density*, work in progress.

[33] M. Ernzerhof, K. Burke, and J.P. Perdew, *Long-range asymptotic behavior of ground-state wavefunctions, one-matrices, and pair densities*, submitted to J. Chem. Phys.

[34] J.P. Perdew, Int. J. Quantum Chem. S **27**, 93 (1993).

[35] D.C. Langreth and J.P. Perdew, Solid State Commun. **17**, 1425 (1975).

[36] A.D. Becke, J. Chem. Phys. **98**, 1372 (1993).

[37] V. Barone, Chem. Phys. Lett. **226**, 392 (1994).

[38] A. Becke, J. Chem. Phys. **104**, 1040 (1996).

[39] C. J. Umrigar and X. Gonze, Phys. Rev. A **50**, 3827 (1994).

[40] C. Filippi, C. J. Umrigar, and M. Taut, J. Chem. Phys. **100**, 1290 (1994).

[41] M. Levy and J.P. Perdew, Phys. Rev. A **32**, 2010 (1985).

[42] K. Burke, J.P. Perdew, and M. Levy, *Improving energies by using exact electron densities*, to appear in Phys. Rev. A, April 1, 1996.

[43] C. J. Umrigar and X. Gonze, in *High Performance Computing and its Application to the Physical Sciences*, Proceedings of the Mardi Gras 1993 Conference, edited by D. A. Browne et al. (World Scientific, Singapore, 1993).

[44] Q. Zhao and R.G. Parr, Phys. Rev. A **46**, 2337 (1992).

[45] A. Görling, Phys. Rev. A **46**, 3753 (1992).

[46] Y. Wang and R.G. Parr, Phys. Rev. A **47**, R1591 (1993).

[47] R. van Leeuwen and E.J. Baerends, Phys. Rev. A **49**, 2421 (1994).

[48] R.C. Morrison and Q. Zhao, Phys. Rev. A **51**, 1980 (1995).

[49] E. Clementi and C. Roetti, At. Data Nucl. Data Tables **14**, 177 (1974).

[50] J.P. Perdew, A. Savin, and K. Burke, Phys. Rev. A **51**, 4531 (1995).

[51] J.P. Perdew, M. Ernzerhof, K. Burke, and A. Savin, *On-top pair-density interpretation of spin-density functional theory, with applications to magnetism*, to appear in Int. J. Quantum Chem.

Author Index

Affiliation Index

Subject Index

Chemical applications of
density-functional theory